Current and Future Developments in Food Science

(*Volume 1*)

Advances in the Determination of Xenobiotics in Foods

Edited by

Belén Gómara

Department of Instrumental Analysis and Environmental Chemistry,
Institute of General Organic Chemistry (IQOG),
Spanish National Research Council (CSIC),
Madrid, Spain

&

María Luisa Marina

Department of Analytical Chemistry,
Physical Chemistry and Chemical Engineering,
University of Alcalá Alcalá de Henares (Madrid), Madrid, Spain

Current and Future Developments in Food Science

Volume # 1

Advances in the Determination of Xenobiotics in Foods

Editors: Belén Gómara & María Luisa Marina

ISBN (Online): 9789811421587

ISBN (Print): 9789811421570

Bentham Science Publishers Pte. Ltd.
80 Robinson Road #02-00
Singapore 068898
Singapore
Email: subscriptions@benthamscience.net

BENTHAM
SCIENCE

CONTENTS

PREFACE

Xenobiotics had been and presently are of great concern, both for the society and the health authorities all over the world. Xenobiotics in food may include a huge variety of compounds of different nature. Nowadays, one of the groups that have caught the attention of researchers and authorities are food chain residues. These compounds are chemicals unintentionally present in the food due to the different procedures of production and preparation methods to which foodstuffs are subjected. Among them, compounds related to food contact materials such as plasticizers and plastic monomers are one of the xenobiotics mostly supervised by the European Food Safety Authority (EFSA) and the Environmental Protection Agency (EPA) and Food and Drug Administration (FDA) in the United States. In addition, pesticides can also be present in the final foodstuff because of their previous use in the field or on the farm. On the other hand, due to the global distribution of environmental pollutants, they are also susceptible to end up in the food chain because of different processes of deposition and/or bioaccumulation. There are several classes of environmental pollutants. Some of them are regulated by local or global legislations such as persistent organic pollutants and heavy metals. There are also many other emerging contaminants that must be controlled such as some halogenated flame retardants and perfluorinated compounds, among others. In addition, some xenobiotics could be present in the final food consumed as a result of food treatments, as is the case of acrylamide and furan which are related to high-temperature cooking processes. Finally, the presence of natural contaminants such as mycotoxins, aflatoxins and biogenic amines in the final foodstuffs must be controlled too.

The control of all these compounds would not be possible without the development of advanced analytical methodologies enabling their unequivocal, precise and accurate determination in foodstuffs. In this regards, one of the most employed methodologies is the separation techniques coupled to mass spectrometry. Depending on the physicochemical properties of each xenobiotic, gas (GC) or liquid (LC) chromatography can be applied for its separation, identification and quantification. Constant research is being carried out in order to develop more sensitive and selective methods for the determination of these xenobiotics at the low concentration levels they use to be present in foodstuffs. Novel analytical approaches in this field are fast GC and ultra-high performance LC (UHPLC) which have been successfully applied to study some of these xenobiotics. In addition, highly sensitive and selective mass analyzers such as triple quadrupole, Orbitrap or other hybrid systems combining some of them and novel developments such as ion mobility equipment are being recently applied to these purposes. These advances in combination with fast and environmentally friendly sample extraction and purification methods provide the society and authorities with the necessary methods for controlling and regulating, if necessary, the presence of all these xenobiotics in food.

Therefore, this book is aimed to present some of the most recent advances and developments achieved in the determination of different xenobiotics in foods. Chapters are organized according to the type of xenobiotic under study.

This book was inspired by the context of the AVANSECAL-CM and AVANSECAL-II-CM research projects funded by the Comunidad of Madrid and European FEDER program and headed by Professor María Luisa Marina from 2014 to nowadays, which was the continuation of two previous research projects (ANALISYC and ANALISYC-II) headed by Professor María José González from 2006 to 2013. Along all these years, a numerous group of researchers made considerable efforts to develop innovative analytical methodologies to control and improve food quality and safety with very relevant results in this field which have been and are being recognized at the international level. The editors are very grateful to these researchers, especially to those who have contributed to this e-book, and dedicated this e-book to Professor María José González in her retirement as a worm acknowledgement for her valuable contribution in the field of xenobiotics analysis.

Experts and researchers in analytical chemistry, food safety and xenobiotic analysis and newcomers in these fields such as Ph.D. students or chemists working in control laboratories or laboratory technicians will find in this e-book updated information including a set of advanced analytical methods used for the analysis of a broad spectrum of xenobiotics revealing the most interesting features and drawbacks to be overcome in this field. PhD students will learn more about novel analytical developments, they will acquire knowledge about xenobiotics and know in depth the field of food contamination. Finally, chemists working in control laboratories or laboratory technicians will have a very useful tool to face the problems arising on food safety.

We are very grateful to all the authors for their relevant contributions to this e-book.

Efforts in this field will be pursued in the next four years, thanks to the fundings from the Comunidad of Madrid (Spain) and European FEDER program through the new AVANSECAL-II-CM research project.

Belén Gómara
Department of Instrumental Analysis and Environmental Chemistry
Institute of General Organic Chemistry (IQOG)
Spanish National Research Council (CSIC)
Madrid
Spain

María Luisa Marina
Department of Analytical Chemistry, Physical Chemistry and Chemical Engineering
University of Alcalá Alcalá de Henares (Madrid)
Madrid
Spain

DEDICATION

Dedicated to Professor Maria José González Carlos in her retirement

List of Contributors

Alberto Ritieni — Università di Napoli Federico II, Department of Pharmacy, Via D. Montesano, 49 -, 80131 Napoli, Italy

Beatriz Gómez-Gómez — Departamento de Química Analítica. Facultad de Ciencias Químicas. Universidad Complutense de Madrid. 28040. Madrid, Spain

Cristina Nerín — I3A – Aragón Institute of Engineering Research, University of Zaragoza, Calle María de Luna 3, 50018 Zaragoza, Spain

Esther Garrido-Gamarro — Food Safety and Quality Officer, Fisheries and Aquaculture, Department, Food and Agriculture Organization of the United Nations (FAO). Viale delle Terme di Caracalla, 00153 Roma RM, Italy

Esteban Abad — Laboratorio de Dioxinas. Departamento de Química Ambiental. IDAEA-CSIC. C/ Jordi Girona, 18-26, 08034 Barcelona, Spain

Ethel Eljarrat — Institute of Environmental Assessment and Water Research, Spanish National Research Council (IDAEA-CSIC). C/ Jordi Girona, 18-26, 08034 Barcelona, Spain

Filomena Silva — ARAID – Agencia Aragonesa para la Investigación y el Desarollo, Av. de Ranillas 1-D, planta 2ª, oficina B, 50018 Zaragoza, Spain
Faculty of Veterinary Medicine, University of Zaragoza, Calle de Miguel Servet 177, 50013 Zaragoza, Spain

Francisca Holgado — Institute of Food Science, Technology and Nutrition, Spanish National Research Council (ICTAN-CSIC), C/ José Antonio Novais, 10, 28040 Madrid, Spain

Francisco J. Morales — Institute of Food Science, Technology and Nutrition, Spanish National Research Council (ICTAN-CSIC), C/ José Antonio Novais, 10, 28040 Madrid, Spain

Gianni Sagratini — University of Camerino, School of Pharmacy, Via S. Agostino 1, 62032 Camerino (MC), Italy

Giovanni Caprioli — University of Camerino, School of Pharmacy, Via S. Agostino 1, 62032 Camerino (MC), Italy

Jordi Parera — Laboratorio de Dioxinas. Departamento de Química Ambiental. IDAEA-CSIC. C/ Jordi Girona, 18-26, 08034 Barcelona, Spain

Kurunthachalam Kannan — Wadsworth Center, New York State Department of Health, and Department of Environmental Health Sciences, University at Albany, State University of New York, Albany, NY 12201, USA

Manuela Ábalos — Laboratorio de Dioxinas. Departamento de Química Ambiental. IDAEA-CSIC. C/ Jordi Girona, 18-26, 08034 Barcelona, Spain

Marta Mesías — Institute of Food Science, Technology and Nutrition, Spanish National Research Council (ICTAN-CSIC), C/ José Antonio Novais, 10, 28040 Madrid, Spain

Massimo Ricciutelli — University of Camerino, School of Pharmacy, Via S. Agostino 1, 62032 Camerino (MC), Italy

Òscar Aznar-Alemany Institute of Environmental Assessment and Water Research, Spanish National Research Council (IDAEA-CSIC). C/ Jordi Girona, 18-26, 08034 Barcelona, Spain

Pilar Fernández-Hernando Departamento de Ciencias Analíticas, Facultad de Ciencias, Universidad Nacional de Educación a Distancia (UNED). Paseo Senda del Rey nº 9, 28040, Madrid, Spain

Qian Wu Wadsworth Center, New York State Department of Health, and Department of Environmental Health Sciences, University at Albany, State University of New York, Albany, NY 12201, USA

Raquel Becerril I3A – Aragón Institute of Engineering Research, University of Zaragoza, Calle María de Luna 3, 50018 Zaragoza, Spain

Rosa Mª Garcinuño-Martínez Departamento de Ciencias Analíticas, Facultad de Ciencias, Universidad Nacional de Educación a Distancia (UNED). Paseo Senda del Rey nº 9, 28040, Madrid, Spain

Sauro Vittori University of Camerino, School of Pharmacy, Via S. Agostino 1, 62032 Camerino (MC), Italy

Vicente Andreu Environmental and Food Safety Research Group of the University of Valencia (SAMA-UV), Desertification Research Center (CIDE), Joint Research Center CSIC-UV-GV, Moncada, Spain

Yolanda Picó Environmental and Food Safety Research Group of the University of Valencia (SAMA-UV), Desertification Research Center (CIDE), Joint Research Center CSIC-UV-GV, Moncada, Spain

Yelko Rodríguez-Carrasco University of Valencia, Department of Food Chemistry and Toxicology, Av/ Vicent A. Estellés, s/n 46100 Burjassot,, Valencia, Spain

Yolanda Madrid Departamento de Química Analítica. Facultad de Ciencias Químicas. Universidad Complutense de Madrid. 28040. Madrid, Spain

Zoyne Pedrero Zayas CNRS/UNIV PAU & PAYS ADOUR, Institut des Sciences Analytiques et de Physicochimie pour l'Environnement et les Matériaux, UMR 5254, 64000, Pau, France

CHAPTER 1

Safety Assessment of Active Food Packaging: Role of Known and Unknown Substances

Filomena Silva[1,2], Raquel Becerril[3] and Cristina Nerín[3,*]

[1] *ARAID – Agencia Aragonesa para la Investigación y el Desarollo, Av. de Ranillas 1-D, planta 2ª, oficina B, 50018 Zaragoza, Spain*

[2] *Faculty of Veterinary Medicine, University of Zaragoza, Calle de Miguel Servet 177, 50013 Zaragoza, Spain*

[3] *I3A – Aragón Institute of Engineering Research, University of Zaragoza, Calle María de Luna 3, 50018 Zaragoza, Spain*

Abstract: Nowadays, consumers are more aware of what they eat and also request, minimally processed foods and they tend to prefer biodegradable or bio-based packaging. One of the most accepted technologies to battle this problematic is active packaging. Active packaging protects the food product by extending its shelf-life while guaranteeing its safety through the addition of antimicrobials or antioxidants that actively interact with the packaging atmosphere or the food product to avoid oxidation processes, microbial growth and other routes responsible for food spoilage. Although yet not fully implemented in Europe, active packaging is expected to reach a compound annual growth rate of 6.9% in 2020. However, in order to get these active packaging solutions into the market, their safety must be ensured and they must comply with the European legislation on the topic, both for the active substances incorporated into the packaging materials as for the packaging material itself. These packaging materials, either plastic or bio-based, can pose food safety risks to consumers due to the migration of compounds from the packaging to the food product. Compounds like plasticizers, additives, polymer monomers/oligomers and even non-intentionally added substances (NIAS) can migrate from the packaging material to the food product at concentrations capable to endanger human health and, therefore, they must be correctly detected and identified, to allow a correct risk assessment and strict monitoring of the packaging materials available.

Keywords: Active packaging, Antioxidant, Antimicrobial, Migration, Release, Food contact materials, Bio-based polymers, Natural compounds, Non-intentionally added substances.

* **Corresponding author Cristina Nerín:** Aragón Institute of Engineering Research, Campus Rio Ebro, Edificio Torres Quevedo, Universidad de Zaragoza, Calle Maria de Luna 3, 50018 Zaragoza, Spain; Tel: +34976761873; E-mail: cnerin@unizar.es

Belén Gómara & María Luisa Marina (Eds.)
Current and Future Developments in Food Science (Vol. 1)

INTRODUCTION

Nowadays, consumers are more aware of what they eat and also request minimally processed foods and they tend to prefer biodegradable or bio-based packaging over the traditional plastic ones. Therefore, there is an urgent real need to develop newer and safer food packaging systems to improve food shelf-life, whether to reduce food waste (packaged food or the package itself) [1] or to distribute products to more distant places. Furthermore, there is a growing need to provide new solutions to ensure the safety and quality of the packaged foods and products. Due to all these demands, there is a growing market for the development of new packaging solutions, called active packaging (AP), to be applied within several fields, such as pharmaceutical, healthcare or food industries [2]. Active packages are based on the incorporation of active agents with the food packages, thus avoiding the direct addition of chemicals to the food. These active agents can be either incorporated directly in the packaging materials, or be included inside a package as pads, trays, sachets or pouches. These active agents include antioxidants, antimicrobials or absorbers that hand over new properties to the pre-existent material such as oxygen or free radical scavenging (antioxidant/absorber) or microbiological control (antimicrobial) [2]. Over the last decades, active packaging has become a reality for the food packaging industry with the constant growth of the global market for active/intelligent packaging, reaching a compound annual growth rate of 6.9% [3]. There is a vast array of AP solutions currently available in the market; with the vast majority of them having the consumer as the final user and a few of them being intended for business-to-business use (Table 1).

Table 1. Examples of worldwide commercially available active packaging solutions.

Tradename	Company	AP Type	Materials	Country	Consumers	B2B
Ageless®	Mitsubishi Gas Chemical Co. Inc.	Oxygen scavenger	Sachets	Japan	X	
シートドライヤー Sheet dryer	Torishige	Dessicant	Laminate papers	Japan	X	
EMAP/AMAP	Perfotec	Modified Atmosphere Packaging (controlled permeability of the packaging and controlled atmosphere)	Plastic trays	The Netherlands	X	X
FreshPaper	Fenugreen	Fiber-based sheets with organic spices	Paper sheet	USA (also export in Europe)	X	X
MegaCO_2	Pomona	Moisture absorbing pads combined with CO_2 scavenger	pads	Poland	X	
EthenAbsorbers/ETENSachets	Pomona	Ethylene scavenger	sachets	Poland	X	
Oxyguard	Clariant	Oxygen scavengers	sachets	USA	X	

(Table 1) cont.....

Tradename	Company	AP Type	Materials	Country	Consumers	B2B
BreatheWay	Landec Corporation	Selective permeability films	films	USA	X	
Darex OST	Darex	Oxygen scavenger	crown	USA		
OxyFresh	STANDA Laboratories (EMCO)	CO_2 scavenger / O_2 emitter	sachets	France	X	
Sanocoat	Mondi Packaging Flexibles AG	Antimicrobial paper	paper	Austria/Germany	X	
NA	Erze Ambalaj/Parx Plastics	Antimicrobial tray	plastic trays	Turkey		X
AntioxidantPack	BTSA	Antioxidant film	plastic films	Spain		X
Supasorb	Thermarite	High capacity moisture absorbing film	pads	Malaysia	X	
Fresh-r-Pax	Maxwell Chase Technologies LLC	Moisture absorbent	trays, pads, pouches	USA	X	
DriFresh® SeaFresh™ Fresh-Hold	Sirane	Absorbing pad (moisture/ice/odor) with CO_2 emitter	pads	UK, available in Poland	X	
Dri-Fresh® SeaFresh™ Ice-Mats	Sirane	Seawater-releasing pad to extend seafood shelf-life	pads, mats	UK, available in Poland	X	
Dri-Fresh® Fresh-Hold™ OA	Sirane	Odour-scavenging pad	pads, labels	UK, available in Poland	X	
Dri-Fresh® Fresh-Hold™ AB	Sirane	Antibacterial pad	pads	UK, available in Poland	X	
NA	Artibal S.A.	Antimicrobial coating	films	Spain	X	
NA	Artibal S.A.	Antioxidant coating	films	Spain	X	
NA	Goglio SpA	Antioxidant film	films	Italy	X	
NA	SAMTACK	Antioxidant adhesive for multilayer	adhesive	Spain	X	
Rycoat F-100, Emulactiv C-1	REPSOL YPF Lubricantes & Especialidades	Antimicrobial/antioxidant coating	paper/cardboard	Spain	X	X
NA	CellComb	CO_2 emitter	pads	Sweden	X	

These packages have been used to preserve all kinds of foods ranging from more perishable goods, as fresh fruit, vegetables, meat, fish and cheese to more processed foods such as breads, cakes and sweets, sauces and jams, processed and dried meat, snacks and even baby food and pet food. These AP solutions include several absorbers such as moisture and odor absorbers, and ethylene and carbon dioxide scavengers. With respect to antioxidant packaging, the two main types of AP available are the use of oxygen and free radical scavengers. When dealing with antimicrobial packaging, there is a broad array of technologies available aiming at reducing microbial growth in the food product, either by changing the atmosphere (selective permeability films for modified atmosphere and carbon dioxide emitters) or by adding antimicrobial substances ranging from chemicals to natural products such as essential oils or herb extracts. The mode of action of each AP depends on the active agent incorporated and the packaging design as well as the characteristics of the packaged food product (Table **2**).

The emergence and development of these new packaging technologies that interact with food triggered a response from the authorities to ensure their safety towards the consumers. In this regard, the European Union adopted specific legislation to regulate the use and application of these new active and intelligent packaging systems, namely European Committee regulations 1935/2004 and 450/2009. Regulation (EC) No 1935/2004 on materials and articles intended to come into contact with food [4] defines for the first time active packaging and authorizes its use by establishing limits to guarantee the safety of consumers. Regulation (EC) No 450/2009 on active and intelligent materials and articles intended to come into contact with food [5] includes general requirements stated in Regulation 1935/2004/EC and provides new specific requirements for the safe use of active and intelligent packaging.

Table 2. Mode of action of different active packaging technologies available on the market.

Active Packaging	Function	AP Material	Mode of Action
Antioxidant	Oxygen scavenger	Iron-based sachets	The iron contained in the sachets is prone to oxidation, absorbing oxygen in the process
	Free radical scavenger	Multi-layer film	This multilayer film contains green tea extract (GTE) and acts by scavenging free radicals formed in the atmosphere of the package, using catechins as free radical acceptor molecules. There is no direct interaction between antioxidant compounds and foods, as the green tea extract is crafted inside the multilayer film, with none or negligible specific migration of green tea compounds from the package
Antimicrobial	Carbon dioxide emitter	Highly absorbent pads	The liquids released by the packaged food activate the CO_2-releasing reaction. As the majority of the bacteria present in foods are aerobic (*e.g.*, require oxygen), the maintenance of a steady CO_2 concentration causes an antimicrobial effect on the food as it inhibits aerobic bacteria growth.
	Antimicrobial releaser	Antimicrobial multi-layer film	These films contain natural volatile extracts The antimicrobial activity of these films is exerted in the vapor phase, taking advantage of the inherent volatility of the extracts used; therefore, no direct contact is required between the package and the food product, which allows for a more homogeneous distribution and action of the antimicrobial agent in the whole food product

(Table 2) cont.....

Active Packaging	Function	AP Material	Mode of Action
Other	Permeability control	Packaging polymeric film	Crystallizable polymer with temperature switch: at low temperatures, the polymer is in a more compact, crystalline structure that has low permeability; meanwhile, when temperature rises, polymer structure changes to an amorphous state which results in an increased permeability
	Ethylene absorber	Absorbing pads	These pads contain substances capable of absorbing ethylene and thus prevent fruit ripening
	Desiccant	Laminate paper	Calcium chloride is used as moisture absorber due to its hygroscopic features The absorber is included in a cellulose matrix to potentiate moisture absorption

According to these regulations, active packaging is allowed to change the composition of the atmosphere around the packaged food, providing that the active substance is an authorized compound and the originated change complies with EU regulation. Besides, AP should be correctly labelled to prevent improper use and misunderstanding by consumers. Another important point of these regulations is the establishment of standardized procedures to assess the safety of both the passive parts of active materials, such as the packaging materials, and the active compounds [6].

Antioxidant Packaging

The presence of an oxygen-rich atmosphere in the packaging headspace is one of the most critical factors affecting both food quality and shelf-life. The presence of residual oxygen can led to product oxidation resulting in changes in its organoleptic properties, such as color, odor, taste, *etc* [7] (Table **3**); in its nutritional value due to oxidation of important vitamins, fatty acids, proteins, among others [8, 9] (Table **3**); and its microbial load as oxygen promotes the growth of aerobic foodborne spoilage and pathogenic bacteria [10]. All the above mentioned detrimental effects of oxygen on food products can have a severe impact on human health, either by the consumption of microbiologically contaminated food or by the ingestion of food oxidation products. These products are formed during food production, processing and storage, being currently described as potentially harmful to humans, as they have been linked to inflammatory and to the onset of carcinogenic processes [11]. Regarding microbial contamination, aerobic pathogens such as *Salmonella spp*, *E. coli* and *L. monocytogenes*, highly prevalent in food products, can lead to severe foodborne illness, with significant morbidity and mortality rates [12]. Additionally, the decrease in consumer's acceptance of oxidized foods with altered organoleptic properties and nutritional value results in great economic losses due to food waste [13]. Consequently, the food industry aims to exclude oxygen from food packaging, which is mainly performed by gas flushing or modified atmosphere packaging (MAP) processes, or more efficiently by using antioxidant packaging capable of controlling residual oxygen [14].

Table 3. Sensorial and nutritional changes caused by oxygen in packaged foods.

Food Product	Organoleptic and Nutritional Changes due to Oxygen Presence	Reference
Meat products	Development of rancidity Potential formation of toxic aldehydes Loss of nutritional quality because of polyunsaturated fatty acid (PUFA) degradation	[15]
Oils and fats	Development of off-flavors and odors (such as tallowy, painty, burned, fishy, grassy) Vitamin E content of vegetable oils and the vitamin A content of fish oils are reduced, if not entirely destroyed Degradation of essential dietary fatty acid components Development of rancidity Darkening of color	[16]
Dairy products	Lipid oxidation and formation of peroxides Development of rancid flavor Off flavors (tallowy, oily, and fishy) in butters Tallowy or oxidized flavor in ice cream Tallowness in sweetened condensed milk Formation of toxic aldehyde and ketone compounds	[17]
Baked products	Degradation of vitamins (A, C, D, E, and folate), essential fatty acids (linoleic acid), and essential amino acids like methionine, cysteine, histidine, lysine, and tryptophan Development of rancidity Changes in texture	[18]
Fruits and vegetables	Phenolic browning of fruit/vegetables Changes in texture (softening) Loss of vitamin C (ascorbic acid) Acceleration of respiration Discoloration	[19]

Types of Antioxidant Packaging

Overall, we can distinguish between two main types of antioxidant compounds: primary antioxidants, such as free-radical scavengers, which target oxidation products; and secondary antioxidants, such as metal chelators, UV absorbers, singlet oxygen ($^1O_2$) quenchers and oxygen scavengers, which prevent oxidation.

Oxygen Scavengers

Oxygen may be accessible in the package through several routes such as entrance through the packaging material, minor leakage due to poor sealing and material barrier properties, on account of the residual oxygen in the air encased in the container, deficient gas flushing or oxygen evacuation from the container head-space [20]. Oxygen scavenging compounds react with the oxygen present in the package to reduce its concentration and make it unavailable for deteriorative reactions [21]. The oxygen scavenger market is expected to reach 2.41 billion dollars by 2022, at a compound annual growth rate (CAGR) of 5.1% from 2017 to 2022 [21]. The most commonly used oxygen scavenger is ferrous oxide, but other compounds include catechol, ascorbic acid, sulfites, photosensitive dyes, unsaturated hydrocarbons and enzymes such as glucose oxidase and catalase [22, 23].

Free Radical Scavengers

When UV light, oxygen or metals cannot be effectively removed from the packaging atmosphere or food product, oxidation reactions start to take place in the food product and its atmosphere. This results in the formation of free radicals, such as oxo, hydroxo or peroxo that initiate oxidative chain reactions [24]. Free radical scavengers react with these free radicals to convert them into more stable compounds that are not able to engage in further oxidative chain reactions. There is a multitude of available antioxidant compounds that are able to scavenge free radicals, ranging from the synthetic ones such as butylated hydroxytoluene (BHT), butylated hydroxyanisole (BHA) and tert -butylhydroquinone (TBHQ) to the natural ones such as natural compounds, plant and fruit extracts and essential oils from herbs and spices [2]. There are also some developments based on nanoparticles such as selenium (Se) nanoparticles, incorporated behind a layer of LDPE in a multilayer, using the adhesive as vehicle [25]. This new antioxidant packaging material has been recently approved by EFSA. The current tendency is to move from synthetic antioxidants towards the natural alternatives, to circumvent the safety issues with the synthetic ones, namely their potential harmfulness to humans [26]. Although natural antioxidants look promising and have already demonstrated their effectiveness as free radical scavengers [25, 27], usually they are required in large quantities in the packaging to attain the same antioxidant activity as the synthetic ones in the food product.

There is a tremendous amount of research on the development of new antioxidant packaging solutions; therefore, for the interest of this book chapter, only research from the last five years with proof-of-concept (assays in food systems) will be considered (Table **4**). Regarding the polymeric matrices used for the packaging material, research has driven from plastic materials, either as a monolayer or multi-layer assembly, such as polyethylene (PE) or polyvinyl alcohol (PVA), to more environmental-friendly and sustainable materials as paper and board, bioplastics such as poly(lactic acid) and biopolymers such as the ones based on starch, proteins and chitosan. In these matrices, the most preferred antioxidants incorporated are of natural origin, such as natural extracts from fruits, green tea, olive leaf; food by-products such as polyvinylpolypyrrolidone washing solution; compounds from natural origin such as resveratrol, α-tocopherol and lycopene; and essential oils such as the ones from cinnamon and clove. All of these strategies were able to effectively inhibit lipid oxidation in mushrooms, beef, pork and poultry meat, fish, butter and oils, and processed foods such as chocolate peanuts and cereals. The decrease in food oxidation was measured through the determination of peroxide values (PV), thiobarbituric acid reactive substances (TBARS), formation of dienes and trienes, enzymatic assays as the tyrosinase assay, measurement of oxidation products such as metmyoglobin, hexanal and several ketones and aldehydes, and also degradation products of fatty acids.

Table 4. Examples of antioxidant packaging strategies described over the last five years. AP-active packaging; NR-not mentioned.

Matrix	Origin (Synthetic or Natural)	Antioxidant Compound	Dosage	Packaging Type	Main Findings	Ref.
Molecularly imprinted hydrogels	Natural	Ferrulic acid	11 µmol g^{-1}	Cover film for can	Protection of butter from oxidation evaluated by the TBARS test 50% decrease in TBARS values in AP packaged butter Oxidation was reduced by 25% after four weeks of butter storage	[28]

(Table 4) cont.....

Matrix	Origin (Synthetic or Natural)	Antioxidant Compound	Dosage	Packaging Type	Main Findings	Ref.
Chitosan	Natural	Kombucha tea	1-3% (w/w polymer)	Cover film	Inhibition of lipid oxidation in minced beef patties after 6 days of storage evaluated by the TBARS test	[29]
Poly(lactic acid) (PLA) nanofibers	Natural	α -Tocopherol	5% (w/w polymer)	Cover film	Inhibition of lipid oxidation in raw beef evaluated by the TBARS test	[30]
Polyurethane	Natural and synthetic	α-tocopherol and butylated hydroxytoluene (BHT)	1-5% (w/w polymer)	Film	Preclusion of soybean oil oxidation, evaluated by FT-IR, during 60 days at 60 °C in the dark	[31]
LDPE	Natural	Polyvinylpolypyrrolidone washing solution (PVPP-WS) extract	3-20% (w/w)	Wrapping film	Reduction of lipid oxidation in raw beef, with reductions in TBARS values higher than 60% at the 9th day of cold storage At the highest concentration applied (20%),	[32]

(Table 4) cont.....

Matrix	Origin (Synthetic or Natural)	Antioxidant Compound	Dosage	Packaging Type	Main Findings	Ref.
					the onset of oxidation was delayed until the ninth day	
Linear LDPE	Natural	Resveratrol	1% (w/w)	Bags	Maximum oxidation reduction of 34.7%, in raw beef evaluated by the TBARS test	[33]
Fish protein isolate (FPI)/fish skin gelatin (FSG)	Natural	Zinc oxide nanoparticles (ZnONP) Basil leaf essential oil (BEO)	3% ZnONP (w/w, based on protein content) and 100% BEO (BEO) (w/w, based on protein content)	Wrapping film	Reduction of lipid oxidation in sea bass slices as evaluated by the PV and TBARS tests AP wrapped fish showed a 30-40% and 50-60% decrease in PV and TBARS, respectively	[34]
NM (plastic)	Natural	Green tea extract	NM	Cover film and bag	Reduction of oxidation in chocolate peanuts and cereals Lower pyrazine and fatty acid oxidation Shelf-life extension from 9 to 18 months Decreased rancidity	[35]

(Table 4) cont.....

Matrix	Origin (Synthetic or Natural)	Antioxidant Compound	Dosage	Packaging Type	Main Findings	Ref.
Paraffin coated paper	Natural	Cinnamon essential oil	100 g of cinnamon EO in paraffin emulsion " m-2"	Pad	Decreased sliced mushrooms browning by decreasing tyrosinase activity No significant colour alterations (browning) after 9 days of storage	[36]
Cassava starch	Natural	Lycopene	2-8% (w/w filmogenic solution)	Pouches	During storage, sunflower oil showed higher stability when stored in AP films with PV inside the limit for vegetable oil	[37]
Cassava starch	Natural	Lycopene	2-8% (w/w filmogenic solution)	Pouches	No significant differences in conjugate dienes and trienes formation between AP and control package	[37]

(Table 4) cont.....

Matrix	Origin (Synthetic or Natural)	Antioxidant Compound	Dosage	Packaging Type	Main Findings	Ref.
PVA-cyclodextrin-gelatin film	Natural	Mango peel extract	5% (v organic extract/v filmogenic solution)	Pouches	Irradiation together with AP decreased lipid oxidation in minced chicken meat as evaluated by the TBARS test	[38]
Soy protein isolate	Natural	Clove essential oil	0.5% (v/v filmogenic solution)	Wrapping film	Decreased tuna oxidation as observed by the decrease in TBARS values and aldehydes and ketones contents	[39]
Chitosan	Natural	Ginger and rosemary EO	2% (v/v filmogenic solution)	Wrapping films	Oxidation inhibition in minced poultry meat throughout storage time No significant differences in the results observed between the two EOs	[40]

(Table 4) cont.....

Matrix	Origin (Synthetic or Natural)	Antioxidant Compound	Dosage	Packaging Type	Main Findings	Ref.
Polyethylene film	Natural	Olive leaf extract	2-15% (w/w adhesive)	Cover film	Reduction of oxidation as seen by a 25% decrease in TBARS values and the decrease in oxidative product bands in Raman spectroscopy	[41]
PVA	Natural	Seed cover extract of *Zanthoxylum rhetsa*	1% (w/v film forming solution)	Packet	Reduction of lipid peroxidation in ground chicken meat during the entire refrigerated storage period as observed by the lower TBARS values obtained Improvement in shelf-life up to 12 days at chilled condition	[42]

(Table 4) cont.....

Matrix	Origin (Synthetic or Natural)	Antioxidant Compound	Dosage	Packaging Type	Main Findings	Ref.
PE	Natural	Green tea extract	NM	Bags	Improves the shelf-life of minced pork meat duet to the reduction of Met myoglobin formation, an oxidation product of myoglobin Maintenance of meat redness for a longer period	[43]
LDPE-PET multilayer	Synthetic	Se nanoparticles	Adhesive	Multilayer film	Improves the shelf-life of cooked ham, potato chips, fresh meat, fresh chicken, vegetable salads and more	[44]

Antimicrobial Packaging

Antimicrobials

The growth of microorganisms in food is one of the most important causes of food spoilage and an important cause of human disease. It has been estimated that about 25% of all foods produced globally are lost due to microbial spoilage [45]. Moreover, in the European Union, 4,786 foodborne outbreaks (including waterborne outbreaks) were reported in 2016 [12]. Microbiological food spoilage is caused by the growth of microorganisms which produce enzymes responsible for biochemical reactions that yield undesirable chemical compounds and cause changes in the original characteristics of the food product. These changes reduce the quality of the food product and decreased its shelf-life [46]. In addition, some of these microorganisms can also result in severe human illnesses (Fig. **1**) if contaminated food is ingested. For instance, in 2016, EFSA (European Food Safety Authority) reported more than 400 death caused by the foodborne pathogens *Campylobacter*, *Salmonella*, *Yersinia*, Sigha toxin-producing *E. coli* and *Listeria* [12].

Foodborne microorganisms can come both from the natural flora of the product itself or from the contamination produced during harvesting, food processing, and distribution [47]. Therefore, there are a huge range of microorganisms that can be founded in food including bacteria, molds and yeasts. The type and load of predominant microorganisms in one specific food product depends mainly on their intrinsic characteristics (pH, water activity, composition, structure) and also on the storage conditions (temperature and atmosphere conditions mainly) [46]. For example, psychotropic *Pseudomonas* are commonly found in fresh meat stored aerobically at cold temperatures [48], while *Brochothrix termosphacta* grows mainly in fresh meat stored at cold temperatures and low levels of oxygen [49]. Considering the variability of existing foodborne microorganisms, the development of new strategies to reduce or inhibit the growth of microorganisms in food products represents a challenge even to developed countries. Moreover, the increasing consumer and producer demands for products with higher quality and longer shelf-life requires more efficient antimicrobial solutions [50].

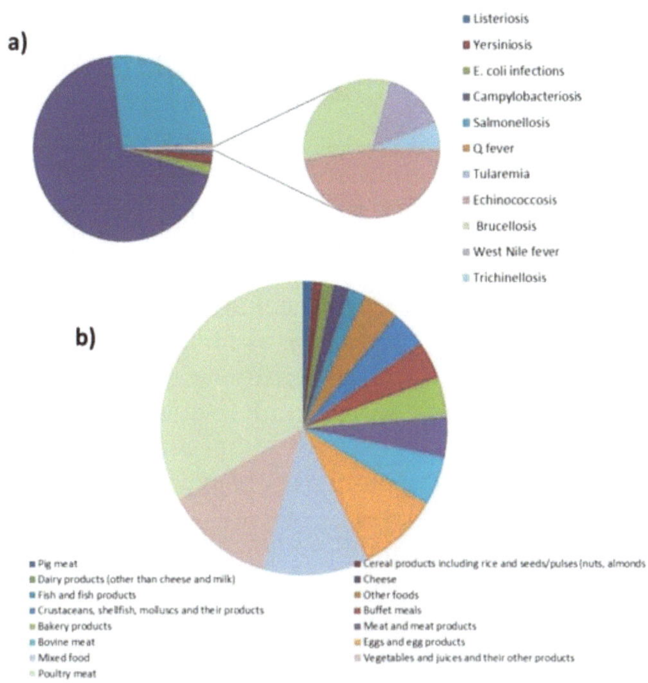

Fig. (1). 2016 data on zoonosis and foodborne outbreaks in Europe: a) reported number of confirmed human zoonosis; b) Distribution of strong-evidence foodborne outbreaks, by type of vehicle (excluding waterborne outbreaks). Source: The European Union summary report on trends and sources of zoonosis, zoonotic agents and food-borne outbreaks in 2016 [12].

In this context, the research dealing with the developments of new antimicrobial active packaging has increased as a viable alternative to reduce microbial growth in food [14, 51]. It is important to point out that each product type will need a specific antimicrobial packaging solution since its intrinsic characteristic and its microflora will be different. Given this, the type of antimicrobial substance and polymer selection will depend on the use of the material. For example, an active material intended for packaging refrigerated fresh meat stored aerobically should contain an antimicrobial active against *Pseudomonas* and a polymer impermeable to water. The number of active substances and polymers used in antimicrobial AP has increased over the last years, especially regarding those of natural origin that are the ones preferred by the consumer [14]. In addition, new technologies to control the release rate of the antimicrobial have also been incorporated into the material in order not only to prolong the release of the active compound, but also as a means to predict and yield more reproducible release rates [52].

Although the amount of research on the development of new antimicrobial AP technologies is huge, most of them do not include proof-of-concept tests with real food. This may be due on one hand, to the novelty of some of these technologies that are not fully developed and, on the other hand, to the difficulties encountered when applying the antimicrobial materials to real systems. It has been demonstrated that the efficiency in food systems is lower probably due to the interactions of the active compounds with the food matrix [53, 54], higher organic acid and trace metal contents, and greater availability of nutrients for microbial cell repair and survival [47]. Similar to antioxidant AP, in the case of antimicrobial AP, focus was given to relevant new antimicrobial packaging developments in last five years tested in food matrices (Table **5**).

Concerning polymeric matrices, research includes classical synthetic polymers such as polyethylene or polyester used alone as monolayer material or combined to compose multilayer materials. Additionally, a high number of bio-based polymers have been investigated, namely poly(lactic acid) and those obtained from proteins, cellulose, polysaccharides (sodium alginate), chitosan or starch. Regarding the antimicrobial compounds evaluated, two main groups can de distinguished: synthetic nanoparticles (especially zinc oxide and silver) and antimicrobial compounds of natural origin. These include bacteriophages, antimicrobial peptides such as sakacin-A, ε-Poly-l-lysine, bacteriocins such as nisin, nanocellulose, essential oils (especially cinnamon essential oil) and natural extracts or compounds derived from plants such as pinosylvin or pomegranate peel extract. Chitosan is a particular case, since chitosan polymers have intrinsic antimicrobial properties. To assess the antimicrobial action of the active packaging, the reduction in *Staphylococcus aureus*, *Salmonella typhimurium, Listeria monocytogenes*, *Escherichia coli*, *Campylobacter jejuni*, total bacteria, moulds and yeast counts has been studied in fresh or read to eat meat (poultry, beef or turkey). Meat and meat products were the main food targets used probably due to their short shelf-life and high microbial load including both spoilage and pathogenic microorganisms. Other antimicrobial materials were also tested in fresh vegetables, bread, milk, fish or cheese. It is important to point out that the antimicrobial action achieved is not very high in all cases and, in some studies, the microbial reduction obtained is less than 2 log CFU.

Table 5. Examples of antimicrobial packaging strategies described over the last five years. MIC-Minimal Inhibitory Concentration. CFU- Colony Forming Unit. PFU-Plaque Forming Unit.

Matrix	Origin (Synthetic or Natural)	Antioxidant Antimicrobial Compound	Dosage	Packaging Type	Main Findings	Ref.
Sodium alginate film	Synthetic	Zinc oxide nanoparticles	3 mg mL^{-1}	Wrapping film	Reduction of *S. aureus* and *S.* Typhimurium in poultry sausages at 8 °C (2 log CFU mL^{-1} in 24 h)	[55]
Low-density polyethylene	Synthetic	Combinations of silver, copper oxide and zinc oxide nanoparticles	1% (w/w polymer)	Wrapping film	Reduction of coliforms in ultra-filtrated cheese at 4 °C (4.21 log CFU g^{-1} after 4 weeks)	[56]
Poly(lactic acid) / poly-ethylenglicol blend	Synthetic and natural	Zinc oxide and silver/copper nanoparticles and cinnamon, garlic and clove oil	1-4% (wt) nanoparticles 100% (w/w, based on PLA content) Essential oils 10-100% (w/w, based on PLA content)	Wrapping film	Films with cinnamon oil reduced *L. monocytogenes* and *S.* Typhimurium in cheese at 4 °C (4.47 log CFU g^{-1} and 3.17 log CFU g^{-1} respectively after 11 days)	[57]
Gelatin coated paper	Synthetic	Citric acid	0.5 and 1.0% relative to the solution mass	Wrapping film	Reduction of total bacterial counts in beef at 4 °C (3 log g^{-1} after 4 days)	[58]

Matrix	Origin (Synthetic or Natural)	Antioxidant Antimicrobial Compound	Dosage	Packaging Type	Main Findings	Ref.
Zein film	Natural	*Zataria multiflora* Boiss. essential oil	10% (w/w dry zein power)	Bags	Reduction of *L. monocytogenes* and *E. coli* in pasteurized cow's milk at 4 °C (0.99 log CFU mL^{-1} after 6 days)	[59]
Multilayer film made of polyester/aluminum/polyethylene	Natural	Cinnamon and oregano essential oil and Benzoic acid	18.5 and 10.2% (w/w dry adhesive)	Bags	Films with cinnamon essential oil reduced *E. coli* O157:H7 (3 log CFU mL^{-1} after 24 h) and *Saccharomyces cerevisiae* (3 log CFU g^{-1} after 48h) in tomato puree	[60]
Cellulose / polypropylene absorbent pads	Natural	Pinosylvin	0.08-0.8 mg cm^{-2} polymer	Absorbent pads	Growth reduction of *C. jejuni* in chicken exudates (56.99-99.99%) Growth reduction of total viable counts, total lactic acid bacteria and psychrotrophs in breast chicken fillets (10-74%)	[61]
Sodium caseinate film	Natural	Pomegranate peel extract	2 x MIC	Cover film	Reduction of total viable bacteria count (1.9 log CFU g^{-1}) and *Staphylococcus aureus* (1.8 log CFU g^{-1}) in ground beef at 4 °C after 9 days	[62]

Matrix	Origin (Synthetic or Natural)	Antioxidant Antimicrobial Compound	Dosage	Packaging Type	Main Findings	Ref.
Cellulose paper	Natural	Bacteriophage cocktails of *Listeria monocytogenes* and *E. coli* O104:H4	2.5x10^{10} PFU mL^{-1}	Pads	Reduction of *Listeria monocytogenes* in cantaloupes (1-3 log g^{-1}) and turkey at 4 °C (0.8 log CFU cm^{-2})	[63]
Poly(lactic acid)	Natural	*Salmonella* phage Felix O1 and *Listeria* phage A511	10^{12} PFU mL^{-1}	Cover film	Reduction of *Salmonella* sp. (0.8 log CFU cm^{-2} at 4 °C and 1.3 log CFU cm^{-2} at 10 °C) and *L. monocytogenes* (6.3 log CFU cm^{-2} at 4 °C and 1.5 log CFU cm^{-2} at 10 °C) in precooked sliced turkey breast	[64]
Low density polyethylene	Natural and synthetic	Nisin, chitosan, potassium sorbate and silver substituted zeolite	2% (w/w polymer)	Multilayer bags	Reduction of mesophilic bacteria, total coliforms and total molds and yeasts in chicken drumsticks stored at 5 °C after 6 days (0.28-1 log CFU g^{-1})	[65]
Poly (lactic acid) / sawdust particle	Natural	Bacteriocin 7293	19.54±2.87 µg cm^{-2}	Packet	Reduction of Gram-positive and Gram negative bacteria (2-5 log CFU cm^{-2}) in chilled pangasius fish fillets	[66]

Matrix	Origin (Synthetic or Natural)	Antioxidant Antimicrobial Compound	Dosage	Packaging Type	Main Findings	Ref.
Polyethylene-coated paper	Natural	Sakacin-A	0.44% (w/w dry weight of coating)	Cover film	Reduction of *L. innocua* in ready-to-eat thin-cut veal meat at 4 °C (1.5 log CFU g^{-1} after 48 h	[67]
Starch	Natural	ε-Poly-l-lysine	1.6, 3.2 and 6.5 mg polylysine cm^{-2} of pad	Cover biofilm	Films with 6.5 mg reduced *A. parasiticus* (2.73 log CFU g^{-1}) and *P. expansum* (1.42 log CFU g^{-1}) in bread	[68]
Chitosan-nanocellulose biocomposites	Natural	Chitosan and nanocellulose	1% (w/0.6 v) chitosan 0.18% (w/w chitosan) nanocellulose	Cover film	Reduction of lactic acid bacteria in ground meat at 4 °C (1.3 log CFU g^{-1}) and 25 °C (3.1 log CFU g^{-1}) after 6 days	[69]
Chitosan	Natural	*Eucalyptus globulus* essential oil	0.5-1.5% (v/v) of essential oil	Wrapping film	Reduction of *L. monocytogenes* in sliced sausages at 23 °C (0.26-1.01 log CFU mL^{-1} after 3 days)	[70]

Safety Assessment

According to European Regulations [4, 5], safety assessment of an active material implies both the safety assessment of the passive parts and the active compounds.

The passive parts of active and intelligent packaging systems refer to the materials or articles in which the active compounds are added or incorporated [71] and are subjected to pre-existing European and national food-contact material regulations such as Regulation (EU) No 10/2011 on plastic materials and articles intended to come into contact with food [72]. It is important to remark that in the case of the releasing systems, the amount of released active substance should not be calculated in the values of overall migration as the active substance is not part of the passive material (article 9(2) (EC) No 450/2009).

The active compounds, as defined in Regulation (EC) No 450/2009, are substances responsible for creating the active and/or intelligent function. These compounds must be subjected to the safety assessment carried out by the European Food Safety Authority before they are authorized for use. In the case of releasing systems, the released active substance should comply with relevant European Community provisions applicable to food as they can be incorporated into the food product. Moreover, active substances which are not in direct contact with the food or the environment surrounding the packaged food product and are separated from food by a functional barrier should not be evaluated if they comply with the requirement establish in article 5(2)(c) of Regulation (EC) No 450/2009.

In agreement with the European Regulations cited, the safety of a specific material implies that all the substances that can be released from the material to the food, either active compounds and substances from the passive parts, comply with the established limits of migration and levels of use in food. These limits have been established by the European Commission on the basis of EFSA scientific opinion on risk assessment made from data on estimated consumer exposure and results from toxicity studies. EFSA [73] indicated that migration data should give realistic account of migration into foodstuffs and mentioned three main approaches: modelling, simulation and direct measurement in foods (indicated when simulation is impossible or not reliable). Regulation (EU) No 10/2011 establishes the specific procedures for migration testing that should be applied to all food contact materials. According to this Regulation, migration from materials not yet in contact with food should be simulated. This means that materials should be exposed to one or several simulants (liquid o solid substances that mimic the behavior of food) during a selected time and temperature to simulate the real food exposure to the material [74]. The time, temperature and food simulant should be selected to simulate the worst case scenario in which the packaging is intended to be used. To a more detail review on active packaging migration please see [52]. After performing all the necessary migration tests, the identification of all compounds released by the new material has to be done, because either the active substances used or their impurities or their interaction products with the packaging material or with the food could contaminate the food with new toxic compounds. For instance, a migration study performed in several active packaging materials have demonstrated that impurities from one of the active agents, citral, were migrating from the packaging materials [75], showing the need for the analysis of compounds at trace levels. This identification procedure includes the qualitative and quantitative analysis of volatile and non-volatile compounds for which high resolution mass spectrometry is usually required.

Active Substances

As it has been explained in the previous section, the active substances that are incorporated into the active packaging material have to comply with the current legislation for approved substances in food. In many cases, they can have migration limits or maximum allowed levels in food and therefore, in this case, migration tests are mandatory.

Table **6** summarizes research carried out within the last five years on active (antioxidant and antimicrobial) packaging that carried out migration assays. As can be seen, it includes active materials based on different types of polymers since the material is a key factor in the migration process [52]. Specifically, synthetic polymers of polyethylene (PE), low and high density polyethylene (LDPE and HDPE), polyethylene terephthalate (PET), polyvinyl alcohol (PVA) or poly(butylene succinate) and other more environmental-friendly such as paper, bioplastics of poly(lactic acid) (PLA) or bio-based polymers of chitosan, proteins, starch, bacterial cellulose or peptides. Regarding the conditions used for the migration assays, in most cases, they only partially fulfilled with European Regulation (EU) No 10/2011 and used established simulants but different temperatures and times of exposure. It is important to clarify that the migration tests are conducted in some studies with the objective of evaluating the release behavior of the active compound from the polymeric matrix and not to assess the safety of the material.

Besides the conditions of the migration tests, the analytical methods used to detect and quantify the active substances in the simulants are essential to guarantee the compliance with the values of migration established, sometimes within the ppm and ppb level. As can be seen in Table **6**, methods depend mainly on the type of substance analyzed. In the case of nanoparticles, the analytical technique mostly used is ICP (inductively coupled plasma) [76] combined with different detectors: MS (mass spectrometry), AES (atomic emission spectrophotometry) or OES (optical emission spectrophotometry) (Table **6**). These techniques detect ultra-trace metals from metallic nanoparticles. To detect and measuring single, individual nanoparticles, SP (single particle) ICP-MS is used [77]. Atomic absorption spectroscopy is also used to detect metallic ions from nanoparticles [73]. Due to their volatility, gas chromatography is the most extensively analytical technique to detect and quantify the main compounds of essential oils [78]. To increase the sensitivity of this technique, a prior extraction technique such as headspace single-drop microextraction HS-SDME (single-drop microextraction) [79, 80] or solid phase microextraction SPME [60, 81] can be used to concentrate the sample. HPLC (high performance liquid chromatography) or UPLC (ultra-high performance liquid chromatography) coupled to different detectors as DAD (Photodiode Array Detection) [82] or MS [83] among others, occupies a leading position in the analysis of phenolic compounds [84] as thymol, quercetin or α-tocopherol (Table **6**). Then, these techniques are used to quantify the migration of compounds of natural substances that are rich in phenolic compounds as essential oils such as cinnamon essential oil or natural extracts such as rosemary (Table **6**).

Although less sensitive, spectrophotometric analysis is also employed with natural compounds that can absorb at different wavelengths. In table **6**, this method is used to measure blueberry pomace, green tea extract and carvacrol. In some cases, compounds are previously subjected to chemical reactions to obtain colored compounds that can be measured using spectrophotometry. This is the case of epigallocatechin gallate that is measured after reaction with the Folin-Ciocalteu reagent and sulfur dioxide released by packaging material containing metabisulfite which is analyzed by using the p-rosaniline method (Table **6**).

Up to date, there have been several strategies to mitigate the migration of active substances from packaging films, by providing a controlled-release of the active agent over time. These strategies include alterations in the polymer structure, polymer coating, polymer blends and composites. Over the past years, nanomaterials such as nanoclays, cellulose nanofibers and crystals, carbon nanotubes, technology has been applied to packaging materials either to be used as fillers, and thus alter polymer structure; or as encapsulating agents for the active compounds and thus control active agent release directly. For a more detailed review [52].

Food Contact Materials

Food contact materials (FCM) refer not only to the packaging materials, but to any material or article in direct or indirect contact with the food product. This definition also involves those materials used in food processing, production and distribution, as during any of the food chain processes, the materials could transfer substances to the food product, resulting in food chemical contamination. The European frame regulation establishes that no FCM should transfer to the food any substance capable of endangering human health, being this the main rule to follow. However, the design of an active packaging implies that some substances will be released from the material and arrive at the food, at levels that could be even higher than those established by the regulation 10/2011/EU of plastics in contact with food; therefore, it is important to say that these active substances have to be approved as food additives, as food ingredients or accepted as active ingredients to be incorporated into the polymers. In the case of active

Table 6. Examples of active packaging materials developed in last five years reporting migration assays.

Compound	Polymer	Migration Assay	Analytical Techniques for Migrant Determination	Migration Results	Safety Remarks	Ref.
Thymol	LDPE	Simulants: ethanol 10% (v/v) and ethanol 95% (v/v) Temperature: 40 °C Time: 7 days Method: immersion	HPLC-DAD	Partition coefficients of 15-82% Diffusion coefficients of 7.5×10^{-13}-3.0×10^{-12}	An ADI of 0.03 mg kg^{-1} body weight (bw) day^{-1} is the agreed value stated within the Commission Review report for thymol (SANCO/10581/2013 rev 3) [85]	[82]
	PLA	Simulants: 15% and 95% ethanol Temperature: 30, 40, 50 and 60 °C (ethanol 95%); 60, 65, 75 and 83 °C (ethanol 15%) Time: 10 hours Method: immersion	GC-MS	increased the temperature increases the release rate of thymol, with equilibrium being reached within 1.4 h 87-100% thymol release to 95% ethanol over the temperature range (30-60 °C) Diffusion coefficients of 0.29-$5.75\text{x}10^{12}$ m^2 s^{-1}		[86]
Carvacrol	Starch/ Polyester	Simulants: ethanol 10%, ethanol 50%, isooctane and acetic acid 3% Temperature: 20 °C Time: 160 h Method: Direct contact	Spectrometry UV-visible	Migration of carvacrol was lower in ethanol 10% and acetic acid 3%. The migration of carvacrol was faster in ethanol 50%	Carvacrol is considered a flavoring substance [87] according to European Regulation	[88]
Curcumin	PVA	Simulant: 50% ethanol/water Temperatures: 25, 35 and 45 °C Time: 10 hours Method: immersion	Spectrophotometric detection at 279 nm	Almost 100% of migration after 1 h of incubation	Clinical trials have shown that curcumin is safe, even when consumed at a daily dosage of 12 g for 3 months According to the Joint FAO/WHO Expert Committee on Food Additives, the estimate of temporary acceptable daily intake for man: 0-0.1 mg kg^{-1} bw [89]	[90]

(Table 6) cont.....

Compound	Polymer	Migration Assay	Analytical Techniques for Migrant Determination	Migration Results	Safety Remarks	Ref.
Green Tea extract	PE	Simulants: 10% and 95% ethanol Temperature: 20 °C Time: 10 days Method: immersion	The volatile compounds were analyzed using SPME-GC–MS and non-volatile compounds were analyzed using UHPLC-QTOF-MS (negative and positive mode)	The highest migration rate was observed for epigallocatechin gallate, gallocatechin gallate and epicatechin gallate in 95% ethanol Lowest migration in 95% ethanol was observed for caffeine Gallocatechin, epigallocatechin gallate and catechin gallate were the most migrating compounds to 10% ethanol Lowest migration in 10% ethanol was observed for gallocatechin gallate	Since 2006, green tea extract used as an active agent is considered as a health claim and it can be used as a food additive without limits, as it is food. However, a 2018 EFSA evaluation on tea catechins safety states that doses of epigallocatechin-3-gallate (EGCG) ≥ 800 mg EGCG/day taken as food supplement has been shown to induce a statistically significant increase of serum transaminases in treated subjects compared to control but that further dose-response studies are needed [91]	[43]
	PET	Food product: IV gamma nectarines Temperature: 4 °C Time: 15 days	UHPLC-MS	Specific migration values of tea compounds were for all the cases below to the limit of detection calculated		[92]
	Cellulose	Simulants: 50% v/v ethanol in water Temperatures: 20 and 40 °C Time: 10 days Method: immersion	Spectrophotometric method with detection at 268 nm	GTE migration was lower at 20 °C (≈35%) than at 40 °C (up to 86%)		[93]
α-Tocopherol	Chitosan	Simulants: 50% EtOH (v/v), 95% EtOH (v/v), and isooctane at 4 and 20 °C Method: immersion	HPLC-DAD	Partition coefficients ≥490 Diffusion coefficients of 1.07×10^{-13}-2.24×10^{-12} cm^2 s^{-1} α-tocopherol migration in 95% EtOH (v/v) was 58.8 (4 °C) and 96.8% (20 °C) At 50% EtOH (v/v), a maximum of 9.6% of α-tocopherol No migration of a-tocopherol from the active films into isooctane	2015 EFSA report states that a Joint FAO/WHO Expert Committee on Food Additives (JECFA) derived an ADI of 0.15–2 mg kg^{-1} bw day^{-1} for α-tocopherol The acute oral toxicity of tocopherols is low NOAEL level of 125 mg kg^{-1} bw day^{-1} [94]	[95]

(Table 6) cont.....

Compound	Polymer	Migration Assay	Analytical Techniques for Migrant Determination	Migration Results	Safety Remarks	Ref.
Olive leaf extract	PE/paper	Simulants: ethanol 10% and acetic acid 3% Temperature: 40 °C Time: 10 days Method: Filled pouches	UPLC–MS	The migration values of all the phenolic compounds identified in olive leaf extract were below the limits of detection in the different simulants studied	2014 EFSA claim states that in one human intervention study showed an increase in glucose tolerance without disproportionate increase in insulin concentrations after daily consumption of the olive leaf water extract for 12 weeks under the conditions of use proposed by the applicant. However, no evidence has been provided in relation to the mechanism by which the olive leaf water extract could exert the claimed effect [96]	[41]
Blueberry pomace	Starch	Ethanol (95%) and 3% acetic acid (v/v) Temperature: 40 °C Time: 10 days Method: immersion	UV-Vis spectra	Release of blueberry pomace was greater into acetic acid Release of phenolic compounds from blueberry pomace was higher when films were immersed into acetic acid Lower amount of phenolic compounds was detected when ethanol was used as simulant	No safety claims or toxicological data available	[97]
Rosemary	Protein	Simulants: 95% ethanol (v/v) and 10% ethanol (v/v) Temperature: 5, 20 and 40 °C Time: 10 days and 30 days (5 °C) Method; single side contact	UHPLC	Maximum EO migration is attained after 7 days of testing Increased temperatures yielded higher EO migration to the simulants	According to 2018 EFSA reports (Question No EFSA-Q-2003-140), only rosemary aqueous and organic extracts are approved to be used as antioxidants, not the essential oil 30-1000 mg kg^{-1} of rosemary extracts are allowed, depending on the food product used [98]	[99]
Quercetin	HDPE/LDPE	Simulant: 95% ethanol Temperature: 37 °C Time: 55 days Method: immersion	HPLC-DAD	Diffusion coefficients of 4.90×10^{-14}- 3.22×10^{-11} cm^2 s^{-1} Quercetin release ranging from 20-100%	Under a worst-case scenario of estimating intake, a person would not be exposed to more than 226 mg quercetin/person/day) from the intended addition of quercetin to foods [100] Although no specific report from EFSA on the substance is available, in Japan, quercetin is allowed as a food additive under the List of Existing Food Additives [100]	[101]

(Table 6) cont.....

Compound	Polymer	Migration Assay	Analytical Techniques for Migrant Determination	Migration Results	Safety Remarks	Ref.
Cinnamon, essential oil	Whey protein	Simulants: 95% ethanol (v/v) Temperature: 5, 20 and 40 °C Time: 10 days Method; single side contact	UHPLC	Eucalyptol is the compound with the highest migration rate and α-pinene with the lowest	Flavoring substances and food ingredients with flavoring properties obtained from vegetable can be used in food unless there is a doubt about their safety [102]	[99]
Natamycin	Chitosan or methylcellulose	Simulants: ethanol 95% Temperature: 40 °C Time: 10 days Method: immersion	HPLC-DAD (diode array detection)	Migration of natamycin was slower from chitosan based films than from methylcellulose films	Natamycin is a food additive (E 232) authorized to use in semi-hard and semi-soft cheese and dry, cured sausage at a maximum level of 1 mg dm^{-2} in the outer 5 mm of the surface [87, 103 - 105]	[106]
Metabisulphite/citric acid	Paper	Simulant: food product (mushrooms) Temperature: 4 °C Time: 11 days	Spectrophotometric detection at 550 nm	No sulfur dioxide was detected in the samples of mushrooms (detection limit of the method-3.2 mg sulfur dioxide g^{-1} of mushrooms).	According to EU regulation 10/2011, the specific migration limit for sulfites incorporated into packaging material is 10 mg sulfur dioxide g^{-1} food [72]	[107]

(Table 6) cont.....

Compound	Polymer	Migration Assay	Analytical Techniques for Migrant Determination	Migration Results	Safety Remarks	Ref.
Silver nanoparticles	PE or HDPE	Simulants: Milli-Q water acetic acid 3% ethanol 10%. Temperature: 40 °C Time: 10 days Method: immersion	Digestion followed by of ICP-MS and single particle ICP-MS	Migration below detection limit in Milli-Q water and 10% ethanol Migration of 2.0-3.1 ng cm^{-2} of silver in 3% acetic acid	Silver nanoparticles: According to EFSA toxicity test should be carried out to specific additive containing Ag particles with a dimension below 100 nm to assure their safety [108]. Moreover, EFSA has establish in 0.05 mg Ag kg^{-1} food the limit for the use of some Ag complexes in food contact material [108] and WHO has estimated a NOAEL of 0.39 mg Ag day^{-1} [109] Titanium dioxide nanoparticles: According to EFSA the effects of TiO_2 nanoparticles were of uncertain biological significance and therefore of limited relevance for the risk assessment of the food additive TiO_2 [110] Titanium oxide is food additive (E 171) [104, 111] authorized to fruit and vegetable preparations excluding compote and processed fish and fisheries products including mollusks and crustaceans. Zinc oxide nanoparticles: According to EFSA ZnO nanoparticles does not migrate in nanoform. Therefore the safety evaluation should focus on the migration of soluble ionic zinc [112] The limit established by European Regulation for ZnO is 25 mg kg^{-1} simulant (2011).	[77]
	Starch/PVA	Simulants: ethanol 10%, Ethanol 20%, ethanol 50%, oleic acid and acetic acid 3% Temperature: 20 °C Time: 7 days Method: immersion	Atomic absorption spectroscopy	Migration was higher in aqueous simulants due the hydrophilic nature of the film The release of silver from films containing silver nanoparticles was lower than those observed with films containing ions		[113]
Titanium dioxide nanoparticles	Soybean polysaccharide	Simulants: water Temperature: 70 °C Method; immersion	ICP-optical emission microscopy (OES)	The maximum migration detected was 21.05 ± 0.054 ppm		[114]
Zinc oxide nanoparticles	Poly(butylene succinate)	Simulants: water, ethanol 10% and acetic acid 3% Time:2 hours - 15 days Method: immersion	Atomic absorption spectroscopy	ZnO released increased with higher concentrations of ZnO in the film. Fast and higher migration was detected in acetic acid 3% (16 ppm)		[73]
Silver and zinc oxide nanoparticles	LDPE	Simulants: aqueous food simulant Temperature: 40 °C Time: 10 days Method: immersion	ICP-MS	Migration of 2.44 ± 0.37 ppm of ZnO were detected. Migration of Ag below detection limit (<0.001 ppm).		[115]
Titanium dioxide and silver nanoparticles	PLA	Simulants: ethanol 50% (v/v) and acetic acid 3% (v/v) Temperature: 25 °C Time: 10, 20, 30, 40 days Method: immersion	ICP-AES	For Ti nanoparticles the maximum migration values were between 2.19 and 3.5 µg kg^{-1} in 3% acetic acid and between 0.59 and 0.99µg kg^{-1} for 50% ethanol. For silver nanoparticles were 3 µg kg^{-1} approx. in 3% acetic acid and 0.8 µg approx. for 50% ethanol.		[116]

(Table 6) cont.....

Compound	Polymer	Migration Assay	Analytical Techniques for Migrant Determination	Migration Results	Safety Remarks	Ref.
Selenium nanoparticles	Adhesive in multilayer	Simulants: 50% ethanol and 3% acetic acid Temperature: 60 ºC Time: 10 days Hazelnuts, walnuts and cheap potatoes	ICP-MS	Non migrating material (LOD ≈ 0.46 ng Se mL^{-1} and LOQ ≈ 1.53 ng Se mL^{-1})	There is not migration over LOD.	[25]

ADI-Acceptable Dairy Intake. NOAEL-no observed adverse effect level.

substances that are entrapped within the polymer matrix and do not need to be released to exert their function, as is the case of radical scavengers, as long as it has been proven that there is no migration of the active agent, the packaging is considered safe.

Food contact materials include plastics, biopolymers, paper and board, glass, lacquers and varnishes, printing inks, adhesives, rubber, silicones and metals. All of them are under the frame regulation, but only some of them are harmonized at European level, what means that they have specific regulation at EU level. National legislations apply to the non-harmonized FCM and the Resolution approved by the Council of Europe (RESAP 2005) that applies to paper and board [117].

As above mentioned, food contact materials safety study usually implies the exposure of the specimen, the FCM, to a simulant that mimics the behavior of the food intended to be in contact with it. Six simulants are established in the EU legislation, as will be later explained in detail. After the exposure, the simulants are analyzed in depth to identify every single compound released from the polymer. Although most of the polymers already available in the market for contact with food comply with the legislation, most of the times there are migrating substances from the polymer to the food, as has been reported for plastics such as polypropylene (PP) [118] and PET [119 - 121], as well as bioplastics such as PLA [122]. This means that no packaging material can be considered as totally inert and most of them release chemical compounds at different concentration levels, with the key point being to keep to a minimum the inevitable migration of these compounds.

One of the main problems dealing with the safety of FCM is the specific migration analysis, which implies the deep analysis of every single compound migrating from the material to the food simulant or to the food. This information is required for applying the risk assessment methodology to each FCM. Many publications have shown that a wide range of substances, both intentionally and non-intentionally added substances (IAS and NIAS, respectively) are found in food simulants after being exposed to the FCM. The main problem in this task is the high difficulty in compound identification, as many of these substances are impurities, degradation products and/or new compounds resulting from the interaction between FCM ingredients during the production or manufacturing steps. The presence of unknown compounds makes it impossible to apply risk assessment methodology to the FCM and this represents a very high concern to the industry and legislators. In order to detect trace amounts of these compounds, analytical chemistry tools are continuously evolving towards new and more sensitive techniques. Although the instrumentation available is sensitive enough, quantification is still a problem, since in many cases, pure standards of the compounds detected are not available, thus requiring "de novo" synthesis of pure compounds or the use of similar chemical families [118, 123].

Other problems related with FCM safety is the presence of oligomers, which can be detected in almost any polymer, including biopolymers. Nowadays, there is a great awareness related with the presence of microplastics in food, but we have to take into consideration that these microplastics are not only small pieces of plastics coming from the micronization of plastic bags and plastic articles but can also result from oligomers migrating from the polymers. In fact, when applying the standardized migration tests, several oligomers can be identified in plastic materials [121] and in some cases the evaporation of the simulants and the deep analysis of the residue allows to measure the size of such oligomers, confirming that they are microplastics. A wide list of oligomers have been found in different polymers, as can be seen in previously published literature [124 - 126].

One special case when dealing with FCM safety is the case of multilayer materials, as many of the FCM available on the market contain a compendium of plastics (conventional ones or biopolymers), paper, aluminum, adhesives, varnishes and printing inks. When considering a combination of materials, one must keep in mind that all materials will contribute to the migration. Printing inks, varnishes and adhesives are a compendium of chemical compounds with some polymeric resin and many of these compounds are of low molecular mass. This situation makes the final format of multilayer materials prone to produce more migration than the single materials. This occurs because, once applied on the material, the compounds reach the equilibrium in the system and the packaging has to be considered as one unique material, as the different layers and contributors cannot be separated. Then, migration from all contributors can take place. For this reason, when testing the FCM for safety, the final format, as a multilayer, should be tested, which implies much more complex and challenging specific migration analysis. Then, the pattern of recognition of each compound found in the migration studies is also very important, as it is critical for improving the safety, substitute some ingredient or change completely the supplier or the kind of material responsible for the migration. For this purpose, the most sensitive and powerful analytical tools are required. High resolution mass spectrometry is needed to identify any single compound migrating. Fig. (**2**) shows an example of a migration test in a multilayer material, where the origin of the different compounds found was identified by UPLC-MS-Q-TOF.

Compound	Tenax® (mg/kg)	Isooctane (mg/kg)	From	SML (mg/kg)
Dodecane (I)	0.004-0.40	0.005-0.06	P/A	?
2,4-Di-tert-butylphenol	0.01-0.04	0.005-0.03	P	60
Butylated hydroxytoluene (II)	0.030-0.033	0.02-0.03	P/A	3
Lactone	0.02-0.13	0.01-0.47	P/A	?
Triphenyl phosphate (III)	0.01-0.02	0.01-0.03	A	0.05*
Caprolactam (III)	nd	0.04-0.19	P	15
Diethyl phthalate (I)	nd	nd	P/A	0.01
Unknown (DEHA) (I)	0.02-0.10	0.002-0.05	A	18
1,2,4-Trichlorobenzene (III)	0.003-0.005	nd	?	?

Tenax®:
Alkanes (I) → P/A
2,4,4-Trimethyl-1-pentanol (I) → P
1,6-Dioxacyclododecane-7,12-dione (I) → A

Tenax®:
3-Tridecanone (II) → P
3-Heptadecanone (II) → P
8-Pentadecanone (II) → P

Fig. (2). Comparison of migration results of a multilayer FCM using isooctane and Tenax as food simulants.

CONCLUDING REMARKS AND FUTURE TRENDS

As we have seen in this chapter, food contact materials either conventional or as active materials aimed at extending the shelf-life of packaged food need to be studied in depth before using them as food containers. Besides the active substances, either antioxidant or antimicrobial, many other different chemical substances at a low concentration level can be transferred from the FCM itself to the food. Many of these migrants are unknown substances and non-intentionally added substances (NIAS). To determine every single chemical migrant from the active package, highly advanced analytical chemistry techniques and methodologies are required as, for instance, high resolution mass spectrometry for identification purposes. Therefore, although active packaging is an attractive option to improve food safety and an excellent way to extend the shelf-life of packaged food, one must always bear in mind that the safety of the packaging material developed has to be insured according to EU regulations. The development of new active packaging systems is a very attractive field of research nowadays but, unfortunately, only a few active materials actually reach the market, as the efficiency and safety of the materials have to be demonstrated in real food at industrial level and most of the new materials have been only tested as laboratory scale or *in vitro*, where no real food is involved.

The analysis of packaging materials, food simulants and the food itself are very challenging, complex and difficult tasks. It implies working with blind samples, as the composition of materials is not disclosed. Therefore, in these cases, experience seems to be the key, because the origin of migrants should also be determined, especially when they are surpassing the established limits. For this reason, safety assessment in active food packaging represents a great analytical challenge and is constantly evolving towards lower limits of detection (<ppb level).

CONSENT FOR PUBLICATION

Not applicable.

CONFLICT OF INTEREST

The authors confirm that this chapter contents have no conflict of interest.

ACKNOWLEDGEMENTS

Declared none.

REFERENCES

[1] United States Environmental Protection Agency (US EPA). Advancing Sustainable Materials Management: Facts and Figures. https://www.epa.gov/smm/advancing-sustainable- materials-management-facts-and-figures

[2] Nerin C. Antioxidant active packaging and antioxidant edible films. Oxidation in foods and beverages and antioxidant applications. Cambridge: Uinted Kingdom 2010; Vol. 2: pp. 496-515. [http://dx.doi.org/10.1533/9780857090331.3.496]

[3] Restuccia D, Spizzirri UG, Parisi OI, *et al.* New EU regulation aspects and global market of active and intelligent packaging for food industry applications. Food Control 2010; 21(11): 1425-35. [http://dx.doi.org/10.1016/j.foodcont.2010.04.028]

[4] European Commission (EC). Regulation (CE) Nº 1935/2004 of the European Parliament and of the Council of 27 October 2004 on materials and articles intended to come into contact with food and repealing Directives 80/590/EEC and 89/109/EEC 2004.

[5] European Commission (EC). Commission Regulation (EC) No 450/2009 of 29 May 2009 on active and intelligent materials and articles intended to come into contact with food 2009.

[6] Dainelli D, Gontard N, Spyropoulos D, Zondervan-van den Beuken E, Tobback P. Active and intelligent food packaging: legal aspects and safety concerns. Trends Food Sci Technol 2008; 19: S103-12. [http://dx.doi.org/10.1016/j.tifs.2008.09.011]

[7] Yildirim S, Rocker B, Pettersen MK, *et al.* Active packaging applications for food. Compr Rev Food Sci Food Saf 2018; 17(1): 165-99. [http://dx.doi.org/10.1111/1541-4337.12322]

[8] Silvagni A, Franco L, Bagno A, Rastrelli F. Thermo-induced lipid oxidation of a culinary oil: The effect of materials used in common food processing on the evolution of oxidised species. Food Chem 2012; 133(3): 754-9. [http://dx.doi.org/10.1016/j.foodchem.2012.01.088]

[9] Ferreira VCS, Morcuende D, Madruga MS, Silva FAP, Estévez M. Role of protein oxidation in the nutritional loss and texture changes in ready-to-eat chicken patties. Int J Food Sci Technol 2018;

53(6): 1518-26.
[http://dx.doi.org/10.1111/ijfs.13733]

[10] Lee DS. Packaging and the microbial shelf-life of food. CRC Press 2010.

[11] Estévez M, Li Z, Soladoye OP, Van-Hecke T. Health Risks of Food Oxidation. Adv Food Nutr Res 2017; 82: 45-81.
[http://dx.doi.org/10.1016/bs.afnr.2016.12.005] [PMID: 28427536]

[12] European Food Safety Authority (EFSA) and European Centre for Disease Prevention and Control. (ECDC). The European Union summary report on trends and sources of zoonoses, zoonotic agents and food-borne outbreaks in 2016. EFSA J 2017; 15(12)e05077

[13] Janjarasskul T, Suppakul P. Active and intelligent packaging: The indication of quality and safety. Crit Rev Food Sci Nutr 2018; 58(5): 808-31.
[http://dx.doi.org/10.1080/10408398.2016.1225278] [PMID: 27645262]

[14] Vilela C, Kurek M, Hayouka Z, *et al.* A concise guide to active agents for active food packaging. Trends Food Sci Technol 2018; 80: 212-22.
[http://dx.doi.org/10.1016/j.tifs.2018.08.006]

[15] Fang Z, Zhao Y, Warner RD, Johnson SK. Active and intelligent packaging in meat industry. Trends Food Sci Technol 2017; 61: 60-71.
[http://dx.doi.org/10.1016/j.tifs.2017.01.002]

[16] Matthaus B. Investigations about the autoxidation of fats and oils by a chemiluminescence method. *Fett Wiss. Technol.-*. Fat Sci Technol 1995; 97(9): 327-9.

[17] Gad AS, Sayd AF. Antioxidant properties of rosemary and Its potential uses as Natural Antioxidant in Dairy Products: A Review. Food Nutr Sci 2015; 06(01): 15.
[http://dx.doi.org/10.4236/fns.2015.61019]

[18] Nanditha B, Prabhasankar P. Antioxidants in bakery products: a review. Crit Rev Food Sci Nutr 2009; 49(1): 1-27.
[http://dx.doi.org/10.1080/10408390701764104] [PMID: 18949596]

[19] Cichello SA. Oxygen absorbers in food preservation: a review. J Food Sci Technol 2015; 52(4): 1889-95.
[http://dx.doi.org/10.1007/s13197-014-1265-2] [PMID: 25829570]

[20] Lopez-Rubio A, Almenar E, Hernandez-Munoz P, Lagaron JM, Catala R, Gavara R. Overview of active polymer-based packaging technologies for food applications. Food Rev Int 2004; 20(4): 357-87.
[http://dx.doi.org/10.1081/FRI-200033462]

[21] Gaikwad KK, Singh S, Lee YS. Oxygen scavenging films in food packaging. Environ Chem Lett 2018; 16(2): 523-38.
[http://dx.doi.org/10.1007/s10311-018-0705-z]

[22] Kerry JP, O'Grady MN, Hogan SA. Past, current and potential utilisation of active and intelligent packaging systems for meat and muscle-based products: A review. Meat Sci 2006; 74(1): 113-30.
[http://dx.doi.org/10.1016/j.meatsci.2006.04.024] [PMID: 22062721]

[23] Sharma R, Ghoshal G. Emerging trends in food packaging. Nutr Food Sci 2018; 48(5): 764-79.
[http://dx.doi.org/10.1108/NFS-02-2018-0051]

[24] Nerín C, Vera P, Canellas E. Active and intelligent food packaging. 2017; p. 700.

[25] Vera P, Echegoyen Y, Canellas E, *et al.* Nano selenium as antioxidant agent in a multilayer food packaging material. Anal Bioanal Chem 2016; 408(24): 6659-70.
[http://dx.doi.org/10.1007/s00216-016-9780-9] [PMID: 27497969]

[26] Nieva-Echevarría B, Manzanos MJ, Goicoechea E, Guillén MD. 2,6-di-tert-butyl-hydroxytoluene and its metabolites in food. Compr Rev Food Sci Food Saf 2015; 14(1): 67-80.
[http://dx.doi.org/10.1111/1541-4337.12121]

[27] Carrizo D, Gullo G, Bosetti O, Nerín C. Development of an active food packaging system with antioxidant properties based on green tea extract. Food Addit Contam Part A Chem Anal Control Expo Risk Assess 2014; 31(3): 364-73.
[http://dx.doi.org/10.1080/19440049.2013.869361] [PMID: 24266419]

[28] Benito-Peña E, González-Vallejo V, Rico-Yuste A, *et al.* Molecularly imprinted hydrogels as functional active packaging materials. Food Chem 2016; 190: 487-94.
[http://dx.doi.org/10.1016/j.foodchem.2015.05.128] [PMID: 26213001]

[29] Ashrafi A, Jokar M, Mohammadi Nafchi A. Preparation and characterization of biocomposite film based on chitosan and kombucha tea as active food packaging. Int J Biol Macromol 2018; 108: 444-54.
[http://dx.doi.org/10.1016/j.ijbiomac.2017.12.028] [PMID: 29223753]

[30] Aytac Z, Keskin NOS, Tekinay T, Uyar T. Antioxidant α-tocopherol/γ-cyclodextrin–inclusion complex encapsulated poly(lactic acid) electrospun nanofibrous web for food packaging. J Appl Polym Sci 2017; 134(21)
[http://dx.doi.org/10.1002/app.44858]

[31] Anbinder PS, Peruzzo PJ, Martino MN, Amalvy JI. Effect of antioxidant active films on the oxidation of soybean oil monitored by fourier transform infrared spectroscopy. J Food Eng 2015; 151: 43-50.
[http://dx.doi.org/10.1016/j.jfoodeng.2014.11.008]

[32] Barbosa-Pereira L, Aurrekoetxea GP, Angulo I, Paseiro-Losada P, Cruz JM. Development of new active packaging films coated with natural phenolic compounds to improve the oxidative stability of beef. Meat Sci 2014; 97(2): 249-54.
[http://dx.doi.org/10.1016/j.meatsci.2014.02.006] [PMID: 24598072]

[33] Busolo MA, Lagaron JM. Antioxidant polyethylene films based on a resveratrol containing clay of interest in food packaging applications. Food Packaging Shelf 2015; 6: 30-41.
[http://dx.doi.org/10.1016/j.fpsl.2015.08.004]

[34] Arfat YA, Benjakul S, Vongkamjan K, Sumpavapol P, Yarnpakdee S. Shelf-life extension of refrigerated sea bass slices wrapped with fish protein isolate/fish skin gelatin-ZnO nanocomposite film incorporated with basil leaf essential oil. J Food Sci Technol 2015; 52(10): 6182-93.
[http://dx.doi.org/10.1007/s13197-014-1706-y] [PMID: 26396365]

[35] Carrizo D, Taborda G, Nerín C, Bosetti O. Extension of shelf-life of two fatty foods using a new antioxidant multilayer packaging containing green tea extract. Innov Food Sci Emerg Technol 2016; 33: 534-41.
[http://dx.doi.org/10.1016/j.ifset.2015.10.018]

[36] Echegoyen Y, Nerín C. Performance of an active paper based on cinnamon essential oil in mushrooms quality. Food Chem 2015; 170: 30-6.
[http://dx.doi.org/10.1016/j.foodchem.2014.08.032] [PMID: 25306314]

[37] de Moraes Crizel T, de Oliveira Rios A, Alves V, Bandarra N, Moldão-Martins M, Hickmann Flôres S. Active food packaging prepared with chitosan and olive pomace. Food Hydrocoll 2018; 74: 139-50.
[http://dx.doi.org/10.1016/j.foodhyd.2017.08.007]

[38] Kanatt SR, Chawla SP. Shelf-life extension of chicken packed in active film developed with mango peel extract. J Food Saf 2018; 38(1)e12385
[http://dx.doi.org/10.1111/jfs.12385]

[39] Echeverría I, López-Caballero ME, Gómez-Guillén MC, Mauri AN, Montero MP. Active nanocomposite films based on soy proteins-montmorillonite- clove essential oil for the preservation of refrigerated bluefin tuna (*Thunnus thynnus*) fillets. Int J Food Microbiol 2018; 266: 142-9.
[http://dx.doi.org/10.1016/j.ijfoodmicro.2017.10.003] [PMID: 29216554]

[40] Pires JRA, de Souza VGL, Fernando AL. Chitosan/montmorillonite bionanocomposites incorporated with rosemary and ginger essential oil as packaging for fresh poultry meat. Food Packaging Shelf

2018; 17: 142-9.
[http://dx.doi.org/10.1016/j.fpsl.2018.06.011]

[41] Moudache M, Nerín C, Colon M, Zaidi F. Antioxidant effect of an innovative active plastic film containing olive leaves extract on fresh pork meat and its evaluation by Raman spectroscopy. Food Chem 2017; 229: 98-103.
[http://dx.doi.org/10.1016/j.foodchem.2017.02.023] [PMID: 28372258]

[42] Muppalla SR, Chawla SP. Effect of Gum Arabic-polyvinyl alcohol films containing seed cover extract of *Zanthoxylum rhetsa* on shelf-life of refrigerated ground chicken meat. J Food Saf 2018; 38(4)e12460
[http://dx.doi.org/10.1111/jfs.12460]

[43] Wrona M, Nerín C, Alfonso MJ, Caballero MÁ. Antioxidant packaging with encapsulated green tea for fresh minced meat. Innov Food Sci Emerg Technol 2017; 41: 307-13.
[http://dx.doi.org/10.1016/j.ifset.2017.04.001]

[44] Vera P, Canellas E, Nerín C. New antioxidant multilayer packaging with nanoselenium to enhance the shelf-life of marked food product. Nanomaterials (Basel) 2018; 8(10): 837.
[http://dx.doi.org/10.3390/nano8100837] [PMID: 30332731]

[45] Gram L, Ravn L, Rasch M, Bruhn JB, Christensen AB, Givskov M. Food spoilage interactions between food spoilage bacteria. Int J Food Microbiol 2002; 78(1-2): 79-97.
[http://dx.doi.org/10.1016/S0168-1605(02)00233-7] [PMID: 12222639]

[46] Petruzzi L, Corbo MR, Sinigaglia M, Bevilacqua A. Chapter 1 - Microbial Spoilage of Foods: Fundamentals. In: Bevilacqua A, Corbo MR, Sinigaglia M, Eds. The Microbiological Quality of Food. Woodhead Publishing 2017; pp. 1-21.

[47] Malhotra B, Keshwani A, Kharkwal H. Antimicrobial food packaging: potential and pitfalls. Front Microbiol 2015; 6(611): 611.
[PMID: 26136740]

[48] Stanborough T, Fegan N, Powell SM, Singh T, Tamplin M, Chandry PS. Genomic and metabolic characterization of spoilage-associated *Pseudomonas* species. Int J Food Microbiol 2018; 268: 61-72.
[http://dx.doi.org/10.1016/j.ijfoodmicro.2018.01.005] [PMID: 29335226]

[49] Comi G. Chapter 8 - Spoilage of Meat and Fish. In: Bevilacqua A, Corbo MR, Sinigaglia M, Eds. The Microbiological Quality of Food. Woodhead Publishing 2017; pp. 179-210.

[50] Becerril R, Manso S, Nerin C, Gómez-Lus R. Antimicrobial activity of Lauroyl Arginate Ethyl (LAE), against selected food-borne bacteria. Food Control 2013; 32(2): 404-8.
[http://dx.doi.org/10.1016/j.foodcont.2013.01.003]

[51] Biji KB, Ravishankar CN, Mohan CO, Srinivasa Gopal TK. Smart packaging systems for food applications: a review. J Food Sci Technol 2015; 52(10): 6125-35.
[http://dx.doi.org/10.1007/s13197-015-1766-7] [PMID: 26396360]

[52] Nerin C, Silva F, Manso S, Becerril R. Chapter 6 - The Downside of Antimicrobial Packaging: Migration of Packaging Elements into Food. In: Jorge Barros-Velazquez , Ed. Antimicrobial Food Packaging. San Diego: Academic Press 2016; pp. 81-93.

[53] Burt S. Essential oils: their antibacterial properties and potential applications in foods-a review. Int J Food Microbiol 2004; 94(3): 223-53.
[http://dx.doi.org/10.1016/j.ijfoodmicro.2004.03.022] [PMID: 15246235]

[54] Gutierrez J, Barry-Ryan C, Bourke P. Antimicrobial activity of plant essential oils using food model media: efficacy, synergistic potential and interactions with food components. Food Microbiol 2009; 26(2): 142-50.
[http://dx.doi.org/10.1016/j.fm.2008.10.008] [PMID: 19171255]

[55] Akbar A, Anal AK. Zinc oxide nanoparticles loaded active packaging, a challenge study against *Salmonella* Typhimurium and *Staphylococcus aureus* in ready-to-eat poultry meat. Food Control 2014;

38: 88-95.
[http://dx.doi.org/10.1016/j.foodcont.2013.09.065]

[56] Beigmohammadi F, Peighambardoust SH, Hesari J, Azadmard-Damirchi S, Peighambardoust SJ, Khosrowshahi NK. Antibacterial properties of LDPE nanocomposite films in packaging of UF cheese. Lebensm Wiss Technol 2016; 65: 106-11.
[http://dx.doi.org/10.1016/j.lwt.2015.07.059]

[57] Ahmed J, Hiremath N, Jacob H. Antimicrobial efficacies of essential oils/nanoparticles incorporated polylactide films against *L. monocytogenes* and *S.* Typhimurium on contaminated cheese. Int J Food Prop 2017; 20(1): 53-67.
[http://dx.doi.org/10.1080/10942912.2015.1131165]

[58] Battisti R, Fronza N, Vargas Júnior Á, Silveira SM, Damas MSP, Quadri MGN. Gelatin-coated paper with antimicrobial and antioxidant effect for beef packaging. Food Packaging Shelf 2017; 11: 115-24.
[http://dx.doi.org/10.1016/j.fpsl.2017.01.009]

[59] Kashiri M, Cerisuelo JP, Domínguez I, *et al.* Zein films and coatings as carriers and release systems of *Zataria multiflora Boiss*. essential oil for antimicrobial food packaging. Food Hydrocoll 2017; 70: 260-8.
[http://dx.doi.org/10.1016/j.foodhyd.2017.02.021]

[60] Gherardi R, Becerril R, Nerin C, Bosetti O. Development of a multilayer antimicrobial packaging material for tomato puree using an innovative technology. Lebensm Wiss Technol 2016; 72: 361-7.
[http://dx.doi.org/10.1016/j.lwt.2016.04.063]

[61] Silva F, Domingues FC, Nerín C. Control microbial growth on fresh chicken meat using pinosylvin inclusion complexes based packaging absorbent pads. Lebensm Wiss Technol 2018; 89: 148-54.
[http://dx.doi.org/10.1016/j.lwt.2017.10.043]

[62] Emam-Djomeh Z, Moghaddam A, Yasini Ardakani SA. Antimicrobial activity of pomegranate (*Punica granatum L.*) peel extract, physical, mechanical, barrier and antimicrobial properties of pomegranate peel extract-incorporated sodium caseinate film and application in packaging for ground beef. Packag Technol Sci 2015; 28(10): 869-81.
[http://dx.doi.org/10.1002/pts.2145]

[63] Lone A, Anany H, Hakeem M, *et al.* Development of prototypes of bioactive packaging materials based on immobilized bacteriophages for control of growth of bacterial pathogens in foods. Int J Food Microbiol 2016; 217: 49-58.
[http://dx.doi.org/10.1016/j.ijfoodmicro.2015.10.011] [PMID: 26490649]

[64] Radford D, Guild B, Strange P, Ahmed R, Lim LT, Balamurugan S. Characterization of antimicrobial properties of *Salmonella* phage Felix O1 and *Listeria* phage A511 embedded in xanthan coatings on Poly(lactic acid) films. Food Microbiol 2017; 66: 117-28.
[http://dx.doi.org/10.1016/j.fm.2017.04.015] [PMID: 28576360]

[65] Soysal Ç, Bozkurt H, Dirican E, *et al.* Effect of antimicrobial packaging on physicochemical and microbial quality of chicken drumsticks. Food Control 2015; 54: 294-9.
[http://dx.doi.org/10.1016/j.foodcont.2015.02.009]

[66] Woraprayote W, Pumpuang L, Tosukhowong A, *et al.* Antimicrobial biodegradable food packaging impregnated with Bacteriocin 7293 for control of pathogenic bacteria in pangasius fish fillets. Lebensm Wiss Technol 2018; 89: 427-33.
[http://dx.doi.org/10.1016/j.lwt.2017.10.026]

[67] Barbiroli A, Musatti A, Capretti G, Iametti S, Rollini M. Sakacin-A antimicrobial packaging for decreasing *Listeria* contamination in thin-cut meat: preliminary assessment. J Sci Food Agric 2017; 97(3): 1042-7.
[http://dx.doi.org/10.1002/jsfa.8120] [PMID: 27790709]

[68] Luz C, Calpe J, Saladino F, *et al.* Antimicrobial packaging based on ε-polylysine bioactive film for the control of mycotoxigenic fungi *in vitro* and in bread. J Food Process Preserv 2018; 42(1)e13370

[http://dx.doi.org/10.1111/jfpp.13370] [PMID: 29456275]

[69] Dehnad D, Mirzaei H, Emam-Djomeh Z, Jafari SM, Dadashi S. Thermal and antimicrobial properties of chitosan-nanocellulose films for extending shelf-life of ground meat. Carbohydr Polym 2014; 109: 148-54.
[http://dx.doi.org/10.1016/j.carbpol.2014.03.063] [PMID: 24815411]

[70] Azadbakht E, Maghsoudlou Y, Khomiri M, Kashiri M. Development and structural characterization of chitosan films containing *Eucalyptus globulus* essential oil: Potential as an antimicrobial carrier for packaging of sliced sausage. Food Packaging Shelf 2018; 17: 65-72.
[http://dx.doi.org/10.1016/j.fpsl.2018.03.007]

[71] European Commission (EC). Commission, E EU Guidance to the Commission Regulation (EC) No 450/2009 of 29May 2009 on active and intelligent materials and articles intended tocome into contact with food 2011.

[72] European Commission (EC). Commission Regulation (EU) No 1282/2011 of 28 November 2011 amending and correcting Commission Regulation (EU) No 10/2011 on plastic materials and articles intended to come into contact with food Text with EEA relevance 2011.

[73] Petchwattana N, Covavisaruch S, Wibooranawong S, Naknaen P. Antimicrobial food packaging prepared from poly(butylene succinate) and zinc oxide. Measurement 2016; 93: 442-8.
[http://dx.doi.org/10.1016/j.measurement.2016.07.048]

[74] Nerin C. Plastics and Polymers for Food Packaging Manufacturing. Reference Module in Food Science. 2016.
[http://dx.doi.org/10.1016/B978-0-08-100596-5.03195-4]

[75] Aznar M, Rodriguez-Lafuente A, Alfaro P, Nerin C. UPLC-Q-TOF-MS analysis of non-volatile migrants from new active packaging materials. Anal Bioanal Chem 2012; 404(6-7): 1945-57.
[http://dx.doi.org/10.1007/s00216-012-6247-5] [PMID: 22836481]

[76] Otoni CG, Espitia PJP, Avena-Bustillos RJ, McHugh TH. Trends in antimicrobial food packaging systems: Emitting sachets and absorbent pads. Food Res Int 2016; 83: 60-73.
[http://dx.doi.org/10.1016/j.foodres.2016.02.018]

[77] Mackevica A, Olsson ME, Hansen SF. Silver nanoparticle release from commercially available plastic food containers into food simulants. J Nanopart Res 2016; 18(1): 5.
[http://dx.doi.org/10.1007/s11051-015-3313-x]

[78] Calo JR, Crandall PG, O'Bryan CA, Ricke SC. Essential oils as antimicrobials in food systems – A review. Food Control 2015; 54: 111-9.
[http://dx.doi.org/10.1016/j.foodcont.2014.12.040]

[79] López P, Sánchez C, Batlle R, Nerín C. Development of flexible antimicrobial films using essential oils as active agents. J Agric Food Chem 2007; 55(21): 8814-24.
[http://dx.doi.org/10.1021/jf071737b] [PMID: 17880148]

[80] Rodríguez A, Batlle R, Nerín C. The use of natural essential oils as antimicrobial solutions in paper packaging. Part II. Prog Org Coat 2007; 60(1): 33-8.
[http://dx.doi.org/10.1016/j.porgcoat.2007.06.006]

[81] Pezo D, Salafranca J, Nerín C. Development of an automatic multiple dynamic hollow fibre liquid-phase microextraction procedure for specific migration analysis of new active food packagings containing essential oils. J Chromatogr A 2007; 1174(1-2): 85-94.
[http://dx.doi.org/10.1016/j.chroma.2007.08.033] [PMID: 17765253]

[82] Torres A, Romero J, Macan A, Guarda A, Galotto MJ. Near critical and supercritical impregnation and kinetic release of thymol in LLDPE films used for food packaging. J Supercrit Fluids 2014; 85: 41-8.
[http://dx.doi.org/10.1016/j.supflu.2013.10.011]

[83] López de Dicastillo C, Nerín C, Alfaro P, Catalá R, Gavara R, Hernández-Muñoz P. Development of new antioxidant active packaging films based on ethylene vinyl alcohol copolymer (EVOH) and green

tea extract. J Agric Food Chem 2011; 59(14): 7832-40.
[http://dx.doi.org/10.1021/jf201246g] [PMID: 21682295]

[84] Spácil Z, Nováková L, Solich P. Analysis of phenolic compounds by high performance liquid chromatography and ultra performance liquid chromatography. Talanta 2008; 76(1): 189-99.
[http://dx.doi.org/10.1016/j.talanta.2008.02.021] [PMID: 18585262]

[85] European Food Safety Authority (EFSA). Outcome of the consultation with Member States, the applicant and EFSA on the pesticide risk assessment for thymol in light of confirmatory data. EFSA Supporting Publications 2017; 14(1): 1162E.
[http://dx.doi.org/10.2903/sp.efsa.2017.EN-1162]

[86] Tawakkal ISMA, Cran MJ, Bigger SW. Release of thymol from poly(lactic acid)-based antimicrobial films containing kenaf fibres as natural filler. Lebensm Wiss Technol 2016; 66: 629-37.
[http://dx.doi.org/10.1016/j.lwt.2015.11.011]

[87] European Commission (EC). Commision implementing regulation (EU) No 872/2012of 1 October 2012adopting the list of flavouring substances provided for by Regulation (EC) No 2232/96 of theEuropean Parliament and of the Council, introducing it in Annex I to Regulation (EC) No1334/2008 of the European Parliament and of the Council and repealing Commission Regulation(EC) No 1565/2000 and Commission Decision 1999/217/EC. Off J Eur Union 2012.

[88] Requena R, Vargas M, Chiralt A. Obtaining antimicrobial bilayer starch and polyester-blend films with carvacrol. Food Hydrocoll 2018; 83: 118-33.
[http://dx.doi.org/10.1016/j.foodhyd.2018.04.045]

[89] World Health Organization (WHO). Toxicological Evaluation of Some Food Colours. Enzymes, Flavour Enhancers, Thickening Agents, and Certain Other Food Additives 1975.

[90] Ma QY, Ren YM, Wang LJ. Investigation of antioxidant activity and release kinetics of curcumin from tara gum/polyvinyl alcohol active film. Food Hydrocoll 2017; 70: 286-92.
[http://dx.doi.org/10.1016/j.foodhyd.2017.04.018]

[91] Younes M, Aggett P, Aguilar F, *et al.* Scientific opinion on the safety of green tea catechins. EFSA J 2018; 16(4)e05239

[92] Colon M, Nerin C. Development and application of an analytical procedure for specific migration of green tea compounds in IV gamma nectarine active packaging. Food Control 2015; 57: 419-25.
[http://dx.doi.org/10.1016/j.foodcont.2015.04.035]

[93] Wrona M, Cran MJ, Nerín C, Bigger SW. Development and characterisation of HPMC films containing PLA nanoparticles loaded with green tea extract for food packaging applications. Carbohydr Polym 2017; 156: 108-17.
[http://dx.doi.org/10.1016/j.carbpol.2016.08.094] [PMID: 27842804]

[94] European Food Safety Authority (EFSA). Scientific Opinion on the re-evaluation of tocopherol-rich extract (E 306), α-tocopherol (E 307), γ-tocopherol (E 308) and δ-tocopherol (E 309) as food additives. EFSA J 2015; 13(9): 4247.

[95] Otero-Pazos P, Sendón R, Blanco-Fernandez B, *et al.* Preparation of antioxidant active films based on chitosan: diffusivity study of α-tocopherol into food simulants. J Food Sci Technol 2016; 53(6): 2817-26.
[http://dx.doi.org/10.1007/s13197-016-2256-2] [PMID: 27478238]

[96] European Food Safety Authority (EFSA). Scientific Opinion on the substantiation of a health claim related to olive (*Olea europaea L.*) leaf water extract and increase in glucose tolerance pursuant to Article 13(5) of Regulation (EC) No 1924/2006. EFSA J 2014; 12(5): 3655.

[97] Luchese CL, Garrido T, Spada JC, Tessaro IC, de la Caba K. Development and characterization of cassava starch films incorporated with blueberry pomace. Int J Biol Macromol 2018; 106: 834-9.
[http://dx.doi.org/10.1016/j.ijbiomac.2017.08.083] [PMID: 28823702]

[98] European Food Safety Authority (EFSA). Use of rosemary extracts as a food additive - Scientific

Opinion of the Panel on Food Additives, Flavourings, Processing Aids and Materials in Contact with Food. EFSA J 2008; 6(6): 721.
[http://dx.doi.org/10.2903/j.efsa.2008.721]

[99] Ribeiro-Santos R, Ramos-de-Melo N, Andrade M, Sanches-Silva A. Potential of migration of active compounds from protein-based films with essential oils to a food and a food simulant. Packag Technol Sci 2017; 30(12): 791-8.
[http://dx.doi.org/10.1002/pts.2334]

[100] Harwood M, Danielewska-Nikiel B, Borzelleca JF, Flamm GW, Williams GM, Lines TC. A critical review of the data related to the safety of quercetin and lack of evidence of *in vivo* toxicity, including lack of genotoxic/carcinogenic properties. Food Chem Toxicol 2007; 45(11): 2179-205.
[http://dx.doi.org/10.1016/j.fct.2007.05.015] [PMID: 17698276]

[101] Lu LJ, Lu LX. Preparation and properties of quercetin-incorporated high density polyethylene/low density polyethylene antioxidant multilayer film. Packag Technol Sci 2018; 31(6): 433-9.
[http://dx.doi.org/10.1002/pts.2371]

[102] European Commission (EC). Regulation (EC) No 1334/2008 of the European Parliament and of the Council of 16 December 2008 on flavourings and certain food ingredients with flavouring properties for use in and on foods and amending Council Regulation (EEC) No 1601/91, Regulations (EC) No 2232/96 and (EC) No 110/2008 and Directive 2000/13/EC (Text with EEA relevance). **2008**.

[103] European Commission (EC). Commission regulation (EU) No 1129/2011of 11 November 2011amending Annex II to Regulation (EC) No 1333/2008 of the European Parliament and of theCouncil by establishing a Union list of food additives. Off J Eur Union 2011.

[104] European Food Safety Authority (EFSA). Scientific Opinion on the safety evaluation of the substance zinc oxide, nanoparticles, uncoated and coated with [3-(methacryloxy)propyl] trimethoxysilane, for use in food contact materials. EFSA J 2015; 13(4)

[105] European Commission (EC). Commission regulation (EU) 2015/647 of 24 April 2015 amending and correcting Annexes II and III to Regulation (EC) No 1333/2008 of the European Parliament and of the Council as regards the use of certain food additives Off J Eur Union 2015.

[106] Santonicola S, García Ibarra V, Sendón R, Mercogliano R, Rodríguez-Bernaldo de Quirós A. Antimicrobial films based on chitosan and methylcellulose containing natamycin for active packaging applications. Coatings 2017; 7(10): 177.
[http://dx.doi.org/10.3390/coatings7100177]

[107] Wrona M, Bentayeb K, Nerín C. A novel active packaging for extending the shelf-life of fresh mushrooms (*Agaricus bisporus*). Food Control 2015; 54: 200-7.
[http://dx.doi.org/10.1016/j.foodcont.2015.02.008]

[108] European Food Safety Authority (EFSA). Scientific opinion on the re-evaluation of silver (E 174) as food additive. EFSA J 2006; 14(1): 4364.

[109] World Health Organization (WHO). Silver in Drinking-water. Guidelines for Drinking-Water Quality 2003; p. 2.

[110] European Food Safety Authority (EFSA) Re-evaluation of titanium dioxide (E 171) as a foodadditive. EFSA J 2016; 14(9): 4545.

[111] European Commission (EC). Commission regulation (EU) No 738/2013 of 30 July 2013 amending Annex II to Regulation (EC) No 1333/2008 of the European Parliament and of the Council as regards the use of certain additives in seaweed based fish roe analogues. Off J Eur Union 2013.

[112] European Food Safety Authority (EFSA). Safety assessment of the substance zinc oxide, nanoparticles, for use in food contact materials. EFSA J 2016; 14(3)

[113] Cano A, Cháfer M, Chiralt A, González-Martínez C. Development and characterization of active films based on starch-PVA, containing silver nanoparticles. Food Packaging Shelf 2016; 10: 16-24.
[http://dx.doi.org/10.1016/j.fpsl.2016.07.002]

[114] Salarbashi D, Tafaghodi M, Bazzaz BSF. Soluble soybean polysaccharide/TiO_2 bionanocomposite film for food application. Carbohydr Polym 2018; 186: 384-93. [http://dx.doi.org/10.1016/j.carbpol.2017.12.081] [PMID: 29456001]

[115] Panea B, Ripoll G, González J, Fernández-Cuello Á, Albertí P. Effect of nanocomposite packaging containing different proportions of ZnO and Ag on chicken breast meat quality. J Food Eng 2014; 123: 104-12. [http://dx.doi.org/10.1016/j.jfoodeng.2013.09.029]

[116] Li W, Zhang C, Chi H, *et al.* Development of antimicrobial packaging film made from poly(lactic acid) incorporating titanium dioxide and silver nanoparticles. Molecules 2017; 22(7): 1170. [http://dx.doi.org/10.3390/molecules22071170] [PMID: 28703753]

[117] Baughan JS. 9 - Global legislation for paper and card materials in contact with food. Global Legislation for Food Contact Materials 2015; 201-10.

[118] Vera P, Canellas E, Nerín C. Identification of non volatile migrant compounds and NIAS in polypropylene films used as food packaging characterized by UPLC-MS/QTOF. Talanta 2018; 188: 750-62. [http://dx.doi.org/10.1016/j.talanta.2018.06.022] [PMID: 30029443]

[119] Dreolin N, Aznar M, Moret S, Nerin C. Development and validation of a LC-MS/MS method for the analysis of bisphenol a in polyethylene terephthalate. Food Chem 2019; 274: 246-53. [http://dx.doi.org/10.1016/j.foodchem.2018.08.109] [PMID: 30372934]

[120] Osorio J, Úbeda S, Aznar M, Nerín C. Analysis of isophthalaldehyde in migration samples from polyethylene terephthalate packaging. Food Addit Contam Part A Chem Anal Control Expo Risk Assess 2018; 35(8): 1645-52. [http://dx.doi.org/10.1080/19440049.2018.1465208] [PMID: 29718767]

[121] Ubeda S, Aznar M, Nerín C. Determination of oligomers in virgin and recycled polyethylene terephthalate (PET) samples by UPLC-MS-QTOF. Anal Bioanal Chem 2018; 410(9): 2377-84. [http://dx.doi.org/10.1007/s00216-018-0902-4] [PMID: 29428989]

[122] Aznar M, Ubeda S, Dreolin N, Nerín C. Determination of non-volatile components of a biodegradable food packaging material based on polyester and polylactic acid (PLA) and its migration to food simulants. J Chromatogr A 2019; 1583: 1-8. [http://dx.doi.org/10.1016/j.chroma.2018.10.055] [PMID: 30477716]

[123] Félix JS, Isella F, Bosetti O, Nerín C. Analytical tools for identification of non-intentionally added substances (NIAS) coming from polyurethane adhesives in multilayer packaging materials and their migration into food simulants. Anal Bioanal Chem 2012; 403(10): 2869-82. [http://dx.doi.org/10.1007/s00216-012-5965-z] [PMID: 22526644]

[124] Carrizo D, Maccagnan A, Felix JS, Nerin C, Bosetti O. The barrier effect of EVOH versus 1,4,7-triaxocyclotridecane-8,13-dione, a non-intentionally added compound from polyurethane adhesives in multilayer food packaging. Packag Technol Sci 2015; 28(12): 1039-46. [http://dx.doi.org/10.1002/pts.2182]

[125] Úbeda S, Aznar M, Vera P, *et al.* Overall and specific migration from multilayer high barrier food contact materials - kinetic study of cyclic polyester oligomers migration. Food Addit Contam Part A Chem Anal Control Expo Risk Assess 2017; 34(10): 1784-94. [http://dx.doi.org/10.1080/19440049.2017.1346390] [PMID: 28665763]

[126] Hoppe M, de Voogt P, Franz R. Identification and quantification of oligomers as potential migrants in plastics food contact materials with a focus in polycondensates - A review. Trends Food Sci Technol 2016; 50: 118-30. [http://dx.doi.org/10.1016/j.tifs.2016.01.018]

CHAPTER 2

Microplastics and Nanoplastics in Food

Pilar Fernández-Hernando[1], **Rosa Mª Garcinuño-Martínez**[1,*] and **Esther Garrido-Gamarro**[2]

[1] *Departamento de Ciencias Analíticas, Facultad de Ciencias, Universidad Nacional de Educación a Distancia (UNED), Paseo Senda del Rey nº 9, 28040, Madrid, Spain*

[2] *Food Safety and Quality Officer, Fisheries and Aquaculture, Department, Food and Agriculture Organization of the United Nations (FAO), Viale delle Terme di Caracalla, 00153 Roma RM, Italy*

Abstract: Plastic production has exponentially increased since the 1950s and reached 322 million tons in 2015. It is expected that the production of microplastic will continue increasing to at least double the production of 2015. As documented in laboratory and field studies, marine organisms of commercial importance for fisheries and aquaculture are affected by microplastics ingestion not only due to the additives used in their manufacture but also because microplastics act as absorbents of persistent organic pollutants (POPs) from the environment. The ingestion of microplastics by aquatic organisms pose a risk to marine environment and food safety. Although microplastics are a human health hazard, their effects on seafood is attenuated by the extraction of the gastrointestinal tract. However, shellfish and other species of crustaceous consumed whole pose a particular concern for human exposure. This chapter discusses the problems associated with microplastics ingested by marine organisms. The most common methods used for sampling, identification, and quantification of microplastics are mentioned and some analytical methods to determine plastic additives and POPs adsorbed on the microplastics in different marine environment matrices are described. Microplastic dietary intake and the limitations for food safety risk assessment are also addressed. Since 2004, many types of research have focused on this topic and analyzed microplastics in various environmental matrices. However, the development of standardized methods for the screening, identification, detection, and quantification of microplastics in marine environment remains a challenge.

Keywords: Additives, Analytical methods, Environmental matrices, Food safety, Microplastics, Risk assessment, Seafood.

INTRODUCTION

Today, plastics have become one of the most utilized materials worldwide, being

* **Corresponding author Rosa Mª Garcinuño-Martínez:** Departamento de Ciencias Analíticas, Facultad de Ciencias, Universidad Nacional de Educación a Distancia (UNED), Paseo Senda del Rey nº 9, 28040, Madrid, Spain; Tel: +34913987366; E-mail: rmgarcinuno@ccia.uned.es

Belén Gómara & María Luisa Marina (Eds.)
Current and Future Developments in Food Science (Vol. 1)

an essential constituent of daily life items.

Since their first development in the 1800s, plastics production has suffered changes to meet the needs of a variety of sectors and consumers and has enabled technological improvements and solutions. Due to their functional properties, plastics displaced traditional materials. Plastics are a range of synthetic or semi-synthetic materials delivered from fossil resources and organic products and are usually divided into three categories: thermoplastics (polymers that can be remelted), thermosets (polymers that remain in a permanent solid state once hardened) and elastomers (elastic polymers that return to their original shape). Depending on the intended use of the plastic, polymers with different physical and chemical properties can be mixed among them and with additives such as plasticizers, flame-retardants, colorants, and antioxidants to enhance plastic performance, which can complicate recycling and the evaluation of their impact on the environment and on human health. The increasing production of plastic requires efficient waste management systems that few countries can implement. For this reason, it is estimated that most plastics persist in the environment whole or fragmented, contributing to plastics and microplastics pollution. Microplastics can become a food safety threat when they get into the food chain and have been found in a variety of food commodities such as salt, beer, honey or fish. Seafood is the best-studied species concerning microplastic intake.

MICROPLASTICS AND NANOPLASTICS DEFINITION

There is an ongoing debate about what can be considered microplastics and nanoplastics. One of the most acknowledged definitions describes microplastics as plastic particles composed of a heterogeneous mixture of different shaped materials in the range of 0.1-5000 μm [1, 2]. Nanoplastics are identified as plastic particles whose size is ranging from 0.001 μm to 0.1 μm [3]. Although the size of microplastics is an important factor that determines their impact in the living organisms that ingest them, shape might also be an influencing factor. There are two types of plastics based on how they are produced or generated. Primary microplastics are produced for industrial purposes such as plastic manufacture or cosmetics.

Secondary microplastics are generated by weathering processes and fragmentation of larger plastics [4], which may occur when plastics are disposed of in the environment or when using plastic products such as textiles or tires. Eventually, both types of microplastics will end up polluting the environment and entering food supply chains.

MICROPLASTICS AND NANOPLASTICS COMPOSITION

As mentioned above, plasticizers, flame-retardants or antioxidants are used as additives in plastics. When plastics reach the environment, they can adsorb or absorb contaminants from the surroundings, such as polycyclic aromatic hydrocarbons (PAHs), polychlorinated biphenyls (PCBs), chlorinated pesticides, which are considered persistent organic pollutants (POPs), trace metals, and microorganisms such as pathogenic bacteria or viruses [5].

Polymers

Monomers, such as ethylene, propylene, and styrene are used as building blocks of polymers that lead to the production of a variety of materials. The most commonly used polymers are: acrylonitrile butadiene styrene (ABS), acrylic (AC), epoxy resin (EP), expanded polystyrene (EPS), polyethylene high density (HDPE), polyethylene low density (LDPE), polyethylene linear low density (LLDPE), polyamide (Nylon) 4, 6, 11, 66 (PA), polycarbonate (PC), polycaprolactone (PCL), polyethylene (PE), polyethylene terephthalate (PET), poly (glycolic) acid (PGA), poly(lactide) (PLA), poly(methyl methacrylate) (PMMA), polypropylene (PP), polystyrene (PS), polyurethane (PU), polyvinyl alcohol (PVA), polyvinyl chloride (PVC), styrene-butadiene rubber (SBR) and thermoplastic polyurethane (TPU) [5]. All these polymers can be expected in microplastic pollution and therefore get into different food chains.

Flame-Retardants

At present, more than 175 chemicals are classified as flame-retardants (FRs) [6]. These compounds are commonly added to polymers to reduce the flammability of plastics and some are not normally added to polymers in processing, but can be found in a polymer matrix from leaching out of the contents. Brominated flame retardants contain a wide variety of organic compounds including polybrominated diphenyl ethers (PBDEs) and hexabromocyclododecanes (HBCDs) that are the most used chemicals in the manufacture of plastic. They are commonly added to polystyrene, polyesters, polyolefins, polyamides, epoxies, and ABS. HBCDs and PBDEs are used by simple blending with the polymers, therefore these compounds are most likely to leach out of the final products [7]. This poses an environmental and food safety concern because PBDEs and HBCD are considered POPs by the Stockholm Convention and many studies associate them with endocrine disorders, teratogenicity, and kidney and liver toxicity [8, 9].

Plasticizers

Substances such as phthalates and bisphenols (BPs) are used as plasticizers, usually being added to synthetic polymers such as PE, PP and/or PVC to enhance flexibility and softness, and reduce brittleness. Both phthalates and bisphenol A (BPA) have been found to be endocrine disruptors while the toxicity of other BPs remains unknown [10].

Antioxidants and Stabilizers

Nonylphenols (NPs) are organic compounds widely used as stabilizers in plastic food packaging and act as antioxidant agents in some polymers such as rubber, vinyl, polyolefins, polystyrenes and PVC [11]. NPs are known to be endocrine disruptors and have been reported to have synergistic effects due to co-occurrence with some chemicals such as plasticizers, BPA, and PBDEs [12, 13].

PRESENCE OF MICROPLASTICS AND NANOPLASTICS IN FOOD

Microplastics pollution has become a potential food safety threat that is especially relevant for fishery products as well as other sea products such as table salt. Although microplastics have been reported in products such as honey and sugar [14] or beer [15], sea products and water seem to be the most researched samples of microplastics uptake.

Fisheries and Aquaculture Products

Ingestion of microplastics by aquatic organisms has been reported in numerous studies, being found in twelve out of the twenty-five most consumed species that constitute the world marine fisheries [5]. Small concentrations of microplastics of around 1 to 2 particles per fish have been detected in a large number of important commercial species such as sardine, mackerel, anchovy, herring and sprat from the Pacific, Atlantic and Indian oceans and the Mediterranean Sea [16, 17]. Other marine and freshwater species of local and regional relevance have been observed to contain microplastics particles as well [5]. Most microplastics have been found in the gastrointestinal tract of fish, where particles seem to concentrate after ingestion. However, several studies in marine organisms have shown that smaller microplastics and nanoplastics could be translocated in other organs such as liver, although the translocation pathways are not well understood yet [18]. Bivalve mollusks have also been found to contain microplastics particles. One of the best-studied species is mussels, where their occurrence has been reported in Europe, North America, Brazil, and China. A concentration of microplastics in mussels ranging between 0.5 particle g^{-1} in Europe to 4 particles g^{-1} in China was observed [2].

Microplastics have also been observed in crustaceans and echinoderms, in important food commodities such as common shrimp and Norway lobster in waters from the coast of North Sea and Irish Sea, with concentrations ranging from 0.03 to 1.92 particles g^{-1}, depending on location and sampling date, and mostly found in the digestive tract [19].

Salt

Microplastics inevitably contaminate salt and other sea products and their occurrence has been reported in different countries such as Italy, Croatia, and Spain. The average microplastic content in Italy fluctuated from 1.57 (high-quality salt) to 8.23 (low-quality salt) particles g^{-1}, with a size range from 4 µm to 2100 µm. The results in Croatia fluctuated from 27.13 (high-quality salt) – to 31.68 (low-quality salt) particles g^{-1} with an average size range from 15 to 4628 µm [20]. In Spain, microplastics were found in commercial table salt. The range content went from 50 to 280 particles kg^{-1}, where PET was the most commonly reported plastic, followed by PP and PE [21].

Water

Microplastics have been detected in treated and untreated drinking water. Microplastics particles are found in different concentrations depending on water treatment. In untreated water, levels of microplastics ranging from 1473 ± 34 to 3605 ± 497 particles L^{-1} were observed whereas in treated water, concentration levels were between 338 ± 76 and 628 ± 28 particles L^{-1}. A relevant finding is that some microplastics particles were reported to be down to the size of 1 µm and around twelve different microplastics compounds were identified, being PET, PP and PE the most prominent [22].

METHODS OF CHARACTERIZATION AND DETECTION OF MICROPLASTICS

In 2004, Thompson and his collaborators [23] pointed out the problem of the abundance and generalization of microplastics in the environment. Since then, many researchers have focused on this matter and microplastics have been analyzed in almost all environmental compartments. Neither of methods for the sampling and analysis of micro (nanoplastics) is harmonized yet nor is there an agreement on the limits on the accuracy of the particle size of microplastics that can be found.

There is still a lack of standardization and the methodologies used vary widely. Microplastics monitoring studies require reliable and comparable qualitative and quantitative analytical methods for their determination. However, these methods are still in the development stage because microplastic research is a re-emerging field [24].

Microplastics include a very heterogeneous set of particles that differ in size, shape, color, specific density, and chemical composition among other features. To make significant comparisons and monitoring of microplastics, it is relevant to establish specific methodological criteria to assess their abundance, distribution, and composition.

Although these methodologies are not widespread, there are standardized protocols for the sampling and classification of microplastics established by the Marine Strategy Framework European Directive [25]. However, when finding suitable analytical methods to perform microplastics analysis, there is not one, but many different approaches and methodologies [26 - 30]. Despite these discrepancies, and because there are no standardized microplastics analysis procedures for each type of sample (sediment, water, marine biota, *etc.*), the analytical protocols are similar for identification and quantification [31]. In general, analytical procedures for microplastics in environmental samples consist of sampling, extraction, isolation (or separation), identification and quantification (or classification).

Sampling

Sampling is the most extensive and complex part of the strategy of the scientific methodology, given that reproducible and representative results depend on this analytical stage. However, in marine sciences, field study is limited by several aspects such as spatial scale, time and economic resources, including personnel costs, specific equipment and vessels [32]. In addition, the sampling of microplastics seldom considers the seasonal or interannual variations of the environmental parameters, or even the huge extensions where samples are analyzed [33].

Microplastics can be detected in the entire water column, in a large variety of sediments and in several tissues and cells of different organisms in the aquatic environment [34 - 37]. An important factor in the choice of microplastics sampling methods is the size range of the microplastic particles existing in the samples that will be studied. Published literature proposes a classification for plastic particles taking into account the size [38], which incorporates a new category called microplastic, generally <5 mm.

Hidalgo-Ruz *et al.* [26] indicated that there is a series of general samples of microplastics in the main marine environments (sea surface, water column, sediment, and biota) that require different approaches. Sampling can be:

a. Selective, which consists of the sample extraction directly from elements that are identifiable to the naked eye (particles between 1 and 5 mm Ø), generally on the surface of the sediments.

b. In bulk, where the whole sample (water or sediment) is used without reducing it in the sampling treatment. Bulk samples are more adequate when microplastics cannot be visually identified for several reasons: (i) sediment particles cover microplastics, (ii) small amount of microplastics which requires the classification/filtration of large volumes of sample (sediment/water), or (iii) microplastics particles are too small to be identified with the naked eye [26].
c. Reduced in volume, decreasing the global sample volume (water or sediment) during sampling, conserving only the part of the sample that will be processed later. For sediments, samples can be sifted directly at the sampling site [39], while for water, samples of reduced volume are usually achieved by filtering huge volumes of water using networks of different mesh sizes.

Table **1** shows the sampling methodologies most commonly used for the collection of environmental samples.

Table 1. Sampling processes of microplastics from various matrices [40, 41].

Matrix	Type	Process
Water	Surface water	Collection with a trawl with a rectangular opening and a net connected with a collecting bag/neuston net/catamaran; "grab" sampling (bottles)
	Mid-water level	Collection with bongo nets
	Riverine systems	Gill nets
Sediments	Bottom samples	Collection with a box corer
	Surface samples	Collection with iron spoons or non-plastic sampling spades
	Seabed samples	Collection with core or bottom trawl
Marine organisms	Benthic invertebrate species	Collected in grabs, traps, and creels, or by bottom trawling
	Planktonic and nektonic invertebrates	Collected by way of manta and bongo nets
	Fish species	Recovered in surface, midwater and benthic trawls, depending on their habitats
	Bivalves, crustaceans and annelids and shellfish	Collected from the field by hand or from commercial fish markets

Different studies on beaches in New Zealand [42 - 47] evaluated the value of microplastic densities in sediments and their results varied considerably. Most of the samples were taken from the most recent floating debris deposited on the high tide line. The tidal zone where samples were taken is not specified explicitly in a large number of studies. The uniformity presented by the different studies shows evidence of the requirement to systematically explore the types of sampling to be able to indicate in which beach area microplastics tend to accumulate. In studies carried out on the surface of the sea between 15 cm to 25 cm thick [27, 30, 48], the samples were taken mainly with neuston nets in order to take samples of huge volumes of water in a short time, only keeping the reduced volume sample.

Mesh sizes ranged between 0.053 and 3 mm, and a range of 0.30 to 0.39 mm was observed in a large number of studies. In the case of water column samples which depth ranged from 1-212 m, circular bongo nets were used in most cases and the opening of the network varied between 0.79 and 1.58 m^2 [30, 49]. The length of the network for samples of the sea surface was referenced in 14 investigations [26] to be between 1.0 and 8.5 m, the most of the networks ranging from 3.0 to 4.5 m in length. Ingestion of microplastics by marine species has been reported in a variety of habitats such as sea surface, water column, benthos (community formed by organisms that inhabit the bottom of aquatic ecosystems), estuaries, river waters, beaches and aquaculture [16]. The great variety of organisms of interest and the habitats from which they originate require a wide variety of techniques for sample collection [26]: the method used for sampling will depend on the objective of the investigation, by the resources available, the habitat and the organism itself. In addition, several methodologies can be used for the collection of samples of marine animals susceptible to the ingestion of microplastics. They can be taken in vanes, traps, and creels, or by bottom trawling [50, 51], through manta and bongo networks [52 - 54] (Table **1**).

The collection of different species of fish will depend on their habitats, usually being trawl surface [55]. Some species such as bivalves, crustaceans, and annelids are commonly collected by hand [56 - 59]. During the sampling and handling of the sample, identification of the potential contamination sources of the plastic samples is essential mainly those related to air pollution and synthetic fibers from clothes, equipment, and rain [1]. Storage and preservation can be achieved by drying at 40 °C or by freezing (sediment), or cooling/freezing (water, rich in organic matter). The samples of multiple uses conserved in formalin or alcohol can be useful for the analysis of the contents of the stomach [60].

Pretreatment and Purification

In most environmental samples, the microplastics sampled are not isolated or purified so their analysis cannot be carried out directly. The Guide on the monitoring of marine debris in European seas by the European Commission in 2013 [28] and other standards developed since 2015 [61 - 66] include recommended pretreatment protocols. A cleaning step [52] before analysis [40] without destroying or erroneously characterizing the analyte is another important challenge, since a large number of purification treatments could modify the microplastics themselves. The fundamental idea of purification is to eliminate what is not plastic materials, such as biofilms, sand, wood, or similar, preserving the original microplastics and avoiding the artificial production of secondary microplastics [67]. Sometimes this pretreatment can modify, damage or discolor some types of microplastics affecting the physical (for example, shape and color) and chemical (polymer and adsorbed contaminants) determination [62]. Therefore, it is advisable to verify if the pre-treatment step fulfills its purpose and does not cause false results. A recommended purification process consists in washing samples of sediments or marine organism tissues with water, saline water or using forceps [54, 68].

Separation

During sampling, microplastics should be collected and extracted from the matrix sample (water, sediment, stomach contents, *etc.*) before they can be recognized/confirmed and chemically characterized [40]. In this context, development of more and better standard protocols is necessary [69, 70].

Separation by Density

The separation by density has been used in studies of water samples, sediments, and biotic samples. The values of the specific density of a plastic material will depend on the type of polymer and the manufacturing process, varying between 0.8 and 1.4 g cm^{-3}, while the values of the density of other materials that may be present in the intestine of the biota (sand, other sediments) are approximately 2.65 g cm^{-3}. Due to this density difference, it is possible to separate lighter plastic particles from heavier particles by mixing samples with fresh water [71], seawater [72], a saturated solution of NaCl or NaI (1.2 g cm^{-3}) [73], or sodium polytungstate solution (1.4 g cm^{-3}) [74], after sample agitation for a certain time, from minutes to hours, depending on the size of the sample [23, 48, 75, 76]. Polystyrene foam, high and low density polyethylene and polypropylene float in freshwater and seawater while flexible and rigid polyvinyl chloride (PVC), polyethylene terephthalate (PET) and nylon float in sodium polytungstate solution of high density [49, 71, 73]. The different plastics are separated by settling or by centrifugation [77]. Subsequently, the supernatant containing plastic particles is filtered for further processing.

Filtration

Isolation of plastic particles from the supernatant obtained by density separation is achieved by filtering the solution usually using a vacuum [23, 48]. Filters with pore sizes of 0.45, 1 and 2 μm are enough to collect all floating microplastics. The use of sieves with different mesh sizes also makes possible the isolation of plastic particles and allows distinguishing size categories. Mesh sizes ranging between 0.038 and 4.75 mm are used in a waterfall or alone [46, 72, 73, 78, 79]. In studies of sea surface water, up to five [73, 80] and six [81, 82] sieves were used.

When particles have a size smaller than 1 μm, chromatographic techniques such as hydrodynamic chromatography (HDC) [83] or Field Flow Fractionation (FFF) [84] can be used for filtration.

Digestion

The analysis of microplastics that may be present in the biota, excised tissues from fish species or environmental samples requires prior digestion since biological material, microbial biofilms, algae, and detritus can mask the presence of microplastics in samples [85]. Therefore, methods for extracting and separating microplastics from the biotic material have greater complexity than those used in water samples or sediments. These methods include different stages: dissection, cleaning, homogenization, and digestion of samples using chemical substances or enzymes [1].

Dissection and Cleaning

Most of the studies found in the literature are focused on the analysis of specific tissues, mainly the digestive tract, including stomach and intestine, although in laboratory studies the whole organism is also of interest. Analysis of the intestinal tract is useful to determine the ingestion of microplastics by invertebrates and vertebrates [23, 49, 50, 86 - 92].

Homogenization

Solid samples of tissue require a homogenization that is usually done manually (with a glass or porcelain mortar) or mechanically (by means of a fully metallic mixer).

Digestion

Traditionally, digestion is performed with strong oxidizing chemicals, although synthetic polymers can be degraded or damaged when these treatments are used, mainly at high temperatures. Digestion can be carried out in an acid medium, with nitric acid [56, 68, 93 - 95] or hydrochloric acid at temperatures of 0-100 °C. Research has shown that polymer particles, which include PE and PS, dissolve after exposure to these acids overnight or boiling 30 min with HNO_3 22.5 M, while polyamides, nylon, PET and PC dissolve easily showing poor resistance to acids even in low concentrations.

It is likely that some acids such as nitric, hydrofluoric, perchloric, and sulfuric at high concentrations destroy or deeply damage most of the tested polymers, especially at high temperatures. However, hydrochloric acid, being a non-oxidizing acid, was not efficient enough to achieve digestion of organic material.

Digestion can also be carried out with alkalis. Digestions were tested with 10% KOH and 60 °C overnight [65, 96 - 100] and with NaOH. Results confirmed that PA, PE, and PP are unaffected when 10% KOH was used; however, with 40% NaOH (60 °C) PC and PET are degraded, PA fibers are deformed, and a yellow color of the PVC granules and the fusion of PE particles were observed [52]. Similarly, Dehaut *et al.* [65] pointed out that PC, PA, and PET were degraded when 40% NaOH was used. In particular, chemical resistance data indicated that PE and PVC are resistant to NaOH, even at 50% concentrations at 50 °C [52]. Modifications in the polymers resistant to the treatment were noticed which emphasize the need to carry out exhaustive tests of the treatments applied before their use in biota.

Oxidants were also used for digestions, such as 30% hydrogen peroxide (H_2O_2) and 55-60 °C [101] for digestion of mussels. Avio *et al.* [102] identified that the direct application of H_2O_2, at room temperature with oscillating agitation during 24-48 h resulted in a recovery of 70% of microplastics. Losses were linked to the formation of H_2O_2 foam. Enzymatic digestion has been considered as a specific biological medium to provoke the hydrolysis of proteins and decomposition of tissues [52]. Cole *et al.* [52] established a protocol for sample digestion that uses a serine protease (proteinase K) to eliminate biological material from the samples collected in the field. This study showed that ultrasonication has a detrimental effect on the efficiency of digestion, due to the precipitation of proteins in the media. Fluorescent polystyrene microspheres (20 mm) have been satisfactorily isolated from marine copepods using this Proteinase K methodology, obtaining a digestion efficiency higher than 97%. Recently, Ding *et al.* [103] published a method for sample pretreatment to achieve the isolation of microplastics from the digestive system of bivalves. For qualitative and quantitative analysis of microplastics, micro-Fourier transformed infrared (μ-FTIR) spectroscopy and stereomicroscopy were used. The developed methodology was employed for the analysis of microplastics in the digestive system in *Chlamys farreri* and *Mytilus galloprovincialis*. Recoveries for PP, PE, PS and PVC ranged between 96.7% and 98.6% when 10% KOH was used as high-efficiency digestion system.

Identification

As mentioned above, after preparing and separating the samples, the quantity and types of microplastics present must be determined. Particles are divided into several types, such as polystyrene foam [104], fibers, films, granules and pellets [78, 105] and they are collected from filters using tweezers [106]. Microplastics can also be observed directly on the filter paper. The most used identification and classification methods today are Visual Identification and Scanning Electron Microscopy (SEM).

Visual Identification

The visual examination of isolated microplastics is still a mandatory step. A careful visual classification of the waste is necessary to isolate plastics from organic waste (fragments of shells, parts of animals, dried seaweed or seagrass, *etc.*) and other materials (metal, paint, coatings, tar, glass, *etc.*). This classification is usually carried out by inspecting the sample directly with the naked eye or with the help of a dissecting microscope [107 - 109]. Visual identification is the most common method to carry out a later characterization and/or chemical quantification. It allows the classification of microplastic particles depending on their physical characteristics by direct observation or using optical, polarized or electronic microscopy. In general, this visual identification and classification method is used for microplastics of larger size (from 0.3 to 5 mm) [110, 111]. Visual examination allows the classification of plastic particles according to their size, color, and shape, and also obtaining some information about their provenience. Additionally, it offers some advantages, for instance, it requires no specialized technical personnel or additional consumables and it is also fast and inexpensive. However, when accurate differentiation of microplastics is required, especially in smaller sizes, practice and knowledge are needed.

Identification of the chemical composition of the most common pellets was carried out by using colors in a preliminary way. When microplastics are dispersed among large amounts of marine litter, color can help in separation. Particles presenting striking colors can be isolated more easily and subject to a subsequent identification as possible microplastics. PP shows transparent plastic pellets while PE shows white color [112]. Opaque color is related to low-density PE and almost transparent color is related to ethylvinyl acetate [113]. The degree of photodegradation, the residence time on the surface of the sea and the degree of desquamation or weathering can be estimated by microplastic coloration [114].

Visual identification can be achieved by using staining dyes as a non-expensive method. However, results attained with Nile EGN network, Eosin B, Hostasol Yellow 3G and Rose Bengal were unsatisfactory [97]. Main problems included the dye affinity for plastic materials and the staining of biogenic material containing the sample which would be avoided by a previous step of complete digestion.

However, the results of this method are arbitrary and nonobjective, since they can lend considerable differences among various research works and be time consuming. Andrady [35] put forward a method to make easy the visualization under a microscope, by using a lipophilic fluorescent dye (Nile Red) to dye microplastics found in water samples; however, this methodology has not continued in time.

Maes *et al.* [97] presented a detailed study on a quick screening for the detection of microplastics in samples of sedimentary deposits in the marine environment based on the use of Nile Red dye, which changes its emission of fluorescence according to the polarity of the environment. This property allows that microplastics could be classified by their hydrophobicity, *i.e.*, polyolefin, polyaromatic, polar (polyesters/nylons), and be a useful indicator to determine different environment-produced changes, such as oxidation, corrosion, *etc.* Despite their deficiencies, the spectroscopic methods mentioned above are the most widely used in the majority of microscopic visual detections of plastic particles.

Lusher *et al.* [1] inspected 332 samples of mussels from 15 different sites, using a stereomicroscope with Infinity camera 1-3C and INFINITY ANALYZE and CAPTURE software to take pictures and measure the size of all the particles found above 70 mm. Most microplastics with sizes greater than 70 mm were identified visually, but for the 25% whose size was below 1 mm, it was necessary to identify them by FTIR. In conclusion, if the analysis is carried out in accordance with the protocols [28], the visual identification of microplastics up to 500 μm is a satisfactory tool, since it is cheap, relatively accurate and fast. Even though visual examination is a simple and accessible technique, in our opinion, it should not be used as independent method due to its limitations in the precision and determination of the type of polymer. Thus, it should be used as a tool for additional analytical applications. Many researchers agree with this statement [26, 61, 115]. Visual preselection of particles is often employed to decrease these practical problems, although it can lead to biases [41].

Scanning Electron Microscope (SEM)

The identification of microplastics with SEM facilitates the distinction between different microplastics and organic particles since it provides large and clear images [43]. However, the use of SEM reduces the practicality, since it requires preparations of complex and time-consuming samples. In addition, the colors of the particles cannot be used to characterize in the SEM analysis, being only proposed for certain plastic particles [24]. The indicated restrictions can lead to inaccuracies in the determination of the bounty of microplastics in a given environment [116]. When SEM is coupled to the energy dispersive X-ray spectroscopy, EDS, (SEM-EDS) the elementary composition of the plastic particles (carbon and inorganic particles) is obtained, distinguishing among different types of plastics present [24, 116].

SEM-EDS is a highly specific system used to a great extent in the characterization of micro and nanomaterials. It has developed into an essential tool for microplastics particles identification and characterization. These useful combined techniques allowed the researchers to reach the conclusion that the first steps of the degradation of plastic waste have taken place quite quickly [117]. Recently, optical microscopy and SEM-EDS have been used together for the analysis of microplastics of samples found in oceanic networks and viscera of different fish. These techniques have determined the size, morphology and chemical composition of microplastics [118]. The conclusions drawn from the study of optical images showed plastic particles between 70 and 600 mm, and that chlorinated plastics, such as PVC, could be clearly identified due to the presence of chlorine and their mineral composition. We also detected species that had been wrongly characterized as different kind of plastics by optical microscopy.

Eriksen *et al.* [119, 120] used SEM-EDS and environmental scanning microscopy-energy dispersive X-ray spectroscopy (ESEM-EDS) both to characterize the morphology of microplastics surface and to determine their elemental distribution.

Reisser *et al.* [121] carried out some identifications of microplastics by color and shape.

To obtain a distribution in space and time, as well as the accumulation of organic and inorganic contaminants absorbed in microplastics, a quantification of microplastics was carried out taking into account that the concentration of the microplastics is given by the particles per square meter (particles per m^2), per cubic meter (particles per m^3), per liter (particles per liter) and per kg of dry sediment. Given that SEM allows to observe in detail the surface of the sample and very small fractions [122 - 124], some researchers such as Halle *et al.* [125] used this technique to demonstrate the degradation of microplastic particles in the environment, and to explain changes in their morphology, such as cuts and scratches.

Detection and Quantification

After the identification analyses, the suspicious plastics are subjected to analytical techniques for their detection and quantification. Among the most commonly used techniques, Fourier transform infrared spectrometry (FTIR) in transmittance or reflectance, Attenuated total reflectance (ATR), Raman spectrometry for color pigment spectrometry and gas chromatography (GC) by pyrolysis coupled with mass spectroscopy (Pyr-GC-MS) can be cited.

Identification of microplastics, using the techniques of FTIR and Raman is based on the comparison of the spectra of the sample in the microplastic, with the spectra of known polymers. The Pyr-GC-MS technique analyzes the microplastic particles based on their thermal degradation properties; in addition, this technique can simultaneously analyze the type of polymer and its plastic additives [1, 126].

The most used techniques for the characterization of microplastics are Infrared and Raman spectroscopies and, they are suggested by the Technical Subgroup of the Marine litter in a Garbage Strategy Framework Directive through the Guide on Marine Garbage Monitoring in the European seas [127].

The most important characteristics of these spectroscopic techniques are that they require minimum preparation of the sample and low sample volumes, being suggested for the distinction between plastics and natural particles in marine and soil samples.

In general, FTIR and Raman spectroscopy are highly recommended for the chemical characterization of microplastics. The MSFD Technical Subgroup on Marine Litter [127] recommends using these methods to 10% of microplastics with sizes from 100 to 5000 μm and to all suspicious particles in the range of 20 to 100 μm. Both vibrational spectroscopy techniques are non-destructive and also highly accurate and complementary. FTIR produces an infrared spectrum resulting from the change in dipole moment, based on the interaction of light with molecules whereas Raman provides a molecular spectrum based on the polarizability of chemical bonds [41, 128].

FTIR and ATR

FTIR microscopy is the most widely used technique to analyze microplastics below the limit of 500 μm. Identification is based on the high specificity of the FT-MIR spectrum of each polymer. Thus, due to the variability in the manufacturing process of plastics and the alterations suffered by the environment, microplastics of the same polymer have similar but not identical spectra [129 - 132].

Once the spectra of the microplastics have been registered, results are evaluated. In this analytical stage, the manual interpretation [133, 134] of the elucidation of the most outstanding vibrational bands can be performed using the reference tables [107] or the complete spectra can be compared with a library of reference spectra [108, 109, 118, 135]. There are commercially available databases of infrared reference spectra for polymer analysis, but these databases can also be created with the laboratory's own measurements.

Technical expertise is required for IR microscopy and the identification of individual particles from their spectra leads to errors easily, especially for small size particles (<20 μm) since the resolution of the microscope includes spectral signals of the environment, other adjacent particles or the filter itself [97]. Recently, the attenuated total reflectance (ATR)-FTIR was used in a study to identify the type of polymer of a total of 240 suspected microplastic particles in surface water and sediments. It was shown that most of the polymers were due to the presence of PP (50.9%), low density PE (18.2%), high density PE (26.4%), a mixture of PP and ethylene propylene (3.6%), as well as styrene acrylonitrile (0.9%) [136].

Harrison *et al.* suggested a more complex method for identifying polyethylene particles in a membrane filter [137, 138] based on ATR and reflectance with FT-MIR microscopy and on previous publications by Thompson *et al.* and Ng *et al.* [23, 48]. The simultaneous detection of different types of common polymers (polyethylene, polypropylene, polyamide 6, polystyrene and polyvinyl chloride) in wastewater was carried out by Tagg *et al.* which developed a method based on the approach of Harrison *et al.* [137] using FT-MIR microscopy with the Focal Plane Array technique, previously eliminating biological contaminants with H_2O_2 [63].

A microplastic analysis ingested by fish was carried out by Lusher *et al.* and Foekema *et al.* [139, 140]. Sometimes photodegradation by UV light, biodegradation, and thermal degradation can modify not only the shape and size, but also the original composition of the polymer [35]. On the other hand, bacteria present in the marine environment can also rapidly repopulate the microplastics, forming selective families of bacteria from successive plastispheres.

The set of these degradation processes makes identification and spectroscopic analysis more difficult, given that there are no reference spectra for all polymer degradation products [69].

Raman Spectroscopy

Raman spectroscopy is an alternative to infrared spectroscopy and microscopy for the identification of microplastics, so in recent years many research groups have used this technique. After different comparative studies with both techniques, Käppler *et al.* [141] recommended Raman especially for fragments smaller than 20 mm. While in FTIR, researchers used their own system parameters, such as the most common laser wavelengths: 532 nm [141, 142] and 785 nm [66, 143, 144], in Raman the displacement varied widely in the range from 200-150 to 3000-1000 cm. Raman spectroscopy shows the union of microplastics within the material and the structures of molecules and networks [145]. As in FTIR, no sample preparation is required in Raman spectroscopy.

Sometimes, microplastics with color could be a problem to identify the plastic type, since the pigment can also produce other spectra. In general, it can be said that Raman spectroscopy and FTIR are complementary techniques. FTIR allows a precise identification of the polar groups [146], while Raman spectroscopy gives a better response of the non-polar symmetric links, fit out a broader spectral coverage as well as better resolution and less water interference in comparison to FTIR [147]. However, vibrational spectroscopy is limited in the analysis of degraded or contaminated microplastics for several reasons inherent to these techniques such as the difficult availability of equipment, high cost, the need for specialized personnel and the treatment of complex data [24, 41, 69, 128, 143, 148].

When spectroscopic techniques are simultaneously used, such as Raman or FTIR spectroscopy and SEM-EDS, this provides more exhaustive information on the microplastics studied. Napper *et al.* and Thompson *et al.* [23, 149] studied, under different experimental conditions, the release of microscopic fibers after successive washes of synthetic fabrics, such as nylon. The analysis of each type of textile material was confirmed by micro-FTIR. In addition, the released fibers were analyzed by SEM to complete the morphological study. Li *et al.* [150] identified some transparent spheres as aluminum silicate by micro-FTIR; however, the analysis by SEM showed that they were diatomaceous earth. These findings highlight the need for using complementary techniques in order to make a more accurate identification of suspect microplastics.

Pyr-GC-MS

Other analytical technique to detect microplastics is Pyr-GC MS or thermal extraction and desorption (TED). This technique has improved the possible tools to characterize synthetic polymers and copolymers, which can generate decomposition products of the original polymers under regulated conditions, such as under an inert gas and high temperatures (500-100 °C) [151, 152].

Pyr-GC-MS consists of the chromatographic separation of the products by means of a capillary column of fused silica, carrying out the detection with a mass spectrometer. The polymers are identified by the interpretation of the obtained mass spectra or by the use of mass spectral libraries.

An advantage of the pyrolysis methods is that no pretreatment is needed for the direct analysis of solid polymer samples, analyzing only a small amount of sample in each measurement (5-200 mg) [152]. Pyr-GC-MS has been used for the identification of particles of different types of plastics such as PE, PP, PS, PA, chlorinated and chlorosulfonated PE in marine sediments, along with other polymers containing plastic manufacturing derivatives such as diethylhexyl phthalate, dibutyl phthalate, diethyl phthalate, di-*iso*-butyl phthalate, dimethyl phthalate, benzyl phthalate and benzoditeno-butylphenol [126]. The simultaneous Pyr-GC-MS analysis of PE, PP, PS, PA, PVC, PMMA was carried out by Dehaut *et al.* [65] and Fischer *et al.* [153] by comparing pyrograms with both domestic and commercial databases. Fischer pointed out that there is an uncertainty due to the misinterpretation because different polymers can give similar pyrolysis products.

Quantification of Microplastics

Different units are used to indicate the quantitative amount of microplastics, mainly mass and abundance. For samples such as sand, sediment, and surface water, mass is commonly measured in grams of microplastics m^{-2} and the abundance in particles m^{-2}. For water columns, mass values are expressed in milligrams m^{-3}, and abundance as the number of items m^{-3}. Often, to quantify microplastics in the marine biota (fish, mussel, *etc.*) the results are expressed as item microplastic $animal^{-1}$. These differences in the way of quantifying microplastics make difficult the comparison among different studies.

METHODS OF ANALYSIS FOR CHEMICAL COMPOUNDS RELATED TO MICROPLASTICS

The determination of organic compounds such as plastic additives used for manufacturing or toxic compounds sorbed onto or into the plastic surface in food samples generally requires multistep and time-consuming analytical methods. Firstly, contaminants have to be isolated from the sample through an extraction procedure using mostly organic solvents. The extraction technique employed depends on the specific properties of the analytes and the nature of the matrix sample investigated. Clean up procedures are usually necessary to remove interferences from the sample before analytical determination (separation, detection, identification and quantification of target compounds) by chromatographic techniques. A great number of articles reporting the determination of chemical compounds related to microplastics in food samples has been published in the scientific literature, and it is continuously increasing. Therefore, methods related to seafood analysis published in the period 2013- 2018 are presented in this section.

Methods of Analysis for Plastics Additives

Brominated flame retardants, phthalates, bisphenol A (BPA), nonylphenols and antioxidants are among the most used as plastic additives potentially toxic to humans when they are released to the environment.

Brominated Flame Retardants (BFRs)

Some Brominated Flame Retardants (BFRs) such as polybrominated diphenyl ethers (PBDEs) and hexabromocyclododecanes (HBCDs) have extensively been used for years for industrial purposes to enhance the fire resistance of commercial products. These compounds are most likely to be found in seafood. BFRs lipophilic character makes necessary the use of low-medium polarity or even non-polar extraction solvents, such as n-hexane, dichloromethane, ethyl acetate, acetone, acetonitrile or their mixtures.

Table **2** summarizes some relevant data published on selected analytical methods employed for the determination of BFRs in a broad type of seafood samples.

Table 2. Analytical details of the selected articles concerning BFRs determination in seafood.

Matrix	Extraction Method	Extraction Solvent	Clean Up	Detection Technique	Ref.
Roach, bleak, eel	Soxhlet	Dichloromethane	Acidic silica column GPC	GC-MS	[154]
Zooplankton, shad, whitefish	Soxhlet	Hexane/Acetone	Multilayer silica/Florisil column	GC-MS	[155]
Wild salmon	Soxhlet	Dichloromethane/hexane	GPC, acidic silica, Florisil columns	GC-MS	[156]
Fish, dolphin blubber	PLE	Dichloromethane/hexane	Acid digestion, SPE alumina	GC-MS	[157]
Crustacean, mollusks, fish	PLE	Toluene/Acetone	Acidic silica gel/Florisil	GC-MS	[158]
Fish, crustaceans	PLE	Toluene/Acetone	Multilayer silica columns	GC-MS	[159]
Eel	ASE	Dichloromethane	Silica gel column GPC	GC-MS	[160]
Eel	ASE	Toluene/Acetone	Acidic silica, Florisil, celite/carbon columns	GC-MS	[161]

(Table 2) cont.....

Matrix	Extraction Method	Extraction Solvent	Clean Up	Detection Technique	Ref.
Fish, mussels, clams	ASE	Hexane/Acetone	ACS	GC-MS	[162]
Fish fillet	ASE	Dichloromethane: hexane	Multi column clean-up	GC-MS	[163]
Smoked salmon, cod fillet	Sonication	Acetonitrile/Toluene	Florisil/acidic silica gel columns	GC-MS	[164]
Mussels	Sonication	Hexane/Acetone	Multilayer columns	LC-MS	[165]
Fish, prawn, squid	MSPD	Acetone/hexane	Multilayer silica	GC-MS	[166]
Fish	QuEChERS	Ethyl acetate	Mini silica column	GC-MS	[167]
Croaker and salmon.	QuEChERS	acetonitrile	Filtered during extraction	GC-MS	[168]

ACS, Automatic Cleanup System

Soxhlet extraction is one of the most classical extraction methods. It has been used for the determination of HBCDs and PBDEs in zooplankton, shad, whitefish [155], and PBDEs in wild fish (roach, bleak, eel) [154] and wild salmon [156].

Solvents most commonly used are dichloromethane, hexane, acetone and their binary mixtures.

Pressure Liquid Extraction (PLE; Dionex trade name ASE for accelerated solvent extraction) usually provides efficient and quick sample extraction employing hexane, toluene, acetone, dichloromethane, or their mixtures. PLE has been used to extract simultaneously 19 brominated compounds in different fish samples and dolphin blubber [157], BFRs in eel [160], crustacean, mollusks and fish [158], PBDEs in eel [161], fish (halibut, trout, bluefish), mussels and clams [162]. Recently, Bichon *et al.* [159] determined BFRs in fish and crustaceans and Neugebauer *et al.* [163] in a broad variety of matrices including fish fillet. Poma G. *et al.* [164] extracted PBDs and several emerging halogenated flame retardants in smoked salmon and cod fillet by sonication using an acetonitrile/toluene mixture as extraction solvent. This extraction methodology has been also used for the determination of some BFRs in mussels by Nostbakkena *et al.* [165]. Separation and purification of BFRs from co-extracted compounds (lipids and chemicals with physicochemical similarities) are required to ensure a sensitive and selective quantification of the target compounds. Several approaches have been used to remove interferences from extracts, including gel permeation chromatography (GPC) and solid-phase extraction (SPE) columns filled with different sorbents such as silica, alumina, Florisil or their combinations, depending on the analytes of interest.

QuEChERS (quick, easy, cheap, effective, rugged and safe) and Matrix Solid

Phase Dispersion (MSPD) are sample preparation methods that comprise extraction and clean-up. MSPD has been used for the determination of BFRs in fish, prawn and squid samples employing acetone/hexane as extraction solvent [166]. Kalachova *et al.* [167] developed QuEChERS extraction with ethyl acetate followed by silica minicolumn clean-up for the determination of 14 BFRs in fish. In addition, an acetonitrile QuEChERS extraction was applied for the analysis of BFRs in croaker and salmon. GC-MS is the technique mainly employed for the analytical determination of BFRs. Liquid chromatography coupled to MS (LC-MS) is also used. Analytical separation of analytes is usually achieved on an octadecyl-functionalized silica column, which main disadvantage is the impossibility to separate PBDE isomers.

Phthalates

Determination of phthalic acid esters (PAEs) or phthalates in food samples is a hard analytical task because of the complexity of the sample matrix that contains large amounts of interferences and the high background levels of contamination related to the ubiquity of PAEs in the analytical laboratories. Contamination sources may come from plastic materials used in the laboratory, such as plastic laboratory consumables or even from some parts of the GC injector. Therefore, analytical methodologies for the determination of these compounds must be simple to reduce sample contamination as much as possible. González-Sálamo *et al.* [169] published recently a review dealing with the determination of PAE in food samples. However, a short number of articles related to the analysis of these contaminants in seafood were reported being sonication the preferred method for extraction. Ultrasonic-assisted technique with acetonitrile followed by SPE clean up with primary secondary amine cartridge was worked out by Gu *et al.* [170] for the quantification of PAEs in fish, prawns, and mollusks. This methodology was used lately by Hu *et al.* [171] for the determination of phthalate monoesters as markers of phthalate contamination in the mentioned marine organisms. Phthalates diesters have been detected in seafood (white hake, breaded fish, sliced and smoked salmon, fish stick, cooked prawn) using acetonitrile as extraction solvent followed by a centrifugation step for fat removal [172]. Ultrasonic extraction using dichloromethane coupled with gas purge microsyringe extraction (GP-MSE) clean-up was performed by He *et al.* [173] as a novel sample cleanup for the determination of PAEs in seafood. Analytical determination in all cases was performed by GC-MS/MS.

Bisphenol A and Nonylphenols

NPs are degradation products of nonylphenol ethoxylates (NPEOs), one of the groups of non-ionic surfactants widely used as detergents, solubilizers, emulsifiers and dispersing compounds.

NPs are used as chemical intermediates during NPEOs manufacture and also added to many plastics as additives. NPs and BPA are considered endocrine disruptor compounds (EDCs), so that may simulate natural hormones and produce negative effects in the endocrine system. In the context of food safety, it is of great interest to determine EDCs in food and many investigations are focused on the multiresidue determination of a large number of estrogenic effect substances. Therefore, NPs and BPA determination is treated together in this section. Several review articles have been published in the literature concerning NP and BPA determination in environmental matrices, including food [174 - 176]. A more detailed revision focused on seafood is included in this section. Table **3** shows published data related to the selected analytical methods recently applied for the determination of NPs and BPA in seafood samples.

Table 3. Analytical details of the selected articles concerning NPs and BPA determination in seafood.

Matrix	Extraction Method	Extraction Solvent	Clean Up	Detection Technique	Ref.
Carp fish	Soxhlet	Dichloromethane	SPE, Silica gel	GC-MS	[177]
Wild tilapia	SLE	Methanol/acetonitrile	SPE, Acidic alumina	LC-MS	[178]
Aquatic foods	SLE	Cyclohexane/ethyl acetate	GPC	LC-MS	[179]
Mullet samples	SLE	Methanol	SPE	LC-MS / LC-FLD	[180]
Mussels*	SLE	Methanol/water	SPE, C18	LC-UV	[181]
Fish, mollusk, prawn	Sonication	Acetonitrile	SPE, PSA	LC-MS/MS	[170]
Freshwater fishes	Sonication	Dichloromethane/acetone	SPE, Aminopropyl	LC-MS/MS	[182]
Mussels	Sonication	Methanol/ammonium acetate	SPE Oasis HLB	HPLC-FLD	[183]
Mariculture fishes	Sonication	Methanol	SPE	HPLC-FLD	[184]
Wild freshwater fishes	Sonication	Acetonitrile	Multilayer columns	GC-MS	[185]
Clams	PLE	Acetone/hexane	-	LC-MS/MS	[186]
Carp, bass, walleye, trout	PLE	Methylene chloride	SPE, C18		[187]
Greenland sharks	PLE	Methanol	SPE, C18	HPLC-FLD	[188]
Mediterranean fish	QuEChERS	Acetonitrile	-	LC-MS	[189]
Fish, prawn, mollusk	QuEChERS	Acetonitrile	-	GC-MS	[190]

(Table 3) cont.....

Matrix	Extraction Method	Extraction Solvent	Clean Up	Detection Technique	Ref.
Carp, white bream, catfish	QuEChERS	Acetonitrile	-	LC-MS	[191]

*NPs are not determined.

Conventional extraction techniques for NPs and BPA determination in seafood include Soxhlet and solid-liquid extraction (SLE) followed by additional cleanup steps. Mortazavi *et al.* [177] determined several NPs and BPA residues in carp fish samples using DMC as extraction solvent by Soxhlet. SLE using methanol/acetonitrile was employed for the analysis of target compounds in wild tilapia tissues [178], in aquatic foods using cyclohexane/ethyl acetate as extraction solvent [179], in mullet samples using methanol [180], and in mussels using methanol/water [181]. Ultrasonication using acetonitrile is widely used for the extraction of NPs and BPA from seafood. This technique was used for the determination of the target compounds in fish, mollusk and prawn [170], and in wild freshwater fishes [185]. Other organic solvents were also employed: dichloromethane/acetone for different freshwater fishes [182], methanol/ammonium acetate for mussel samples [183] or methanol for three different species of mariculture fishes [184]. PLE was applied for the extraction and simultaneous cleanup in single or multiple steps. Some authors used PLE to analyze NPs and BPA in clams [186], carp, bass, walleye and lake trout [187], and in Greenland sharks [188]. Several applications of QuEChERS to NPs and BPA residue analysis in seafood were described in the literature. Recently, this methodology was applied in eleven different species of Mediterranean fish [189], in fish, prawn and mollusk [190], and carp, white bream and catfish [191]. Analytical determination of NPs and BPA in seafood can be completed by LC or GC using different detectors such as mass spectrometry (LC-MS, GC-MS), molecular fluorescence (LC-FLD) and ultraviolet detection (LC-UV).

Methods of Analysis for Plastics Contaminants (Dioxins, PCBs...)

An additional concern to the contamination of microplastics in the aquatic environment is the demonstrated fact that polypropylene microplastics easily adsorb hydrophobic substances [192]. There are already many studies that have found that microplastics in the marine environment are polluted with persistent organic pollutants (POPs) by means of adsorption/absorption procedures. These POPs can be accumulated in concentrations of up to 6 orders of magnitude (1 million times higher) more than those of the surrounding water, as shown by Mato *et al.* [193], for a number of organic compounds, such as PCBs, dichlorodiphenyldichloroethylene (DDE) and nonylphenols (NP). The provision of microplastics to attract this type of pollutants, together with their prevalence and capacity for moving to long-distance, is a cause for very serious concern. The MP-POP interaction directly affects the biota. Aquatic organisms usually eat microplastics. The food chain is exposed to the introduction of highly toxic chemical contaminants through the ingestion of microplastics contaminated by organisms. In addition, toxicity can be increased by the interaction of persistent organic pollutants with microplastics particles, which become possible vectors of chemical contaminants. Therefore, it is essential to identify and quantify the chemical products accumulated by the microplastics in diverse environmental matrices, as well as to study the possible pathways and sorption mechanisms of POPs on microplastics. Currently, the knowledge about microplastics-POP interactions is insufficient, so related research is considered a priority by many scientists, both in the environmental and food fields. Different simulation studies were achieved to verify the scope of this interaction. Zhan *et al.* carried out, in simulated water, PCB sorption experiments on PP, with a contact range from 1 to 24 h, and with several alterations in the size of the particles, the temperature and the environment of the solution. Results showed that the equilibrium absorption time was about 8 h and the absorption capability increased when the particle's size and temperature decreased [194]. Recently, Liu *et al.* [195] stated the distribution coefficients for PAH sorption in PS nanoplastic (70 nm) in simulated freshwater.

Plastic waste contaminated with microplastic-sized pieces was found in the North Pacific. This study showed that this garbage was composed of 70% PE from which 40% was contaminated with pesticides, 50% with PCBs and 80% with PAHs [192]. It could happen that polyethylene microplastics contaminated with PCBs subsequently immersed in the benthic regions become incorporated into the sediments, which would be an additional problem, since PCB contamination would be transferred to these benthic regions. They could also be ingested by benthic organisms, which would facilitate the introduction of toxic contaminants of PCBs into the food chain. Additionally, lighter microplastics which float in surface waters can potentially be contaminated with toxic chemicals through the deposition of pollutants from the atmosphere. A study conducted by Crawford *et al.* indicated that lower density plastics are likely to remain floating in surface water for long periods of time, despite the accumulation of embedded material, and, therefore, are more likely to suffer PAH deposition on their surface through atmospheric sedimentation [192]. It was also shown that the increase in salinity affects by increasing the absorption capacity of PCBs by PE microplastics. The standard and official methods for the analysis of POPs are available in the guidelines for air, water, solid waste and food analysis. These methods generally include GC and LC using different detectors (DAD, fluorescence, MS, ECD, *etc.*) with analysis times of more than 30 min. In addition to microplastics-POP interactions, microplastics can adsorb metals (Ag, Cd, Co, Cr, Cu, Hg, Ni, Pb, and

Zn). In freshwater at pH ~ 6.5 metals can be adsorbed onto microplastics due to the formation of modified organic complexes [196].

A more specific study indicated that metals were found in five types of plastic (PET, high-density PE, PVC, low-density PE and PP) in different concentrations, according to space and time [197]. Metals were extracted from the microplastic using 20% Aqua Regia and afterwards analyzed to determine Al, Cr, Mn, Ir, Co, Ni, Zn, Cd, and Pb by Inductively Coupled Plasma Mass Spectrometry (ICP-MS).

Brennecke *et al.* [198] examined the adsorption of zinc and copper in antifouling paint on PS beads and PVC fragments aged in a simulated aqueous environment. Extraction was carried out with modified Aqua Regia (12 M HCl and 16 M HNO_3 at 3:1 ratio) to separate the metals from the surface of the microplastics, determining their concentration in the simulated sea water by flame atomic absorption spectrometry. Concentrations of Zn and Cu in microplastics increased with time and, the accumulation in PVC continued increasing during the 14 days of the experiment [198]. Based on the above, contamination of microplastics by POPs can produce toxic effects in aquatic organisms and as a result of the ingestion of biota could pass to the food chain. This problem is important enough to require a detailed and accurate study to demonstrate these hypotheses and ascertain the health and environmental risks involved in the marine ecosystems and, ultimately, to humans.

MICROPLASTICS DIETARY INTAKE

The dietary intake of microplastics in foods depends on consumption habits. For instance, microplastics mainly accumulate in the gastrointestinal tract of aquatic animals, which minimizes the direct exposure to microplastics for fishery products in areas where these products are degutted. There are other products such as small pelagic fish species, a small-size freshwater fish, crustaceans, echinoderms and bivalves that are commonly eaten whole; therefore, the exposure to microplastics would be higher. One of the most consumed bivalve mollusks are mussels, where the occurrence of microplastics was reported in several studies [2]. In crustaceans such as shrimp, there is a clear difference in the microplastic content of whole and peeled shrimps. Commonly, microplastic particles are not found in abdominal muscle tissue, which is generally the edible part. Also, in the preparation of food, most of the digestive tract, the head and the gills, which is estimated that contains 90% microplastic particles, it is usually thrown away [19]. For other food commodities such as salt or water, the dietary intake of microplastics will also depend on consumption habits, and risk assessment exercises would need to take this into consideration.

LIMITATIONS FOR FOOD SAFETY RISK ASSESSMENT

The location of microplastics in the human body and their potential food safety impact is unknown. Although it is thought that the particles below 1.5 µm can enter into the capillaries of the organs, while larger particles will be eliminated [199], the toxicological data of plastics possibly ingested and their related compounds are nowadays gaps in knowledge. The best studied dietary sources of microplastics are fisheries and aquaculture products, which are important food commodities in certain areas. However, the toxicity of most plastic monomers, polymers, and additives present in microplastics has never been estimated by relevant international expert scientific committees such as the Joint FAO/WHO Expert Committee on Food Additives (JECFA).

JECFA is one of the scientific committees of international experts, which is key to assess the possible toxicity through the scientific data already originated and thus to establish the basis for the exercises of risk analysis (risk assessment, risk management, and risk communication expressed in Fig. (**1**). In 2017, a preliminary risk assessment taking into consideration certain microplastic additives and contaminants was carried out by FAO.

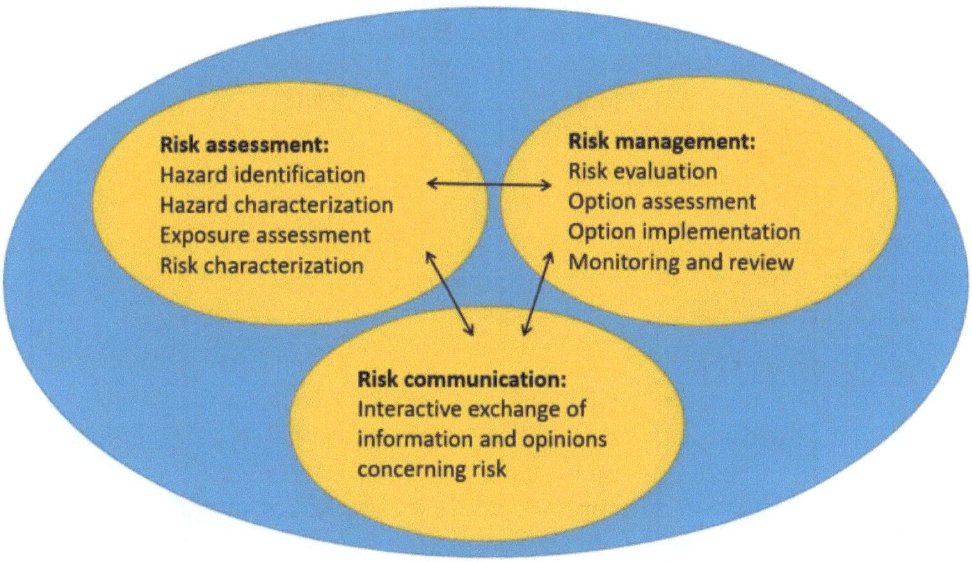

Fig. (1). FAO/WHO risk analysis framework.

Examples of Risk Assessment

As previously mentioned, although microplastics have been found in a variety of food commodities such as salt, beer or honey, seafood is the greatest research source of dietary intake of microplastics.

With plastics and microparticles principally accumulating in the gastrointestinal tract of animals, that are commonly degutted before consumption, direct exposure to microplastics is minimized for several commodities. However, small aquatic animals consumed with viscera are a cause for particular concern. This is the case of mussels, which have been of great research interest. The largest amount of microplastics in mussels appeared in China with contents reaching 4 particles g^{-1} [2]. According to the summary statistics of the individual consumption of chronic foods (CIFOCOss), the highest consumption of mussels described corresponds to the elderly of Belgium (P95), 250 g per day per person is calculated. Taking into account the combination of these data in reference to mussels, it can be estimated that a ration of mussels could contain up to 1000 microplastic particles, that is, around 9 µg, depending on the volume and density of the particles. The exposure evaluation on microplastics in mussels proved that the intake of microplastics per day is 0.15 µg kg^{-1} which indicates that for a person of 60 kg is equal to 9 µg per day. With these values, the most unfavorable cases are selected, covering all the populations, including the most at risk. Although the assessment of mussel exposure was carried out, the tolerable daily intake (TDI) of microplastic particles and most of their compounds has not yet been proven, so whether this grade of exposure is safe or not cannot be defined. However, the TDI for some plastic additives such as phthalates, BPA, alkylphenol, and brominated flame retardant, as well as adhered contaminants such as PCBs, PAHs, dichloro- diphenyl-trichloroethane, has been established and the exposure assessment exercise concluded that the level of the compounds present in microplastics from mussels is significantly lower than the TDI. Therefore, the food safety impact for this specific case appears to be insignificant [5]. Nevertheless, it would be important to consider the exposure to these compounds through other food sources for further conclusions.

INVESTIGATION GAPS

Nowadays, the detection and identification of microplastics in environmental and food samples still present some data gaps for different reasons: first, the risk of losses and contamination of microplastics during sampling, storage and analysis is highlighted. There is a lack of consistency in the results presented in the visual identification of microplastics, as well as a lack of standardized methods for identification, classification, and detection. In addition, sample contamination is frequent and there are few protocols of action to prevent it. Appropriate sampling is complicated and long-time-consuming. There is no regulation regarding the toxicity of microplastics for biota and for humans, *etc*.

Analytical Control and Quality Assurance

The utilization of approved and standardized methods is critical for quality assurance and also to establish the reference techniques to compare the results obtained with developed analytical techniques. However, the normalization of analytical methods for identification, detection, and quantification of microplastics in the marine environment is in its infancy and there are hardly any validation methods.

Sample Contamination

Sample contamination in the analysis of microplastics in environmental and food samples should be avoided. Pollution and cross-contamination are not a one-step problem, but it can occur at every step of the analysis. This contamination can occur due to the existence of samples in the air with synthetic fibers from apparel [200, 201] or atmospheric rain [68, 96, 202 - 205]. If possible, sources of contamination should be eliminated and quantification carried out using environmental filters or procedural targets.

It is essential to have blank spaces to perform the procedures, as well as replicated samples and spiked blank samples. However, all this is blocked by the absence of certified reference materials containing known microplastics concentrations, which are essential to control possible contamination and carry out method validation.

Sampling

A major problem in quality assurance in microplastics determination is the variety of polymers with respect to type, size, color, and morphology join to inhomogeneity within environmental samples. This undoubtedly affects each step of the analytical methodology: sampling, purification, extraction, and quantification.

It is essential to get a representative sample for different marine environmental matrices before the identification and quantification of microplastics. Sadly, there is a need of standardization also for sampling methodology, as well as inter-laboratory studies in global research [206]. Quantitative analytical methods may include validation studies and blanks assays to establish the reliability and evaluation of the method [207]. In addition, microplastics sampling does not cover seasonal or interannual variations of environmental parameters, which may lead to greater inconsistency when giving precise results of specific places and periods of these pollutants [206].

Visual Identification of Microplastics

As mentioned above, the most ordinary approach for the identification and detection of microplastics is the visual observation of microplastic particles apparently followed by confirmation by chemical composition analysis, generally using the combination of optical and spectroscopic techniques [144, 207]. Visual detection can lead to the appearance of mistaken positives and/or false negatives [208] due to confusions of microplastics with biota, deterioration or degradation of microplastics, *etc*. Eriksen *et al.* [119] described the identification of different particles with approximately 20% error. These particles were initially identified as microplastics by visual observation, but later they were characterized as aluminum silicates from carbon ash using SEM. In other investigations, 32% of the particles, with a size smaller than 100 mm, indicated as microplastics by visual identification were confirmed as non-microplastics after the application of micro-Raman [146] and up to 70% of the particles were erroneously identified as microplastics before the FTIR analysis [26]. In addition, it was shown that plastics size varies considerably and higher recoveries are reported for plastics with larger size compared to the fraction of smaller size.

With regard to nanoplastics, nowadays, there are no established rules for their identification and quantification in marine environmental samples. However, some analytical methods have already proved their validity for this purpose, such as electron microscopy, atomic spectrometry, and light scattering techniques [209].

Detection of micro-Plastics by Different Analytical Techniques

Currently, the determination of polymers of microplastics by infrared spectroscopy is critically debated. Although there is a standard analytical practice, the results of the analyses are not comparable or even contradictory with each other in many cases [210]. An example of this discrepancy is found in the results on microplastics classification and detection performed using FTIR by Lusher *et al.* and Foekema *et al.* [139, 140] in the oesophagus, stomach, and intestines of fish that had ingested microplastics.

Lusher *et al.* [139] informed on the ingestion of polyamide 6 with a 36% of positive findings. However, Foekema *et al.* [140] identified PE and PP.

This may be due to the analysis method employed, since most of the spectrum evaluation is done using spectral libraries, which are applied in most FTIR programs. The Impact Quality Index (HQI) describes the similarity of the spectrum of a measured sample compared with the reference spectrum. Although there is a lot of different algorithms to calculate HQI, few investigation groups work on this and there are not any validation parameters in this context.

In general, the development of a greater number of certified reference materials with chemical groups that include absorbed chemical substances and additives of the different types of polymers (for example, PE, PP, PS, PET, and PVC), could greatly improve the accuracy and validity of analytical data related to microplastics.

So far, only certified references of PE and PVC containing BPA and phthalates are available [211]. In addition, the analytical quality control would be strengthened with interlaboratory studies and certified reference materials. An additional gap is found in the fact that the lack of standardized protocols has led to a series of erroneous reports describing the presence of micro and, to a lesser extent, nanoplastics in environments that are not susceptible of direct comparison, because they use different units (for example, mass/volume, number of elements/volume, *etc.*)

As research goes forward, the need to standardize methods is obvious, so that it is essential to be able to assess more accurately the potential health risks of microplastics to human beings as well as their ecological and environmental impact.

These standardized methodologies should anticipate the numerous potential sources of error and contamination, as discussed above, as well as address the need for control and analysis of a large number of samples in a short time. For marine waters, there is a technical note recently published by NOAA on laboratory methods in the analysis of microplastics [212], which could be important for future research on the standardization of methods and for the quantification of microplastics.

The discrepancy in the investigations of identification, detection, and quantification of microplastic samples has led to numerous additional questions regarding the uptake and variation of microplastics within ecosystems. Is it the representative plastic catchment? To what extent is the trophic transfer? Are the negative effects of plastic ingestion observed in laboratory experiments valid in the environment? *etc* [1]. Therefore, it is necessary to design studies of the results of which can contribute to clarify these problems and questions. In conclusion, there is an urgent need to standardize protocols to promote consistency in the collection and analysis of microplastic data. The validation of analytical methods will increase the comparability of studies and will lead to more harmonized results and a thorough assessment of the impacts and risks that microplastics represent for biota.

CONCLUDING REMARKS

The amount of microplastics ingested by humans through the consumption of different food commodities is poorly understood, as well as the size and shape of microplastics that can translocate across the gut epithelium. Although microplastics and nanoplastics may have significant concerns, more secure data to carry out thorough food safety risk assessments are still required. The current lack of harmonized analytical methods for the determination of microplastics, nanoplastics or their derivatives makes extremely difficult to compare analytical results between methods and/or laboratories. The representativeness and reproducibility of the methods of sampling, identification, and quantification of microplastics still pose many unresolved problems. In addition, microplastics are distributed very heterogeneously in the marine environment, so the information on their temporal and geographical distribution is unknown both in the marine environment and in the environments necessary to assess the exposure in the body. Therefore, there is an urgent need to continue researching in this field and to look for validated, quick and simple analysis methodologies. Although food safety risks posed by plastics additives and contaminants found in the microplastics evaluated in food appear to be low so far, it is important to take into account the inevitable growth of microplastics and nanoplastics in the near future as a result of the degradation of plastics released into the environment over the last decades as well as from future sources.

CONSENT FOR PUBLICATION

Not applicable.

CONFLICT OF INTEREST

The author confirms that this chapter contents have no conflict of interest.

ACKNOWLEDGEMENT

This work was funded by the Comunidad of Madrid (Spain) and European funding from FEDER Program (project S2013/ABI-3028, AVANSECAL-CM) and FAO. The views expressed in this publication are those of the authors and do not necessarily reflect the views or policies of the Food and Agriculture Organization of the United Nations.

REFERENCES

[1] Lusher A, Welden N, Sobral P, Cole M. Sampling, isolating and identifying microplastics ingested by fish and invertebrates. Anal Methods 2017; 9: 1346-60. [http://dx.doi.org/10.1039/C6AY02415G]

[2] Presence of microplastics and nanoplastics in food, with particular focus on seafood. EFSA J 2016; 14(6): 4501.

[3] Klaine SJ, Koelmans AA, Horne N, *et al.* Paradigms to assess the environmental impact of manufactured nanomaterials. Environ Toxicol Chem 2012; 31(1): 3-14. [http://dx.doi.org/10.1002/etc.733] [PMID: 22162122]

[4] Sources, fate and effects of microplastics in the marine environment: a global assessment (Kershaw Ed.). (IMO/FAO/UNESCO-IOC/UNIDO/WMO/ IAEA/UN/UNEP/UNDP Joint Group of Experts on the Scientific Aspects of Marine Environmental Protection). *Report and Studies GESAMP*, 2015, 90, 96.

[5] Microplastics in fisheries and aquaculture: status of knowledge on their occurrence and implications for aquatic organisms and food safety. FAO Fish Aquac Tech Pap 2017; 615. [Rome, Italy].

[6] Alaee M, Arias P, Sjödin A, Bergman A. An overview of commercially used brominated flame retardants, their applications, their use patterns in different countries/regions and possible modes of release. Environ Int 2003; 29(6): 683-9. [http://dx.doi.org/10.1016/S0160-4120(03)00121-1] [PMID: 12850087]

[7] Hutzinger O, Thoma H. Polybrominated dibezo-*p*-dioxins and dibenzofurans: the flame retardant issue. Chemosphere 1987; 16: 1877-80. [http://dx.doi.org/10.1016/0045-6535(87)90181-0]

[8] Muirhead EK, Skillman AD, Hook SE, Schultz IR. Oral exposure of PBDE-47 in fish: toxicokinetics and reproductive effects in Japanese Medaka (*Oryzias latipes*) and fathead minnows (*Pimephales promelas*). Environ Sci Technol 2006; 40(2): 523-8. [http://dx.doi.org/10.1021/es0513178] [PMID: 16468398]

[9] Yogui GT, Sericano JL. Polybrominated diphenyl ether flame retardants in the U.S. marine environment: a review. Environ Int 2009; 35(3): 655-66. [http://dx.doi.org/10.1016/j.envint.2008.11.001] [PMID: 19100622]

[10] Chen Y, Shu L, Qiu Z, *et al.* BPA-Substitute Bisphenol S causes unique alterations of germline function. PLoS Genet 2016; 12(7)e1006223 [http://dx.doi.org/10.1371/journal.pgen.1006223] [PMID: 27472198]

[11] USEPA (United States Environmental Protection Agency). Nonylphenol (NP) and Nonylphenol Ethoxylates (NPEs) Action Plan. United States Environmental Protection Agency 2010.

[12] Vethaak AD, Lahr J, Schrap SM, *et al.* An integrated assessment of estrogenic contamination and biological effects in the aquatic environment of The Netherlands. Chemosphere 2005; 59(4): 511-24. [http://dx.doi.org/10.1016/j.chemosphere.2004.12.053] [PMID: 15788174]

[13] Soares A, Guieysse B, Jefferson B, Cartmell E, Lester JN. Nonylphenol in the environment: a critical review on occurrence, fate, toxicity and treatment in wastewaters. Environ Int 2008; 34(7): 1033-49. [http://dx.doi.org/10.1016/j.envint.2008.01.004] [PMID: 18282600]

[14] Liebezeit G, Liebezeit E. Non-pollen particulates in honey and sugar. Food Addit Contam Part A Chem Anal Control Expo Risk Assess 2013; 30(12): 2136-40. [http://dx.doi.org/10.1080/19440049.2013.843025] [PMID: 24160778]

[15] Liebezeit G, Liebezeit E. Synthetic particles as contaminants in German beers. Food Addit Contam Part A Chem Anal Control Expo Risk Assess 2014; 31(9): 1574-8. [http://dx.doi.org/10.1080/19440049.2014.945099] [PMID: 25056358]

[16] Lusher A. Microplastics in the marine environment: distribution, interactions and effects. In: Bergmann, Gutow Klages, Ed. Marine anthropogenic litter. Springer International Publishing 2015; pp. 245-307.

[17] Sources, fate and effects of microplastics in the marine environment: part two of a global assessment. In: Eds. (IMO/FAO/ UNESCO-IOC/UNIDO/WMO/IAEA/UN/UNEP/UNDP Joint Group of Experts

on the Scientific Aspects of Marine Environmental Protection) Report and Studies GESAMP. 2016; 93: p. 220.

[18] Collard F, Gilbert B, Compère P, *et al.* Microplastics in livers of European anchovies (*Engraulis encrasicolus, L.*). Environ Pollut 2017; 229: 1000-5. [http://dx.doi.org/10.1016/j.envpol.2017.07.089] [PMID: 28768577]

[19] Devriese LI, van der Meulen MD, Maes T, *et al.* Microplastic contamination in brown shrimp (*Crangon crangon*, Linnaeus 1758) from coastal waters of the Southern North Sea and Channel area. Mar Pollut Bull 2015; 98(1-2): 179-87. [http://dx.doi.org/10.1016/j.marpolbul.2015.06.051] [PMID: 26456303]

[20] Renzi M, Blašković A. Litter & microplastics features in table salts from marine origin: Italian *versus* Croatian brands. Mar Pollut Bull 2018; 135: 62-8. [http://dx.doi.org/10.1016/j.marpolbul.2018.06.065] [PMID: 30301081]

[21] Iñiguez ME, Conesa JA, Fullana A. Microplastics in Spanish table salt. Sci Rep 2017; 7(1): 8620. [http://dx.doi.org/10.1038/s41598-017-09128-x] [PMID: 28819264]

[22] Pivokonsky M, Cermakova L, Novotna K, Peer P, Cajthaml T, Janda V. Occurrence of microplastics in raw and treated drinking water. Sci Total Environ 2018; 643: 1644-51. [http://dx.doi.org/10.1016/j.scitotenv.2018.08.102] [PMID: 30104017]

[23] Thompson RC, Olsen Y, Mitchell RP, *et al.* Lost at sea: where is all the plastic? Science 2004; 304(5672): 838. [http://dx.doi.org/10.1126/science.1094559] [PMID: 15131299]

[24] Shim WJ, Hong SH, Eo S. Identification methods in microplastic analysis: a review. Anal Methods 2017; 9: 1384-91. [http://dx.doi.org/10.1039/C6AY02558G]

[25] Galgani F, Fleet D, van Franeker J, *et al.* Marine Strategy Framework Directive Task group 10 report Marine Litter. European Commission Joint Research Center 2010.

[26] Hidalgo-Ruz V, Gutow L, Thompson RC, Thiel M. Microplastics in the marine environment: a review of the methods used for identification and quantification. Environ Sci Technol 2012; 46(6): 3060-75. [http://dx.doi.org/10.1021/es2031505] [PMID: 22321064]

[27] Norén F. Small plastic particles in Coastal Swedish waters. https://www.researchgate.net/publication/284312290 [accessed 30/11/2018];

[28] Galgani F, Hanke G, Werner S, *et al.* Guidance on monitoring of marine litter in European seas Report. Institute for Environment and Sustainability 2013.

[29] Dekiff JH, Remy D, Klasmeier J, Fries E. Occurrence and spatial distribution of microplastics in sediments from Norderney. Environ Pollut 2014; 186: 248-56. [http://dx.doi.org/10.1016/j.envpol.2013.11.019] [PMID: 24448461]

[30] Doyle MJ, Watson W, Bowlin NM, Sheavly SB. Plastic particles in coastal pelagic ecosystems of the Northeast Pacific ocean. Mar Environ Res 2011; 71(1): 41-52. [http://dx.doi.org/10.1016/j.marenvres.2010.10.001] [PMID: 21093039]

[31] Hollman PCH, Bouwmeester H, Peter RJB. Microplastics in the aquatic food chain; Sources, measurement, occurrence and potential risks. Rikilt Wageninger UR 2013.

[32] Costa MF, Duarte AC. Microplastics sampling and sample Handling (Chapter 2). In: Rocha-Santos and Duarte, Ed. Comprehensive Analytical Chemistry Characterization and analysis of microplastics. Elsevier 2017.

[33] Lima ARA, Barletta M, Costa MF. Seasonal distribution and interactions between plankton and microplastics in a tropical estuary. Estuar Coast Shelf Sci 2015; 165: 213-25. [http://dx.doi.org/10.1016/j.ecss.2015.05.018]

[34] Wright SL, Thompson RC, Galloway TS. The physical impacts of microplastics on marine organisms:

a review. Environ Pollut 2013; 178: 483-92.
[http://dx.doi.org/10.1016/j.envpol.2013.02.031] [PMID: 23545014]

[35] Andrady AL. Microplastics in the marine environment. Mar Pollut Bull 2011; 62(8): 1596-605. [http://dx.doi.org/10.1016/j.marpolbul.2011.05.030] [PMID: 21742351]

[36] von Moos N, Burkhardt-Holm P, Köhler A. Uptake and effects of microplastics on cells and tissue of the blue mussel *Mytilus edulis L.* after an experimental exposure. Environ Sci Technol 2012; 46(20): 11327-35.
[http://dx.doi.org/10.1021/es302332w] [PMID: 22963286]

[37] Anderson JC, Park BJ, Palace VP. Microplastics in aquatic environments: Implications for Canadian ecosystems. Environ Pollut 2016; 218: 269-80.
[http://dx.doi.org/10.1016/j.envpol.2016.06.074] [PMID: 27431693]

[38] Ivar do Sul JA, Costa MF. Plastic pollution risks in an estuarine conservation unit. J Coast Res 2013; 65: 48-53.
[http://dx.doi.org/10.2112/SI65-009.1]

[39] Frias JP, Sobral P, Ferreira AM. Organic pollutants in microplastics from two beaches of the Portuguese coast. Mar Pollut Bull 2010; 60(11): 1988-92.
[http://dx.doi.org/10.1016/j.marpolbul.2010.07.030] [PMID: 20800853]

[40] Rocha-Santos T, Duarte AC. A critical overview of the analytical approaches to the occurrence, the fate and the behavior of microplastics in the environment, TrAC, Trend Anal Chem 2014; 65: 47-53.

[41] Qiu Q, Tan Z, Wang J, Peng J, Li M, Zhan Z. Extraction, enumeration and identification methods for monitoring microplastics in the environment. Estuar Coast Shelf Sci 2016; 176: 102-9.
[http://dx.doi.org/10.1016/j.ecss.2016.04.012]

[42] Costa MF, Ivar do Sul JA, Silva-Cavalcanti JS, Araújo MCB, Spengler A, Tourinho PS. On the importance of size of plastic fragments and pellets on the strandline: a snapshot of a Brazilian beach. Environ Monit Assess 2010; 168(1-4): 299-304.
[http://dx.doi.org/10.1007/s10661-009-1113-4] [PMID: 19680758]

[43] Cooper DA, Corcoran PL. Effects of mechanical and chemical processes on the degradation of plastic beach debris on the island of Kauai, Hawaii. Mar Pollut Bull 2010; 60(5): 650-4.
[http://dx.doi.org/10.1016/j.marpolbul.2009.12.026] [PMID: 20106491]

[44] Endo S, Takizawa R, Okuda K, *et al.* Concentration of polychlorinated biphenyls (PCBs) in beached resin pellets: variability among individual particles and regional differences. Mar Pollut Bull 2005; 50(10): 1103-14.
[http://dx.doi.org/10.1016/j.marpolbul.2005.04.030] [PMID: 15896813]

[45] Abu-Hilal AH, Al-Najjar TH. Plastic pellets on the beaches of the northern Gulf of Aqaba, Red Sea. Aquat Ecosyst Health 2009; 12: 461-70.
[http://dx.doi.org/10.1080/14634980903361200]

[46] Van A, Rochman CM, Flores EM, *et al.* Persistent organic pollutants in plastic marine debris found on beaches in San Diego, California. Chemosphere 2012; 86(3): 258-63.
[http://dx.doi.org/10.1016/j.chemosphere.2011.09.039] [PMID: 22014466]

[47] Shiber JG. Plastic pellets on Spain's "Costa del Sol" beaches. Mar Pollut Bull 1982; 13: 409-12.
[http://dx.doi.org/10.1016/0025-326X(82)90014-5]

[48] Ng KL, Obbard JP. Prevalence of microplastics in Singapore's coastal marine environment. Mar Pollut Bull 2006; 52(7): 761-7.
[http://dx.doi.org/10.1016/j.marpolbul.2005.11.017] [PMID: 16388828]

[49] Carpenter EJ, Anderson SJ, Harvey GR, Miklas HP, Peck BB. Polystyrene spherules in coastal waters. Science 1972; 178(4062): 749-50.
[http://dx.doi.org/10.1126/science.178.4062.749] [PMID: 4628343]

[50] Murray F, Cowie PR. Plastic contamination in the decapod crustacean *Nephrops norvegicus* (Linnaeus, 1758). Mar Pollut Bull 2011; 62(6): 1207-17. [http://dx.doi.org/10.1016/j.marpolbul.2011.03.032] [PMID: 21497854]

[51] Welden NAC, Cowie PR. Environment and gut morphology influence microplastic retention in langoustine, *Nephrops norvegicus.* Environ Pollut 2016; 214: 859-65. [http://dx.doi.org/10.1016/j.envpol.2016.03.067] [PMID: 27161832]

[52] Cole M, Webb H, Lindeque PK, Fileman ES, Halsband C, Galloway TS. Isolation of microplastics in biota-rich seawater samples and marine organisms. Sci Rep 2014; 4: 4528. [http://dx.doi.org/10.1038/srep04528] [PMID: 24681661]

[53] Cole M, Lindeque P, Fileman E, Halsband C, Galloway TS, Galloway TS. The impact of polystyrene microplastics on feeding, function and fecundity in the marine copepod *Calanus helgolandicus.* Environ Sci Technol 2015; 49(2): 1130-7. [http://dx.doi.org/10.1021/es504525u] [PMID: 25563688]

[54] Desforges JPW, Galbraith M, Ross PS. Ingestion of microplastics by zooplankton in the northeast Pacific Ocean. Arch Environ Contam Toxicol 2015; 69(3): 320-30. [http://dx.doi.org/10.1007/s00244-015-0172-5] [PMID: 26066061]

[55] Pedà C, Caccamo L, Fossi MC, *et al.* Intestinal alterations in European sea bass Dicentrarchus labrax (Linnaeus, 1758) exposed to microplastics: Preliminary results. Environ Pollut 2016; 212: 251-6. [http://dx.doi.org/10.1016/j.envpol.2016.01.083] [PMID: 26851981]

[56] Van Cauwenberghe L, Claessens M, Vandegehuchte MB, Janssen CR. Microplastics are taken up by mussels (*Mytilus edulis*) and lugworms (*Arenicola marina*) living in natural habitats. Environ Pollut 2015; 199: 10-7. [http://dx.doi.org/10.1016/j.envpol.2015.01.008] [PMID: 25617854]

[57] Watts AJ, Lewis C, Goodhead RM, *et al.* Uptake and retention of microplastics by the shore crab *Carcinus maenas.* Environ Sci Technol 2014; 48(15): 8823-30. [http://dx.doi.org/10.1021/es501090e] [PMID: 24972075]

[58] Watts AJ, Urbina MA, Corr S, Lewis C, Galloway TS. Ingestion of plastic microfibers by the crab *Carcinus maenas* and its effect on food consumption and energy balance. Environ Sci Technol 2015; 49(24): 14597-604. [http://dx.doi.org/10.1021/acs.est.5b04026] [PMID: 26529464]

[59] Canesi L, Ciacci C, Bergami E, *et al.* Evidence for immunomodulation and apoptotic processes induced by cationic polystyrene nanoparticles in the hemocytes of the marine bivalve Mytilus. Mar Environ Res 2015; 111: 34-40. [http://dx.doi.org/10.1016/j.marenvres.2015.06.008] [PMID: 26115607]

[60] Costa MF, da Costa JP, Duarte AC. Sampling of micro(nano)plastics in environmental compartments: How to define standard procedures? Curr Opin Environ Sci Health 2018; 1: 36-40. [http://dx.doi.org/10.1016/j.coesh.2017.10.001]

[61] Song YK, Hong SH, Jang M, *et al.* A comparison of microscopic and spectroscopic identification methods for analysis of microplastics in environmental samples. Mar Pollut Bull 2015; 93(1-2): 202-9. [http://dx.doi.org/10.1016/j.marpolbul.2015.01.015] [PMID: 25682567]

[62] Wagner J, Wang ZM, Ghosal S, Rochman C, Gassel M, Wall S. Novel method for the extraction and identification of microplastics in ocean trawl and fish gut matrices. Anal Methods 2017; 9: 1479-90. [http://dx.doi.org/10.1039/C6AY02396G]

[63] Tagg AS, Sapp M, Harrison JP, Ojeda JJ. Identification and quantification of microplastics in wastewater using focal plane array-based reflectance micro-FTIR imaging. Anal Chem 2015; 87(12): 6032-40. [http://dx.doi.org/10.1021/acs.analchem.5b00495] [PMID: 25986938]

[64] Crichton EM, Noel M, Gies EA, Ross PS. A novel, densityindependent and FTIR-compatible

approach for the rapid extraction of microplastics from aquatic sediments. Anal Methods 2017; 9(9): 1419-28.
[http://dx.doi.org/10.1039/C6AY02733D]

[65] Dehaut A, Cassone AL, Frère L, *et al.* Microplastics in seafood: Benchmark protocol for their extraction and characterization. Environ Pollut 2016; 215: 223-33.
[http://dx.doi.org/10.1016/j.envpol.2016.05.018] [PMID: 27209243]

[66] Karami A, Golieskardi A, Choo CK, Romano N, Ho YB, Salamatinia B. A high-performance protocol for extraction of microplastics in fish. Sci Total Environ 2017; 578: 485-94.
[http://dx.doi.org/10.1016/j.scitotenv.2016.10.213] [PMID: 27836345]

[67] Bergmann M, Gutow L, Klages M. Marine anthropogenic litter. Springer 2015.
[http://dx.doi.org/10.1007/978-3-319-16510-3]

[68] Davidson K, Dudas SE. Microplastic ingestion by wild and cultured Manila clams (*Venerupis philippinarum*) from Baynes Sound, British Columbia. Arch Environ Contam Toxicol 2016; 71(2): 147-56.
[http://dx.doi.org/10.1007/s00244-016-0286-4] [PMID: 27259879]

[69] Löder MGJ, Gerdts G. Methodology used for the detection and identification of microplastics-A critical appraisal.Marine Anthropogenic Litter. Springer International Publishing 2015; pp. 201-27.
[http://dx.doi.org/10.1007/978-3-319-16510-3_8]

[70] Underwood AJ, Chapman PM, Browne MA. Some problems and practicalities in design and interpretation of samples of microplastic wastes. Anal Methods 2017; 9: 1332-45.
[http://dx.doi.org/10.1039/C6AY02641A]

[71] Zucher N. Small plastic derbis on beaches in Hong Kong: An Inithial Investigation Masther Tesis. University of Hong Kong 2009.

[72] Kusui T, Noda M. International survey on the distribution of stranded and buried litter on beaches along the Sea of Japan. Mar Pollut Bull 2003; 47(1-6): 175-9.
[http://dx.doi.org/10.1016/S0025-326X(02)00478-2] [PMID: 12787616]

[73] Lattin GL, Moore CJ, Zellers AF, Moore SL, Weisberg SB. A comparison of neustonic plastic and zooplankton at different depths near the southern California shore. Mar Pollut Bull 2004; 49(4): 291-4.
[http://dx.doi.org/10.1016/j.marpolbul.2004.01.020] [PMID: 15341821]

[74] Corcoran PL, Biesinger MC, Grifi M. Plastics and beaches: a degrading relationship. Mar Pollut Bull 2009; 58(1): 80-4.
[http://dx.doi.org/10.1016/j.marpolbul.2008.08.022] [PMID: 18834997]

[75] Reddy MS, Basha S, Adimurthy S, Ramachandraiah G. Description of the small plastics fragments in marine sediments along the Alang-Sosiya ship-breaking yard, India. Estuar Coast Shelf Sci 2006; 68: 656-60.
[http://dx.doi.org/10.1016/j.ecss.2006.03.018]

[76] Browne MA, Galloway TS, Thompson RC. Spatial patterns of plastic debris along Estuarine shorelines. Environ Sci Technol 2010; 44(9): 3404-9.
[http://dx.doi.org/10.1021/es903784e] [PMID: 20377170]

[77] Collard F, Gilbert B, Eppe G, Parmentier E, Das K. Detection of anthropogenic particles in fish stomachs: an isolation method adapted to identification by Raman Spectroscopy. Arch Environ Contam Toxicol 2015; 69(3): 331-9.
[http://dx.doi.org/10.1007/s00244-015-0221-0] [PMID: 26289815]

[78] Claessens M, De Meester S, Van Landuyt L, De Clerck K, Janssen CR. Occurrence and distribution of microplastics in marine sediments along the Belgian coast. Mar Pollut Bull 2011; 62(10): 2199-204.
[http://dx.doi.org/10.1016/j.marpolbul.2011.06.030] [PMID: 21802098]

[79] Ivar do Sul JA, Spengler Â, Costa MF. Here, there and everywhere. Small plastic fragments and pellets on beaches of Fernando de Noronha (Equatorial Western Atlantic). Mar Pollut Bull 2009;

58(8): 1236-8.
[http://dx.doi.org/10.1016/j.marpolbul.2009.05.004] [PMID: 19486997]

[80] Moore CJ, Moore SL, Leecaster MK, Weisberg SB. A comparison of plastic and plankton in the North Pacific Central Gyre. Mar Pollut Bull 2001; 42(12): 1297-300.
[http://dx.doi.org/10.1016/S0025-326X(01)00114-X] [PMID: 11827116]

[81] Shaw DG, Day RH. Colour- and form-dependent loss of plastic micro-debris from the North Pacific Ocean. Mar Pollut Bull 1994; 28: 39-43.
[http://dx.doi.org/10.1016/0025-326X(94)90184-8]

[82] Moore CJ, Moore SL, Weisberg SB, Lattin GL, Zellers AF. A comparison of neustonic plastic and zooplankton abundance in southern California's coastal waters. Mar Pollut Bull 2002; 44(10): 1035-8.
[http://dx.doi.org/10.1016/S0025-326X(02)00150-9] [PMID: 12474963]

[83] Striegel AM, Brewer AK. Hydrodynamic chromatography. Annu Rev Anal Chem (Palo Alto, Calif) 2012; 5: 15-34.
[http://dx.doi.org/10.1146/annurev-anchem-062011-143107] [PMID: 22708902]

[84] Loeschner K, Navratilova J, Legros S, *et al.* Optimization and evaluation of asymmetric flow field-flow fractionation of silver nanoparticles. J Chromatogr A 2013; 1272: 116-25.
[http://dx.doi.org/10.1016/j.chroma.2012.11.053] [PMID: 23261297]

[85] Europe P. Plastic Europe, Plastics – the Facts 2014/2015. An analysis of European plastics production, demand and waste data. Available from: http://issuu.com/plasticseuropeebook/does/final_plastics_the_facts_2014_19122 [accessed 30/11/2018];

[86] Goldstein MC, Goodwin DS. Gooseneck barnacles (*Lepas spp.*) ingest microplastic debris in the North Pacific Subtropical Gyre. PeerJ 2013; 1e184
[http://dx.doi.org/10.7717/peerj.184] [PMID: 24167779]

[87] Kartar S, Milne RA, Sainsbury M. Polystyrene waste in the Severn Estuary. Mar Pollut Bull 1973; 4(9): 144.
[http://dx.doi.org/10.1016/0025-326X(73)90010-6]

[88] Dos Santos J, Jobling M. A model to describe gastric evacuation in cod (*Gadus morhua L.*) fed natural prey. ICES J Mar Sci 1992; 49(2): 145-54.
[http://dx.doi.org/10.1093/icesjms/49.2.145]

[89] Cannon SME, Lavers JL, Figueiredo B. Plastic ingestion by fish in the Southern Hemisphere: A baseline study and review of methods. Mar Pollut Bull 2016; 107(1): 286-91.
[http://dx.doi.org/10.1016/j.marpolbul.2016.03.057] [PMID: 27058965]

[90] Bellas J, Martínez-Armental J, Martínez-Cámara A, Besada V, Martínez-Gómez C. Ingestion of microplastics by demersal fish from the Spanish Atlantic and Mediterranean coasts. Mar Pollut Bull 2016; 109(1): 55-60.
[http://dx.doi.org/10.1016/j.marpolbul.2016.06.026] [PMID: 27289284]

[91] Lusher AL, O'Donnell C, Officer R, O'Connor I. Microplastic interactions with North Atlantic mesopelagic fish. ICES J Mar Sci 2016; 73(4): 1214-25.
[http://dx.doi.org/10.1093/icesjms/fsv241]

[92] Rummel CD, Löder MG, Fricke NF, *et al.* Plastic ingestion by pelagic and demersal fish from the North Sea and Baltic Sea. Mar Pollut Bull 2016; 102(1): 134-41.
[http://dx.doi.org/10.1016/j.marpolbul.2015.11.043] [PMID: 26621577]

[93] Van Cauwenberghe L, Janssen CR. Microplastics in bivalves cultured for human consumption. Environ Pollut 2014; 193: 65-70.
[http://dx.doi.org/10.1016/j.envpol.2014.06.010] [PMID: 25005888]

[94] Braid HE, Deeds J, De Greasse SL, Wilson JJ, Osborne J, Hanner RH. Preying on commercial fisheries and accumulating paralytic shellfish toxins: A dietary analysis of invasive Dosidicus gigas (*Cephalopoda Ommastrephidae*) stranded in Pacific Canada. Mar Biol 2012; 159(1): 25-31.

[http://dx.doi.org/10.1007/s00227-011-1786-4]

[95] Lu Y, Zhang Y, Deng Y, *et al.* Uptake and accumulation of polystyrene microplastics in zebrafish (*Danio rerio*) and toxic effects in liver. Environ Sci Technol 2016; 50(7): 4054-60. [http://dx.doi.org/10.1021/acs.est.6b00183] [PMID: 26950772]

[96] Rochman CM, Tahir A, Williams SL, *et al.* Anthropogenic debris in seafood: Plastic debris and fibers from textiles in fish and bivalves sold for human consumption. Sci Rep 2015; 5: 14340. [http://dx.doi.org/10.1038/srep14340] [PMID: 26399762]

[97] Maes T, Jessop R, Wellner N, Haupt K, Mayes AG. A rapid-screening approach to detect and quantify microplastics based on fluorescent tagging with Nile Red. Sci Rep 2017; 7: 44501. [http://dx.doi.org/10.1038/srep44501] [PMID: 28300146]

[98] Catarino AI, Thompson R, Sanderson W, Henry TB. Development and optimization of a standard method for extraction of microplastics in mussels by enzyme digestion of soft tissues. Environ Toxicol Chem 2017; 36(4): 947-51. [http://dx.doi.org/10.1002/etc.3608] [PMID: 27583696]

[99] Munno K, Helm PA, Jackson DA, Rochman C, Sims A. Impacts of temperature and selected chemical digestion methods on microplastic particles. Environ Toxicol Chem 2018; 37(1): 91-8. [http://dx.doi.org/10.1002/etc.3935] [PMID: 28782833]

[100] Prata JC, da Costa JP, Duarte AC, Rocha-Santos T. Methods for sampling and detection of microplastics in water and sediment: a critical review. environmental. TrAC Trend Analyt Chem 2019; 110: 150-9. [http://dx.doi.org/10.1016/j.trac.2018.10.029]

[101] Li J, Yang D, Li L, Jabeen K, Shi H. Microplastics in commercial bivalves from China. Environ Pollut 2015; 207: 190-5. [http://dx.doi.org/10.1016/j.envpol.2015.09.018] [PMID: 26386204]

[102] Avio CG, Gorbi S, Regoli F. Experimental development of a new protocol for extraction and characterization of microplastics in fish tissues: First observations in commercial species from Adriatic Sea. Mar Environ Res 2015; 111: 18-26. [http://dx.doi.org/10.1016/j.marenvres.2015.06.014] [PMID: 26210759]

[103] Ding JF, Li JX, Sun CJ, *et al.* Separation and identification of microplastics in digestive system of bivalves. Chin J Anal Chem 2018; 46(5): 690-7. [http://dx.doi.org/10.1016/S1872-2040(18)61086-2]

[104] Jayasiri HB, Purushothaman CS, Vennila A. Quantitative analysis of plastic debris on recreational beaches in Mumbai, India. Mar Pollut Bull 2013; 77(1-2): 107-12. [http://dx.doi.org/10.1016/j.marpolbul.2013.10.024] [PMID: 24210947]

[105] Lee J, Hong S, Song YK, *et al.* Relationships among the abundances of plastic debris in different size classes on beaches in South Korea. Mar Pollut Bull 2013; 77(1-2): 349-54. [http://dx.doi.org/10.1016/j.marpolbul.2013.08.013] [PMID: 24054782]

[106] Nuelle MT, Dekiff JH, Remy D, Fries E. A new analytical approach for monitoring microplastics in marine sediments. Environ Pollut 2014; 184: 161-9. [http://dx.doi.org/10.1016/j.envpol.2013.07.027] [PMID: 24051349]

[107] Hummel DO. Atlas of plastics additives Analysis by spectrometric methods. Springer Science & Business Media 2012.

[108] Mani T, Hauk A, Walter U, Burkhardt-Holm P. Microplastics profile along the Rhine River. Sci Rep 2015; 5: 17988. [http://dx.doi.org/10.1038/srep17988] [PMID: 26644346]

[109] Lusher AL, Hernandez-Milian G, O'Brien J, Berrow S, O'Connor I, Officer R. Microplastic and macroplastic ingestion by a deep diving, oceanic cetacean: the True's beaked whale *Mesoplodon mirus*. Environ Pollut 2015; 199: 185-91.

[http://dx.doi.org/10.1016/j.envpol.2015.01.023] [PMID: 25667115]

[110] McDermid KJ, McMullen TL. Quantitative analysis of small-plastic debris on beaches in the Hawaiian Archipelago. Mar Pollut Bull 2004; 48(7-8): 790-4. [http://dx.doi.org/10.1016/j.marpolbul.2003.10.017] [PMID: 15041436]

[111] Heo NW, Hong SH, Han GM, *et al.* Distribution of small plastic debris in cross-section and high strandline on Heungnam beach, South Korea. Ocean Sci J 2013; 48(2): 225-33. [http://dx.doi.org/10.1007/s12601-013-0019-9]

[112] Ismail A, Adilah NMB, Nurulhudha MJ. Plastic pellets along Kuala Selangor-Sepang coastline. Malaysian Appl Biol J 2009; 38: 85-8.

[113] Eriksson C, Burton H. Origins and biological accumulation of small plastic particles in fur seals from Macquarie Island. Ambio 2003; 32(6): 380-4. [http://dx.doi.org/10.1579/0044-7447-32.6.380] [PMID: 14627365]

[114] Turner A, Holmes L. Occurrence, distribution and characteristics of beached plastic production pellets on the island of Malta (central Mediterranean). Mar Pollut Bull 2011; 62(2): 377-81. [http://dx.doi.org/10.1016/j.marpolbul.2010.09.027] [PMID: 21030052]

[115] Wesch C, Barthel AK, Braun U, Klein R, Paulus M. No microplastics in benthic eelpout (*Zoarces viviparus*): An urgent need for spectroscopic analyses in microplastic detection. Environ Res 2016; 148: 36-8. [http://dx.doi.org/10.1016/j.envres.2016.03.017] [PMID: 27198628]

[116] Dehghani S, Moore F, Akhbarizadeh R. Microplastic pollution in deposited urban dust, Tehran metropolis, Iran. Environ Sci Pollut Res Int 2017; 24(25): 20360-71. [http://dx.doi.org/10.1007/s11356-017-9674-1] [PMID: 28707239]

[117] Girao AV, Caputo G, Ferro MF. Application of Scanning Electron Microscopy-Energy Dispersive X-Ray Spectroscopy (SEM-EDS) (Chapter 6). In: Eds. Comprehensive Analytical Chemistry Characterization and Analysis of Microplastics. Elsevier 2017.

[118] Wang ZM, Wagner J, Ghosal S, Bedi G, Wall S. SEM/EDS and optical microscopy analyses of microplastics in ocean trawl and fish guts. Sci Total Environ 2017; 603-604: 616-26. [http://dx.doi.org/10.1016/j.scitotenv.2017.06.047] [PMID: 28646780]

[119] Eriksen M, Mason S, Wilson S, *et al.* Microplastic pollution in the surface waters of the Laurentian Great Lakes. Mar Pollut Bull 2013; 77(1-2): 177-82. [http://dx.doi.org/10.1016/j.marpolbul.2013.10.007] [PMID: 24449922]

[120] Eriksen M, Maximenko N, Thiel M, *et al.* Plastic pollution in the South Pacific subtropical gyre. Mar Pollut Bull 2013; 68(1-2): 71-6. [http://dx.doi.org/10.1016/j.marpolbul.2012.12.021] [PMID: 23324543]

[121] Reisser J, Shaw J, Wilcox C, *et al.* Marine plastic pollution in waters around Australia: characteristics, concentrations, and pathways. PLoS One 2013; 8(11)e80466 [http://dx.doi.org/10.1371/journal.pone.0080466] [PMID: 24312224]

[122] Anderson PJ, Warrack S, Langen V, Challis JK, Hanson ML, Rennie MD. Microplastic contamination in Lake Winnipeg, Canada. Environ Pollut 2017; 225: 223-31. [http://dx.doi.org/10.1016/j.envpol.2017.02.072] [PMID: 28376390]

[123] Ter Halle A, Ladirat L, Gendre X, *et al.* Understanding the fragmentation pattern of marine plastic debris. Environ Sci Technol 2016; 50(11): 5668-75. [http://dx.doi.org/10.1021/acs.est.6b00594] [PMID: 27138466]

[124] Courtene-Jones W, Quinn B, Murphy F, Gary SF, Narayanaswamy BE. Optimisation of enzymatic digestion and validation of specimen preservation methods for the analysis of ingested microplastics. Anal Methods 2017; 9(9): 1437-45. [http://dx.doi.org/10.1039/C6AY02343F]

[125] Ter Halle A, Ladirat L, Martignac M, Mingotaud AF, Boyron O, Perez E. To what extent are microplastics from the open ocean weathered? Environ Pollut 2017; 227: 167-74. [http://dx.doi.org/10.1016/j.envpol.2017.04.051] [PMID: 28460234]

[126] Fries E, Dekiff JH, Willmeyer J, Nuelle MT, Ebert M, Remy D. Identification of polymer types and additives in marine microplastic particles using pyrolysis-GC/MS and scanning electron microscopy. Environ Sci Process Impacts 2013; 15(10): 1949-56. [http://dx.doi.org/10.1039/c3em00214d] [PMID: 24056666]

[127] European Commission. MSFD Technical Subgroup on Marine Litter. Guidance on Monitoring of Marine Litter in European Seas A guidance document within the Common Implementation Strategy for the Marine Strategy Framework Directive 2013.

[128] Elert AM, Becker R, Duemichen E, *et al.* Comparison of different methods for MP detection: What can we learn from them, and why asking the right question before measurements matters? Environ Pollut 2017; 231(Pt 2): 1256-64. [http://dx.doi.org/10.1016/j.envpol.2017.08.074] [PMID: 28941715]

[129] Fotopoulou KN, Karapanagioti HK. Surface properties of beached plastic pellets. Mar Environ Res 2012; 81: 70-7. [http://dx.doi.org/10.1016/j.marenvres.2012.08.010] [PMID: 23013745]

[130] Chen Z, Hay J, Jenkins M. FTIR spectroscopic analysis of poly(ethylene terephthalate) on crystallization. Eur Polym J 2012; 48(9): 1586-610. [http://dx.doi.org/10.1016/j.eurpolymj.2012.06.006]

[131] Sarnmon C, Yarwood L, Everall N. An FT-IR study of the effect of hydrolytic degradation on the structure of thin PET films. Polym Degrad Stabil 2000; 67(1): 149-58. [http://dx.doi.org/10.1016/S0141-3910(99)00104-4]

[132] Galimberti D, Quarti C, Milani A, Brambilla L, Civalleri B, Castiglioni C. IR spectroscopy of crystalline polymers from ab initio calculations: Nylon 6,6. Vib Spectrosc 2013; 66: 83-92. [http://dx.doi.org/10.1016/j.vibspec.2013.02.005]

[133] Neves D, Sobral P, Ferreira JL, Pereira T. Ingestion of microplastics by commercial fish off the Portuguese coast. Mar Pollut Bull 2015; 101(1): 119-26. [http://dx.doi.org/10.1016/j.marpolbul.2015.11.008] [PMID: 26608506]

[134] Frias JP, Gago J, Otero V, Sobral P. Microplastics in coastal sediments from southern Portuguese shelf waters. Mar Environ Res 2016; 114: 24-30. [http://dx.doi.org/10.1016/j.marenvres.2015.12.006] [PMID: 26748246]

[135] Watermann BT, Löder M, Herlyn M, Daehne B, Thomsen A, Gall K. Long-term 2007-2013 monitoring of reproductive disturbance in the dun sentinel Assiminea grayana with regard to polymeric materials pollution at the coast of Lower Saxony, North Sea, Germany. Environ Sci Pollut Res Int 2017; 24(4): 3352-62. [http://dx.doi.org/10.1007/s11356-016-8058-2] [PMID: 27864739]

[136] Tsang YY, Mak CW, Liebich C, Lam SW, Sze ET, Chan KM. Microplastic pollution in the marine waters and sediments of Hong Kong. Mar Pollut Bull 2017; 115(1-2): 20-8. [http://dx.doi.org/10.1016/j.marpolbul.2016.11.003] [PMID: 27939688]

[137] Harrison JP, Ojeda JJ, Romero-González ME. The applicability of reflectance micro-Fourie--transform infrared spectroscopy for the detection of synthetic microplastics in marine sediments. Sci Total Environ 2012; 416: 455-63. [http://dx.doi.org/10.1016/j.scitotenv.2011.11.078] [PMID: 22221871]

[138] Renner G, Schmidt TC, Schram J. Characterization and quantification of microplastics by Infrared Spectroscopy (Chapter 4). In: Rocha-Santos T, Duarte A, Eds. Wilson and Wilson's Comprehensive Analytical Chemistry, Characterization and Analysis of Microplastics. Amsterdam: Elsevier 2017; 75.

[139] Lusher AL, McHugh M, Thompson RC. Occurrence of microplastics in the gastrointestinal tract of

pelagic and demersal fish from the English Channel. Mar Pollut Bull 2013; 67(1-2): 94-9. [http://dx.doi.org/10.1016/j.marpolbul.2012.11.028] [PMID: 23273934]

[140] Foekema EM, De Gruijter C, Mergia MT, van Franeker JA, Murk AJ, Koelmans AA. Plastic in north sea fish. Environ Sci Technol 2013; 47(15): 8818-24. [http://dx.doi.org/10.1021/es400931b] [PMID: 23777286]

[141] Käppler A, Fischer D, Oberbeckmann S, *et al.* Analysis of environmental microplastics by vibrational microspectroscopy: FTIR, Raman or both? Anal Bioanal Chem 2016; 408(29): 8377-91. [http://dx.doi.org/10.1007/s00216-016-9956-3] [PMID: 27722940]

[142] Sruthy S, Ramasamy EV. Microplastic pollution in Vembanad Lake, Kerala, India: The first report of microplastics in lake and estuarine sediments in India. Environ Pollut 2017; 222: 315-22. [http://dx.doi.org/10.1016/j.envpol.2016.12.038] [PMID: 28041839]

[143] Zhao S, Danley M, Ward JE, Li D, Mincer TJ. An approach for extraction, characterization and quantitation of microplastic in natural marine snow using Raman microscopy. Anal Methods 2017; 9(9): 1470-8. [http://dx.doi.org/10.1039/C6AY02302A]

[144] Zhang K, Xiong X, Hu H, *et al.* Occurrence and characteristics of microplastic pollution in Xiangxi bay of three gorges reservoir, China. Environ Sci Technol 2017; 51(7): 3794-801. [http://dx.doi.org/10.1021/acs.est.7b00369] [PMID: 28298079]

[145] Van Cauwenberghe L, Vanreusel A, Mees J, Janssen CR. Microplastic pollution in deep-sea sediments. Environ Pollut 2013; 182: 495-9. [http://dx.doi.org/10.1016/j.envpol.2013.08.013] [PMID: 24035457]

[146] Lenz R, Enders K, Stedmon CA, Mackenzie DMA, Nielsen TG. A critical assessment of visual identification of marine microplastic using Raman spectroscopy for analysis improvement. Mar Pollut Bull 2015; 100(1): 82-91. [http://dx.doi.org/10.1016/j.marpolbul.2015.09.026] [PMID: 26455785]

[147] Ribeiro-Claro P, Nolasco MM, Araújo C. Characterization of microplastics by Raman spectroscopy (Chapter 5). In: Rocha-Santos T, Duarte A, Eds. Comprehensive Analytical Chemistry. Elsevier 2017; 1: pp. 119-51.

[148] Dümichen E, Barthel AK, Braun U, *et al.* Analysis of polyethylene microplastics in environmental samples, using a thermal decomposition method. Water Res 2015; 85: 451-7. [http://dx.doi.org/10.1016/j.watres.2015.09.002] [PMID: 26376022]

[149] Napper IE, Thompson RC. Release of synthetic microplastic plastic fibres from domestic washing machines: Effects of fabric type and washing conditions. Mar Pollut Bull 2016; 112(1-2): 39-45. [http://dx.doi.org/10.1016/j.marpolbul.2016.09.025] [PMID: 27686821]

[150] Li J, Qu X, Su L, *et al.* Microplastics in mussels along the coastal waters of China. Environ Pollut 2016; 214: 177-84. [http://dx.doi.org/10.1016/j.envpol.2016.04.012] [PMID: 27086073]

[151] Kusch P. Pyrolysis-gas chromatography/mass spectrometry of polymeric materials. Advanced Gas Chromatography, Progress in Agricultural, Biomedical and Industrial Applications. InTech, Rijeka 2012; pp. 343-63.

[152] Kusch P. Application of Pyrolysis-Gas Chromatography/Mass Spectrometry (Py-GC/MS) (Chapter 7). Comprehensive Analytical Chemistry Characterization and Analysis of Microplastics. Elsevier 2017.

[153] Fischer M, Scholz-Böttcher BM. Simultaneous trace identification and quantification of common types of microplastics in environmental samples by pyrolysis-gas chromatography–mass spectrometry. Environ Sci Technol 2017; 51(9): 5052-60. [http://dx.doi.org/10.1021/acs.est.6b06362] [PMID: 28391690]

[154] Jürgens MD, Johnson AC, Jones KC, Hughes D, Lawlor AJ. The presence of EU priority substances mercury, hexachlorobenzene, hexachlorobutadiene and PBDEs in wild fish from four English rivers.

Sci Total Environ 2013; 461-462: 441-52.
[http://dx.doi.org/10.1016/j.scitotenv.2013.05.007] [PMID: 23747559]

[155] Poma G, Volta P, Roscioli C, Bettinetti R, Guzzella L. Concentrations and trophic interactions of novel brominated flame retardants, HBCD, and PBDEs in zooplankton and fish from Lake Maggiore (Northern Italy). Sci Total Environ 2014; 481: 401-8.
[http://dx.doi.org/10.1016/j.scitotenv.2014.02.063] [PMID: 24614155]

[156] Zacs D, Rjabova J, Bartkevics V. Occurrence of brominated persistent organic pollutants (PBDD/DFs, PXDD/DFs, and PBDEs) in Baltic wild salmon (Salmo salar) and correlation with PCDD/DFs and PCBs. Environ Sci Technol 2013; 47(16): 9478-86.
[http://dx.doi.org/10.1021/es401852d] [PMID: 23883417]

[157] Barón E, Eljarrat E, Barceló D. Gas chromatography/tandem mass spectrometry method for the simultaneous analysis of 19 brominated compounds in environmental and biological samples. Anal Bioanal Chem 2014; 406(29): 7667-76.
[http://dx.doi.org/10.1007/s00216-014-8196-7] [PMID: 25260412]

[158] Bichon E, Guiffard I, Vénisseau A, *et al.* Simultaneous determination of 16 brominated flame retardants in food and feed of animal origin by fast gas chromatography coupled to tandem mass spectrometry using atmospheric pressure chemical ionisation. J Chromatogr A 2016; 1459: 120-8.
[http://dx.doi.org/10.1016/j.chroma.2016.07.003] [PMID: 27425757]

[159] Bichon E, Guiffard I, Vénisseau A, *et al.* Simultaneous analysis of historical, emerging and novel brominated flame retardants in food and feed using a common extraction and purification method. Chemosphere 2018; 205: 31-40.
[http://dx.doi.org/10.1016/j.chemosphere.2018.04.070] [PMID: 29679786]

[160] Sühring R, Möller A, Freese M, *et al.* Brominated flame retardants and dechloranes in eels from German Rivers. Chemosphere 2013; 90(1): 118-24.
[http://dx.doi.org/10.1016/j.chemosphere.2012.08.016] [PMID: 22985592]

[161] Couderc M, Poirier L, Zalouk-Vergnoux A, Kamari A, Blanchet-Letrouvé I, Marchand P, Vénisseau A, Veyrand B, Mouneyrac C, Le Bizec B. Occurrence of POPs and other persistent organic contaminants in the European eel (*Anguilla anguilla*) from the Loire estuary, France. Sci Total Environ 2015; 505: 199-215.
[http://dx.doi.org/10.1016/j.scitotenv.2014.09.053]

[162] Martellini T, Diletti G, Scortichini G, *et al.* Occurrence of polybrominated diphenyl ethers (PBDEs) in foodstuffs in Italy and implications for human exposure. Food Chem Toxicol 2016; 89: 32-8.
[http://dx.doi.org/10.1016/j.fct.2015.12.026] [PMID: 26746818]

[163] Neugebauer F, Dreyer A, Lohmann N, Koschorreck J. Determination of halogenated flame retardants by GC-API-MS/MS and GC-EI-MS: a multi-compound multi-matrix method. Anal Bioanal Chem 2018; 410(4): 1375-87.
[http://dx.doi.org/10.1007/s00216-017-0784-x] [PMID: 29236134]

[164] Poma G, Malarvannan G, Voorspoels S, *et al.* Determination of Halogenated Flame Retardants in Food: Optimization and validation of a method based on a two-step clean-up and gas chromatography-mass spectrometry. Food Control 2016; 65: 168-76.
[http://dx.doi.org/10.1016/j.foodcont.2016.01.027]

[165] Nøstbakken OJ, Duinker A, Rasinger JD, *et al.* Factors influencing risk assessments of brominated flame-retardants; evidence based on seafood from the North East Atlantic Ocean. Environ Int 2018; 119: 544-57.
[http://dx.doi.org/10.1016/j.envint.2018.04.044] [PMID: 30077002]

[166] Portolés T, Sales C, Gómara B, *et al.* Novel analytical approach for brominated flame retardants based on the use of gas chromatography-atmospheric pressure chemical ionization-tandem mass spectrometry with emphasis in highly brominated congeners. Anal Chem 2015; 87(19): 9892-9.
[http://dx.doi.org/10.1021/acs.analchem.5b02378] [PMID: 26354040]

[167] Kalachova K, Pulkrabova J, Cajka T, Drabova L, Stupak M, Hajslova J. Gas chromatography-triple quadrupole tandem mass spectrometry: a powerful tool for the (ultra)trace analysis of multiclass environmental contaminants in fish and fish feed. Anal Bioanal Chem 2013; 405(24): 7803-15. [http://dx.doi.org/10.1007/s00216-013-7000-4] [PMID: 23657455]

[168] Sapozhnikova Y, Simons T, Lehotay SJ. Evaluation of a fast and simple sample preparation method for polybrominated diphenyl ether (PBDE) flame retardants and dichlorodiphenyltrichloroethane (DDT) pesticides in fish for analysis by ELISA compared with GC-MS/MS. J Agric Food Chem 2015; 63(18): 4429-34. [http://dx.doi.org/10.1021/jf505651g] [PMID: 25644932]

[169] González-Sálamo J, Socas-Rodríguez B, Hernández-Borges J. Analytical methods for the determination of phthalates in food. Curr Opin Food Sci 2018; 22: 122-36. [http://dx.doi.org/10.1016/j.cofs.2018.03.002]

[170] Gu YY, Yu XJ, Peng JF, *et al.* Simultaneous solid phase extraction coupled with liquid chromatography tandem mass spectrometry and gas chromatography tandem mass spectrometry for the highly sensitive determination of 15 endocrine disrupting chemicals in seafood. J Chromatogr B Analyt Technol Biomed Life Sci 2014; 965: 164-72. [http://dx.doi.org/10.1016/j.jchromb.2014.06.024] [PMID: 25023229]

[171] Hu X, Gu Y, Huang W, Yin D. Phthalate monoesters as markers of phthalate contamination in wild marine organisms. Environ Pollut 2016; 218: 410-8. [http://dx.doi.org/10.1016/j.envpol.2016.07.020] [PMID: 27435611]

[172] Cariou R, Larvor F, Monteau F, *et al.* Measurement of phthalates diesters in food using gas chromatography-tandem mass spectrometry. Food Chem 2016; 196: 211-9. [http://dx.doi.org/10.1016/j.foodchem.2015.09.045] [PMID: 26593485]

[173] He M, Yang C, Geng R, *et al.* Monitoring of phthalates in foodstuffs using gas purge microsyringe extraction coupled with GC-MS. Anal Chim Acta 2015; 879: 63-8. [http://dx.doi.org/10.1016/j.aca.2015.02.066] [PMID: 26002478]

[174] Salgueiro-González N, Castiglioni S, Zuccato E, Turnes-Carou I, López-Mahía P, Muniategui-Lorenzo S. Recent advances in analytical methods for the determination of 4-alkylphenols and bisphenol A in solid environmental matrices: A critical review. Anal Chim Acta 2018; 1024: 39-51. [http://dx.doi.org/10.1016/j.aca.2018.02.081] [PMID: 29776546]

[175] Acir IH, Guenther K. Endocrine-disrupting metabolites of alkylphenol ethoxylates - A critical review of analytical methods, environmental occurrences, toxicity, and regulation. Sci Total Environ 2018; 635: 1530-46. [http://dx.doi.org/10.1016/j.scitotenv.2018.04.079] [PMID: 29874777]

[176] Li C, Jin F, Snyder SA. Recent advancements and future trends in analysis of nonylphenol ethoxylates and their degradation product nonylphenol in food and environment. TrAC, Trend Anal Chem 2018; 107: 78-90.

[177] Mortazavi S, Bakhtiari AR, Sari AE, Bahramifar N, Rahbarizadeh F. Occurrence of endocrine disruption chemicals (bisphenol a, 4-nonylphenol, and octylphenol) in muscle and liver of *Cyprinus Carpino* Common, from Anzali wetland, Iran. Bull Environ Contam Toxicol 2013; 90(5): 578-84. [http://dx.doi.org/10.1007/s00128-013-0964-0] [PMID: 23429963]

[178] Chen WL, Gwo JC, Wang GS, Chen CY. Distribution of feminizing compounds in the aquatic environment and bioaccumulation in wild tilapia tissues. Environ Sci Pollut Res Int 2014; 21(19): 11349-60. [http://dx.doi.org/10.1007/s11356-014-3062-x] [PMID: 24878555]

[179] Niu Y, Zhang J, Duan H, Wu Y, Shao B. Bisphenol A and nonylphenol in foodstuffs: Chinese dietary exposure from the 2007 total diet study and infant health risk from formulas. Food Chem 2015; 167: 320-5. [http://dx.doi.org/10.1016/j.foodchem.2014.06.115] [PMID: 25148994]

[180] Errico S, Nicolucci C, Migliaccio M, Micale V, Mita DG, Diano N. Analysis and occurrence of some phenol endocrine disruptors in two marine sites of the northern coast of Sicily (Italy). Mar Pollut Bull 2017; 120(1-2): 68-74.
[http://dx.doi.org/10.1016/j.marpolbul.2017.04.061] [PMID: 28477989]

[181] Ocharoen Y, Boonphakdeea C, Boonphakdeea T, Shinnc AP, Moonmangmeed S. High levels of the endocrine disruptors bisphenol A and 17β-estradiol detected in populations of green mussel, *Perna viridis*, cultured in the Gulf of Thailand. Aquaculture 2018; 497: 348-56.
[http://dx.doi.org/10.1016/j.aquaculture.2018.07.057]

[182] Lee CC, Jiang LY, Kuo YL, *et al.* Characteristics of nonylphenol and bisphenol A accumulation by fish and implications for ecological and human health. Sci Total Environ 2015; 502: 417-25.
[http://dx.doi.org/10.1016/j.scitotenv.2014.09.042] [PMID: 25268571]

[183] Staniszewska M, Graca B, Sokołowski A, Nehring I, Wasik A, Jendzul A. Factors determining accumulation of bisphenol A and alkylphenols at a low trophic level as exemplified by mussels *Mytilus trossulus.* Environ Pollut 2017; 220(Pt B): 1147-59.
[http://dx.doi.org/10.1016/j.envpol.2016.11.020] [PMID: 27843016]

[184] Ismail NAH, Wee SY, Aris AZ. Bisphenol A and alkylphenols concentrations in selected mariculture fish species from Pulau Kukup, Johor, Malaysia. Mar Pollut Bull 2018; 127: 536-40.
[http://dx.doi.org/10.1016/j.marpolbul.2017.12.043] [PMID: 29475693]

[185] Zhou X, Yang Z, Luo Z, Li H, Chen G. Endocrine disrupting chemicals in wild freshwater fishes: Species, tissues, sizes and human health risks. Environ Pollut 2019; 244: 462-8.
[http://dx.doi.org/10.1016/j.envpol.2018.10.026] [PMID: 30366293]

[186] Casatta N, Mascolo G, Roscioli C, Viganò L. Tracing endocrine disrupting chemicals in a coastal lagoon (Sacca di Goro, Italy): sediment contamination and bioaccumulation in Manila clams. Sci Total Environ 2015; 511: 214-22.
[http://dx.doi.org/10.1016/j.scitotenv.2014.12.051] [PMID: 25546459]

[187] Barber LB, Loyo-Rosales JE, Rice CP, Minarik TA, Oskouie AK. Endocrine disrupting alkylphenolic chemicals and other contaminants in wastewater treatment plant effluents, urban streams, and fish in the Great Lakes and Upper Mississippi River Regions. Sci Total Environ 2015; 517: 195-206.
[http://dx.doi.org/10.1016/j.scitotenv.2015.02.035] [PMID: 25727675]

[188] Ademollo N, Patrolecco L, Rauseo J, Nielsen J, Corsolini S. Bioaccumulation of nonylphenols and bisphenol A in the Greenland shark Somniosus microcephalus from the Greenland seawaters. Microchem J 2018; 136: 106-12.
[http://dx.doi.org/10.1016/j.microc.2016.11.009]

[189] Jakimska A, Huerta B, Bargańska Ż, Kot-Wasik A, Rodríguez-Mozaz S, Barceló D. Development of a liquid chromatography-tandem mass spectrometry procedure for determination of endocrine disrupting compounds in fish from Mediterranean rivers. J Chromatogr A 2013; 1306: 44-58.
[http://dx.doi.org/10.1016/j.chroma.2013.07.050] [PMID: 23890552]

[190] Luo L, Yang Y, Wang Q, *et al.* Determination of 4-noctylphenol, 4-n-nonylphenol and bisphenol A in fish samples from lake and rivers within Hunan Province, China. Microchem J 2017; 132: 100-6.
[http://dx.doi.org/10.1016/j.microc.2017.01.012]

[191] Luo Z, Lu J, Li H, Tu Y, Wan Y, Yang Z. Air-assisted liquid-liquid microextraction integrated with QuEChERS for determining endocrine-disrupting compounds in fish by high-performance liquid chromatography-tandem mass spectrometry. Food Chem 2018; 260: 174-82.
[http://dx.doi.org/10.1016/j.foodchem.2018.04.007] [PMID: 29699659]

[192] Crawford CB, Quinn B. The interactions of microplastics and chemical pollutants. In: Crawford CB, Quinn B, Eds. Microplastic Pollutants. Elsevier 2017; 6: pp. 131-57.
[http://dx.doi.org/10.1016/B978-0-12-809406-8.00006-2]

[193] Mato Y, Isobe T, Takada H, Kanehiro H, Ohtake C, Kaminuma T. Plastic resin pellets as a transport

medium for toxic chemicals in the marine environment. Environ Sci Technol 2001; 35(2): 318-24. [http://dx.doi.org/10.1021/es0010498] [PMID: 11347604]

[194] Zhan Z, Wang J, Peng J, Xie Q, Huang Y, Gao Y. Sorption of 3,3′,4,4′-tetrachlorobiphenyl by microplastics: A case study of polypropylene. Mar Pollut Bull 2016; 110(1): 559-63. [http://dx.doi.org/10.1016/j.marpolbul.2016.05.036] [PMID: 27230985]

[195] Liu L, Fokkink R, Koelmans AA. Sorption of polycyclic aromatic hydrocarbons to polystyrene nanoplastic. Environ Toxicol Chem 2016; 35(7): 1650-5. [http://dx.doi.org/10.1002/etc.3311] [PMID: 26588181]

[196] Turner A, Holmes LA. Adsorption of trace metals by microplastic pellets in fresh water. Environ Chem 2015; 12: 600-10. [http://dx.doi.org/10.1071/EN14143]

[197] Rochman CM, Hentschel BT, Teh SJ. Long-term sorption of metals is similar among plastic types: implications for plastic debris in aquatic environments. PLoS One 2014; 9(1)e85433 [http://dx.doi.org/10.1371/journal.pone.0085433] [PMID: 24454866]

[198] Brennecke D, Duarte B, Paiva F, Caçador I, Canning-Clode J. Microplastics as vector for heavy metal contamination from the marine environment. Estuar Coast Shelf Sci 2016; 178: 189-95. [http://dx.doi.org/10.1016/j.ecss.2015.12.003]

[199] Yoo JW, Doshi N, Mitragotri S. Adaptive micro and nanoparticles: temporal control over carrier properties to facilitate drug delivery. Adv Drug Deliv Rev 2011; 63(14-15): 1247-56. [http://dx.doi.org/10.1016/j.addr.2011.05.004] [PMID: 21605607]

[200] Ramos JAA, Barletta B, Costa MF. Ingestion of nylon threads by Gerreidae while using a tropical estuary as foraging grounds. Aquat Biol 2012; 17: 29-34. [http://dx.doi.org/10.3354/ab00461]

[201] Browne MA, Crump P, Niven SJ, *et al.* Accumulation of microplastic on shorelines worldwide: sources and sinks. Environ Sci Technol 2011; 45(21): 9175-9. [http://dx.doi.org/10.1021/es201811s] [PMID: 21894925]

[202] Dris R, Gasperi J, Saad M, Mirande C, Tassin B. Synthetic fibers in atmospheric fallout: A source of microplastics in the environment? Mar Pollut Bull 2016; 104(1-2): 290-3. [http://dx.doi.org/10.1016/j.marpolbul.2016.01.006] [PMID: 26787549]

[203] De Witte B, Devriese L, Bekaert K, *et al.* Quality assessment of the blue mussel (*Mytilus edulis*): comparison between commercial and wild types. Mar Pollut Bull 2014; 85(1): 146-55. [http://dx.doi.org/10.1016/j.marpolbul.2014.06.006] [PMID: 24969855]

[204] Mathalon A, Hill P. Microplastic fibers in the intertidal ecosystem surrounding Halifax Harbor, Nova Scotia. Mar Pollut Bull 2014; 81(1): 69-79. [http://dx.doi.org/10.1016/j.marpolbul.2014.02.018] [PMID: 24650540]

[205] Santana MFM, Ascer LG, Custódio MR, Moreira FT, Turra A. Microplastic contamination in natural mussel beds from a Brazilian urbanized coastal region: Rapid evaluation through bioassessment. Mar Pollut Bull 2016; 106(1-2): 183-9. [http://dx.doi.org/10.1016/j.marpolbul.2016.02.074] [PMID: 26980138]

[206] Silva AB, Bastos AS, Justino CIL, da Costa JP, Duarte AC, Rocha-Santos TAP. Microplastics in the environment: Challenges in analytical chemistry - A review. Anal Chim Acta 2018; 1017: 1-19. [http://dx.doi.org/10.1016/j.aca.2018.02.043] [PMID: 29534790]

[207] Hanvey JS, Lewis PJ, Lavers JL, Crosbie ND, Pozo K, Clarke BO. A review of analytical techniques for quantifying microplastics in sediments. Anal Methods 2017; 9: 1369-83. [http://dx.doi.org/10.1039/C6AY02707E]

[208] da Costa JP, Duarte AC, Rocha-Santos TAP. Microplastics: occurrence, fate and behaviour in the environment.Comprehensive Analytical Chemistry. Elsevier 2017; 1: pp. 1-24. [http://dx.doi.org/10.1016/bs.coac.2016.10.004]

[209] Koelmans AA, Besseling E, Shim WJ. Nanoplastics in the aquatic environment. Critical Review. In: Bergmann M, Gutow l, Klages M, Eds. Marine Anthropogenic Litter. Springer International Publishing 2015; pp. 325-40.

[210] Filella M. Questions of size and numbers in environmental research on microplastics: methodological and conceptual aspects. Environ Chem 2015; 12(5): 527-38. [http://dx.doi.org/10.1071/EN15012]

[211] Hong SH, Shim WJ, Hong L. Methods of analyzing chemicals associated with microplastics: a review. Anal Methods 2017; 9: 1361-8. [http://dx.doi.org/10.1039/C6AY02971J]

[212] Masura J, Baker J, Foster G, Arthur C. Laboratory methods for the analysis of microplastics in the marine environment: Recommendations for quantifying synthetic particles in waters and sediments. NOAA Technical Memorandum NOS-OR&R-48 2015. https://marinedebris.noaa.gov/sites/default/files/publications files/noaa_microplastics_methods_manual.pdf (accessed 30/11/2018).

CHAPTER 3

Nanotechnology in the Food Field: Application of Metal-Based Nanoparticles

Beatriz Gómez-Gómez and **Yolanda Madrid***

Departamento de Química Analítica, Facultad de Ciencias Químicas, Universidad Complutense de Madrid, 28040, Madrid, Spain

Abstract: Nanotechnology offers a wide range of applications in the food sector such as development of new tastes and textures, nanoencapsulation of bioactive food components, design of nutrient delivery systems, nanosensors to detect spoilage or contamination, and the design of new food packaging materials. Although metal-based nanoparticles (AgNPs, SiO_2NPs, TiO_2NPs, ZnONPs...) have extensively been applied due to their antimicrobial, antioxidant and UV-blocking properties, there is limited knowledge about the impact of nanoparticles on human health and environment. For safety reasons, the EU has issued regulations requiring labelling of the nanomaterials in the ingredients list. Therefore, new analytical methods should be used to characterize nanomaterials but, since there is no single and universal method that can be applied to fully characterize nanoparticles, the need for multimethod approaches is widely acknowledged. This chapter focuses primarily on the application of metal-based nanoparticles in the food sector and the analytical methodologies used for nanoparticle characterization. Regarding the applications of nanoparticles, special attention should be paid to their antimicrobial properties and their use for developing active food packaging materials. Since the characterization of nanoparticles in complex matrices is troublesome, a detailed description of the prospects and difficulties of the analytical techniques commonly employed is given. Similarly, factors affecting nanoparticles stability such as sample preparation, interaction with food matrices, food stimulants, and chemicals used in "in vitro" gastric digestion procedures are also described. Finally, EU regulatory guidelines on nanomaterials are included and discussed.

Keywords: Analytical methodologies, Current EU directives, Food, Metal-based nanoparticles, Nanoparticles stability, Nanotechnology, Sample treatment.

INTRODUCTION

Nanotechnology has a huge impact on our daily life. It has revolutionized the industrial sector due to large-scale production of nanosized materials and the

* **Corresponding author Yolanda Madrid:** Departamento de Química Analítica, Facultad de Ciencias Químicas, Universidad Complutense de Madrid, 28040, Madrid, Spain; Tel/Fax: +34 913945149; E-mail: ymadrid@quim.ucm.es

Belén Gómara & María Luisa Marina (Eds.)
Current and Future Developments in Food Science (Vol. 1)

growing investment in this field from governments and industry worldwide. According to the European Commission recommendation, a nanomaterial can be defined as "*a natural, incidental, or manufactured material containing particles, in an unbound state or as an aggregate or as an agglomerate and where, for 50% or more of the particles in the number size distribution, one or more external dimensions are in the size range 1-100 nm. In specific cases and where warranted by concerns for the environment, health, safety, or competitiveness, the number size distribution threshold of 50% may be replaced by a threshold between 1% and 50%*" [1]. The impact of nanotechnology is such that the term "nanotechnology" has become synonymous with promising innovative products.

Nanosized materials have gained special attention due to their singular properties compared to bulk materials. First, they have a greater surface area per unit mass in contrast to larger particles that makes them more reactive. Second, the dominance of quantum effects at the nanoscale significantly affects the optical, magnetic, and electrical properties of the material. Among nanostructures, nanoparticles (NPs) are a category with important applications on sectors like medicine and medical devices, engineering and communication technologies, and in some industrial areas such as electronics, photonics, textile, pharmaceutical, food, and cosmetics. A nanoparticle is defined as a nanoform that has one or more dimensions of the order of 100 nm or less [2, 3]. This group comprises a heterogeneous variety of materials that are classified based on their composition into different categories: carbon-based nanoparticles (nanotubes, fullerenes, NPs of latex and graphenes), metal/metalloid-based nanoparticles (Fe, Au, Ag, TiO_2NPs, ZnONPs, SiO_2NPs, CeO_2NPs, quantum dots) and aluminium silicates (zeolites, clays). Nowadays, there are more than 1600 nanotechnology-based consumer products on the market [4]. Among all, metal and metal oxide nanoparticles have gained great research attention due to their relevant antioxidant, antimicrobial and blocking UV properties.

The use of nanomaterials in the food sector has exponentially grown in recent years. Nanotechnology offers a wide range of applications (Fig. **1**) in food processing such as the development of new tastes and textures, and the encapsulation of food components and additives to control the release of flavors and deliver nutraceuticals. Nanomaterials in the food industry have been mostly applied to develop food packaging materials for extending the shelf-life of foodstuffs. Nanoparticles are added to the packaging materials to improve their mechanical strength and barrier properties. Moreover, smart and intelligent food packages including nanoparticles as active components and nanosensors have been manufactured for food quality parameters control such as moisture, oxygen and carbon dioxide contents, microbial surface contamination, freshness and food conditions during transport and storage. Although most of the aforesaid applications are still in progress, the prospects of nanotechnology in the food technology sector are extraordinary.

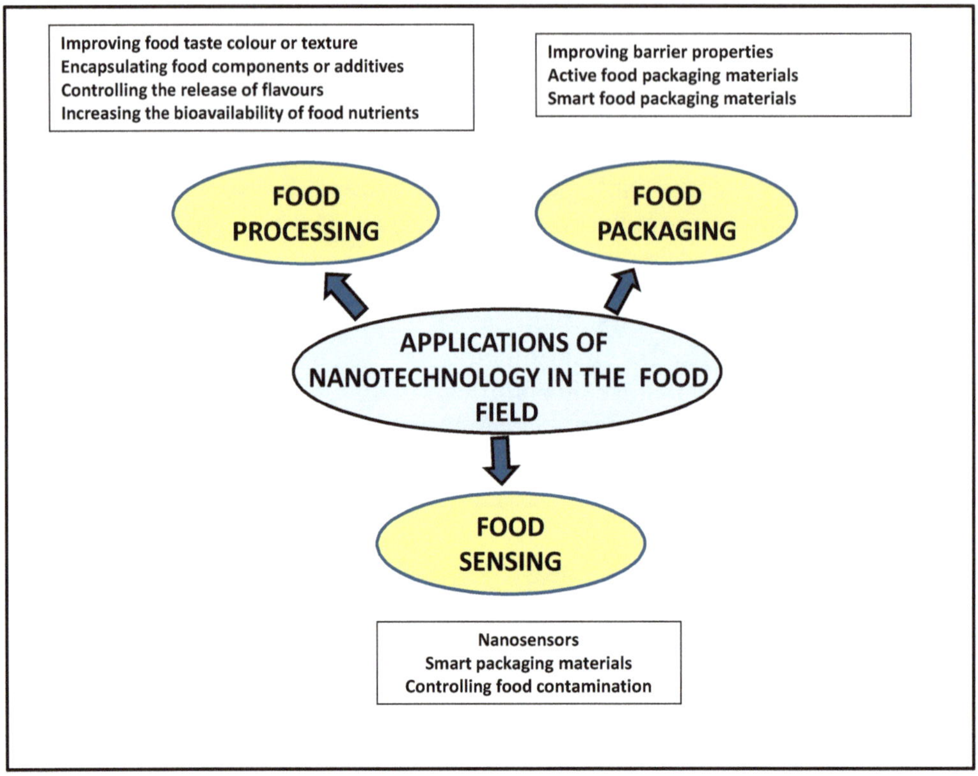

Fig. (1). Main applications of nanotechnology in the food field.

However, the widespread use of nanotechnology in our daily life is a matter of concern due to the increasing exposure of humans and ecosystems to nanoparticles which makes necessary a thorough assessment of their toxicological impact. For safety reasons, the EU has issued regulations requiring labelling of the nanomaterials in the ingredients list [5]. Consequently, there is an urgent need for developing analytical methods to identify nanoparticles in consumer products and to enable scientists to detect, identify and quantify them in complex matrices such as food and environmental samples. Unfortunately, since there is no single and universal method that can be applied to fully characterize nanoparticles, the need for multimethod approaches is widely accepted. The most common techniques for characterizing nanoparticles are Dynamic Light Scattering (DLS), Multi-angle Light Scattering (MALS) and nanoparticles tracking analysis (NTA); classical electron microscopy (TEM/SEM); analytical separation techniques such as ultrafiltration and asymmetric field flow fractionation coupled to ICP-MS or MALS detectors, and single particle ICPMS (sp-ICPMS). NPs stability is a key step when characterizing NPs. Several factors such as sample treatment, dilution, pH modifications, interaction with mobile phases, and food matrix components should be carefully controlled as they can alter the initial morphology of NPs. Along with developing analytical methods, the production of reference materials (RMs) is also required for method validation. Additionally, methods for testing nanomaterials safety including *in vitro and vivo* assays should be applied.

This chapter provides a detailed overview on the application of metal-based nanoparticles in the food sector. The first part is focused on their main applications and regulatory issues. The second outlines the analytical methodologies used for their detection, identification and quantification, and the difficulties encountered in their determination. Particular attention will be paid to those factors affecting nanoparticles stability (sample treatment, interaction with food matrix, and food simulants).

METAL-BASED NANOPARTICLES IN THE FOOD SECTOR

As mentioned earlier, the application of nanotechnology in the food field is growing tremendously. Metal-based nanoparticles are one of the most extensively applied nanostructures, especially as antimicrobial and antioxidant agents to increase the shelf-life of food products. This section will place greater emphasis on the most important applications of this type of nanoparticles.

Nanoparticles in Food Processing

Nanotechnology is used in food processing to improve color, texture and food taste; for encapsulating food components or additives, controlling the release of flavors, increasing the bioavailability of food components, enhancing functionality and nutritional values, and expanding the shelf-life of food products [6]. These applications are restricted in the case of metal and metal oxide nanoparticles due to their greater toxicity compared to other nano-systems such as polymeric nanoparticles, nanoemulsions, liposomes or nanoscale edible coatings that are often based on natural food ingredients [7]. However, several studies including metallic NPs have been reported in the literature. For example, Shah and co-workers produced polysaccharide based edible coatings with silver nanoparticles (AgNPs) in order to extend the shelf-life and preserve the quality of kinnow mandarin during long storage periods. For this purpose, they mixed 1:1 carboxymethyl cellulose (CMC) or guar gum with 20 nm spherical AgNPs to achieve the edible coating CMC-Ag and guar gum-Ag, respectively. Kinnow mandarins were further spread with the edible coatings and subsequently storage during 120 days at 4 °C and 10 °C. The results evidenced that the coated CMC-Ag and guar gum-Ag prevented losses from the main volatile components responsible for the aroma (limonene, linalool and c-terpinene) and preserved the sensorial characteristics of the fruit [8]. Xu and co-workers employed ZnO nanoparticles combined with radio frequency heating as an effective method for performing pasteurization. The procedure was firstly applied to carrots and they reported that the shelf-life of treated carrots was extended up to 60 days. Losses of hardness, color and carotenoids were also reduced [9]. The same treatment was later applied to a twice-cooked pork dish to control microorganism proliferation. The use of this pasteurization method allowed the authors to get heating uniformity and to reduce the loss of texture and flavor when compared with the conventional high pressure steam sterilization [10].

Nanoparticles as Antimicrobial Agents: Food Safety

Foodborne diseases are serious threats to global health. Food contamination may occur at any step of the food chain production or even during food consumption due to food spoilage and food contamination with human pathogen microorganisms [11]. It has been documented that many bacterial infections are resistant to at least one of the antimicrobial agents frequently used to fight infection [12]. The incidence of persistent bacterial infections has notably increased over the past few decades. In this line, the production of bacterial biofilms has been considered in several studies as the major factor linked with this issue. A biofilm is an assemblage of microbial cells that grow attached to a surface and embedded in a matrix composed of extracellular polymeric substances (EPS) mainly polysaccharides, proteins, phospholipids, and nucleic acids [13, 14]. The presence of microbial biofilms is becoming one of the major causes of food contamination in industrial food processing sectors such as brewing and seafood, dairy, poultry, and meat. Microbial biofilms are formed on the equipment surfaces of production lines originating hygiene problems, food spoilage, and post-processing contamination. As a consequence, the shelf-life of the food products decreases along with the appearance of infections and economic losses [14]. For instance, microbial biofilms of *Bacillus cereus, E. coli, Shigella spp., and Staphylococcus aureus* are commonly found in dairy and egg processing industries. Similarly, *Listeria spp., Staphylococcus spp., and Vibrio spp.* have been frequently isolated from surfaces of industrial equipment used in seafood processing plants [15]. Bacterial cells within the confines of a biofilm have the ability to survive in nutrient deficient conditions, to acquire new genes for virulence and even to protect the bacteria community against currently employed disinfectant and antibiotics [16]. Hence, biofilm associated infections are 100-1000 times less susceptible to antibiotic treatment compared to the planktonic mode of microbial growth [17]. Consequently, there is an urgent demand for alternative antibacterial and antibiofilm agents to fight against antibiotic resistance and biofilm-associated infections. In this sense, the use of nanoparticles as new tools to overcome such problems is gaining interest within the scientific community. Nanoparticles may not only increase drug efficacy but also reduce antibiotic resistance due to their direct interaction with the bacterial cell [18].

The antimicrobial activity of nanoparticles is highly dependent on their physicochemical properties such as size, charge, zeta potential, surface morphology, and crystal structure. Additionally, factors like the type of bacterial strain, the exposure time, and the environmental conditions like pH or UV-Vis irradiation play a key role when evaluating the antimicrobial activity of nanoparticles [18]. The influence of many different parameters makes difficult to entirely rule out the mechanisms governing the antimicrobial activity of nanoparticles that still remain under discussion [19]. However, several mechanisms have been proposed in the literature (Fig. **2**).

The production of reactive oxygen species (ROS) has been postulated as one of the main modes of antimicrobial action of metal-based NPs. ROS comprise a group of molecules and reactive intermediates that mainly include the following compounds: superoxide radical (O^{2-}), the hydrogen peroxide (H_2O_2), the hydroxyl radical ($OH^{\cdot}$), and singlet oxygen. ROS produce changes in the permeability of the cell membrane resulting in bacterial cell membrane damage [20]. Moreover, ROS interact with DNA, therefore accelerating bacterial cell apoptosis [18].

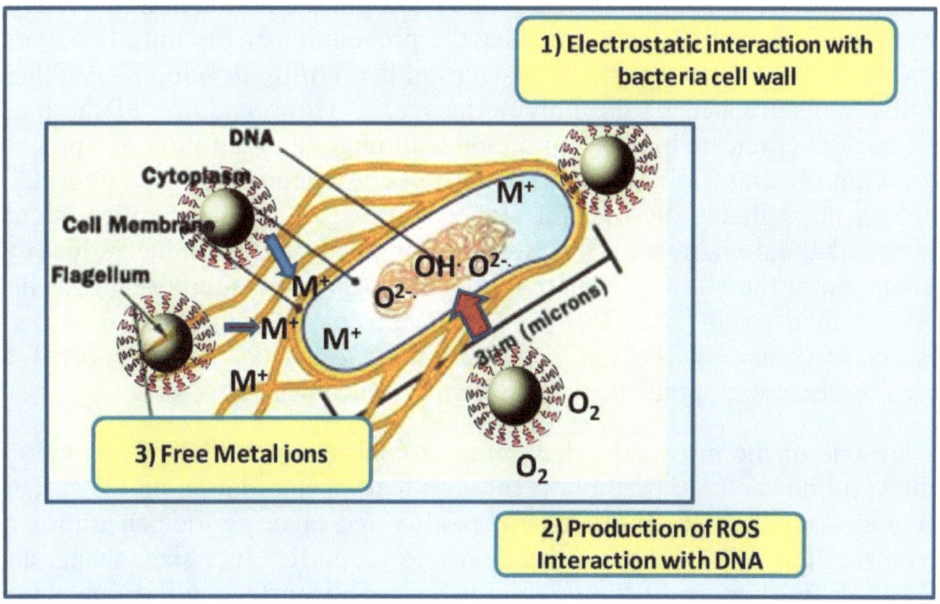

Fig. (2). Antimicrobial mechanisms of nanoparticles.

The release of metal ions from nanoparticles to the surroundings of the bacteria has also been proposed as an antimicrobial mechanism of NPs. The resulting free metal ions are absorbed through the cell membrane and interact with the functional groups of biomolecules disturbing metabolic function. Another mechanism reported is through electrostatic interaction between electronegative groups on bacterial membrane and free metal cations that lead to nanoparticle accumulation on the cell surface, therefore altering the structure and permeability of the cell membrane. Other postulated mechanisms include protein and enzyme dysfunction, genotoxicity, and photokilling [21].

Knowledge of the mechanism of nanoparticles to control pathogenic biofilm growth is still in its infancy stage. The interaction between NPs and biofilm matrix involves three steps. The first comprises the transport of the nanoparticle to the surroundings of the biofilm, followed by the attachment of nanoparticles onto biofilm surface, and the subsequent migration of nanoparticles to the biofilm confines. The EPS where bacterial cells are embedded contribute to the initial deposition and subsequent accumulation. Once the NPs have penetrated into the biofilm, they move within the biofilm mainly by diffusion. NPs diffusion is dependent on the pore size of the biofilm matrix, the charge of the NPs and EPS, the hydrophobicity of the surrounding environment, and the chemical gradient. Once NPs get the bacterial cells, they induce the bacteria dead [22].

Several strategies have been proposed in order to fight the production of bacterial biofilms (Fig. **3**). Some of them consider the prevention of the initial bacterial adhesion on the food surface at the first stages of the biofilm development. Others deal with the disturbance of the polymeric matrix. However, one of the most effective tools to eradicate biofilm formation is through the inhibition of a process called quorum sensing (QS), a bacterial cell-to-cell communication system [23]. QS mechanism regulates the biofilm mode of growth through chemical signal molecules called autoinducers. Thus, the disruption of the autoinducers leads to the interruption of the biofilm formation process [24]. Some microorganisms that occur in the food sector like *Bacillus cereus, E. coli, Staphylococcus aureus, Listeria monocytogen, and Pseuodomona aeruginos,* have been reported to produce, release and respond to the QS signal molecules [25 - 29].

AgNPs are one of the most extended antimicrobial nanomaterials. They induce the damage of the bacterial membrane through their accumulation on the surface producing changes in the membrane permeability and favoring the penetration of the NPs inside the cell. Therefore, nanoparticle properties like size, shape, and surface charge and nature of the coating agent highly influence the interaction between bacterial membrane and nanoparticles [30]. For example, Demirbas and co-workers synthesized AgNPs using anthocyanins-rich berries obtained from

Rubus fruticosus L. (blackberry), *Fragaria vesca L.* (strawberry) and *Rubus idaeus L.* (raspberry) extracts. The antimicrobial properties of the resulting AgNPs were evaluated against *E. coli, S. enterica, S. aureus and B. Cereus.* Although all nanoparticles showed inhibitory effects against the microorganisms selected, those produced from strawberry extracts were the most effective. These results evidenced that differences in the composition of the surface of the capping agent lead to different results in the antimicrobial tests [31]. Similarly, Kujda and co-workers produced AgNPs with different surface properties, and therefore, with different antimicrobial properties [32]. Nanoparticle size has also shown an important effect on antimicrobial activity of AgNPs. Small AgNPs have been reported to be more effective against microorganisms. Dogru and co-workers tested the antimicrobial effectiveness of AgNPs of different diameter size (70, 50 and 30 nm) against *E. coli and S. aureus*. Growth inhibition was obtained with all AgNPs tested at a level of NPs exposure of 50 mg L^{-1}; however, the smallest ones (30 nm) showed the best antimicrobial properties. The authors attributed the major inhibitory properties of the smallest AgNPs to their highest surface to volume ratio and their highest capability to release Ag^{+} [33]. Similarly, Raza and co-workers evidenced that nanoparticle shape also influences the antimicrobial activity of AgNPs against *P. aeruginosa* and *E. coli.* Spherical nanoparticles offered the best biocidal behavior compared to other morphologies [34].

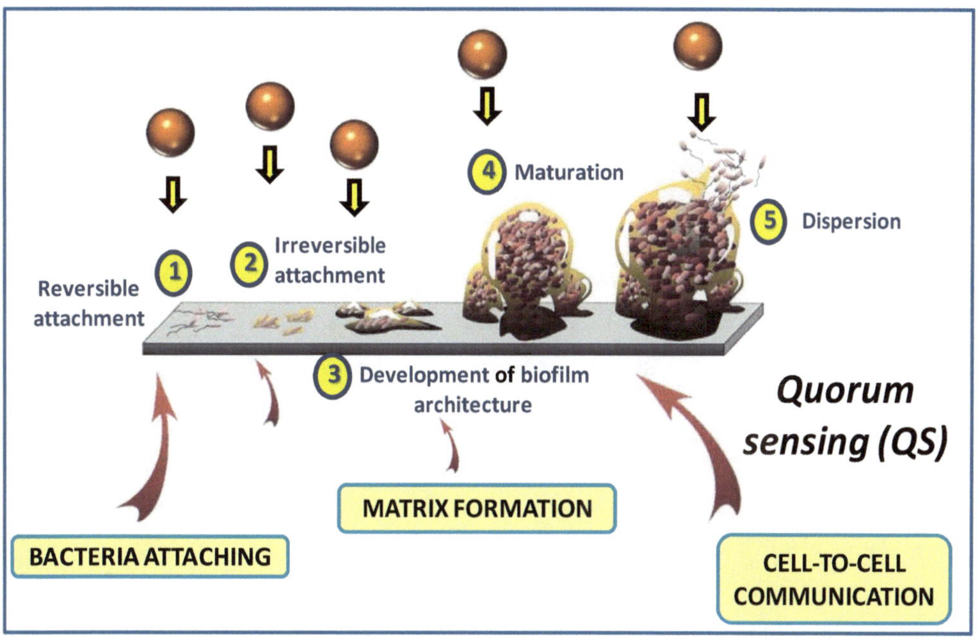

Fig. (3). Mechanisms of nanoparticles against microbial biofilms.

The results described so far are related to bacteria growing in planktonic cultures. However, very few studies have appeared in the literature concerning the impact of AgNPs against bacterial biofilm. Lotha and co-workers have demonstrated that AgNPs synthesized by using purified plant extracts were able to inhibit the biofilm formation of some foodborne pathogens such as *Enterococcus faecalis*, *Bacillus subtilis*, *Staphylococcus aureus*, *Escherichia coli*, *Salmonella enterica serovar Typhi*, *Pseudomonas aeruginosa* and *Enterobacter cloacae* [35]. Similarly, Anjugam and co-workers showed the antibiofilm potential against *E. faecalis* and *P. aeruginosa* of AgNPs manufactured from crustacean immune molecule β-1,3 glucan binding protein [36].

The control of *quorum sensing* mechanism is a promising tool to eradicate biofilm formation. Singh and co-workers evidenced that AgNPs were able to attenuate *Pseudomona aeruginosa* QS systems [37]. In this line, Waght and co-workers and Naik and co-workers studied the potential of silver nanowire and AgCl-TiO_2NPs to interrupt QS- regulated virulence and biofilm formation. In both cases they were able to inhibit the violacein production in the *Chromobacterium violaceum* strain. This bacterium produces the pigment violacein in response to the autoinducer N-acyl homoserine lactone, the QS signal molecule of Gramm negative bacteria. Therefore, a decrease in the violacein production in presence of AgNPs suggests that nanoparticles affect the bacterial QS system [38, 39].

Besides AgNPs, other metallic nanoparticles have been evaluated as antimicrobial agents against food borne pathogens. Copper nanoparticles (CuNPs) could damage the bacterial membrane integrity leading to the death of the microorganisms. Additionally, CuNPs increase the production of ROS affecting the normal cell function in multiple ways such as lipid peroxidation, protein oxidation or DNA degradation [40]. Tovar-Corono and co-workers indicated that CuNPs produced with cyanidine-0-3-glucoside were able to inhibit the growth of *Pseudomona aeruginosa*, *Staphylococcus saprophyticus* and *Staphylococcus epidermis* [41]. Moreover, CuNPs have also showed antimicrobial activity against *B. subtilis, E. coli* and *S. aureus* [42]. A few papers reported the antimicrobial activity of gold nanoparticles. These nanoparticles are able to interact with bacterial membrane, which facilitates their attachment to the internal structure of the membrane, and therefore, activate the mechanism through ROS production [43].

ZnONPs, TiO_2NPs or CuONPs are also considered as antimicrobial agents. Their antibacterial activity mechanism has been largely discussed in the literature but it still remains unclear. ZnONPs are able to interact with bacterial membrane affecting directly the phospholipid bilayer, thus originating cell membrane disruption and nanoparticle internalization. Moreover, the photocatalitic properties of ZnONPs favor the production of ROS species (O^{2-}, OH^-, H_2O_2) on nanoparticle surface under UV irradiation. Furthermore, ZnONPs tend to release Zn^{2+} ions when they are dispersed in aqueous solutions. Consequently, a decrease of cell viability has been observed due to the presence of Zn^{2+} [44 - 46]. TiO_2NPs also act as a photocatalyst agent in presence of UV-light producing ROS species that induce bacteria death. In addition, TiO_2NPs cause lipid peroxidation that originates changes on membrane fluidity favoring membrane integrity disruption. Similarly, CuONPs are able to penetrate through bacterial cell causing cell functions disturbance [47].

The antibacterial action of metal oxide nanoparticles against food borne pathogens has attracted the attention of an important number of researchers [21]. For example, Duffy and co-workers recently reported the effectiveness of ZnONPs and CuONPs against a *Salmonella strains* isolated from poultry [48]. Furthermore, Shakerimoghaddam and co-workers found that ZnONPs were able to reduce biofilm formation of *E. coli* isolated from urine samples [49]. Additionally, Long and co-workers studied the effect of UV irradiation time on the photocatalytic disinfection process of TiO_2NPs against food borne pathogens such as *Salmonella typhimurium* and *Listeria monocytogenes* [50]. Furthermore, Jalvo and co-workers proposed TiO_2NPs as additive for making self-cleaning antimicrobial flat and porous glass surfaces. The resulting functionalized surfaces were irradiated with UV-A during 2 h. Their antibacterial activity was tested against *Staphylococcus aureus* and *Pseudomona putida* afterwards. Although cell viability was clearly affected when bacteria were grown in the self-cleaning surface, the formation of biofilm was not affected [51].

Nanoparticles in Food Packaging

Food packaging materials provide a way of preserving the quality and safety of products during both distribution and storage. Generally, the packaging material should exhibit suitable mechanical, optical, and thermal properties. For instance, packaging materials should act as a barrier against chemical contaminants, permeation of water vapor, oxygen, carbon dioxide or flavors as well as microorganisms [52]. A great variety of materials have been used for this purpose including ceramics, metals or paper. However, synthetic polymers are the most popular ones since they offer attractive advantages such as thermostability, thermal and mechanical properties, lightness, and low prices. Polymers usually employed in food packaging are polyethylene, polypropylene, polystyrene, polyvinyl chloride (PVC), and polyethylene terephthalate (PET), all derived from non-renewable petroleum resources with a potential risk factor for health and environment [53]. Consequently, the development of biopolymers in the field of food packaging is an active research topic. In this sense, a wide range of biodegradable polymers has been developed. Some of them are derived from plant and animal resources such as polysaccharides, polyesters or proteins while others are synthetic biodegradable polymers like poly(L-lactide) (PLA) or poly(butylene succinate) (PBS); others are produced by microbial fermentation, for example, poly(β-hydroxybutyrate) (PHB). Unfortunately, they offer high gas and vapor permeability and poor resistance which limit their applications [53 - 55]. The addition of nanoparticles to these materials may improve their mechanical strength and barrier properties and provide new characteristics to the material (Fig. **4**).

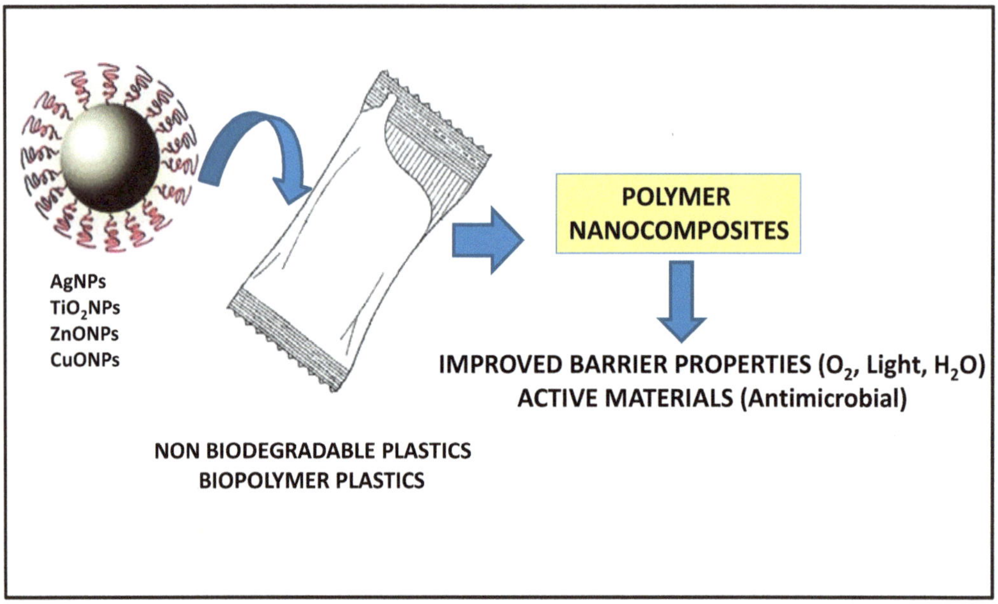

Fig. (4). Application of nanoparticles in developing food packaging materials.

Following the recommendations addressed by the European Union Guidance to the Commission Regulation No 450/2009 [56], a packaging material is designed as active when it performs some desired function other than merely providing a barrier to the external environment. In general, the active material is obtained by adding some compounds into the packaging material that absorb or release substances in order to improve the quality of packaged food or to extend its shelf-life [57 - 59]. In recent years, the use of nanosized compounds has been of great interest in order to develop active food packaging. The resulting nanocomposites can release specific compounds like antioxidant and/or flavors with the aim of improving the sensory characteristics of food as well as preserving the quality of fresh and processed products by scavenging moisture, oxygen and carbon dioxide inside the package. Furthermore, they would offer antibacterial and antioxidant effects, and even alert consumers of food safety conditions through nanosensor systems which identify specific microbial and/or chemical contaminants or environmental factors [2, 7, 60].

Due to their antimicrobial properties, AgNPs have been widely studied for developing biopolymers packaging coatings. For example, Singh and Sahareen [61] proposed a low cost and eco-friendly cellulosic material impregnated with AgNPs to extend the shelf-life of vegetables (cabbage and tomato). The nutritional value of the vegetables stored in the packaging containing nanoparticles remained unaltered. The developed material also exhibited antimicrobial properties and it was able to keep moisture content that makes vegetables look fresher for longer time. Similarly, a low-density polyethylene nanocomposite with AgNPs has been designed to protect chicken breast fillets during their storage. The study revealed an increase of the product shelf-life due to the presence of nanoparticles [62].

Bimetallic (Ag-Cu) alloy nanoparticles have been also evaluated to enhance the properties of agar-based films. Agar has been suggested as an alternative source for designing renewable and biodegradable plastic packaging materials. Unfortunately, it offers low thermal stability and very poor oxygen and water diffusion barriers. The Agar/Ag-Cu nanocomposite film significantly improved thermo-mechanical, oxygen barrier and optical properties of the agar. Moreover, the material exhibited an important antibacterial activity against two food-borne pathogenic bacteria, *L. monocytogenes* and *S. typhimurium.* Recently, the Food and Drug Administration (FDA) has approved the use of agar nanocomposite films as material for food packaging [63]. In this line, Li and co-workers employed AgNPs and TiO_2NPs for improving mechanical and thermal stability and antimicrobial properties of PLA film. Moreover, the authors examined the migration of the nanoparticles tested from the film to different food simulants [64].

Among the natural polymers used to develop biodegradable food packaging, chitosan offers attractive features due to its excellent ability to form films, high mechanical strength and antimicrobial properties [65]. Moreover, chitosan has been classified as safe by the Food and Drug Administration (FDA) to be used as a food preservative and has been widely employed for preserving meat and fruits [66, 67]. A chitosan-TiO_2NPs composite coating has been developed by Youssef and co-workers. The presence of TiO_2NPs enhances antibacterial activity against gram positive and gram negative bacteria and fungi. The developed material was successfully applied for storing soft cheese during 30 days [68]. Recently, several examples have appeared in the literature on the use of biodegradable food packaging modified with nanoparticles. Zhang and co-workers employed

TiO_2NPs to improve the mechanical characteristics and expand new functionalities of chitosan-based nanocomposite films [69]. A polysaccharides-based polymer like soluble soybean has been recently investigated in combination with TiO_2NPs for developing transparent, water-soluble and colorless food packaging materials with enhanced mechanical and antimicrobial properties [70, 71]. In addition, a bilayer membrane constituted by PLA, cellulose nanofibers, and TiO_2 nanoparticles was proposed as innovative food container with optical transparency and gas barrier performances. It was found that the presence of TiO_2NPs sharply decreased the transmittance in the UV range and therefore enhanced the optical transparency of PLA [72].

ZnONPs have also been considered as an active agent to be incorporated into a polymeric matrix due to their antimicrobial and UV-light barrier properties without impairing polymer transparency [73]. ZnO bulk is defined by the FDA and the European Food Safety Authority (EFSA) as a "Generally Recognized as Safe (GRAS)" substance, and therefore, their use as an additive for plastic materials is allowed [74]. The number of studies focused on the application of ZnONPs in food packaging has increased in recent years [75 - 77]. Zhang and co-workers, for instance, incorporated ZnO nanoparticles in PLA in order to elaborate a paper-based packaging material to avoid microbiological contamination in foods. Antimicrobial test revealed that the active material was effective in inactivating *E. coli* and *S. aureus* [78]. A blend of ZnO nanorods with the natural mineral kaolin nanoclay was incorporated as nanofillers to provide functionality to semolina-based edible films for food applications. Semolina is a type of wheat flour with high gluten content and with antioxidant activity. However, semolina films offer poor water diffusion barrier and irregular mechanical characteristics. The reinforcement of this material with ZnO nanorods and nano-kaolin endowed the material with low oxygen permeability and water solubility and high UV barrier properties. Additionally, the resulting bio-nanocomposite showed antimicrobial activity against *E. coli* bacteria [79]. Hydrogels have been widely used as water or drip adsorption pads in the food packaging industry [80]. Among them, carrageenan, a water soluble polysaccharide extracted from various species of red marine algae, has been preferred due to its strong gel forming ability along with high water holding capacity [80]. Oun and Rhim prepared carrageenan-based hydrogels film using KCl as a cross-linker and ZnO and CuO nanoparticles as functional fillers. The integration of nanoparticles furnished the material with water holding capacity and thermal stability along with strong antimicrobial activity against *E. coli* and *L. monocytogenes* [81].

Nanoparticles for Food Sensing

Food contamination may appear at any step of the food production chain. Therefore, the development of analytical methodologies enabling a fast and reliable identification and quantification of food components and contaminants is of great importance in order to ensure food quality and safety. In this respect, nanotechnology combined with suitable analytical methodologies and detection techniques is very useful for assessing food quality and safety [82]. For instance, metal and metal oxide nanoparticles are used to develop nanosensors incorporated into the food packaging in the so called "smart packaging" with the aim of monitoring in real time storage conditions. Nanosensors monitor freshness, display information on the quality of the product, and improve safety. Moreover, nanosensors can be inserted in measurement devices for determining parameters affecting food quality and safety [83]. All of these advantages justify the growing interest in developing nanosensors. Du and co-workers, for example, marketed a colorimetric biosensor using single-stranded DNA conjugated with flower-shaped AuNPs for the rapid detection of *Listeria monocytogenes* [84]. AuNPs have extensively been studied in the field of nanotechnology due to their attractive properties such as high chemical and physical stability, biocompatibility and great affinity with biological molecules that makes their functionalization easy. Moreover, they exhibit exceptional optical properties related to surface plasmons [85]. Consequently, AuNPs can produce different colorimetric signals depending on their size, shape, and surface morphology. In this line, color change has been induced in AuNPs by varying their aggregation states, which enables the performance of colorimetric sensing assays. Based on that, a colorimetric sensing assay has been used to detect selectively *Listeria monocytogenes* [84]. Similarly, Lim and co-workers developed a time-temperature indicator using gelatin-templated gold nanoparticles in order to monitor changes in temperature during storage and distribution of food products. An increment in temperature induces alterations in the stability of gelatin matrix, and therefore, changes in shape, size, and color of AuNPs [86].

Metal oxide nanoparticles have been also used to produce sensor systems. In this context, Ansari and co-workers produced an electrochemical immunosensor assisted by ZnONPs for the immobilization of immunoglobulin antibodies and bovine serum albumin (BSA) for ochratoxin-A (OTA) detection [87]. The mycotoxin OTA can be found in a wide variety of food such as cereals, coffee, wines or dried fruits being hazardous to health through food contamination. The results obtained in this work evidenced that ZnONPs amplified the electrochemical signal and constituted a good platform for attaching biomolecules [87, 88]. Similarly, Shao and co-workers developed an aptasensor based on nano-gapped structures and Fe_3O_4 magnetic nanoparticles for the detection of OTA. The constructed nanosensor was able to successfully detect OTA in water and food samples and it was described as an easy operated, low-priced, ultrasensitive and highly selective sensor system by the authors [89]. Finally, Nesakumar and co-workers immobilized lactate dehydrogenase on ZnO nanorods in order to fabricate an electrochemical biosensor for the detection of lactate. Lactate is used for extending the shelf-life of food products. However, it has been reported that lactate induces adverse effects in children under the age of three years [90].

Nano-food Regulatory Issues in the European Union

As mentioned previously, nanotechnology offers tremendous benefits in many food industry related areas including food processing, food packaging or food safety and quality control. Some of these applications are already on the market but most of them are still being developed. However, the arrival of these new products and ingredients into the market has led to the urgent need of addressing regulatory frameworks. The risks of nanomaterials to human health and environment still remain unclear. In recent years, the European Union has addressed specific legislation for food in which nanomaterials are explicitly or implicitly included [91]. However, sometimes it is difficult to draw the line between what is allowed and not allowed in the legislative frame. For example, regulations have not been developed for all food applications. Additionally, the lack of indications on how to apply the regulatory framework makes difficult its implementation. It is worth noting that most efforts have been directed towards elaborating soft law measurements such as communications, resolutions, recommendations, opinions and guidelines rather than implementing legally-binding regulations, directives, and decisions [92].

Food additives are legislated according to the EC No 1333/2008. This directive highlights that when a food additive is used with characteristics different from the initial approved it shall be considered as a different additive and therefore a re-evaluation is needed in terms of potential adverse effects [93]. TiO_2 (E171) and SiO_2 (E551) are listed as food additives approved for their use in food in EC No 1129/20011, which amended Annex II of EC No 1333/2008. SiO_2 is used in the food industry as anti-caking agent to maintain flow properties of powdered mixes, seasonings and coffee whiteners but also to preserve color or durability as well as to improve handling [94]. TiO_2 is employed as whitening or brightening agent, and therefore, it is mainly found in cheese and sauces, skimmed milk, ice-creams and pastries. In addition, it can be added to the coating of sweets and chewing-gum [95]. Unfortunately, legislation does not specify if these compounds can be used as additives when they are present as nanoforms. The re-evaluation of TiO_2 (E171) performed by EFSA concluded that E171 is not a nanomaterial [96]. However, some publications in the literature described the existence of nanoparticles in the food additive E171. For example, Peters and co-workers found particles in the nanometer size when monitoring 7 food products containing the additive E171. However, only around 15% of the particles had a size below 100 nm [97]. According to the definition given by the European Commission, a material is considered a nanomaterial only when 50% of the particles are smaller than 100 nm [1]. Similarly, EFSA re-evaluated the use of SiO_2 (E551) as food additive. The document specifies that E551 is a material composed of aggregated nanosized particles with a diameter size greater than 100 nm but it cannot be totally excluded that some of the aggregates are not smaller than 100 nm [98]. In both scientific opinions, EFSA committee has warned that EU specifications for silicon dioxide (E551) and titanium dioxide (E171) are insufficient to properly characterize the use of these materials as food additives. EFSA also advises of the need to select appropriate statistical data when particle size distribution is measured. Moreover, the need for determining the percentage of nanoparticles (in number and in mass) present in these additives is emphasized [96, 98].

EC No 2015/2283 Regulation on novel food established the conditions that a new product has to fulfill to be placed on the EU market. Novel food is defined as “any food that was not used for human consumption to a significant degree within the Union before 15 May 1997, irrespective of the dates of accession of Member States to the Union” [99]. In this context, foods containing or consisting of engineered nanomaterials is listed as one of these food categories and considered as novel food. Textually, engineered nanomaterial is described as “any intentionally produced material that has one or more dimensions of the order of 100 nm or less or that is composed of discrete functional parts, either internally or at the surface, many of which have one or more dimensions of the order of 100 nm or less, including structures, agglomerates or aggregates, which may have a size above the order of 100 nm” [99]. Nevertheless, in this definition, the percentage of particles with a size below 100 nm that is allowed to be considered a nanomaterial is not mentioned. Therefore, differences between what is considered a nanomaterial and what is not in the area of food law may lead to misunderstandings in its application. To the best of our knowledge, none nanomaterial has been included in the list of authorized novel foods although they are regulated and mentioned in EC No 2015/2283 on novel food [100]. Regarding labeling, EC No 1169/2011 Regulation [5] indicates that any nanomaterial present in a food product shall be included in the list of ingredients with the word “nano” after the name of such ingredient. Finally, Regulation No 2015/2283 stipulates that risk assessment of novel food should be carried out by EFSA. Currently, only calcium carbonate (E170) and vegetal carbon (E153) are allowed in foods as nanoforms. Other potential nanoingredients based on titanium dioxide (E171), iron (hydr)oxides (E172), silver (E174), gold (E175), silicon dioxide (E551), calcium silicate (E552), magnesium silicate (E553a) and talc (E553b), are under study.

Concerning the presence of nanomaterials in food packaging, two legislative documents refer to nanomaterials added or used on food contact materials: "Regulation (EC) 450/2009 on active and intelligent materials and articles to come into contact with food" and "Regulation (EU) No 10/2011 on plastic materials and articles intended to come into contact with food". Regulation (EC) 450/2009 indicates that "intelligent packaging systems provide the user with information on the conditions of the food and should not release their constituents into the food. Intelligent systems should be positioned on the outer surface of the package and separated from the food by a functional barrier with the aim of preventing the migration of substances from behind that barrier into the food". Generally, the use of non-authorized substances behind this functional barrier is permitted; however, their migration should remain below 0.01 mg kg^{-1} level. Interestingly, the regulation explicitly indicates that nanomaterials are not considered to be positioned after any functional barrier and this is the only mention to materials with a size range below 100 nm [101]. Additionally, Regulation (EU) No 10/2011 addressed that the use of nanosized substances on plastic materials and articles intended to come into contact with food is only possible when they are in the list of allowed substances to be used as additive or polymer production. Nowadays, Titanium nitride nanoparticles are the only nanomaterial allowed in PET up to 20 mg kg^{-1}. Silicon dioxide also appears in this list but without including any specification on size ranges [102].

As evidenced, nanotechnology is growing and developing without the existence of clear rules and regulations. With the increasing use of these nanomaterials, there is a wide debate on the environmental impact and unknown risks of these materials. Consequently, it is of great importance to assess the potential risks of nanomaterials before their use in the food industry [103]. EFSA has recently launched a guide that provides an overview on how to perform risk assessment of nanomaterials intended to be used in the food and feed area [104]. The main key points of the guide are compiled in Fig. (**5**). This document highlights the importance to carry out a complete characterization of the physicochemical characteristics of a nanomaterial in order to elucidate its safety. According to the EFSA recommendation, the parameters that need to be properly measured are: material composition and purity, particle size, agglomeration or aggregation state, particle shape, structure, surface chemical composition (composition of the functional groups or coating on the particle surface), surface area, surface charge, appearance, melting point, boiling point, density, porosity, dustiness, pH, formulation (nanopowder, dispersed in solution and the nature of stabilizers).

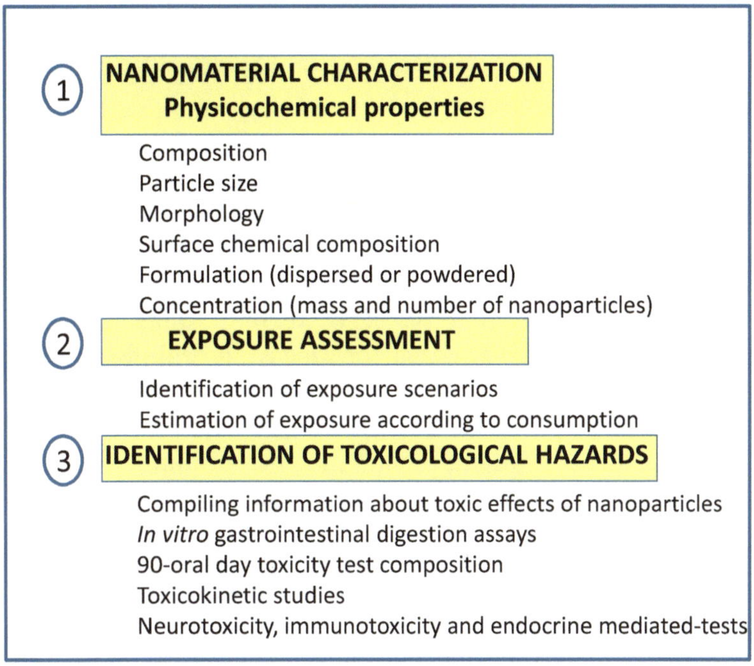

Fig. (5). Main points addressed by EFSA [104] for stabilizing NPs risk assessment.

Concentration of the nanomaterial in terms of both mass (g kg^{-1}) and particle number (n kg^{-1}), and the mass of the material present as ionic form, are requested. Information on physical and chemical stability of the nanomaterial and coating, as well as the solubility, the dispersibility and chemical reactivity, is also required by the EFSA recommendation. Moreover, Appendix C of this guidance includes a list of suitable analytical techniques to be used to determine the parameters described above [104]. The document also recommends the use of standardized method and certified RMs to control method performance in terms of sensitivity, detection limits, and precision. If no RMs are available, EFSA recommends the use of self-generated test materials properly characterized. Additionally, it is pointed out that particle size parameters must always be measured by at least two independent methods (one being electron microscopy) and also the acquisition of other parameters should be performed with more than one method paying special care to the sampling, sample preparation, and dispersion of particles.

Nanoparticles risk assessment implies not only the characterization of their physicochemical properties but also the determination of levels of exposure and hazard identification. EFSA guidelines point out the need of providing data on exposure estimation and scenarios with the aim of properly evaluating the effect of nanosized materials on human health and environment. According to the guidance, information on the presence of the nanomaterial or its degradation product(s) in the food matrix should also be provided. When exposure data are not available, the worst-case model shall be adopted assuming that the material added in the food matrix and ingested and absorbed is present as nanoform. Finally, the identification of toxicological hazards should be carried out by performing the following steps: First step shall include information on the carcinogenic, mutagenic, and neurotoxic properties of the nanomaterial along with the performance of degradation test under simulated lysosomal conditions and *in vitro* toxicity tests including genotoxicity. The second stage shall include the performance of 90-day oral toxicity test for assessing oral absorption and tissue distribution at different times. Finally, the third step will be dedicated to toxicokinetic studies to explore the extent of accumulation of the nanomaterial during long-term exposure along with the performance of neurotoxicity, immunotoxicity or endocrine-mediated tests [104].

CHARACTERIZATION OF METAL-BASED NANOPARTICLES: THE NEED FOR USING A MULTI-TECHNIQUE APPROACH

The use of nanomaterials in the food field implies the development of analytical methodologies for their detection and quantification. To ensure consumer safety, all ingredients present as nanoforms shall be clearly compiled in the label of the product. Consequently, it is of prime importance to develop analytical methods that can help determine whether products contain nanomaterials or not as a means of supporting nanoparticles regulations.

Unfortunately, there is no single standardized and validated methodology fulfilling the requirements needed for performing a proper characterization of metal-based nanoparticles in terms of size, morphology, identity, and concentration either in mass or particle number. Consequently, nanoparticles characterization implies application of multi-technique platforms. Nowadays, there are a plethora of techniques available (Fig. **6**) for characterizing nanoparticles such as those based on light scattering measurements like Dynamic Light Scattering (DLS), Multiangle Light Scattering (MALS) and nanoparticles tracking analysis (NTA); classical electron microscopy (TEM/SEM), and separation techniques such as ultrafiltration, hydrodynamic chromatography (HDC), and asymmetrical flow field flow fractionation (AF^4). In the last few years, ICPMS in the single particle mode (sp-ICPMS) has emerged as a powerful technique for simultaneously detecting and quantifying metallic nanoparticles without the need of applying previous separation steps. Additionally, ICPMS is commonly used as detector in the separation techniques mentioned above. These techniques have both advantages and drawbacks that must be taken into account when applied to nanoparticles characterization. For instance, light scattering measurements in DLS mode is widely applied for estimating average size nanoparticles in dispersions; however, the results are not accurate when applying to polydispersive suspensions.

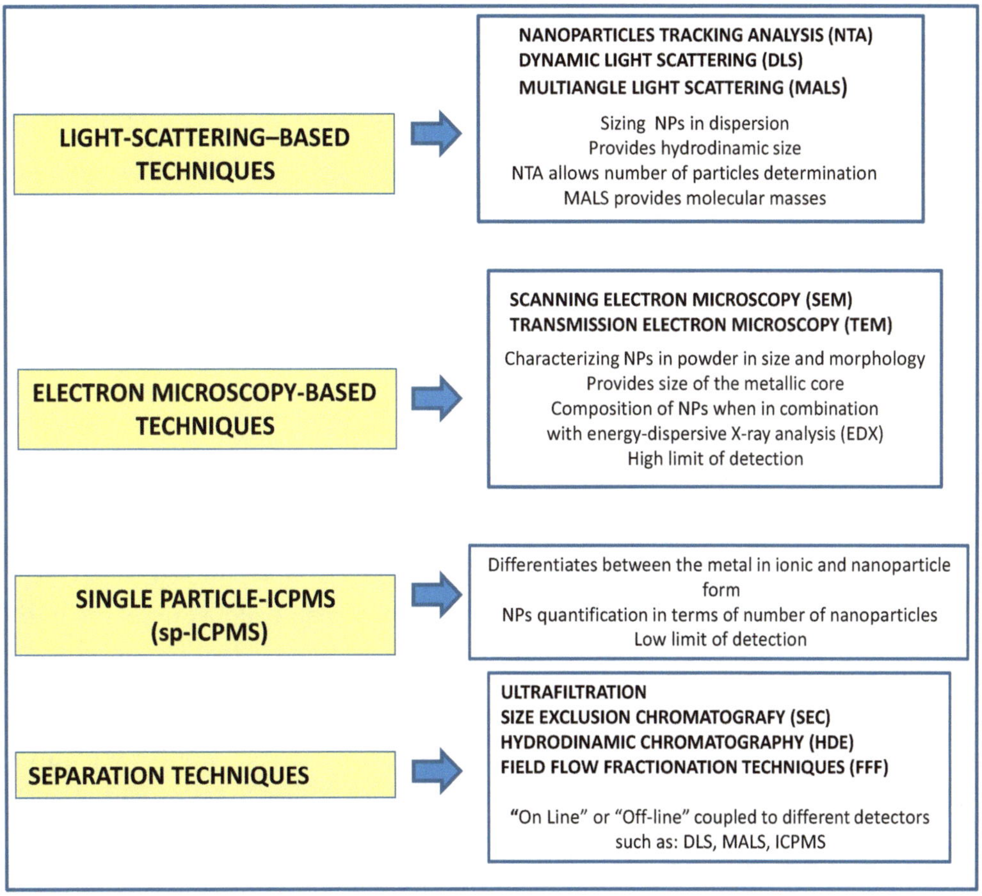

Fig. (6). Techniques most commonly employed for metallic NPs characterization.

DLS provides information on hydrodynamic size and zeta-potential values. The hydrodynamic radius characterizes the overall size of the particle, including particle core, stabilizing agents, and other molecules, *e.g.*, proteins attached to the particle's surface. Zeta potential estimates the stability of the NPs and the degree of aggregation. MALS is used for determining molecular masses and size of macromolecules. NTA allows counting nanoparticles directly in solution in agreement with EC recommendation (2011/696EU). This recommendation indicates that nanoparticles quantification should be done in terms of particle number concentration rather than in mass concentration [105]. Light scattering techniques are used for sizing NPs in dispersion while electron microscopy-based techniques (SEM/TEM) and atomic force microscopy (AFM) are able to characterize NPs in size and morphology. SEM/TEM techniques are usually the first choice for characterizing NPs since a picture conveys information more effectively than words; however, a significant number of nanoparticles needs to be measured to obtain representative results and additionally, solvent removal is required that may alter the original morphology of the nanoparticles [106]. The above-mentioned techniques can be divided into two groups: those based on light scattering measurements providing information on the hydrodynamic size, and SEM/TEM measurements that provide the size of the metallic core. Therefore, the methodology employed for sizing NPs should always be specified.

The properties of NPs are dependent also on their chemical composition. SEM/TEM in combination with energy-dispersive X-ray analysis (EDX) enables the determination of nanoparticles composition; however, its detection limit is not high enough for detecting nanoparticles at low concentration levels. ICPMS is the method of choice for detecting metal-based NPs as low concentration levels since the limits of detection achieved by the technique reach the ng L^{-1} levels. ICPMS is a very suitable technique for facing the issues derived from NPs characterization due to its broad dynamic range and its capability of performing multielemental and isotopic analysis. Moreover, ICPMS can characterize NPs by using the single particle-ICPMS (sp-ICPMS) mode. ICPMS has been also used as detector in the separation techniques employed in characterizing nanoparticles. In the last few years, the interest of using sp-ICPMS for characterizing NPs has grown considerably. sp-ICPMS is a technique that allows: 1) discrimination between the metal in ionic and nanoparticle form without the need to apply separation techniques, 2) determination of the size distribution of the NPs, and 3) NPs quantification in terms of number of nanoparticles at low limits of detection. The use of sp-ICPMS implies the continuous nebulization of diluted NPs suspensions. During sp-ICPMS analysis, ionic metallic form is detected as a constant signal while metallic nanoparticle form appeared as a transient signal or pulse with a typical duration of 200-500 μs, therefore, ICPMS signals need to be recorded at high time resolution. The intensity of the pulse is proportional to the mass of the detected NPs while the number of pulses is related to particle number concentration [107]. Although the principles behind sp-ICPMS are very straightforward, accurate determination of NP sizes and nanoparticle concentration is not an easy task. Calibrating NPs mass requires the determination of the absolute detection efficiency (counts/atom or counts/mass). For this purpose, a calibration curve of the intensity against NPs of different sizes needs to be constructed; however, the availability of well-defined dispersed NPs solution is very limited. The size detection limit is also dependent on the resolution of the mass analyzer to distinguish between particulate signals and background noise. In this respect, the technique provides lower detection limits for nanoparticles of less abundant elements, such as silver. The use of sp-ICPMS with quadrupole or sector field-mass analyzer only allows measuring one isotope per particle. In recent years, the use of ICP-MS with time-of-flight, ToF, a simultaneous mass analyzer, has appeared as a promising tool for simultaneously detecting mixtures of nanoparticles of different composition, NPs with multi-elemental composition and untargeted nanoparticles [108]. Therefore, the use of multi-element detection (time-of-flight) sp-ICP-ToFMS is a powerful analytical tool for detecting untargeted NPs and overcoming problems related with background signals due to colloids and dissolved elements [109]. Unfortunately, ICP-ToFMS provides limits of detection 10-100 times higher than ICPMS with a quadrupole as mass analyzer and it is not economically affordable for most laboratories [110].

Legislation also requires the quantification of nanoparticles in the sample. The amount of NPs fraction can be expressed as mass-based concentration or particle-based concentration. Both terms are not directly comparable and the one cannot be converted to the other. Only for those NPs with spherical morphology and known density it is possible to transform NPs concentration in mass units into particle mass-units and *vice versa*. In 2011, the EU commission (2011/696/EU) [1] recommended that NPs quantification should be expressed as particle number concentration rather than as mass concentration. In fact, a small mass of NP fraction can contain a big amount of NPs of small size which are the most harmful. Among the above-mentioned techniques, only NTA and sp-ICPMS can count directly nanoparticles in solution.

Separation techniques have been extensively applied for characterizing NPs according to their size. Ultrafiltration (cut off with a range of 1-100 nm) has been applied for purifying and fractionating NPs. The method provides a first screening of NPs size distribution and it allows easily differentiating between ionic and particulate forms. Chromatographic techniques such as size exclusion chromatography (SEC) and hydrodynamic chromatography (HDC) have been also used for the fractionation of nanoparticles. The former is based on the separation of nanoparticles or a mixture of macromolecules on a stationary phase with a distribution of pore size in the range of the nanoparticles of interest, while in HDC the separation is achieved in narrow capillary tubes. The separation of NPs by chromatographic techniques is challenging. One of the problems when employing SEC is NPs adsorption onto the column packaging material, leading to poor recovery of NPs along with modifications in their hydrodynamic diameter. Several approaches have been applied to avoid such problems, like using columns with large pore size or mobile phases containing additives (surfactants and molecules commonly used as capping agents for stabilizing NPs as for instance citrate for AuNPs) [111]. Hydrodynamic chromatography is similar to SEC, except that solid-phase interactions are decreased due to the non-porous nature of the stationary phase [112]. Both techniques have been coupled to different detectors such as Light dispersion detectors, MALS, ICPMS, and even single particle-ICPMS.

Field-flow fractionation (FFF) techniques have strongly re-emerged for fractionating and sizing a great variety of compounds of different sizes, from a few nanometers to a few micrometers (nanoparticles, macromolecules, supramolecular assemblies, colloids, and particles). The technique has been applied in many areas such as biomedical, environmental, food, and consumer products. FFF was invented by Giddings in 1966 [113]. The idea behind this separation approach is the use of an external field applied perpendicularly to the direction of sample flowing through an empty and thin channel. Depending on the type of field applied, several FFF techniques are described: flow (FlFFF), thermal (ThFFF), electrical (ElFFF), sedimentation (SdFFF), gravitational (GrFFF), dielectrophoretic (DEP-FFF), acoustic (AcFFF), and magnetic (MgFFF) field-flow fractionation, respectively. However, only flow FFF has application in the laboratories. In flow FFF, a cross flow is perpendicularly applied to the direction of the main flow. The cross flow drives analytes to the accumulation wall, which is covered by an ultrafiltration membrane through which the carrier solution passes while retaining the analytes. Flow FFF separates analytes on the basis of their diffusion coefficient and, therefore according to their hydrodynamic diameter. The main advantages of using flow FFF for NPs fractionation is the absence of stationary phases, the use of mobile phases composed of different pHs and ionic strengths (water is commonly employed as mobile phase) and the capability of on-line coupling with different detectors such as UV-Vis, MALS, and ICPMS [114]. However, the performance of this fractionation technique is affected by a great variety of experimental parameters (nature of the membrane, channel dimensions, mobile phase composition, cross-flow, time, and volume of injection and temperature). The optimization of these parameters is not an easy task which means that the technique is sometimes difficult to apply [115, 116].

As previously mentioned, there is not a single technique that fulfills EFSA requirements and EU guidelines to characterize NPs in food and food technology sectors as far as sizing, composition, morphology, bioaccesibility and toxicology are concerned. Difficulties are greater when NPs need to be characterized in complex matrices such as food where NPs′ stability is an important issue. Moreover, the lack of RMs to perform method validation, supporting legislation, and consumer confidence is also an additional problem. Attempts have been made to produce RMs. For instance, two RMs (SiO_2NPs-containing tomato soup and AgNPs-containing breast chicken) were prepared within the frame of NanoLyse project sponsored by the European Union Seventh Framework Programme (FP7/2007-2013) (www.nanolyse.eu). Production of RMs of NPs-containing food is not an easy task because NPs transformation due to food matrix, sample treatment and storage. Additionally, lack of homogeneity of the resulting material is also an added problem. Owing to the absence of RMs, Linsinger and co-workers suggested that method validation should include the following steps [117]:

1. Detection of the presence of the nanoparticle (size identity). If this is the case, the type of size must be clarified (hydrodynamic size or core size) as well as the particle size distribution.
2. Identification of the type of nanoparticle (chemical identity and morphology).
3. Quantification of the fraction of nanoparticle present in the sample (mass or number fraction).

Factors Affecting Stability of Nanoparticles

Stability of NPs is a key factor for correctly assessing the impact and potential toxicity of NPs. Dissolution and aggregation/agglomeration of NPs can take place during synthesis, storage, dilution, variations in pH, effect of mobile phases during separation, sample treatment and so on (Fig. 7).

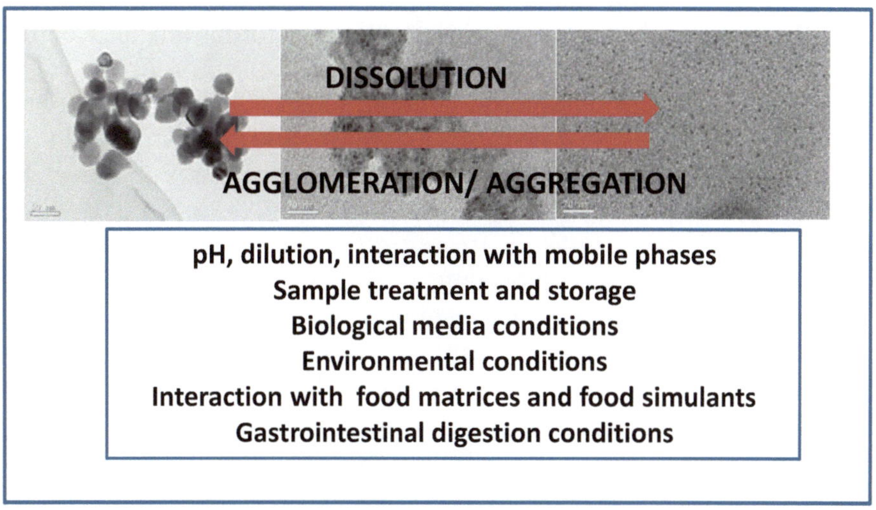

Fig. (7). Factors affecting nanoparticles stability.

Several procedures are used for keeping NPs stability in solution. The use of stabilizers (surfactants, polymers, organic acids, proteins...) is the method of choice for ensuring short and long term stability or even during NPs synthesis.

Selection of the stabilizer or capping agent is crucial since it is accepted that NPs′ fate and toxicity are largely influenced by the physical interaction between the surfaces of NPs and cells used in cytotoxicity studies. Several authors further indicated that the nature of the stabilizer is the most important factor in evaluating NPs toxicity [118]. Moreover, the use of a stabilizer may hinder the normal utilization of synthesized nanoparticles in clinical applications because of its toxicity. Palomo-Siguero and co-workers evaluated the toxicity of three types of SeNPs against *Lactobacillus Bulgaricus* [119]. Spherical SeNPs were synthesized using three types of coating materials (chitosan (CS-SeNPs), hydroxyethyl cellulose (HEC-SeNPs), and a non-ionic surfactant (ethoxylated-SeNPs) [120]. The presence of HEC-SeNPs or CS-SeNPs barely affects the viability of the exposed bacteria even at a level of exposure of 10 µg Se mL^{-1} while the presence of 10 µg Se mL^{-1} in form of ethoxylated-SeNPs produced a dramatic decrease of bacterial cell viability, suggesting the surfactant as the primary cause of the drastic decrease observed in bacterial viability. These results evidenced the importance of the capping agent when establishing the true toxicity of NPs [119]. Unfortunately, this crucial information is missing in most reported toxicity studies.

Similarly, in biological media, NPs can interact with proteins resulting in the formation of a biological corona on the NPs surface. As a result, the NPs surface induces conformational changes in the adsorbed proteins, which may affect the impact of NPs fate and behavior *in vivo* [120]. The formation of a corona is a dynamic process and it is influenced by the physicochemical properties of the NPs (size, morphology, surface chemistry, coating, and so on) [121]. For instance, AuNPs of different sizes (*i.e.*, 5, 10, 20, 30, 60, 80, and 100 nm) were incubated sequentially with blood proteins, such as albumin, fibrinogen, γ-globulin, histone, and insulin. Results from fluorescence measurements evidenced that NPs with a diameter less than 10 nm presented a greater interaction with blood proteins [122]. The shape of the NP is also important as confirmed for TiO_2 nanorods and nanotubes that were differentially adsorbed by plasma proteins [123]. NPs surface charge can also influence protein adsorption. Studies have demonstrated that NPs surfaces with neutral charge bind less proteins than their negatively charged (COOH functionalized) or positively charged (NH_2 functionalized) counterparts [124]. For example, polyethyleneimine-functionalization of ZnONPs improves their interaction with albumin as compared to non-functionalized NPs.

Once released in the environment, NPs can suffer dissolution, oxidation, adsorption and aggregation, depending on the physicochemical characteristics of the receiving media and nanoparticles. Therefore, once emitted, NPs can exist as free ions, adsorbed on particulate matter or forming complexes with ligands. For instance, AgNPs suffer several processes one released in the aquatic environment such as photooxidative dissolution, photoreduction, sulfidation and chloridation [125]. All these transformations need to be considered to provide a correct assessment of the impact of NPs on human health and environmental risk.

Food is also a complex matrix where metallic/metalloid nanoparticles can be intentionally or unintentionally added. For instance, NPs can migrate from packaging to food and once migrated the stability of the NPs can be affected by the food matrix. In this line, Grombe and co-workers [126] highlighted a decrease in the size of SiO_2NPs when added to tomato soup for preparing a certified reference material. SEM measurements evidenced that tomato soup matrix produced an increase in agglomeration of SiO_2NPs, but still individual NPs were detected. However, SiO_2NPs present a smaller size (30-50 nm) than pure solutions of SiO_2NPs (above 60 nm). Similar results were described for AgNPs spiked chicken meat [127, 128]. A decrease in particle size was observed along with a 60% decrease of particle mass concentration. These observations were explained by a combination of AgNPs dissolution, the appearance of chemical transformations such as the reaction between silver and sulphur (such as Ag_2S) and the occurrence of processes of agglomeration/aggregation. Size decrease and even NPs dissolution was also detected when ZnONPs were in contact with fruit juice and chicken meat. Results provided by ultrafiltration followed by AAS measurements evidenced that the percentage of dissolved Zn in ZnONPs enriched fruit juice was around 60%. Additionally, chicken breast matrix provided an important decrease in ZnONPs size. All these results corroborated the importance of food matrix on NPs stability, and therefore, in their potential toxicity [129].

Once ingested, nanoparticles can also modify their size and morphology as they pass through different conditions before being absorbed by intestinal cells. From a nutritional point of view, the term bioaccesibility has been defined as the fraction of a compound that is released from the food matrix in the gastrointestinal tract, and becomes available for intestinal absorption to enter the blood stream. Otherwise, bioavailability is referred to the fraction of the nutrient ingested that is available for use in physiologic function. Usually, bioaccesibility is assessed by reproducing *in vitro* the physiological conditions occurring during human digestion and by considering the three areas of the digestive system: mouth, stomach and intestines. The main parameters monitored during gastrointestinal *in vitro* assays are temperature, pH, agitation, and enzymatic and chemical composition of the saliva, gastric juice, and intestinal juice. For instance, the physiological conditions occurring during digestion (presence of enzymes and drastic changes of pH) could alter NPs stability, and therefore, affect their absorption in the intestine walls. Böhmer and co-workers [130] reported the effect of the experimental conditions involved in *in vitro* gastrointestinal assay on AgNPs stability. DLS measurements evidenced that artificial saliva has no effect on the size and aggregation state of the nanoparticles. In contrast, the low pH conditions of the stomach and intestinal fluids produced an increase in the hydrodynamic particle radius as determined by DLS measurements. This increment was attributed to the appearance of conformational changes of the capping agent and/or the adsorption of proteins. Other authors [131, 132] described a release of free silver ions and the formation of chlorine interparticle bridges, and the precipitation of AgCl when AgNPs were exposed to a synthetic stomach fluid because of its low pH (nearly 2) and high electrolyte concentration. However, the absence of food components during the application of *in vitro* gastrointestinal assays could lead to wrong conclusions. Therefore, it is of paramount importance to perform these type of studies in presence of food matrices. Peters and co-workers [133] evaluated the stability of 32 nm SiO_2NPs when added to hot water, coffee with powdered cream milk, soup, and pancake samples. Subsequently, the SiO_2NPs-enriched matrices were submitted to *in vitro* gastrointestinal digestion. SEM and DLS data showed the presence of nanosized SiO_2 during salivar stage in a relative amount (5 to 40% depending on the food matrix) whereas during the subsequent gastric step, SiO_2NPs almost disappeared in soup and coffee. Authors attributed this effect to the low pH (lower than 2) of the gastric step. After intestinal digestion, when the pH increased up to neutral values, SiO_2NPs appeared again and the percentage of silica as SiO_2NPs was around 80% for coffee and 15% for soup and pancake [133]. The influence of the digestion process on AgNPs with and without the presence of food components (proteins, carbohydrates and fatty acids) has been also evaluated by Small-angle X-ray scattering (SAXS) measurements. Particles were stable during the *in vitro digestion* procedure either in presence or in absence of food components [134].

The results of these different studies suggest that nanoparticles reach the intestine. However, very few data on the capability of nanoparticles to pass through the intestinal barrier has been reported. The human colon adenocarcinoma Caco-2 cell line has been extensively used due to its ability to undergo spontaneous differentiation, leading to the formation of a cell monolayer. This *in vitro* system simulates the mature small intestine enterocytes, both morphologically and functionally. However, this model is far away from the complexity of the *in vivo* intestinal barrier that is composed of other cells such as mucus secretory and globet cells. However, the results provided by Caco 2-cells correlated well with *in vivo* absorption in humans which makes it quite appropriate for establishing NPs risk assessment according to EFSA regulation. Recently, Vila and co-workers reported the effects caused by the exposure of AgNPs [135]. Several parameters such as transcellular and paracellular transport, as well as the cellular uptake of AgNPs with an average size of 7.74 nm were assessed with different techniques such as confocal microscopy, transmission electron microscopy complemented with energy dispersive X-ray and ICP-MS, respectively. Additionally, the induction of oxidative DNA damage and changes in gene expression by qPCR were also assessed. Caco-2 cell viability was not significantly affected by AgNPs showing a decrease of only 20%, after being exposed during 24 h to 100 μg mL^{-1}. Interestingly, cytotoxicity of AgNPs in differentiated cells was significantly lower than in undifferentiated cells. No significant effects were observed on the monolayer's integrity/permeability after exposure to AgNPs. The important uptake of AgNPs was not accompanied by translocation through the monolayers of Caco-2 cells. Similarly, the impact of Al containing nanoparticles on Caco-2 cells evidenced that aluminum uptake by cells occurred predominantly in particulate form, whereas ionic Al was not taken up by intestinal cell lines. None of the Al species showed cytotoxic effects up to 200 μg Al mL^{-1}. The transcriptome analysis indicated effects on oxidative stress pathways. Moreover, the study showed that intestinal cellular uptake of Al occurs mainly in the nanosize form whereas toxicological effects appear to be related to the ionic aluminum [136].

As mentioned before, food matrix or food components could alter the interaction between nanoparticles and Caco-2 cells. Lichtenstein and co-workers reported that digested and undigested AgNPs in presence of food components were comparably taken up by Caco-2 cells, whereas the uptake of particles digested without food components was decreased by 60%. The results obtained may indicate that NPs can pass through the intestinal barrier and may enter in the blood stream [134]. Munger and colleagues reported the presence of silver in serum samples after ingestion of particles with diameters between 5 and 10 nm [137].

Sample preparation is a main challenge when determining NPs in food. Most of the analytical techniques employed for characterizing nanoparticles require the presence of nanoparticles in solution, therefore nanoparticles need to be firstly isolated from the complex matrix where they are suspended or embedded. Several extractants have been reported in the literature to characterize NPs in food samples [138]. Enzymatic hydrolysis has been widely used for degrading matrix and isolating NPs. The enzymes used have mainly been proteases such as Proteinase K [139], but also amylase and Macerozyme R-10 (mixture of enzymes cellulose, hemicelluloses and pectinase) [128, 139, 140]. Others extractants such as tetramethylammonium hydroxide have been employed for isolating AuNPs from rat livers [141]. In order to prevent aggregation, serum albumin was added to the homogenates. NPs stability can also be compromised during sample storage. In this regard, the short and long-term stability of SiO_2NPs of AgNPs was checked with the purpose of preparing NPs-based RMs (SiO_2NPs enriched tomato soup and AgNPs enriched chicken meat). Several procedures for preserving NPs and food matrix integrity were tested including sterilization, γ-irradiation, freezing, and lyophilization. Thermal sterilization produced a decrease of SiO_2NPs size while lyophilization led to an irreversible sedimentation.

Consequently, γ- irradiation was selected for preserving SiO_2NPs tomato soup as NPs size and morphology remained unaltered. With respect to AgNPs, autoclaving favored NPs dissolution while freezing with liquid N_2 maintained NPs size and morphology

Maintaining NPs integrity is not an easy task and consequently NPs size and morphology should be checked during storage, pretreatment and sample preparation, otherwise the results obtained will not reflect the characteristics of the NPs originally present in the samples.

Nanoparticles Migration from Packaging to Food

As described above, metallic and metalloid nanoparticles (AgNPs, SiO_2NPs, ZnONPs and TiO_2NPs) have been mainly employed for developing food packaging materials, primarily with antimicrobial, antioxidant and blocking UV-Vis radiation properties. Therefore, the implementation of migration studies to assess the impact of these active packaging materials on the consumers is of great relevance. Most migration studies appeared in the literature have been performed according to European Regulation 10/2011 (EU, 2011) which sets out the rules for performing migration testing packaging plastic materials intended to come into contact with the food. Following the recommendations, a piece of plastic material is immersed in a vial containing a food simulant. The food simulants used are those listed in Table II, annex III, in particular: (A) 10% (v/v) ethanol; (B) 3% (w/v) acetic acid; (C) 20% (v/v) ethanol; and (D) vegetable oil. Aqueous food simulants A, B, and C are used to mimic foods that have a hydrophilic character and which are able to extract hydrophilic substances. Food simulant B is used in food with pH below 5. In this line, Echegoyen and Nerín [142] evaluated the release of AgNPs from different nanosilver antimicrobial food containers by using food simulants. Silver migration was observed for all samples studied, with the total silver migration values ranging between 1.66 and 31.46 ng cm^{-2} (lower values than the permissible limits).

Metak, Nabhani and Connolly [143] studied the migration of AgNPs from nanosilver impregnated polymer containers and coated nanosilver films by using both food and food simulants (water and 3% acetic acid) for 10 days at 40 °C. Nano-silver coated films demonstrated a slightly higher level of migration compared with food samples stored in nanosilver containers. The highest level of migration was obtained in orange juice with maximum values of 28.92 ± 0.01 $\mu g L^{-1}$. TEM measurements evidenced the presence of AgNPs in food as a consequence of the migration from packaging. A study of the stability of AgNPs in 3% v/v acetic acid by AF^4-MALS was achieved by Ntim and co-workers [144]. Their results showed that after 1 h of incubation, some Ag NPs were dissolved and after 4 h, all AgNPs were transformed to Ag ions. Although most studies reported on the literature have been performed by using food simulants, migration studies using food matrices have also been completed. In this respect, the migration of silver from plasticized polyvinyl chloride (PVC) nanocomposites to chicken meat was evaluated by Cushen and co-workers [145]. For this purpose, meat was wrapped with the plastic film and stored at 5-20 °C for 1-4 days. The silver content of the chicken was measured by ICPMS. Silver was found in the meat at a concentration level of 0.03-8.4 mg kg^{-1} that is below the limits of migration set for conventional migrants (no more than 60 mg kg^{-1}, or 10 mg dm^{-2}). As previously stated, nanomaterials are novel substances and, consequently, they should be used with precaution. The use of nanomaterials in the food field is increasing, especially for developing active food packaging materials. Risk assessment on nanomaterials are underway, especially those concerning the evaluation of the exposure from food packaging. Accordingly, migration studies are of great importance for evaluating food packaging safety involving nanomaterials.

CONCLUDING REMARKS

The application of nanoparticles in the food sector is infinite with novel uses being continuously explored. However, their impact on consumer safety requires analytical methodologies able to fulfill the requirements established by current EU directives on nanomaterials. This chapter has put forward that determination of metal-based nanoparticles is not an easy task and a plethora of techniques with their specific advantages and disadvantages needs to be applied. Moreover, nanoparticles stability has to be preserved during the analytical steps to obtain accurate results. Quantitative extraction of nanoparticles, while keeping their original properties, remains still a challenge. Hence, it is crucial to develop robust and valid analytical methods. To this end, producing Certified Reference Materials is a key issue to support method validation and legislation to increase and assure consumer confidence.

CONSENT FOR PUBLICATION

Not applicable.

CONFLICT OF INTEREST

The authors confirm that this chapter contents have no conflict of interest.

ACKNOWLEDGEMENTS

The authors thank the Spanish Commission of Science and Technology (CTQ2017-83569-C2-1-R) and the Comunidad of Madrid (Spain) and European funding from FEDER Program (project S2013/ABI-3028, AVANSECAL-CM) for funding.

REFERENCES

[1] European Commission. Commission recommendation of 18 October 2011 on the definition of nanomaterial (text with EEA relevance) (2011/696/EU) Off J EurUnion 2011; 275/38-40.

[2] Cushen M, Kerry J, Morris M, Crux-Romero M, Cummins E. Nanotechnologies in the food industry-Recent developments, risks and regulation. Trends Food Sci Technol 2012; 24: 30-46. [http://dx.doi.org/10.1016/j.tifs.2011.10.006]

[3] Som C, Berges M, Chaudhry Q, *et al.* The importance of life cycle concepts for the development of safe nanoproducts. Toxicology 2010; 269(2-3): 160-9. [http://dx.doi.org/10.1016/j.tox.2009.12.012] [PMID: 20025922]

[4] Vance ME, Kuiken T, Vejerano EP, *et al.* Nanotechnology in the real world: Redeveloping the nanomaterial consumer products inventory. Beilstein J Nanotechnol 2015; 6: 1769-80. [http://dx.doi.org/10.3762/bjnano.6.181] [PMID: 26425429]

[5] REGULATION (EU) No 1169/2011 OF THE EUROPEAN PARLIAMENT AND OF THE COUNCIL of 25 October 2011 on the provision of food information to consumers, amending Regulations (EC) No 1924/2006 and (EC) No 1925/2006 of the European Parliament and of the Council, and repealing Commission Directive 87/250/EEC, Council Directive 90/496/EEC, Commission Directive 1999/10/EC, Directive 2000/13/EC of the European Parliament and of the Council, Commission Directives 2002/67/EC and 2008/5/EC and Commission Regulation (EC) No 608/2004.

[6] Adeyeye SAO, Fayemi OE. Nanotechnology and food processing: between innovations and consumer safety. J Culin Sci Technol 2018; 1-18. [http://dx.doi.org/10.1080/15428052.2018.1476276]

[7] Singh T, Shukla S, Kumar P, Wahla V, Bajpai VK, Rather IA. Application of Nanotechnology in food science: perception and overview. Front Microbiol 2017; 8: 1501. [http://dx.doi.org/10.3389/fmicb.2017.01501] [PMID: 28824605]

[8] Shah SWA, Qaisar M, Jahangir M, *et al.* Influence of CMC-and guar gum-based silver nanoparticle coatings combined with low temperature on major aroma volatile component and the sensory quality of kinnow (*Citur reticulate*). Int J Food Sci Technol 2016; 51: 2345-52. [http://dx.doi.org/10.1111/ijfs.13213]

[9] Xu J, Zhang M, Bhandari B, Kachele R. ZnO nanoparticles combined radio frequency heating: A novel method to control microorganism and improve product quality of prepared carrots. Innov Food Sci Emerg Technol 2017; 44: 46-53. [http://dx.doi.org/10.1016/j.ifset.2017.07.025]

[10] Xu J, Zhang M, Bhandari B, Cao P. Microorganism control and product quality improvement of Twice-cooked pork dish using ZnO nanoparticles combined radio frequency pasteurization. Lebensm Wiss Technol 2018; 95: 65-71. [http://dx.doi.org/10.1016/j.lwt.2018.04.067]

[11] Food safety and foodborne illness. World Health Organization 2017.

[12] Zazo H, Colino CI, Lanao JM. Current applications of nanoparticles in infectious diseases. J Control Release 2016; 224: 86-102.

[http://dx.doi.org/10.1016/j.jconrel.2016.01.008] [PMID: 26772877]

[13] Kumar A, Alam A, Rani M, Ehtesham NZ, Hasnain SE. Biofilms: Survival and defense strategy for pathogens. Int J Med Microbiol 2017; 307(8): 481-9. [http://dx.doi.org/10.1016/j.ijmm.2017.09.016] [PMID: 28950999]

[14] Srey S, Jahid IK, Ha SD. 2013 Biofilm formation in food industries: A food safety concern. Food Control 2013; 31: 572-85. [http://dx.doi.org/10.1016/j.foodcont.2012.12.001]

[15] Speranza B, Corbo MR. The Impact of Biofilms on Food Spoilage.The Microbiological Quality of food: foodborne spoilers. Elsevier Ltd. 2017; Vol. 3: pp. 259-82.

[16] Mu H, Tang J, Liu Q, Sun C, Wang T, Duan J. Potent antibacterial nanoparticle against biofilm and intracellular bacteria. Sci Rep 2016; 6: 18877. [http://dx.doi.org/10.1038/srep18877] [PMID: 26728712]

[17] Natan M, Banin E. From Nano to Micro: using nanotechnology to combat microorganisms and their multidrug resistance. FEMS Microbiol Rev 2017; 41(3): 302-22. [http://dx.doi.org/10.1093/femsre/fux003] [PMID: 28419240]

[18] Wang L, Hu C, Shao L. The antimicrobial activity of nanoparticles: present situation and prospects for the future. Int J Nanomedicine 2017; 12: 1227-49. [http://dx.doi.org/10.2147/IJN.S121956] [PMID: 28243086]

[19] Hajipour MJ, Fromm KM, Ashkarran AA, *et al.* Antibacterial properties of nanoparticles. Trends Biotechnol 2012; 30(10): 499-511. [http://dx.doi.org/10.1016/j.tibtech.2012.06.004] [PMID: 22884769]

[20] Slavin YN, Asnis J, Häfeli UO, Bach H. Metal nanoparticles: understanding the mechanisms behind antibacterial activity. J Nanobiotechnology 2017; 15(1): 65. [http://dx.doi.org/10.1186/s12951-017-0308-z] [PMID: 28974225]

[21] Raghunath A, Perumal E. Metal oxide nanoparticles as antimicrobial agents: a promise for the future. Int J Antimicrob Agents 2017; 49(2): 137-52. [http://dx.doi.org/10.1016/j.ijantimicag.2016.11.011] [PMID: 28089172]

[22] Ikuma K, Decho AW, Lau BLT. When nanoparticles meet biofilms-interactions guiding the environmental fate and accumulation of nanoparticles. Front Microbiol 2015; 6: 591. [http://dx.doi.org/10.3389/fmicb.2015.00591] [PMID: 26136732]

[23] Defoirdt T. Quorum-Sensing systems as targets for antivirulence therapy. Trends Microbiol 2018; 26(4): 313-28. [http://dx.doi.org/10.1016/j.tim.2017.10.005] [PMID: 29132819]

[24] Verbeke F, De Craemer S, Debunne N, *et al.* Peptides as Quorum Sensing Molecules: measurement techniques and obtained levels *in vitro* and *in vivo.* Front Neurosci 2017; 11(83): 183. [http://dx.doi.org/10.3389/fnins.2017.00183] [PMID: 28446863]

[25] Huillet E, Tempelaars MH, André-Leroux G, *et al.* PlcRa, a new quorum-sensing regulator from *Bacillus cereus*, plays a role in oxidative stress responses and cysteine metabolism in stationary phase. PLoS One 2012; 7(12)e51047 [http://dx.doi.org/10.1371/journal.pone.0051047] [PMID: 23239999]

[26] Torres-Cerna CE, Morales AJ, Alanis AY, Hernández-Vargas EA. Discrete-time neural network identification of Quorum Sensing *Escherichia coli* regulator s. IFAC-paper Online 2018; 13: 120-4. [http://dx.doi.org/10.1016/j.ifacol.2018.07.265]

[27] Wang B, Muir TW. Regulation of Virulence in *Staphylococcus aureus*: Molecular Mechanisms and Remaining Puzzles. Cell Chem Biol 2016; 23(2): 214-24. [http://dx.doi.org/10.1016/j.chembiol.2016.01.004] [PMID: 26971873]

[28] Kalaiarasan E, Thirumalaswamy K, Harish BN, Gnanasambandam V, Sali VK, John J. Inhibition of

quorum sensing-controlled biofilm formation in *Pseudomonas aeruginosa* by quorum-sensing inhibitors. Microb Pathog 2017; 111: 99-107.
[http://dx.doi.org/10.1016/j.micpath.2017.08.017] [PMID: 28818490]

[29] Garmyn D, Gal L, Lemaitre JP, Hartmann A, Piveteau P. Communication and autoinduction in the species Listeria monocytogenes: A central role for the agr system. Commun Integr Biol 2009; 2(4): 371-4.
[http://dx.doi.org/10.4161/cib.2.4.8610] [PMID: 19721895]

[30] Zheng K, Setyawati M, Leong D. Antimicrobial silver nanomaterials. Coord Chem Rev 2018; 357: 1-17.
[http://dx.doi.org/10.1016/j.ccr.2017.11.019]

[31] Demirbas A, Yilmaz V, Ildiz N, Baldemir A, Ocsoy I. Anthocyanins-rich Berry extracts directed formation of AgNPs with the investigation of their antioxidant and antimicrobial activities. J Mol Liq 2017; 248: 1044-9.
[http://dx.doi.org/10.1016/j.molliq.2017.10.130]

[32] Kujda M, Oćwieja M, Adamczyk Z, *et al.* Charge stabilized silver nanoparticles applied as antibacterial agents. J Nanosci Nanotechnol 2015; 15(5): 3574-83.
[http://dx.doi.org/10.1166/jnn.2015.9727] [PMID: 26504979]

[33] Dogru E, Demirbas A, Altinsoy B, Duman F, Ocsoy I. Formation of Matricaria chamomilla extract-incorporated Ag nanoparticles and size-dependent enhanced antimicrobial property. J Photochem Photobiol B 2017; 174: 78-83.
[http://dx.doi.org/10.1016/j.jphotobiol.2017.07.024] [PMID: 28756155]

[34] Raza MA, Kanwal Z, Rauf A, Sabri AN, Riaz S, Naseem S. Size- and shape-dependent antibacterial studies of silver nanoparticles synthesized by wet chemical routes. Nanomaterials (Basel) 2016; 6(4): 74.
[http://dx.doi.org/10.3390/nano6040074] [PMID: 28335201]

[35] Lotha R, Sundaramoorthy NS, Shamprasad BR, Nagarajan S, Sivasubramanian A. Plant nutraceuticals (Quercetrin and Afzelin) capped silver nanoparticles exert potent antibiofilm effect against food borne pathogen Salmonella enterica serovar Typhi and curtail planktonic growth in zebrafish infection model. Microb Pathog 2018; 120: 109-18.
[http://dx.doi.org/10.1016/j.micpath.2018.04.044] [PMID: 29715535]

[36] Anjugam M, Vaseeharan B, Iswarya A, Divya M, Prabhu NM, Sankaranarayanan K. Biological synthesis of silver nanoparticles using β-1, 3 glucan binding protein and their antibacterial, antibiofilm and cytotoxic potential. Microb Pathog 2018; 115: 31-40.
[http://dx.doi.org/10.1016/j.micpath.2017.12.003] [PMID: 29208541]

[37] Singh BR, Singh BN, Singh A, Khan W, Naqvi AH, Singh HB. Mycofabricated biosilver nanoparticles interrupt Pseudomonas aeruginosa quorum sensing systems. Sci Rep 2015; 5: 13719.
[http://dx.doi.org/10.1038/srep13719] [PMID: 26347993]

[38] Wagh Nee Jagtap MS, Patil RH, Thombre DK, Kulkarni MV, Gade WN, Kale BB. Evaluation of anti-quorum sensing activity of silver nanowires. Appl Microbiol Biotechnol 2013; 97(8): 3593-601.
[http://dx.doi.org/10.1007/s00253-012-4603-1] [PMID: 23224498]

[39] Naik K, Kowshik M. Anti-quorum sensing activity of AgCl-TiO_2 nanoparticles with potential use as active food packaging material. J Appl Microbiol 2014; 117(4): 972-83.
[http://dx.doi.org/10.1111/jam.12589] [PMID: 24965598]

[40] Mahmoodi S, Elmi A, Hallaj-Nezhadi S. Copper nanoparticles as antibacterial agents. J Mol Pharm Org Process Res 2018; 6: 1-7.
[http://dx.doi.org/10.4172/2329-9053.1000140]

[41] Tovar-Corona A, Lobo-Sánchez MA, Herrera-Perez JL, Zanella R, Rodriguez-Mora JI, Vazquez-Cuchillo O. Green Synthesis of copper (0) nanoparticles with cyanidine-0-3-glucoside and its strong antimicrobial activity. Mater Lett 2018; 211: 266-9.

[http://dx.doi.org/10.1016/j.matlet.2017.10.020]

[42] Chakra CS, Rao KV, Pavani T. Antimicrobial activity of pure Cu nano particles synthesized by surfactant varied chemical reduction method. Environ Nanotechnol Monit Manag 2016; 6: 88-94. [http://dx.doi.org/10.1016/j.enmm.2016.06.004]

[43] Kundu S. Gold Nanoparticles: Their application as antimicrobial agents and vehicles of gene delivery. Adv Biotechnol Microbiol 2017; 5: 1-4. [http://dx.doi.org/10.19080/AIBM.2017.05.555658]

[44] Happy Agarwal , Soumya Menon , Venkat Kumar S, Rajeshkumar S. Mechanistic study on antibacterial action of zinc oxide nanoparticles synthesized using green route. Chem Biol Interact 2018; 286: 60-70. [http://dx.doi.org/10.1016/j.cbi.2018.03.008] [PMID: 29551637]

[45] Sirelkhatim A, Mahmud S, Seeni A, *et al.* Review on zinc oxide nanoparticles: Antibacterial activity and toxicity mechanism. Nano-Micro Lett 2015; 7(3): 219-42. [http://dx.doi.org/10.1007/s40820-015-0040-x] [PMID: 30464967]

[46] Kumar R, Umar A, Kumar G, Nalwa HS. Antimicrobial properties of ZnO nanomaterials: A review. Ceram Int 2017; 43: 3940-61. [http://dx.doi.org/10.1016/j.ceramint.2016.12.062]

[47] Dizaj SM, Lotfipour F, Barzegar-Jalali M, Zarrintan MH, Adibkia K. Antimicrobial activity of the metals and metal oxide nanoparticles. Mater Sci Eng C 2014; 44: 278-84. [http://dx.doi.org/10.1016/j.msec.2014.08.031] [PMID: 25280707]

[48] Duffy LL, Osmond-McLeod MJ, Judy J, King T. Investigation into the antibacterial activity of silver, zinc oxide and copper oxide nanoparticles against poultry-relevant isolates of Salmonella and Campylobacter. Food Control 2018; 92: 293-300. [http://dx.doi.org/10.1016/j.foodcont.2018.05.008]

[49] Shakerimoghaddam A, Ghaemi EA, Jamalli A. Zinc oxide nanoparticle reduced biofilm formation and antigen 43 expressions in uropathogenic *Escherichiacoli.* Iran J Basic Med Sci 2017; 20(4): 451-6. [PMID: 28804616]

[50] Long M, Wang J, Zhuang H, Zhang Y, Wu H. Performance and mechanism of standard nano-TiO_2 (P-25) in photocatalytic disinfection of foodborne microorganisms-*Salmonella typhimurium* and *Listeria monocytogenes.* Food Control 2014; 39: 68-74. [http://dx.doi.org/10.1016/j.foodcont.2013.10.033]

[51] Jalvo B, Faraldos M, Bahamonde A, Rosal R. Antimicrobial and antibiofilm efficacy of self-cleaning surfaces functionalized by TiO_2 photocatalytic nanoparticles against Staphylococcus aureus and Pseudomonas putida. J Hazard Mater 2017; 340: 160-70. [http://dx.doi.org/10.1016/j.jhazmat.2017.07.005] [PMID: 28715739]

[52] Sarkar P, Choudhary R, Panigrahi S, Syed I, Sivapratha S, Dhumal CV. Nano-inspired systems in food technology and packaging. Environ Chem Lett 2017; 15: 607-22. [http://dx.doi.org/10.1007/s10311-017-0649-8]

[53] Ghanbarzadeh B, Oleyaei SA, Almasi H. Nanostructured materials utilized in biopolymer-based plastics for food packaging applications. Crit Rev Food Sci Nutr 2015; 55(12): 1699-723. [http://dx.doi.org/10.1080/10408398.2012.731023] [PMID: 24798951]

[54] Rhim JW, Park HM, Ha CS. Bio-nanocomposites for food packaging applications. Prog Polym Sci 2013; 38: 1629-52. [http://dx.doi.org/10.1016/j.progpolymsci.2013.05.008]

[55] Hottle TA, Bilec MM, Landis AE. Sustainability assessments of bio-based polymers. Polym Degrad Stabil 2013; 98: 1898-907. [http://dx.doi.org/10.1016/j.polymdegradstab.2013.06.016]

[56] EU. Guidance to the commission regulation (EC) No 450/2009 of 29 May 2009 on active and

intelligent materials and articles intended to come into contact with food.Version 10 European commission health and consumers directorate-general directorate e-safety of the food chain E6-innovation and sustainability. 2009.

[57] Biji KB, Ravishankar CN, Mohan CO, Srinivasa Gopal TK. Smart packaging systems for food applications: a review. J Food Sci Technol 2015; 52(10): 6125-35. [http://dx.doi.org/10.1007/s13197-015-1766-7] [PMID: 26396360]

[58] McMillin KW. Advancements in meat packaging. Meat Sci 2017; 132: 153-62. [http://dx.doi.org/10.1016/j.meatsci.2017.04.015] [PMID: 28465018]

[59] Ahmed I, Lin H, Zou L, *et al.* A comprehensive review on the application of active packaging technologies to muscle foods. Food Control 2017; 82: 168-73. [http://dx.doi.org/10.1016/j.foodcont.2017.06.009]

[60] Ariyarathna IR, Rajakaruna RMPI, Karunaratne DN. The rise of inorganic nanomaterial implementation in food applications. Food Control 2017; 77: 251-9. [http://dx.doi.org/10.1016/j.foodcont.2017.02.016]

[61] Singh M, Sahareen T. Investigation of cellulosic packets impregnated with silver nanoparticles for enhancing shelf-life of vegetables. Lebensm Wiss Technol 2017; 86: 116-22. [http://dx.doi.org/10.1016/j.lwt.2017.07.056]

[62] Azlin-Hasim S, Cruz-Romero MC, Morris MA, Cummins E, Kerry JP. Effects of a combination of antimicrobial silver low density polyethylene nanocomposite films and modified atmosphere packaging on the shelf life of chicken breast fillets. Food Packaging Shelf 2015; 4: 26-35. [http://dx.doi.org/10.1016/j.fpsl.2015.03.003]

[63] Arfat YA, Ahmed J, Jacob H. Preparation and characterization of agar-based nanocomposite films reinforced with bimetallic (Ag-Cu) alloy nanoparticles. Carbohydr Polym 2017; 155: 382-90. [http://dx.doi.org/10.1016/j.carbpol.2016.08.097] [PMID: 27702525]

[64] Li W, Zhang C, Chi H, *et al.* Development of antimicrobial packaging film made from Poly(Lactic Acid) incorporating titanium dioxide and silver nanoparticles. Molecules 2017; 22(7)E1170 [http://dx.doi.org/10.3390/molecules22071170] [PMID: 28703753]

[65] Youssef AM, Abou-Yousef H, El-Sayed SM, Kamel S. Mechanical and antibacterial properties of novel high performance chitosan/nanocomposite films. Int J Biol Macromol 2015; 76(76): 25-32. [http://dx.doi.org/10.1016/j.ijbiomac.2015.02.016] [PMID: 25709016]

[66] Friedman M, Juneja VK. Review of antimicrobial and antioxidative activities of chitosans in food. J Food Prot 2010; 73(9): 1737-61. [http://dx.doi.org/10.4315/0362-028X-73.9.1737] [PMID: 20828484]

[67] Hernández-Muñoz P, Almenar E, Ocio M, Gavara R. Effect of calcium dips and chitosan coatings on postharvest life of strawberries (*Fragaria x ananassa*). Postharvest Biol Technol 2006; 39(3): 247-53. [http://dx.doi.org/10.1016/j.postharvbio.2005.11.006]

[68] Youssef AM, El-Sayed SM, Salama HH, El-Sayed HS, Dufresne A. Evaluation of bionanocomposites as packaging material on properties of soft white cheese during storage period. Carbohydr Polym 2015; 132: 274-85. [http://dx.doi.org/10.1016/j.carbpol.2015.06.075] [PMID: 26256350]

[69] Zhang X, Xiao G, Wang Y, Zhao Y, Su H, Tan T. Preparation of chitosan-TiO_2 composite film with efficient antimicrobial activities under visible light for food packaging applications. Carbohydr Polym 2017; 169(169): 101-7. [http://dx.doi.org/10.1016/j.carbpol.2017.03.073] [PMID: 28504125]

[70] Salarbashi D, Tafaghodi M, Bazzaz BSF, Jafari B. Characterization of soluble soybean (SSPS) polysaccharide and development of eco-friendly SSPS/TiO_2 nanoparticle bionanocomposites. Int J Biol Macromol 2018; 112: 852-61. [http://dx.doi.org/10.1016/j.ijbiomac.2018.01.182] [PMID: 29410370]

[71] Salarbashi D, Tafaghodi M, Bazzaz BSF. Soluble soybean polysaccharide/TiO_2 bionanocomposite film for food application. Carbohydr Polym 2018; 186: 384-93. [http://dx.doi.org/10.1016/j.carbpol.2017.12.081] [PMID: 29456001]

[72] Roilo D, Maestri CA, Scarpa M, Bettotti P, Checchetto R. Gas barrier and optical properties of cellulose nanofiber coatings with dispersed TiO_2 nanoparticles. Surf Coat Tech 2017; 343: 131-7. [http://dx.doi.org/10.1016/j.surfcoat.2017.10.015]

[73] El-Feky OM, Hassan EA, Fadel SM, Hassan ML. Use of ZnO nanoparticles for protecting oil paintings on paper support against dirt, fungal attack, and UV aging. J Cult Herit 2014; 15(2): 165-72. [http://dx.doi.org/10.1016/j.culher.2013.01.012]

[74] EFSA Panel on Food Conact Materials, Enzymes, Flavourings and Processing Aids. Scientific opinion on the safety evaluation of the substance zinc oxide, nanoparticles, uncoated and coated with [3-(methacryloxy)propyl] trimethoxysilane, for use in food contact materials. EFSA J 2015; 13(4): 4063-72.

[75] Castro-Mayorga JL, Fabra MJ, Pourrahimi AM, Olsson RT, Lagaron JM. The impact of zinc oxide particle morphology as an antimicrobial and when incorporated in poly(3-hydroxybutyrate-c--3-hydroxyvalerate) films for food packaging and food contact surfaces applications. Food Bioprod Process 2017; 101: 32-44. [http://dx.doi.org/10.1016/j.fbp.2016.10.007]

[76] Rahman PM, Mujeeb VMA, Muraleedharan K. Flexible chitosan-nano ZnO antimicrobial pouches as a new material for extending the shelf life of raw meat. Int J Biol Macromol 2017; 97(97): 382-91. [http://dx.doi.org/10.1016/j.ijbiomac.2017.01.052] [PMID: 28104367]

[77] Tankhiwale R, Bajpai SK. Preparation, characterization and antibacterial applications of ZnO-nanoparticles coated polyethylene films for food packaging. Colloids Surf B Biointerfaces 2012; 90: 16-20. [http://dx.doi.org/10.1016/j.colsurfb.2011.09.031] [PMID: 22015180]

[78] Zhang H, Hortal M, Jordá-Beneyto M, Rosa E, Lara-Lledo M, Lorente I. ZnO-PLA nanocomposite coated paper for antimicrobial packaging application. Lebensm Wiss Technol 2017; 78: 250-7. [http://dx.doi.org/10.1016/j.lwt.2016.12.024]

[79] Jafarzadeh S, Ariffin F, Mahmud S, Alias AK, Hosseini SF, Ahmad M. Improving the physical and protective functions of semolina films by embedding a blend nanofillers (ZnO-nr and nano-kaolin). Food Packag Shelf Life 2017; 12: 66-75. [http://dx.doi.org/10.1016/j.fpsl.2017.03.001]

[80] Roy N, Saha N, Kitano T, Saha P. Biodegradation of PVP-CMC hydrogel film: a useful food packaging material. Carbohydr Polym 2012; 89(2): 346-53. [http://dx.doi.org/10.1016/j.carbpol.2012.03.008] [PMID: 24750729]

[81] Oun AA, Rhim JW. Carrageenan-based hydrogels and films: Effect of ZnO and CuO nanoparticles on the physical, mechanical, and antimicrobial properties. Food Hydrocoll 2017; 67: 45-53. [http://dx.doi.org/10.1016/j.foodhyd.2016.12.040]

[82] Liu JM, Hu Y, Yang YK, *et al.* Emerging functional nanomaterials for the detection of food contaminants. Trends Food Sci Technol 2018; 71: 94-106. [http://dx.doi.org/10.1016/j.tifs.2017.11.005]

[83] Fuertes G, Soto I, Carrasco R, Vargas M, Sabattin J, Lagos C. Intelligent packaging systems: sensors and nanosensors to monitor food quality and safety. J Sens 2016; 2: 1-8. [http://dx.doi.org/10.1155/2016/4046061]

[84] Du J, Sing H, Dong W, Bai YH, Yi TH. Colorimetric detection of Listeria monocytogenes using one-pot biosynthesized lower-shaped gold nanoparticles. Sens Actuators B Chem 2018; 265: 285-92. [http://dx.doi.org/10.1016/j.snb.2018.03.067]

[85] Amendola V, Pilot R, Frasconi M, Maragò OM, Iatì MA. Surface plasmon resonance in gold

nanoparticles: a review. J Phys Condens Matter 2017; 29(20)203002
[http://dx.doi.org/10.1088/1361-648X/aa60f3] [PMID: 28426435]

[86] Lim S, Gunasekaran S, Imm JY. Gelatin-templated gold nanoparticles as novel time-temperature indicator. J Food Sci 2012; 77(9): N45-9.
[http://dx.doi.org/10.1111/j.1750-3841.2012.02872.x] [PMID: 22900571]

[87] Ansari AA, Kaushik A, Solanki PR, Malhotra BD. Nanostructured zinc oxide platform for mycotoxin detection. Bioelectrochemistry 2010; 77(2): 75-81.
[http://dx.doi.org/10.1016/j.bioelechem.2009.06.014] [PMID: 19648064]

[88] Cabañes FJ, Bragulat MJ. Black aspergilli and ochratoxin A-producing species. Curr Opin Food Sci 2018; 23: 1-10.
[http://dx.doi.org/10.1016/j.cofs.2018.01.006]

[89] Shao B, Ma X, Zhao S, *et al.* Nanogapped $Au_{(core)}$ @ $Au\text{-}Ag_{(shell)}$ structures coupled with Fe_3O_4 magnetic nanoparticles for the detection of Ochratoxin A. Anal Chim Acta 2018; 1033: 165-72.
[http://dx.doi.org/10.1016/j.aca.2018.05.058] [PMID: 30172322]

[90] Nesakumar N, Thandavan K, Sethuraman S, Krishnan UM, Rayappan JB. An electrochemical biosensor with nanointerface for lactate detection based on lactate dehydrogenase immobilized on zinc oxide nanorods. J Colloid Interface Sci 2014; 414: 90-6.
[http://dx.doi.org/10.1016/j.jcis.2013.09.052] [PMID: 24231089]

[91] Rauscher H, Rasmussen K, Sokull-Kluttgen B. Regulatory aspects of nanomaterials in the EU. Chemieingenieurtechnik (Weinh) 2017; 89(3): 224-31.
[http://dx.doi.org/10.1002/cite.201600076]

[92] Jain A, Ranjan S, Dasgupta N, Ramalingam C. Nanomaterials in food and agriculture: An overview on their safety concerns and regulatory issues. Crit Rev Food Sci Nutr 2018; 58(2): 297-317.
[http://dx.doi.org/10.1080/10408398.2016.1160363] [PMID: 27052385]

[93] European Commission. Regulation (EC) No 1333/2008 of the European Parliament and of the council of 16 December 2008 on food additives (text with EEA relevance). Off J Eur Union 2008; 354/16-33.

[94] Yusoff R, Nguyen LTH, Chiew P, Wang ZM, Ng KW. Comparative differences in the behaviour of TiO_2 and SiO_2 food additives in food ingredient solutions. J Nanopart Res 2018; 20: 76.
[http://dx.doi.org/10.1007/s11051-018-4176-8]

[95] Dudefoi W, Moniz K, Allen-Vercoe E, Ropers MH, Walker VK. Impact of food grade and nano-TiO_2 particles on a human intestinal community. Food Chem Toxicol 2017; 106(Pt A): 242-9.
[http://dx.doi.org/10.1016/j.fct.2017.05.050] [PMID: 28564612]

[96] Re-evaluation of titanium dioxide (E171) as a food a food additive. EFSA Panel on Food Additives and Nutrient Sources added to Food (ANS). EFSA J 2016; 14(9): 4545.

[97] Peters RJB, van Bemmel G, Herrera-Rivera Z, *et al.* Characterization of titanium dioxide nanoparticles in food products: analytical methods to define nanoparticles. J Agric Food Chem 2014; 62(27): 6285-93.
[http://dx.doi.org/10.1021/jf5011885] [PMID: 24933406]

[98] Re-evaluation of silicon dioxide (E 551) as a food additive. EFSA J 2018; 16(1): 5088.

[99] European Commission. Regulation (EU) 2015/2283 OF THE EUROPEAN PARLIAMENT AND OF THE COUNCIL of 25 November 2015 on novel foods, amending Regulation (EU) No 1169/2011 of the European Parliament and of the Council and repealing Regulation (EC) No 258/97 of the European Parliament and of the Council and Commission Regulation (EC) No 1852/ 2001. Off J Eur Union 2015; 327/1-327/22.

[100] European Commission. Commision implementing regulation (EU) 2017/2470 of December 2017 establishing the Union list of novel food in accordance with Regulation (EU) 2015/2283 of the European Parliament and of the Council no novel foods. Off J Eur Union 2017; 351/72-351/201.

[101] European Commission. Commision regulation (EC) No 450/2009 on active and intelligent materials and articles intended to come into contact with food. Off EurUnion 2009; 135/3-135/11.

[102] European Commission. Commission regulation (EU) No 10/2011 of 14 January 2011 on plastic materials and articles intended to come into contact with food. (Text with EEA relevance). Off J EurUnion 2011; 12/1-12/89.

[103] Oomen AG, Steinhäuser KG, Bleeker EAJ, *et al.* Risk assessment frameworks for nanomaterials: scope, link to regulations, applicability, and outline for future directions in view of needed increase in efficiency. NanoImpact 2018; 9: 1-13.
[http://dx.doi.org/10.1016/j.impact.2017.09.001]

[104] Guidance on risk assessment of the nanoscience and nanotechnologies in the food and feed chain: Part 1, human and animal health. EFSA J 2018; 16(7): 5327.

[105] European Commission. Communication from the commission to the European Parliament, the Council and the European Economic and Social Committee. Second regulatory review on nanomaterials. (Text with EEA relevance). Off J EurUnion 2012; 12/1-12/89.

[106] Dudkiewicz A, Tiede K, Loeschner LHS, Jensen E, Wierzbicki R, Molhav K. Characterization of nanomaterials in food by electron microscopy. Trends Analyt Chem 2011; 30(1): 28-43.
[http://dx.doi.org/10.1016/j.trac.2010.10.007]

[107] Laborda F, Bolea E, Jiménez-Lamana J. Single particle inductively coupled plasma mass spectrometry: a powerful tool for nanoanalysis. Anal Chem 2014; 86(5): 2270-8.
[http://dx.doi.org/10.1021/ac402980q] [PMID: 24308527]

[108] Naasz S, Weigel S, Borovinskaya O, *et al.* Multi-element analysis of single nanoparticles by ICP-MS using quadrupole and time-of-flight technologies. J Anal At Spectrom 2018; 33: 835-45.
[http://dx.doi.org/10.1039/C7JA00399D]

[109] Praetorius A, Gundlach-Graham A, Goldberg E, *et al.* Single-particle multi-element fingerprinting (spMEF) using inductively-coupled plasma time-of-flight mass spectrometry (ICP-TOFMS) to identify engineered nanoparticles against the elevated natural background in soils. Environ Sci Nano 2017; 4: 307-14.
[http://dx.doi.org/10.1039/C6EN00455E]

[110] Hendriks L, Gundlach-Graham A, Hattendorf B, Günther D. Characterization of a new ICP-TOFMS instrument with continuous and discrete introduction of solution. J Anal At Spectrom 2017; 32: 548-61.
[http://dx.doi.org/10.1039/C6JA00400H]

[111] Pitkänen L, Striegel AM. Size-exclusion chromatography of metal nanoparticles and quantum dots. Trends Analyt Chem 2016; 80: 311-20.
[http://dx.doi.org/10.1016/j.trac.2015.06.013] [PMID: 27335508]

[112] Pitkänen L, Montoro Bustos AR, Murphy KE, Winchester MR, Striegel AM. Quantitative characterization of gold nanoparticles by size-exclusion and hydrodynamic chromatography, coupled to inductively coupled plasma mass spectrometry and quasi-elastic light scattering. J Chromatogr A 2017; 1511: 59-67.
[http://dx.doi.org/10.1016/j.chroma.2017.06.064] [PMID: 28697934]

[113] Giddings JC. A new separation concept based on a coupling of concentration and flow non uniformities. Sep Sci 1996; 1: 123.
[http://dx.doi.org/10.1080/01496396608049439]

[114] Von der Kammer F, Legros S, Larsen EH, Loeschner K, Hofmann T. Separation and characterization of nanoparticles in complex food and environmental samples by field-flow fractionation. Trends Analyt Chem 2011; 30: 425-36.
[http://dx.doi.org/10.1016/j.trac.2010.11.012]

[115] Gigault J, Pettibone JM, Schmitt C, Hackley VA. Rational strategy for characterization of nanoscale

particles by asymmetric-flow field flow fractionation: a tutorial. Anal Chim Acta 2014; 809: 9-24. [http://dx.doi.org/10.1016/j.aca.2013.11.021] [PMID: 24418128]

[116] Loeschner K, Navratilova J, Legros S, *et al.* Optimization and evaluation of asymmetric flow field-flow fractionation of silver nanoparticles. J Chromatogr A 2013; 1272: 116-25. [http://dx.doi.org/10.1016/j.chroma.2012.11.053] [PMID: 23261297]

[117] Linsinger TP, Chaudhry Q, Dehalu V, *et al.* Validation of methods for the detection and quantification of engineered nanoparticles in food. Food Chem 2013; 138(2-3): 1959-66. [http://dx.doi.org/10.1016/j.foodchem.2012.11.074] [PMID: 23411331]

[118] von Moos N, Slaveykova VI. Oxidative stress induced by inorganic nanoparticles in bacteria and aquatic microalgae--state of the art and knowledge gaps. Nanotoxicology 2014; 8(6): 605-30. [http://dx.doi.org/10.3109/17435390.2013.809810] [PMID: 23738945]

[119] Palomo-Siguero M, Madrid Y. Exploring the behavior and metabolic transformations of SeNPs in exposed Lactic Acid Bacteria. Effect of nanoparticles coating agent. Int J Mol Sci 2017; 18(8): 1712. [http://dx.doi.org/10.3390/ijms18081712] [PMID: 28783048]

[120] Saptarshi SR, Duschl A, Lopata AL. Interaction of nanoparticles with proteins: relation to bio-reactivity of the nanoparticle. J Nanobiotechnology 2013; 11: 26-38. [http://dx.doi.org/10.1186/1477-3155-11-26] [PMID: 23870291]

[121] Mahmoudi M, Lynch I, Ejtehadi MR, Monopoli MP, Bombelli FB, Laurent S. Protein-nanoparticle interactions: opportunities and challenges. Chem Rev 2011; 111(9): 5610-37. [http://dx.doi.org/10.1021/cr100440g] [PMID: 21688848]

[122] Lacerda SH, Park JJ, Meuse C, *et al.* Interaction of gold nanoparticles with common human blood proteins. ACS Nano 2010; 4(1): 365-79. [http://dx.doi.org/10.1021/nn9011187] [PMID: 20020753]

[123] Deng ZJ, Mortimer G, Schiller T, Musumeci A, Martin D, Minchin RF. Differential plasma protein binding to metal oxide nanoparticles. Nanotechnology 2009; 20(45)455101. [http://dx.doi.org/10.1088/0957-4484/20/45/455101] [PMID: 19822937]

[124] Deng ZJ, Liang M, Toth I, Monteiro M, Minchin RF. Plasma protein binding of positively and negatively charged polymer-coated gold nanoparticles elicits different biological responses. Nanotoxicology 2013; 7(3): 314-22. [http://dx.doi.org/10.3109/17435390.2012.655342] [PMID: 22394123]

[125] Zhang W, Xiao B, Fang T. Chemical transformation of silver nanoparticles in aquatic environments: Mechanism, morphology and toxicity. Chemosphere 2018; 191: 324-34. [http://dx.doi.org/10.1016/j.chemosphere.2017.10.016] [PMID: 29045933]

[126] Grombe R, Charoud-Got J, Emteborg H, *et al.* Production of reference materials for the detection and size determination of silica nanoparticles in tomato soup. Anal Bioanal Chem 2013; 405: 8185-95. [PMID: 24390460]

[127] Peters RJB, Rivera ZH, van Bemmel G, Marvin HJP, Weigel S, Bouwmeester H. Development and validation of single particle ICP-MS for sizing and quantitative determination of nano-silver in chicken meat. Anal Bioanal Chem 2014; 406(16): 3875-85. [http://dx.doi.org/10.1007/s00216-013-7571-0] [PMID: 24390462]

[128] Loeschner K, Navratilova J, Købler C, *et al.* Detection and characterization of silver nanoparticles in chicken meat by asymmetric flow field flow fractionation with detection by conventional or single particle ICP-MS. Anal Bioanal Chem 2013; 405(25): 8185-95. [http://dx.doi.org/10.1007/s00216-013-7228-z] [PMID: 23887279]

[129] Gomez-Gomez B, Perez-Corona MT, Madrid Y. Effect of food matrix on ZnONPs stability. 9th Franco-Spanish workshop on Bio-inorganic Analytical Chemistry. Huelva. 2017.

[130] Böhmert L, Girod M, Hansen U, *et al.* Analytically monitored digestion of silver nanoparticles and their toxicity on human intestinal cells. Nanotoxicology 2014; 8(6): 631-42.

[http://dx.doi.org/10.3109/17435390.2013.815284] [PMID: 23763544]

[131] Mwilu SK, El Badawy AM, Bradham K, *et al.* Changes in silver nanoparticles exposed to human synthetic stomach fluid: effects of particle size and surface chemistry. Sci Total Environ 2013; 447: 90-8.
[http://dx.doi.org/10.1016/j.scitotenv.2012.12.036] [PMID: 23376520]

[132] Walczak AP, Fokkink R, Peters R, *et al.* Behaviour of silver nanoparticles and silver ions in an *in vitro* human gastrointestinal digestion model. Nanotoxicology 2013; 7(7): 1198-210.
[http://dx.doi.org/10.3109/17435390.2012.726382] [PMID: 22931191]

[133] Peters R, Kramer E, Oomen AG, *et al.* Presence of nano-sized silica during in vitro digestion of foods containing silica as a food additive. ACS Nano 2012; 6(3): 2441-51.
[http://dx.doi.org/10.1021/nn204728k] [PMID: 22364219]

[134] Lichtenstein D, Ebmeyer J, Knappe P, *et al.* Impact of food components during *in vitro* digestion of silver nanoparticles on cellular uptake and cytotoxicity in intestinal cells. Biol Chem 2015; 396(11): 1255-64.
[http://dx.doi.org/10.1515/hsz-2015-0145] [PMID: 26040006]

[135] Vila L, García-Rodríguez A, Cortés C, Marcos R, Hernández A. Assessing the effects of silver nanoparticles on monolayers of differentiated Caco-2 cells, as a model of intestinal barrier. Food Chem Toxicol 2018; 116(Pt B): 1-10.
[http://dx.doi.org/10.1016/j.fct.2018.04.008] [PMID: 29626574]

[136] Sieg H, Braeuning C, Kunz BM, *et al.* Uptake and molecular impact of aluminum-containing nanomaterials on human intestinal caco-2 cells. Nanotoxicology 2018; 12(9): 992-1013.
[http://dx.doi.org/10.1080/17435390.2018.1504999] [PMID: 30317887]

[137] Munger MA, Radwanski P, Hadlock GC, *et al. In vivo* human time-exposure study of orally dosed commercial silver nanoparticles. Nanomedicine (Lond) 2014; 10(1): 1-9.
[http://dx.doi.org/10.1016/j.nano.2013.06.010] [PMID: 23811290]

[138] De la Calle I, Menta M, Sébe F. Current trends and challenges in sample preparation for metallic nanoparticles analysis in daily products and environmental samples: A review. Spectrochim Acta B At Spectrosc 2016; 125: 66-96.
[http://dx.doi.org/10.1016/j.sab.2016.09.007]

[139] Beltrami D, Calestani D, Maffini M, *et al.* Development of a combined SEM and ICP-MS approach for the qualitative and quantitative analyses of metal microparticles and sub-microparticles in food products. Anal Bioanal Chem 2011; 401: 1401-9.
[http://dx.doi.org/10.1007/s00216-011-5149-2] [PMID: 21660413]

[140] Dan Y, Zhang W, Xue R, Ma X, Stephan C, Shi H. Characterization of gold nanoparticle uptake by tomato plants using enzymatic extraction followed by single-particle inductively coupled plasma-mass spectrometry analysis. Environ Sci Technol 2015; 49(5): 3007-14.
[http://dx.doi.org/10.1021/es506179e] [PMID: 25633589]

[141] Schmidt B, Loeschner K, Hadrup N, *et al.* Quantitative characterization of gold nanoparticles by field-flow fractionation coupled online with light scattering detection and inductively coupled plasma mass spectrometry. Anal Chem 2011; 83(7): 2461-8.
[http://dx.doi.org/10.1021/ac102545e] [PMID: 21355549]

[142] Echegoyen Y, Nerín C. Nanoparticle release from nano-silver antimicrobial food containers. Food Chem Toxicol 2013; 62: 16-22.
[http://dx.doi.org/10.1016/j.fct.2013.08.014] [PMID: 23954768]

[143] Metak AM, Nabhani F, Connolly SN. Migration of engineered nanoparticles from packaging into food products. Lebensm Wiss Technol 2015; 64: 781-7.
[http://dx.doi.org/10.1016/j.lwt.2015.06.001]

[144] Addo Ntim S, Thomas TA, Begley TH, Noonan GO. Characterisation and potential migration of silver

nanoparticles from commercially available polymeric food contact materials. Food Addit Contam Part A Chem Anal Control Expo Risk Assess 2015; 32(6): 1003-11. [http://dx.doi.org/10.1080/19440049.2015.1029994] [PMID: 25831019]

[145] Cushen M, Kerry J, Morris M, Cruz-Romero M, Cummins E. Migration and exposure assessment of silver from a PVC nanocomposite. Food Chem 2013; 139(1-4): 389-97. [http://dx.doi.org/10.1016/j.foodchem.2013.01.045] [PMID: 23561122]

CHAPTER 4

Halogenated and Organophosphorus Flame Retardants

Òscar Aznar-Alemany and **Ethel Eljarrat***

Institute of Environmental Assessment and Water Research, Spanish National Research Council (IDAEA-CSIC), C/ Jordi Girona, 18-26. 08034 Barcelona, Spain

Abstract: Flame retardants are applied to a wide range of materials to improve their fire resistance. However, they leak from those materials into the environment. There are many compounds used as flame retardants, the most relevant organic ones being polybrominated diphenyl ethers (PBDEs), hexabromocyclododecane (HBCD), other halogenated flame retardants (HFRs) and organophosphorus flame retardants (OPFRs). Exposition to flame retardants can be through ingestion, inhalation and skin permeation. Different studies report that food account for most of the exposition to PBDEs. Data indicates that seafood is the main contributor to PBDE intake in Europe and Japan, while meat is the main contributor in the United States and Canada. For this reason, it is one of the main public health interests that food be innocuous. This chapter compares seventeen publications that apply methods suitable for the analysis of flame retardants in food. Some publications include different methods targeting different groups of compounds. PBDEs and most HFRs are commonly analyzed together by GC. HBCD tends to be extracted separately and analyzed by LC. OPFRs are also extracted and analyzed independently, but few methods target them currently. The present text presents and compares the sample treatment, the instrumental analysis and the quality parameters for the listed methods. A final comment on levels of flame retardants in food and dietary intake is provided.

Keywords: BFRs, Clean-up, Dechloranes, Dietary intake, Extraction techniques, Flame retardants, Food analysis, Food safety, Gas chromatography, HBCD, HFRs, Instrumental analysis, Lipid removal, Liquid chromatography, Mass spectrometry, Methods comparison, OPFRs, PBDEs, Quality parameters, Sample treatment.

INTRODUCTION

Flame retardants (FRs) are a group of compounds that are added to materials to

* **Corresponding author Ethel Eljarrat:** Institute of Environmental Assessment and Water Research, Spanish National Research Council (IDAEA-CSIC), C/ Jordi Girona, 18-26. 08034 Barcelona, Spain; Tel: +34 934006100 (extension 1627); E-mail: eeeqam@cid.csic.es

Belén Gómara & María Luisa Marina (Eds.)
Current and Future Developments in Food Science (Vol. 1)

increase their resistance to fire. The search for flame retarding compounds produced, among others, halogenated flame retardants (HFRs), mainly brominated and chlorinated.

In the gas phase, HFRs capture hydroxyl and hydrogen radicals produced in the first steps of combustion and which allows the propagation of the reaction [1].

At the beginning of the 20th century, the electrical industry needed a dielectric insulator that acted as a FR. Production of polychlorinated biphenyls (PCBs) started in the United States in 1929, then in Europe and later in Japan in 1954. A Swedish biologist detected PCBs in fish in 1966 [2]. Two years later, a thousand of Japanese people were found intoxicated with PCB-contaminated rice oil. PCBs were banned in Japan in 1972 and their production in the United States was stopped in 1976 [3].

Hexacyclopentadiene was described first in 1930 and was later considered as an insecticide and FR [4]. Hooker Electrochemical commercialized its dimer, mirex, in the 1960s calling it Dechlorane. However, mirex was banned in 1977 due to its degradation forming a carcinogenic compound [3]. In 1964 Hooker Electrochemical -currently Occidental Chemical (OxyChem)- had already developed a derivate from hexacyclopentadiene which they named Dechlorane Plus.

Some brominated flame retardants (BFRs) available in the 1950s were pentabrominated diphenyl ether (pentaBDE), tris(2,3-dibromopropyl) phosphate (Tris) and tetrabromobisphenol A (TBBPA). As BFRs were more effective than chlorinated FRs, BFRs allowed for smaller amount of additives in the materials, thus not compromising so much their physical properties and BFRs became popular very fast. On the other hand, some of them had to be banned do to their toxicity, as was the case of Tris in 1977 [3].

HFRs are used for years until the scientific community gathers enough data to assess their adequacy. Being persistent, the impact of banned HFRs on the environment and the organisms can last long after their ban.

Not all FRs are halogenated; there are also organophosphorus flame retardants (OPFRs). OPFRs accounted for 20% of the FRs used in Europe in 2006 —which doubles the amount of PBDEs used that year— and have been increasingly applied after the restrictions on PBDEs [5]. Apart from FRs, OPFRs are applied as plasticizers as well.

The fact that FRs are found in the environment and accumulate in organisms, some of which serve to feed humans, implies that these contaminants are likely to be present in food to some extent. Humans are at the top of the food web and, thus, the final recipients of the biomagnification effect.

Exposition to FRs can be through ingestion, inhalation and skin permeation. A study performed in Vietnam considering different types of exposition to polybrominated diphenyl ethers (PBDEs) estimated that fish consumption accounted for 70% of the total exposition to PBDEs and 80% of the exposition to BDE-209 [6]. Some preliminary studies in all kinds of foods from Canada, Japan, the United States of America and Europe concluded that the average daily intake of PBDEs was between 13 and 113 ng day^{-1} [7]. Their data also showed that seafood is the main contributor to PBDE intake in Europe and Japan, while meat is the main contributor in the United States and Canada. The average daily intake of PBDEs per kilogram of body weight in Europe was 2.2 ng bw^{-1} day^{-1}. A Swedish study calculated that fish accounted for more than 60% of the total intake of PBDEs and more than 80% of the intake of BDE-47 [8].

Seafood production has grown to 3.2% every year since 1961 [9]. Nowadays, aquaculture provides half the seafood consumed worldwide. On average, a person consumes 20 kg of fish per year and 17% of the world intake of protein comes from fish.

Not only do organisms accumulate contaminants present in the environment, but the feed used in farms and fish farms contains animal parts with no commercial value, which might add their accumulated contaminants into the diet of new animals.

For this reason, it is one of the main public health interests that food —and especially seafood be innocuous.

FLAME RETARDANTS

Polybrominated Diphenyl Ethers and Hexabromocyclododecane

PBDEs and hexabromocyclododecane (HBCD) are some of the most popular FRs. They can be found in a broad variety of elements such as plastics, furniture, vehicles and electronic appliances [10].

There are 209 PBDE congeners depending on their degree of bromination and the position of the bromine atoms (Fig. **1**). As their structures are analogous to those of PCBs, the same nomenclature by Ballschmiter and Zell is used [11].

The three commercial mixtures are PentaBDE (0-1% triBDE, 24-37% tetraBDE, 50-60% pentaBDE and 4-8% hexaBDE), OctaBDE (10-12% hexaBDE, 43-44% heptaBDE, 31-35% octaBDE, 10-11% nonaBDE and 0-1% decaBDE) and

DecaBDE (0,3-3% nonaBDE and 97-98% decaBDE) [12].

There are 16 possible isomers of HBCD-more specifically, 1,2,5,6,9,10-hexabromocyclododecane. The most common isomers and those that are found in commercial mixtures are diastereoisomers α-HBCD, β-HBCD and γ-HBCD (Fig. **2**). They are found in approximate proportions of 1-12%, 10-13% and 75-89%, respectively [13]. Each diastereoisomer presents two possible enantiomers, for example (+)-α-HBCD and (−)-α-HBCD. In biological samples, δ-HBCD has also been detected [14 - 16], but there is no data of ε-HBCD in the environment so far.

PBDEs and HBCD are blended into materials, not chemically bonded. This makes them easy to leak into the environment [10, 17]. This fact added to their physicochemical and toxicological properties required them to be banned and alternatives to be found.

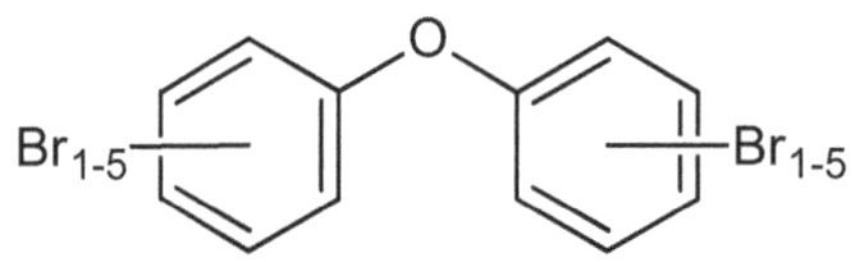

Fig. (1). General structure of PBDEs.

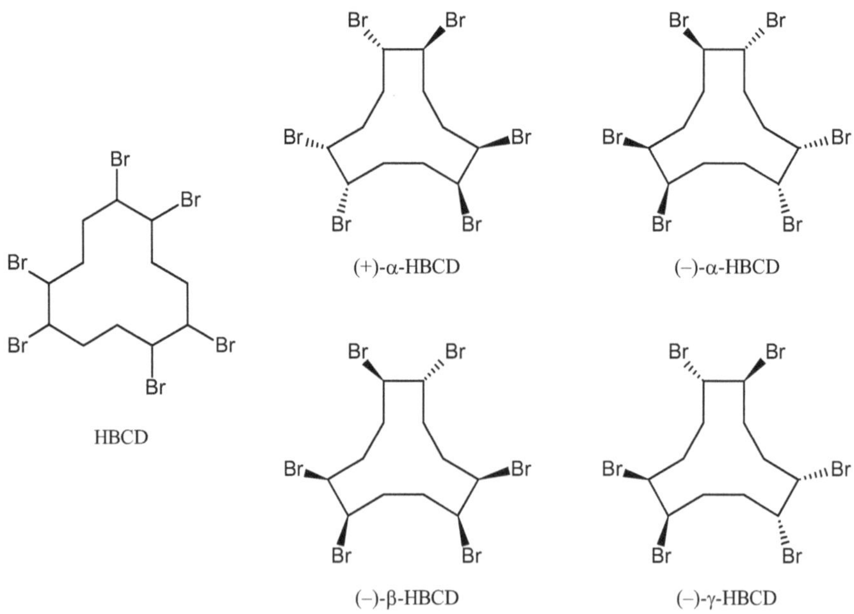

Fig. (2). Structures of some HBCD isomers.

Emerging Flame Retardants

The new FRs that are produced as alternatives are called emerging flame retardants (EFRs), as more data still needs to be gathered to learn the details about their properties and behavior. Some EFRs are BFRs. Hexabromobenzene (HBB), pentabromoethylbenzene (PBEB), decabromodiphenyl ethane (DBDPE) and 1,2-bis-(2,4,6-tribromophenoxy)ethane (BTBPE) are some examples (Fig. **3**). Production of these EFRs has increased in the last three decades and they have been included in monitoring programs [18].

HBB is an additive FR historically used in plastics, textiles and woods [19], but also in paper and electric manufactured goods [20]. HBB is produced mainly in Japan and China [21]. Apart from being commercially produced, there are other possible sources for HBB to enter into environment. HBB was found to be the major product from pyrolysis of OctaBDE and DecaBDE technical products [22].

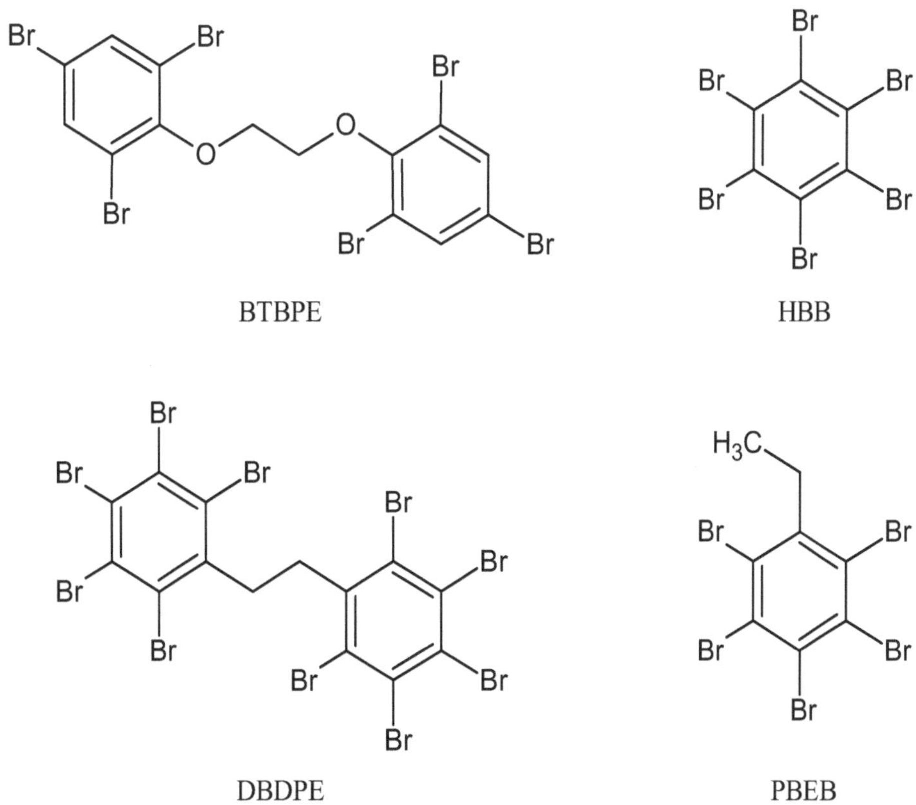

Fig. (3). Structures of brominated EFRs.

The main applications of PBEB, used also as an additive FR, have been in thermoset polyester resins, circuit boards, textiles, adhesives, wire and cable coatings, and polyurethane foam [23]. PBEB was produced in the United States in the 1970s and 1980s, but it is now a low volume production FR in France [23].

DBDPE became available on the market in the mid-1980s. It has been marketed under the trade names Saytex 8010 (Albemarle, Arkansas, USA) and Firemaster 2100 (Chemtura). DBDPE is used as a substitute for BDE-209. Unlike BDE-209, DBDPE does not generate furans or dioxins when exposed to sunlight [24].

BTBPE is an additive FR marketed as FF-680 (Great Lakes Chemical Corporation, Arkansas, USA). BTBPE has been produced since the mid-1970s and is now being used as a replacement for OctaBDE [23].

Dechloranes or halogenated norbornenes (HNs)— are chlorinated EFRs. Dechloranes include Dechlorane Plus (DP), Dechlorane 602 (Dec 602), Dechlorane 603 (Dec 603) and Dechlorane 604 (Dec 604) (Fig. **4**). Despite having been used for years, the first data about DP and other dechloranes in the environment was not published until 2006 and 2010, respectively [25, 26]. This lack of data qualifies them as EFRs, even if they are not new compounds.

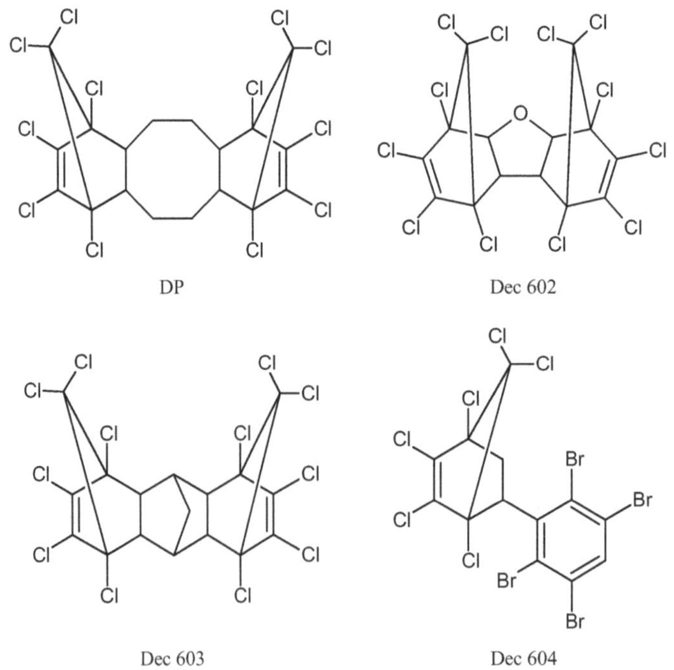

Fig. (4). Structures of dechloranes.

The two known producers of DP are OxyChem (New York, united States) and Anpon Electrochemical (Huai'an, China) [27]. DP presents isomers *syn*-DP and *anti*-DP, which are in commercial mixtures at an approximated 1:2 ratio [28]. DP can account for up to 35% of the weight of the plastics that contain it [29]. In the European Union (EU), DP is considered a low volume production FR, but the United States Environmental Protection Agency (EPA) considers DP a high volume production compound, implying a production or importation higher than 450,000 kg per year [30]. OxyChem patented Dec 602, Dec 603 and Dec 604 in the 1960s. They are less restricted than DP and are sometimes used where DP does not meet the legal requirements [31].

Organophosphorus Flame Retardants

Apart from HFRs, there are also OPFRs (Fig. **5**). They accounted for 20 % of the FR use in 2006 in Europe —twice as much as brominated FRs and have been increasingly applied after the ban on PBDEs [5]. However, inorganic FRs were preferred, accounting for the remaining 70%. Like PBDEs, OPFRs leak from materials and can access environmental matrices through deposition, washout, infiltration, *etc.* [32, 33]. Additionally, OPFRs are used as plasticizers; hence, they can be released from the tones of plastic present in seas and oceans. OPFRs have been found in sediments, fish and water [34 - 36]. OPFRs show toxic effects on the reproductive and endocrine systems, as well as systemic and carcinogenic effects [5, 37].

Physicochemical Properties

FRs have more or less similar physicochemical properties (Table **1**) [5, 10, 18, 30, 38 - 44]. FRs have a high relative molecular mass, with the maximum at 972 g mol^{-1} for DBDPE. HFRs are very hydrophobic with logarithms of their octanol-water partition coefficient (K_{ow}) between 5.6 and 13.5 and solubility in water between 16.8 pg l^{-1} and 210 µg l^{-1}. OPFRs are more soluble and have a bigger range of log K_{ow}, from −0.65 for trimethyl phosphate to 9.49 for TEHP.

Half-life ($t_{1/2}$) of BDE-209 in aerobic conditions in sludge is between 6.0 and 8.2 months. In anaerobic conditions, the amount of BDE-209 can decrease to a 70% [45]. Half-life of HBCD in anaerobic conditions has been estimated to 0.66 or 5 days, which go up to 35 days for sterilised sludge [46 - 48]. Digestion of α-HBCD takes twice as much time as digestion of β-HBCD and γ-HBCD, which is consistent with the fact that environmental samples contain a higher proportion of the α isomer than the commercial product.

More studies are required to properly assess the physicochemical properties and the stability of EFRs. Some models predict persistence greater than 16 months for many EFRs, including HBB and PBEB [49]. A different estimation calculates half-lives of FRs other than PBDEs in air, water and soil [39]. All compounds meet the condition of persistence for persistent organic pollutants (POPs), set by the Stockholm Convention at $t_{1/2} > 2$ days (Table **1**).

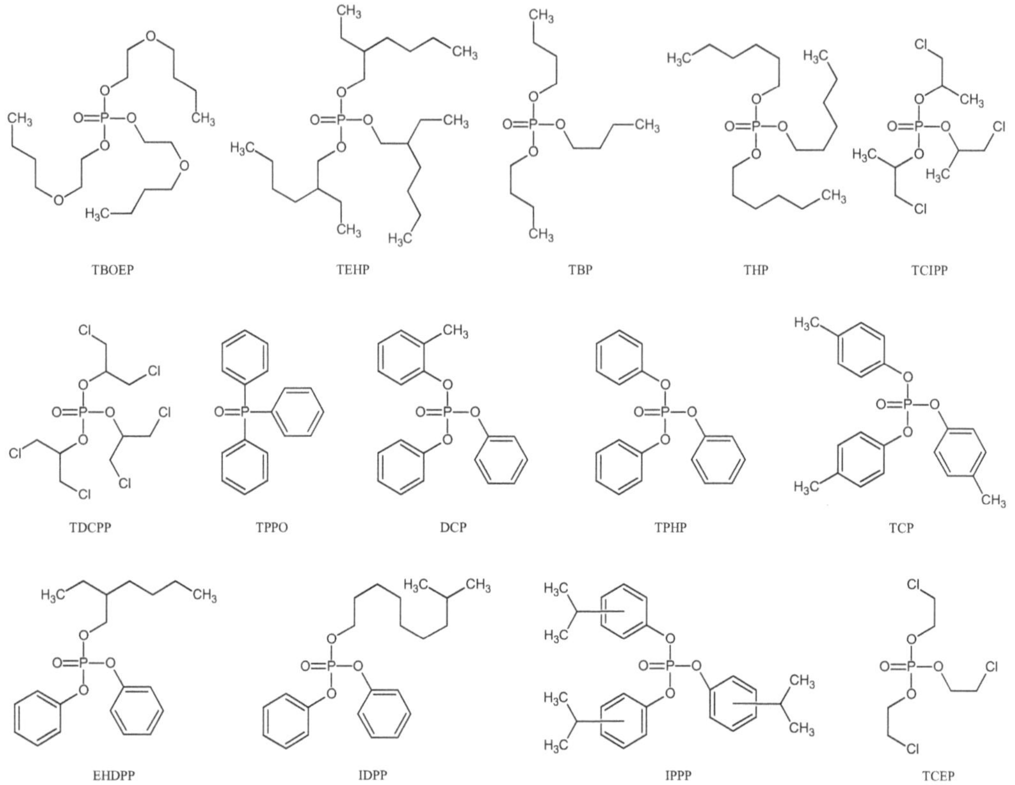

Fig. (5). Structures of some OPFRs: tris(2-butoxyethyl) phosphate (TBOEP), tris(2-ethylhexyl) phosphate (TEHP), tributyl phosphate (TBP), trihexyl phosphate (THP), tris(chloroisopropyl) phosphate (TCIPP), tris(1,3-dichloropropyl) phosphate (TDCPP), triphenylphosphine oxide (TPPO), diphenyl cresyl phosphate (DCP), triphenyl phosphate (TPHP), tricresyl phosphate (TCP), 2-ethylhexyl diphenyl phosphate (EHDPP), isodecyl diphenyl phosphate (IDPP), isopropyl phenyl phosphate (IPPP), tris(2-chloroethyl) phosphate (TCEP).

Table 1. Properties of FRs.

Compound	**Formula**	**Relative Molecular Mass (g mol^{-1})**	**log K_{ow}**	**Solubility in Water (25 °C, mg l^{-1})**	$t_{1/2,air}$	$t_{1/2,water}$	$t_{1/2,soil}$
PBDEs	$C_{12}H_xOBr_y$	248-950	5.88-12.1	4.4×10^{-3} – 14×10^{-3}	1-15.9 months		
HBCD	$C_{12}H_{18}Br_6$	642	5.6	2.1×10^{-3} – 49×10^{-3}			

(Table 1) cont.....

Compound	Formula	Relative Molecular Mass (g mol^{-1})	log K_{ow}	Solubility in Water (25 °C, mg l^{-1})	$t_{1/2,air}$	$t_{1/2,water}$	$t_{1/2,soil}$
DBDPE	$C_{14}H_4Br_{10}$	972	11.1	–	2.23 days	2 years	4 years
HBB	C_6Br_6	552	6.11	1.6×10^{-2}	15.6 months	2 years	4 years
PBEB	$C_8H_5Br_5$	501	6.76	21×10^{-2}	4.65 days	8 months	16 months
BTBPE	$C_{14}H_8Br_6O_2$	688	8.9	2.23×10^{-4}			
Dechloranes	(see Fig. 4)	614-693	8.05-11.5	1.7×10^{-8} – 1.8×10^{-5}			8-11 years
DCP	$C_{19}H_{17}O_4P$	340	4.51	0.24			
EHDPP	$C_{20}H_{27}O_4P$	362	4.21	1.9			
IDPP	$C_{22}H_{31}O_4P$	390	5.44	0.75			
IPPP	$C_{27}H_{33}O_4P$	453	5.31	2.2			
TBOEP	$C_{18}H_{39}O_7P$	398	3.75				
TBP	$C_{12}H_{27}O_4P$	266	4.00	280			
TCEP	$C_6H_{12}Cl_3O_4P$	285	1.47	7×10^3	17.5 hours		
TCIPP	$C_9H_{18}Cl_3PO_4$	328	2.59	1.6×10^3	8.6 hours		
TCP	$C_{21}H_{21}O_4P$	368	5.48				
TDCPP	$C_9H_{15}Cl_6O_4P$	431	3.28	1.5	21.3 hours		
TEHP	$C_{24}H_{51}O_4P$	435	9.49	0.6			
THP	$C_{18}H_{39}O_4P$	350	7.45				
TPHP	$C_{18}H_{15}O_4P$	326	4.59	1.9			
TPPO	$C_{18}H_{15}OP$	278	2.83				

Bioaccummulation and Biomagnification

PBDEs were first detected in environmental samples in 1970 [50]. However, it was not until a couple of decades later that the scientific community became concerned when increasing concentrations of PBDEs were detected in human breast milk [51].

There are two key terms regarding concentrations in high trophic levels of the food web, bioaccumulation and biomagnification [52]. Bioaccumulation is the incorporation of contaminants in an organism through ingestion, skin permeation or inhalation —or through the gills in aquatic organisms. Biomagnification is the increase of concentration as contaminants are transferred up the food web. Bioaccumulation and biomagnification are two important factors in marine food webs, especially since most of the PBDE intake of humans comes from seafood, as mentioned before.

Lower brominated PBDEs have higher bioaccumulation factor (BAF) and biomagnification factor (BMF) than higher brominated congeners [53]. BDE-47, BDE-99 and BDE-100 represent an 80% of the PBDE profiles of biota samples [54]. The European Food Safety Authority (EFSA) published BAF values for HBB in aquatic species comparable to those published for PBDEs and other models predict similar BAFs for other EFRs [49].

Another important term is the biota-sediment accumulation factor (BSAF). The BSAF is a relation between the contamination in an organism in lipid weight (lw) and the contamination in the corresponding sediment normalized to organic carbon. For hydrophobic compounds as FRs, the BSAF is relevant as sediments can act as a source of contamination by resuspending the compounds attached to them.

Table **2** shows BMF values for some FRs in a food web of Lake Winnipeg (Canada) and BSAF values from several studies. The organisms included in those studies were fish, bivalves, plankton and worms, some of them from fish farms. The table shows that the BMF for the emerging DBDPE is in the same range of BMFs for PBDEs and HBCD. On the other hand, if BSAFs for emerging HBB and DP are somewhat lower than for PBDEs, they are higher than for HBCD, which is also a restricted POP. This highlights the importance of monitoring EFRs in food as well as legacy POPs.

Toxicity

PBDEs affect hormonal regulation as well as neuronal, thyroid and liver activity [55 - 57].

Table 2. BMF and BSAF values for some FRs in aquatic environments.

Compound	BMF [58]	BSAF	BSAF Reference
BDE-28		0.05	[59]
		5.89	[60]
BDE-47	0.1-8.9	3.4	[59]
		5	[61]
		10.0	[62]
		31.8	[62]
BDE-99	0.1-9.5		

(Table 2) cont.....

Compound	BMF [58]	BSAF	BSAF Reference
BDE-100	0.1-4.9	3.3	[59]
		4.74	[60]
BDE-153	0.1-5.5		
BDE-209	0.1-34		
α-HBCD	0.1-8.2	0.10	[62]
		0.15	[62]
β-HBCD	0.3-5	0.01	[62]
		0.01	[62]
γ-HBCD	0.1-6.3	0.01	[62]
		0.02	[62]
DBDPE	0.2-9.2		
HBB		0.4	[59]
		2.3-6.1	[63]
anti-DP		0.3	[42]
syn-DP		0.2	[59]
		0.8	[42]

PBDEs are endocrine disruptors as they affect hormonal regulation [55]. It has been proved *in vitro* that PBDEs affect thyroid homeostasis. In female rats, oral administration of PentaBDE and OctaBDE mixtures for four days produce a decrease of thyroxine in serum [64]. Exposure to DecaBDE and other mixtures did not show those effects. A study on seals of England found a directly proportional correlation between total concentration of PBDEs in fat and the amount of thyroid hormones in blood, especially during the first year of life [65].

Some PBDE congeners act as estrogen receptor agonists, especially BDE-100, BDE-75 and BDE-51 [66, 67]. PBDEs have also been found to be cytotoxic [67]. Additionally, the PentaBDE mixture showed antiandrogenic effects in rats *in vivo* and *in vitro* [68].

Neonatal exposure of mice to PBDEs has a negative effect on their spontaneity, learning and memory [69 - 72]. These neurotoxic effects became apparent and increased with the age of the mice.

The no observed adverse effect level (NOAEL) in rats for DecaBDE is > 1.000 mg kg^{-1} (mg of DecaBDE per kg of body weight), while OctaBDE shows effects at 100 mg kg^{-1} in a 90-day study and PentaBDE has a no observed effect concentration (NOEC) of 1 mg kg^{-1} in a 30-day study [73].

HBCD affects the liver, thyroidal homeostasis and the reproductive, nerve and immune systems [74]. On the other hand, it does not seem to be genotoxic.

Some experimental models attribute carcinogen and genotoxic properties to some EFRs [30, 49]. Although more data is required, here are some recent studies. HBB caused oxidative stress in goldfish liver, inhibited sodium and potassium channels and could inhibit enzymatic processes in rat liver [75, 76]. DBDPE showed similar toxicological potential [77, 78]. Dechloranes could also affect rat and fish livers and interfere in protein expression and calcium channels [79, 80]. Dechlorane studies tend to focus on DP and suggest toxicity similar to PBDEs and HBCD.

Exposure to OPFRs has been associated with adverse reproductive, neurologic, and other systemic effects [81]. Besides neurotoxicity, some OPFRs are also known for being endocrine disruptors [81]. Moreover, several works have been carried out to study the carcinogenicity of OPFRs. TCEP and TDCPP have been found to be carcinogenic, affecting the liver, kidney and testes tissue [82].

Legislation

PentaBDE and OctaBDE were removed from the markets of Europe and the United States in 2004 and in 2006 the regulation for Registration, Evaluation, Authorization and Restriction of Chemicals (REACH regulation) was adopted by the EU [83, 84]. The REACH regulation established at 0.1% in mass the maximum amount of PentaBDE or OctaBDE that a material could have in order to be possible to commercialise it. In 2010, DecaBDE, HBCD and TBBPA were included in the REACH regulation. DecaBDE and HBCD are considered substances of very high concern since 2010 and 2008, respectively. HBCD and DecaBDE have been restricted by this regulation in recent years. The Stockholm Convention on POPs banned the production of most PBDE congeners in 2009 and of DecaBDE in 2017 and limited production of HBCD in 2013 [85]. It also bans recycling materials containing any of these compounds from 2030.

Regarding EFRs, DBDPE is registered in the REACH regulation, HBB and PBEB are preregistered and the three are included in monitoring programs [18].

Other legislation sets criteria for the elimination of FRs. With the increasing concern about the intake of FRs and its effect in humans, it would be convenient that reference values were defined eventually. Some directives regarding BFRs are Directive 2000/60/EC (protection of aquatic environment, including periodic controls of biota), Directive 2008/56/EC (protection of marine environment, including a remark on seafood containing safe levels of contaminants), Directive 2011/65/EU and Directive 2012/19/EU (both about electronic equipment).

Dechloranes have not been regulated so far; however, they —and DBDPE— are included in the Chemical Management Plan.

Finally, the European Commission has adopted specific limit values for OPFRs in toys, reaching at 5 mg kg^{-1} [86]. Recently, various environmental organizations have submitted a petition to the United States EPA requiring further tests to determine if TCEP, TClPP and TDCPP present a concerning health or environmental risk due to possible toxicity and persistence [87].

ANALYTICAL METHODOLOGIES

This section lists methods of analysis of FRs that have been or can be applied to analysis of food samples. The concept *food samples* include any kind of edible item; that is meat, liver, seafood (fish, mollusks, crustaceans, *etc.*), eggs, dairy products, vegetables, *etc.*

Table **3** shows a simplified description of methods used in articles published between 2012 and 2018. Most of these articles do indeed describe studies on food samples. A reduced few applied the methods to environmental samples; muscle, liver or eggs of species that are not consumed by humans. However, regardless human eating habits, these are the same biological matrices analyzed with the other methods and are thus included in this text. The first column of Table **3** shows a practical code that will be used in this section to refer to each method and the reference to the literature. The referenced articles are not the original method development, but studies that apply the method to real samples. The reason for that is that some methods suffer changes or adaptations with time and that this chapter aims at showing the methods that are currently used in real studies. In the compounds column of Table **3**, PBDEs are shown separated from other BFRs; therefore, in the table —but not in the text that follows— BFRs stands for 'brominated flame retardants other than PBDEs'. In the table, the different steps of the analytical method are summarized including techniques and sorbents or solvents when relevant. When different aliquots of the extracts undergo different clean-ups, this is shown in separate lines in the clean-up column. When within the clean-up the purified extracts divide into different fractions, these fractions are identified with a capital F followed by a number. These complex clean-ups and instrumental analysis with several runs of the same final extract have numbers between brackets to better distinguish the different steps. Further details are given in the specific sections below.

Main Compounds

All the methods for HFRs include PBDEs, except for one that was a study on EFRs and HBCD. The number of PBDE congeners studied varied from 7 to 39.

The eight included in almost all methods were BDE-28, -47, -99, -100, -153, -154, -183 and -209. Regarding other BFRs, HBB, BTBPE, PBEB and DBDPE are the most studied —in this order— and included in almost all methods listed. Tetrabromobenzoates, tetrabromophthalates and polybromobenzenes are examples of other BFRs analyzed by some methods. As for chlorinated FRs, *syn-* and *anti*-DP are included in all studies except for one that focuses on BFRs. Dechloranes 601 through 604 and DP analogues are monitored in only a few studies.

All compounds mentioned in the previous paragraph are extracted together and into the same final chromatography vial. On the other hand, HBCD and TBBPA, which are analyzed together, commonly undergo different extractions or a fraction of the extract requires a different clean-up for the analysis of this pair of compounds. OPFR analysis also differs in some steps in most cases.

There are 23 different OPFRs included in all these methods. Method S-13 targets only four, which are included in all the others as well. These compounds are TCEP, TCIPP, TDCPP and TPHP. Four other OPFRs are targeted in all methods but S-13; these are EHDPP, TBOEP, TBP TEHP.

Extraction

While most methods use freeze-dried samples, a few do not. When the samples are not freeze-dried, anhydrous sodium sulfate is added to the homogenized sample in order to absorb humidity.

For **BFRs** (excluding HBCD and TBBPA) and **chlorinated FRs**, the most common extraction technique is **pressurized liquid extraction** (PLE). The extracting solvent can either be dichloromethane (DCM) alone (M-18, S-16, W-16) or DCM:hexane (1:1) (methods A-18, P-17 and V-18). A-18 also obtains HBCD from this PLE. All matrices are able to be extracted using PLE, and either freeze-dried samples (A-18 1.5 g, M-18 5-10 g, W-16 1-3 g) or fresh samples (P-17 2.5 g, S-16 1 g, V-18 1 g) are used.

Table 3. Methods for analyzing flame retardants in food matrices.

Code*	Samples	Compounds	Extraction	Lipid Removal	Clean-up	Instrumental Analysis
P-16 [88]	Fish, chicken meat, eggs, milk, olive oil	10 PBDEs, 4 BFRs, DP	UAE. ACN:toluene (9:1). Frit filtered		SPE 1: Florisil. Hexane: DCM (4:1), hexane: DCM (1:1). SPE 2: acidified silica. Hexane: DCM (1:1)	GC-ECNI-MS(SIM)

(Table 3) cont.....

Code*	Samples	Compounds	Extraction	Lipid Removal	Clean-up	Instrumental Analysis
P-18 [89]	Composite food samples: seafood, meat, dairy, fat, grains, eggs, potatoes, vegetables	10 PBDEs, 5 BFRs, DP	UAE. ACN:toluene (9:1). Frit filtered		SPE 1: Florisil. Hexane:DCM (4:1), hexane:DCM (1:1). SPE 2: acidified silica. Hexane:DCM (1:1)	GC-ECNI-MS(SIM)
		4 BFRs, TBBPA, HBCD	UAE. DCM:hexane (1:1). Supernatant		SPE acid silica. DCM	UPLC-ESI(−)-MS/MS. Mobile phases: H_2O:ACN (95:5) 0.1% CH_3COOH and ACN:H_2O (95:5) 0.1% CH_3COOH. MRM
X-15 [90]	Meat, seafood, eggs	9 PBDEs, 4 BFRs, DP	Vacuum UAE. ACN:toluene (9:1). Frit filtered with vacuum		(1) SPE: Florisil. Hexane (F1), ACN (F2). (2) F1, SPE: acidified silica. Hexane:DCM (1:1). (3) F2, dSPE: DSC18/Z-SEP. (4) F2 supernatant + F1 = F3. SPE: aminopropyl silica. Hexane (F4), Hexane:DCM 1:1 (F5)	F4, GC-ECNI-MS(SIM)
		9 OPFRs (with isomers)				F5, GC-EI-MS(SIM)
X-17 [91]	Blended mix of all foods and drinks consumed over 24 h	9 PBDEs, 4 BFRs, DP	UAE. ACN:toluene (9:1). Syringe filtration		(1) SPE: Florisil. Hexane (F1), Acetone, ACN (F2). (2) F2, d-SPE: aminopropyl silica. Hexane:DCM (9:1) (F3), DCM (F4). (3) F1 + F3 = F5, SPE: acid silica. (4) F4, dSPE: Z-SEP	F5, GC-ECNI-MS(SIM)
		8 OPFRs (with isomers)				F4, GC-EI-MS(SIM)
T-16 [92]	Chicken (muscle, liver, skin), eggs, pork, fish	5 BFRs, DP	Rapid solvent extractor. Acetone:hexane (1:1), toluene.	GPC: SX-3 Bio-Beads	SPE 1: deactivated silica. 50% diethyl ether in hexane. SPE 2: Florisil. Hexane, DCM.	GC-EI-MS(SIM)

(Table 3) cont.....

Code*	Samples	Compounds	Extraction	Lipid Removal	Clean-up	Instrumental Analysis
		HBCD			Partitioned: concentrated H_2SO_4	LC-ESI(−)-MS/MS(MRM)
A-18 [93]	Muscle, liver, seafood, eggs	8 PBDEs, 3 BFRs, 4 Decs, HBCD	PLE. DCM:hexane (1:1)	Concentrated H_2SO_4	SPE: neuter alumina. Hexane:DCM (1:2)	(1) GC-EI-MS/MS(MRM) (2) GC-ECNI-MS/MS(MRM) (3) GC-ECNI-MS(SIM) (4) LC-MS/MS(MRM)
		14 OPFRs	UAE. Acetone:hexane (1:1)			TFC-LC-MS/MS(MRM)
M-18 [94]	Muscle, liver, mushrooms, vegetation	28 PBDEs, 8 BFRs, DP	PLE. DCM.	GPC. DCM:hexane (1:1)	SPE 1: silica. DCM:MeOH (9:1). SPE 2: silica. Hexane, DCM:hexane (1:1), 5% MeOH in DCM	GC-ECNI-LRMS(SIM) (1) Rtx®-1614 fused silica column (2) Rxi®-5sil MS fused silica column
S-15 [95]	Seafood muscle	15 PBDEs, 3 BFRs, DP	Soxhlet. DCM:hexane (1:1), 48 h	Concentrated H_2SO_4	SPE: Florisil, neutral silica, acid silica, anhydrous Na_2SO_4. Hexane, DCM	GC-ECNI-MS(SIM) (1) DB-XLB capillary column (2) DB-5HT capillary column
P-17 [96]	Fish, shrimp, whole amphipods	19 PBDEs, 2 BFRs, DP	PLE. DCM:hexane (1:1)	GPC	SPE: Florisil. Hexane, hexane:DCM (85:15)	HRGC-ECNI-MS(SIM)
Z-15 [97]	Fish	7 PBDEs, 8 BFRs, DP	UAE. DCM:hexane (1:1)	GPC. DCM:hexane (1:1)	SPE: Florisil. Hexane:DCM (1:1)	GC-EI-MS/MS(MRM)
		TBBPA, HBCD			PTFE syringe filter	LC-ESI-MS/MS(MRM). Mobile phases: MilliQ water and ACN:MeOH (30:70)
S-13 [98]	Fish	7 PBDEs, 6 BFRs, DP, 4 OPFRs	QuEChERS: anhydrous $MgSO_4$ and NaCl		dSPE: Z-Sep with anhydrous $MgSO_4$	LP-GC-EI-MS/MS(MRM)
W-16 [99]	Fish	20 PBDEs, 9 BFRs, 10 Decs	PLE. DCM	GPC	SPE: silica gel. Hexane, hexane:DCM (6:4), DCM	GC-ECNI-MS(SIM)
S-16 [59]	Fish	8 PBDEs, 24 BFRs, 9 Decs, DBCD, DBALD, HCCPD, HCPN	PLE. DCM	GPC: Bio-Beads S-X3. DCM:hexane (1:1)	SPE: deactivated silica gel. Hexane	(1) GC-ECNI-MS(SIM) (2) GC-EI-MS/MS(MRM)

(Table 3) cont.....

<table>
<tr><th>Code*</th><th>Samples</th><th>Compounds</th><th>Extraction</th><th>Lipid Removal</th><th>Clean-up</th><th>Instrumental Analysis</th></tr>
<tr><td rowspan="2">V-18 [100]</td><td rowspan="2">Liver</td><td>38 PBDEs, 4 BFRs, DP</td><td>PLE. DCM:hexane (1:1)</td><td></td><td>SPE 1: acid-basic-neutral silica. SPE 2: neutral alumina. DCM:hexane (1:1)</td><td>GC-ECNI-MS(SIM)</td></tr>
<tr><td>17 OPFRs</td><td>PLE. DCM:hexane (1:1)</td><td></td><td>SPE: NH2 aminopropyl silica gel. DCM:hexane (2:8), DCM, DCM:MeOH (9:1)</td><td>HPLC-ESI(+)-MS/MS(MRM)</td></tr>
<tr><td rowspan="2">Z-16 [101]</td><td rowspan="2">Eggs</td><td>17 PBDEs, 4 BFRs</td><td rowspan="2">Soxhlet. Acetone:hexane (1:1), 24 h</td><td></td><td>SPE: Florisil, neutral silica, sulfuric acid silica, anhydrous Na_2SO_4. Hexane</td><td>GC-ECNI-MS(SIM) (1) DB-XLB capillary column (2) DB-5HT capillary column</td></tr>
<tr><td>TBBPA, HBCD</td><td>Concentrated H_2SO_4</td><td>SPE: neutral silica, anhydrous Na_2SO_4. Hexane, DCM:hexane (1:1)</td><td>LC-ES-MS/MS(SIM) Mobile phases: MeOH:water (90:10) and ACN</td></tr>
<tr><td rowspan="2">H-18 [102]</td><td rowspan="2">Eggs</td><td>18 PBDEs, 3 BFRs, DP</td><td rowspan="2">Soxhlet. Acetone:hexane (1:1)</td><td rowspan="2">Concentrated H_2SO_4</td><td>SPE: multilayer Florisil/silica gel</td><td>GC-ECNI-MS(SIM) (1) DB-XLB capillary column, (2) DB-5HT capillary column</td></tr>
<tr><td>HBCD</td><td>SPE: silica gel. Hexane, DCM:hexane (1:1)</td><td>LC-APCI-MS/MS(MRM) Mobile phases: MeOH:water (90:10) and ACN</td></tr>
<tr><td>V-17 [103]</td><td>Eggs</td><td>39 PBDEs, 7 BFRs, TBBPA, HBCD, DP</td><td>Hot extraction. Cyclohexane:toluene (1:1)</td><td></td><td>SPE 1: deactivated silica gel. Hexane. SPE 2: activated silica gel. Hexane, hexane:EtAc (9:1)</td><td>GC/ECNI-MS(SIM)</td></tr>
</table>

*The code for the methods consists of the initial of the first author and the year of publishing.

Three other methods are based on **ultrasound assisted extraction** (UAE). P-16 and P-18, by the same authors, use 5 mL of acetonitrile:toluene (9:1) to extract BFRs and DP from 0.2 g dw of fatty matrices or 2.5 g dw of non-fatty matrices. Method X-17 is a modification of X-15 to extract BFRs, DP and OPFRs together from 2 and 2.5 g dw, respectively, of all food matrices. Both methods also use 5 mL acetonitrile:toluene (9:1). X-15 pumped the extract through an empty solid phase extraction (SPE) cartridge with a frit and topped-up with 1 g of magnesium sulfate on a vacuum manifold. X-17 replaces this step with syringe filtration to make the process significantly simpler and faster. Method Z-15 extracts 5-10 g ww of fish using 50 mL of DCM:hexane (1:1) three times. Z-15 obtains HBCD and TBBPA in the same extract.

There are some other **fast extraction techniques** used in the literature. S-13 extracts BFRs, DP and OPFRs from 10 g ww of fish with unbuffered QuEChERS. This extraction uses 10 mL of acetonitrile shaking for 30 s, then adds anhydrous magnesium sulfate and sodium chloride shaking for 1 min. T-16 uses a rapid solvent extractor to extract BFRs other than PBDEs, DP and HBCD from 5 g dw of all food matrices. This extraction runs 2 mL min^{-1} of acetone:hexane (1:1) at 35 ºC for 40 min and 2 mL min^{-1} of toluene at 80 ºC for 40 min. V-17 extracts BFRs, including HBCD and TBBPA, and DP from whole freeze-dried eggs using cyclohexane:toluene (1:1) and an aliquot representing about 80 mg of extracted egg is redissolved to 20 mL.

Soxhlet extractions are more time-consuming. S-15 extracts BFRs and DP from 3 g dw of seafood using 200 mL of DCM:hexane (1:1) for 48 h. Z-16 and H-18 perform a Soxhlet extraction on eggs using 200 mL of acetone:hexane (1:1) for 24 h and 48 h, respectively. In the Soxhlet extraction, Z-16 obtains BFRs, HBCD and TBBPA while H-18 obtains BFRs, DP and HBCD.

There is one extraction that extracts exclusively **HBCD, TBBPA** and four bromophenols with one to three bromine atoms. P-18 uses DCM:hexane (1:1) and collects the supernatant after centrifugation. This is a different extraction from the one P-18 uses for BFRs and DP. In contrast, it has been explained how other methods extract HBCD, TBBPA or both along with the other BFRs: A-18 by PLE, Z-15 by UAE, Z-16 and H-18 by Soxhlet extraction and T-16 and V-17 by other techniques.

There are two methods that extract **OPFRs** separately. A-18 extracts 14 OPFRs from 0.5 g dw of all biota matrices by UAE using acetone:hexane (1:1). The extract is redissolved in 5 mL of hexane:methanol (1:3) and, unlike any other method listed, an aliquot of 200 µl is used for the instrumental analysis. V-18 extracts 17 OPFRs from 0.6 g ww liver by PLE using DCM:hexane (1:1). As seen above, OPFRs are also extracted with the rest of the compounds in some methods: X-15 and X-17 extract 9 and 6 OPFRs by UAE and S-13 extracts 4 OPFRs using QuEChERS. While a separate extraction of OPFRs require extra work, it seems to allow for the analysis of more congeners.

Lipid Removal

Almost half of the methods for analysis of FRs include a step identified as lipid removal. The methods that skip this step tend to perform a two-step clean-up;

usually two SPEs, one of them including acid silica. In ten cases there is a lipid removal step; four consist in using concentrated sulfuric acid and six perform gel permeation chromatography (GPC) with DCM:hexane (1:1).

There is no correlation between the type of sample and the performance of a specific step for removing lipids as all kind of foodstuff is analyzed with or without that step in the different methods listed. However, as already mentioned, the methods that skip this step tend to compensate with a two-step clean-up.

None of the methods for OPFRs includes this step.

Clean-up

For the analysis of **HFRs**, clean-up usually includes an **SPE** for methods that include lipid removal and two for methods that do not. Florisil and silica are the most commonly used sorbents for SPE. Out of 22 clean-up routes for analysis of HFRs listed here, 10 include Florisil and 16 include silica. Eluting solvents are hexane and DCM —with a few some exceptions. There is a growing proportion of DCM through elutions when more than one are performed, DCM:hexane (1:1) being the most recurring elution mixture.

Both methods P-17 and Z-15 perform one only SPE with a **Florisil** cartridge to extract HFRs from seafood, not including HBCD or TBBPA. P-17 and Z-15 extract the compounds with DCM:hexane (1:1) followed by GPC to remove lipids. While Z-15 elutes HFRs from the Florisil cartridges with 160 mL of the same solvent mixture, P-17 first elutes *ortho*-PCBs with 35 mL of hexane and then uses 38 mL of DCM:hexane (15:85) to obtain the fraction that contains HFRs.

Other methods that are applied to all kinds of foodstuff use a combination of **Florisil and silica**. After an extraction with acetonitrile:toluene (9:1), P-16 (and P-18, as they are the same method) performs a first SPE with 3 mL Florisil cartridges eluting with 6 mL of DCM:hexane (1:4) followed by 6 mL of DCM:hexane (1:1). A second SPE uses 6 mL cartridges containing 1.5 g of acid silica eluting with 12 mL of DCM:hexane (1:1). T-16 inverts the order using first a 2 g deactivated silica gel column eluting with 10 mL of diethyl ether:hexane (1:1) followed by a 8 g Florisil column eluting with 50 mL of hexane and 50 mL of DCM.

X-15 and X-17 are two variations of a more complex method that extracts HFRs and OPFRs together from any foodstuff and separates different fractions through clean-up. Both X-15 and X-17 start with a Florisil cartridge and use 8 or 10 mL of hexane, respectively, to elute fraction 1. Fraction 2 is eluted with 5 mL of acetonitrile in X-15 and with 5 mL of acetone and 2 mL of acetonitrile in X-17. X-15 has an extra clean-up step for fraction 1 consisting of elution through an acid silica cartridge with 12 mL of DCM:hexane (1:1). X-15 performs a dispersive SPE (dSPE) with 200 mg of DSC18/Z-SEP mixture with fraction 2 and combines the supernatant of this dSPE with fraction 1. The combined purified extract is divided in an aminopropyl silica cartridge eluting with 10 mL of hexane for the HFRs fraction and then 12 mL of DCM:hexane (1:1) for the OPFRs fraction. X-17 divides fraction 2 through dSPE with aminopropyl silica eluting with 10 mL of DCM:hexane (1:9) for HFRs and 12 mL of DCM for OPFRs. X-17 combines this HFRs fraction with fraction 1 and performs an SPE with acid silica. The authors argue that this change from X-15 better removes interferences and reduces degradation of some tetrabromobenzoate and tetrabromophtalate. The OPFRs fraction undergoes a dSPE with Z-SEP; as opposed to DSC18/Z-SEP, for increased sensitivity for TBOEP and TCEP. Method S-13 also extracts HFRs and OPFRs from fish using QuEChERS and dSPE with 50 mg of Z-SEP and 150 mg of anhydrous magnesium sulfate. While this method is much simpler, four OPFRs are analyzed in contrast with the nine and six OPFRs and some of their isomers targeted by X-15 and X-17, respectively.

S-15, Z-16 and H-18 use multilayer Florisil/silica gel after Soxhlet extraction of seafood or eggs and lipid removal with sulfuric acid to elute HFRs, not including HBCD and TBBPA. The 30 cm gel column (inner diameter = 1.0 cm) consists of 14 g of Florisil, 2 g of neutral silica and 7 g of acid silica, topped with a 2-g layer of anhydrous sodium sulfate. Hexane is the eluting solvent for methods Z-16 (80 mL) and H-18 (55 mL), while S-15 uses 80 mL of hexane followed by 60 mL of DCM.

Other methods rely only on **silica** for the SPE in the analysis of HFRs. After PLE and GPC lipid removal, W-16 purifies the fish extract with a 2-g silica SPE cartridge. The first 3 mL eluted with hexane are discarded. The eluate produced with 6.5 mL of DCM:hexane (4:6) and 7 mL of DCM is kept for analysis. S-16 uses 2.5 g of deactivated silica gel and elutes the purified fish extract with 20 mL of hexane. M-18 extracts different food matrices with PLE and removes lipids with GPC before performing two SPEs. The extracts are first eluted on 0.75 g of silica topped with sodium sulfate using 10 mL of DCM:methanol (9:1) and then eluted on 20 cm (wet-packed height) of silica with 70 mL of hexane, 80 mL of DCM:hexane (1:1) and 100 mL of 5% methanol in DCM. Finally, V-17 extracts HBCD and TBBPA along with the other HFRs from egg samples. V-17 first elutes on 3 g of deactivated silica gel with 60 mL of hexane and then elutes on 8 g of activated silica gel to distribute polyhalogenated compounds into fractions. Fraction 1 is eluted with 48 mL of hexane (for PCB analysis), fraction 2 contains the HFRs and is eluted with 50 mL of ethyl acetate:hexane (1:9).

Two methods use **alumina** as sorbent for SPE. After PLE and removal of lipids with sulfuric acid, method A-18 purifies the extracts of several food matrices with neutral alumina cartridges eluting with 20 mL of DCM:hexane (2:1). These extracts are also used to analyze HBCD. V-18 extracts liver with PLE using DCM:hexane (1:1) and uses the same solvent mixture to elute from an acid-basi--neutral silica column first and a neutral alumina column.

It can be noticed that A-18 obtained HBCD in the purified extract of HFRs on neuter alumina and V-17 obtained **HBCD and TBBPA** in the same extract purified on deactivated silica and activated silica. Other methods have separate clean-ups from an aliquot of the extract and some are a completely different method of extraction and clean-up.

T-16 uses an **aliquot of the extract** (50%) after removing lipids with GPC. Instead of undergoing an SPE, the aliquot is re-dissolved in 2 mL of hexane and then simply partitioned against 3-4 mL of concentrated sulfuric acid for the analysis of HBCD. Z-15 takes a quarter of the fish extract after GPC, re-dissolves it in methanol and filters it using a 0.2-µm polytetrafluoroethylene (PTFE) syringe filter for the analysis of HBCD and TBBPA. Z-16 takes half of the Soxhlet egg extract, removes its lipids with sulfuric acid and performs an SPE for the analysis of HBCD and TBBPA. The extract is eluted through a silica gel column (inner diameter = 1.0 cm), containing 12 cm of neutral silica topped with a 2-cm layer of anhydrous sodium sulfate, using 13 mL of hexane, 5 mL of DCM:hexane (1:1) and 45 mL of DCM:hexane (1:1). The first and second fractions are discarded. H-18 is a variation of Z-16. Differences are that the aliquot is taken after removing the lipids with sulfuric acid and only HBCD is analyzed.

Method P-18 performs a **separate extraction** of HBCD, TBBPA and four bromophenols. After extraction with DCM:hexane (1:1), the extract undergoes an SPE with 8 g of acid silica eluting with 25 mL of DCM.

It has been explained how method S-13 obtains **OPFRs** in the same final purified extract as HFRs and that X-15 and X-17 divide the food extracts in different fractions during the clean-up process to obtain a separate final extract with OPFRs. Method V-18 uses the same PLE extraction for HFRs and OPFRs, but starting with different amounts of sample. The clean-up also differs. The OPFRs extract undergoes an SPE with a 1-g aminopropyl silica gel cartridge, eluting with 3 mL of DCM:hexane (2:8), 8 mL of DCM, and 4 mL of DCM:methanol (9:1).

The remaining method for OPFRs, A-18, performs a separate UAE and clean-up is carried out on-line in the instrumental analysis with **turbulent flow chromatography** (TFC) [36]. TFC is used for on-line clean-up of biological samples in liquid chromatography coupled to mass spectrometry (LC-MS). It was developed in the late 1990s combining size-exclusion chromatography and traditional stationary phase column chemistry to separate macromolecules —such as lipids— from smaller molecules and target analytes. TFC allows direct injection of extracts onto a column packed with large particles avoiding time-consuming clean-up steps. A high flow rate (1.5-5.0 mL min^{-1}) generates a turbulent flow inside the TFC column and the small molecules are retained in the particle pores via diffusion, whereas macromolecules are washed away. Then the target compounds are eluted from the TFC column into the analytical column. This technique had previously been applied to the analysis of pesticides, perfluoroalkyls or drugs [104, 105]. Method A-18 combined two TFC columns: CycloneTM-P (50 × 0.5 mm) and C18-XL (50 × 0.5 mm) from ThermoScientific. As OPFRs are compounds with a wide range of relative molecular mass and K_{ow} values, some of the larger OPFRs require a second column in order to be retained. Mobile phases are (A) water (0.1% formic acid) and (B) methanol (0.1% formic acid) at 0.75 mL min^{-1}. Gradient is 98% of A to 100% of B during the first two min and it is held for the remaining 34 min. Transfer time for the analytes to be eluted from the TFC column onto the analytical column is 120 s.

Instrumental Analysis

Table **4** adds detail to the instrumental analysis of the methods listed in Table **3**. In Table **4**, HBCD and TBBPA are specified in the *compounds* column and not included under HFRs. The reason for this is that HBCD and TBBPA commonly undergo an instrumental analysis different from the rest of the HFRs; that is, HBCD and TBBPA are usually analyzed by liquid chromatography (LC) instead of gas chromatography (GC). Table **4** is structured presenting first the analysis of HFRs, followed by the analysis of HBCD and TBBPA and, finally, the analysis of OPFRs. In two cases, HBCD and TBBPA or OPFRs are analyzed in the same GC run as HFRs and said compounds are marked in bold to make their identification out of the logic structure easier. Within the HFRs group, methods with the same ionization technique are presented following each other. Boxes have been merged when different methods share chromatographic columns or mobile phases. Methods that, additionally to that, use the same instrument are presented in the same line. When the instrumental analysis has been developed and published before being applied in the studies listed in Table **3**, references to the original method publications are provided in the *techniques and instruments* column, even if minor changes have been made to the original methods. The absence of a reference means that the instrumental method was developed in the same study or that the original instrumental method is not clearly referenced in the publication.

According to the listed methods, the analysis of **HFRs**, excluding HBCD and TBBPA, is always performed by GC. HFRs are volatile compounds that can resist the high temperatures in the injector and remain unaltered for the analysis, which makes them suitable for GC. Negative ion chemical ionization (NICI), and more precisely, **electron capture negative ionization** (ECNI) is the most common form of ionization. A few methods use **electron ionization** (EI). Methods using EI use multiple reaction monitoring (MRM), while methods using ECNI use single ion monitoring (SIM). While ECNI provides better sensitivity than EI for halogenated compounds, EI is more selective [106]. When using MRM, transitions for PBDEs obtained with ECNI may not prove acceptable regarding sensitivity. EI also allows quantification by isotope dilution, relating the native compounds to labelled congeners, which is more reliable than quantification with non-labelled standards. In SIM methods, commonly ions with a mass-to-charge ratio (*m*/*z*) of 79 and 81, corresponding to bromine, are monitored for PBDEs.

Over half of the methods use DB-5 **columns** for the GC analysis of HFRs. Most methods targeting ten or fewer PBDE congeners and five to eleven other HFRs use a 15 m column (either DB-5 or any other). Methods including 15 or more PBDE congeners use a 30 m column and may add a different run with a 15 m column. Methods V-17 (39 PBDEs and 10 HFRs, including HBCD and TBBPA) and S-16 (8 PBDEs and 37 HFRs) use a 30 m HP-5 column. S-15, Z-16 and H-18 (15-18 PBDEs, 4 HFRs) perform a run on a 30 m DB-XLB column for di- to hepta-BDEs and some other HFRs and another run on a 15 m DB-5HT column for octa- to deca-BDE and some other HFRs. It is known that the instrumental analysis of PBDEs with a higher level of bromination performs worse in longer runs [38, 107]. Hence the separation of congeners with two to seven bromine atoms in the longer column and congeners with eight to ten bromine atoms in the shorter one. S-15, Z-16 and H-18 also agree in analyzing HBB and PBEB in the shorter column and BTBPE and DBDPE in the longer column. However, DP is analyzed in either or depending on the method.

Table 4. Instrumental analysis of flame retardants in food matrices.

Compounds	Code	Techniques and Instruments	Columns and Mobile Phases	
HFRs	P-16, P-18	GC-ECNI-MS(SIM)	DB-5 capillary column (15 m × 0.25 mm, 0.10 μm)	
HFRs	X-15, X-17	GC-ECNI-MS(SIM) [108]	DB-5 column (15 m × 0.25 mm, 0.10 μm), with a deactivated silica retention gap (1.0 m × 0.25 mm)	
HFRs	P-17	HRGC-ECNI-MS(SIM) [109]	DB-5 capillary column (10 m × 0.25 mm, 0.25 μm)	
HFRs	W-16	GC-ECNI-MS(SIM)	DB-5HT column (15 m × 0.25 mm, 0.1 μm)	

(Table 4) cont.....

Compounds	Code	Techniques and Instruments	Columns and Mobile Phases	
HFRs	V-18	GC-ECNI-MS(SIM) [110]	DB-5HT column (15 m × 0.25 mm, 0.1 μm)	
HFRs	S-15, Z-16, H-18	GC-ECNI-MS(SIM) [101, 111]	DB-XLB capillary column (30 m × 250 μm, 0.25 μm). Di- to hepta-BDEs, HBB, PBT, PBEB, DP	DB-5HT capillary column (15 m × 250 μm, 0.10 μm). Octa- to deca-BDEs, BTBPE, DBDPE, DP
HFRs	M-18	GC-ECNI-LRMS(SIM)	Rtx®-1614 fused silica column (15 m × 250 μm, 0.10 μm). PBDEs	Rxi®-5sil MS fused silica column (30 m × 250 μm, 0.25 μm). HFRs
HFRs, **HBCD, TBBPA**	V-17	GC-ECNI-MS(SIM) [112]	HP-5 ultra-inert column (30 m × 0.25 mm, 0.25 μm)	
HFRs	S-16	GC-ECNI-MS(SIM)	HP-5MS column (30 m × 0.25 mm, 0.25 μm), restriction capillary (0.8 m × 0.1 mm, deactivated) with a backflush system	
		GC-EI-QqQ(MRM) [113]	Restek 1614 column (15 m × 0.25 mm, 0.10 μm), restriction capillary (0.8 m × 0.1 mm, deactivated) with a backflush system	
Decs	A-18	GC-ECNI-QqQ(MRM) [114]	DB-5ms column (15 m × 0.25 mm, 0.1 μm)	
BDE-209, DBDPE	A-18	GC-ECNI-MS(SIM) [115]		
BFRs	A-18	GC-EI-QqQ(MRM) [106]		
HFRs	Z-15	GC-EI-QqQ(MRM)	HT-5MS column (15 m × 0.25 mm, 0.10 μm)	
HFRs	T-16	GC-EI-MS(SIM) [116]	DB-1HT capillary column (15 m × 0.25 mm, 0.1 μm)	
HFRs, **OPFRs**	S-13	LP-GC-EI-QqQ(MRM) [117]	Rti®-5 ms GC column (15 m × 0.53 mm, 1 μm) connected to a Restek non-coated guard column (5 m × 0.18 mm)	

(Table 4) cont.....

Compounds	Code	Techniques and Instruments	Columns and Mobile Phases	
HBCD, TBBPA, bromophenols	P-18	UPLC-ESI-QqQ(MRM) [118]	ACQUITY UPLC™ BEH C18 column (2.1 × 100 mm, 1.7 µm) with an ACQUITY UPLC™ BEH C18 VanGuard pre-column (2.1 × 5 mm, 1.7 µm) at 45 °C	Mobile phase. (A) ACN:water (5:95) and (B) ACN:water (95:5), both with 0.1% acetic acid, at 0.3 mL min^{-1}. 1.5 min 50% A, 0.5 min 50>10% A; held for 3.5 min, 0.1 min 10>50% A, held for 2.9 min
HBCD	T-16	LC-ESI-QqQ(MRM) [119]	Varian Pursuit XRS3 C18 analytical column (150 mm × 2 mm, 3 µm)	Mobile phase. (A) MeOH:water (1:1) with 2 mM ammonium acetate and (B) MeOH at 0.12 mL min^{-1}. 3 min 50>100% B, held for 5 min, 2.5 min 100>65% B, held for 5.5 min
HBCD	A-18	LC-ESI-QTRAP(SRM) [120]	Symmetry C18 (150 mm × 2.1 mm, 5 µm) with C18 guard column (10 mm × 2.1 mm)	Mobile phase. (A) MeOH:water (3:1) and MeOH (B) at 0.25 mL min^{-1}. 17 min 100>10% A, 3 min 10>100% A, held for 15 min
HBCD, TBBPA	Z-15	LC-ESI-QqQ(MRM)	ZORBAX Eclipse Plus C18 column (2.1 mm × 100 mm, 1.8 µm)	Mobile phase. (A) MilliQ water and (B) ACN:MeOH (3:7) at 0.4 mL min^{-1}. 4 min 70>80% B, 5 min 80>88% B, 0.1 min 88>100% B, held for 4.9 min, 0.1 min 100>70% B, held for 5.9 min

(Table 4) cont.....

Compounds	Code	Techniques and Instruments	Columns and Mobile Phases	
HBCD, TBBPA	Z-16	LC-ESI-QqQ(SRM) [121]	XDB-C_{18} column (50 mm × 4.6 mm, 1.8 μm)	Mobile phase. (A) MeOH:water (9:1) and (B) ACN at 0.25 mL min^{-1}. 1 min 90>60% A, 4 min 60>30% A, 9 min 30>90% A
HBCD	H-18	LC-APCI-QqQ(MRM)		
OPFRs	X-15, X-17	GC-EI-MS(SIM)	HT-8 column (25 m × 0.22 mm, 0.25 μm) with deactivated silica retention gap (1.0 m × 0.22 mm)	
OPFRs	A-18	TFC-LC-ESI-QqQ(MRM) [36]	TFC: CycloneTM-P (50 × 0.5 mm) and C18-XL (50 × 0.5 mm). Mobile phase. (A) water with 0.1% formic acid and (B) MeOH with 0.1% formic acid at 0.75 mL min^{-1}. 98% A for sample loading, 100% B for analyte transfer and column cleaning	LC: Purosphere Star RP-18 column (125 × 0.2 mm). Mobile phase. (A) Water with 0.1% formic acid and (B) ammonium acetate in MeOH at 0.25 mL min^{-1}. 1 min 50>80% B, held for 1 min, 1 min 80>90% B, held for 5 min, 8 min 90>100% B, held for 5 min, 9 min 100>50% B
OPFRs	V-18	HPLC-ESI-QqQ(MRM) [122]	Waters Xterra® phenyl column (100 mm × 2.1 mm, 3.5 μm)	Mobile phase. (A) Water and (B) methanol, both with 0.1% formic acid, at 0.2 mL min^{-1}. 3 min 5>70% B, 12 min 70>80% B, 3 min 80>95% B, held for 12 min, 1 min 95>5% B, held for 15 min

There are four exceptions to the tendency of using 15 min columns for a reduced number of PBDE congeners or for just the lower-brominated ones. P-17 analyses 19 PBDE congeners, HBB, PBEB and DP and uses a 10 m DB-5 column. In this case, the total amount of analytes is similar to other methods including fewer PBDEs but more of the other HFRs. However, W-16 analyses 20 PBDE congeners, 9 BFRs and 10 dechloranes with a 15 m DB-5HT column. V-18 targets

38 PBDE congeners, 4 BFRs and DP using a 15 m DB-5HT column. And M-18 analyses separately 28 PBDE congeners on a 15 m Rtx-1614 fused silica column and 8 BFRs and DP on a 30 m Rxi-5sil MS fused silica column.

Regarding the inclusion of **different groups of compounds** in the same run, V-17 analyses 39 PBDEs, HBCD, TBBPA, 7 other BFRs and DP in one run and S-13 includes 7 PBDEs, 6 other BFRs, DP and 4 OPFRs. Methods S-15, Z-16, H-18 and M-18 —in increasing order— divide 19 to 37 HFRs into two runs with different columns in the same instrument with the same ionization technique. Finally, A-18 performs a GC-EI-MS/MS run for 9 PBDEs and 2 BFRs, excluding BDE-209 and DBDPE, which are analyzed by GC-ECNI-MS, and a third GC-ECNI-MS/MS run for Dec 602, Dec 603, Dec 604, *anti*-DP and *syn*-DP. Compensating for the time-consuming different runs, A-18 uses the same extract for the analysis of HBCD, which (except in the case of method V-17) always requires a different instrumental analysis and a division of the extract during clean-up or even a complete separate extraction method.

The analysis of **HBCD and TBBPA** is always performed by LC-MS/MS, (excluding V-17). The reason for the preference of LC is that the different diastereoisomers of HBCD isomerize to α-HBCD in GC due to the heat in the injector, making them impossible to be differentiated [123]. Five of the six methods in Table **4** use electrospray ionization (ESI), the other uses atmospheric pressure chemical ionization (APCI). Method H-18 updates method Z-16's, moving from a LC-ESI-MS/MS to a LC-APCI-MS/MS and keeping the same column and mobile phase gradient. All columns are C18, however, they are all from different commercial houses, length ranges from 50 to 150 mm, inner diameter ranges from 2 to 4.6 mm and particle size ranges from 1.7 to 5 µm. Longer columns (150 mm) have bigger particles (3-5 µm) and shorter columns (50 mm) have bigger diameters (4.6 mm). While mobile phase B is always acetonitrile or methanol, mobile phase A ranges from 100% water (Z-15) to 10% water (Z-16, H-18). This is relevant as A-18, Z-16 and H-18 use much lower percentages of B in their gradients, but when corrected with the organic component of A, all gradients use mostly organic solvent the whole time (Fig. **6**). This makes sense, as these compounds are hydrophobic. In Fig. **6**, it can be appreciated that three methods have runs lasting 14-20 min, while P-18 analyses α-, β- and γ-HBCD, TBBPA and four bromophenols in 8.5 min and A-18 analyses only α-, β- and γ-HBCD in 35 min.

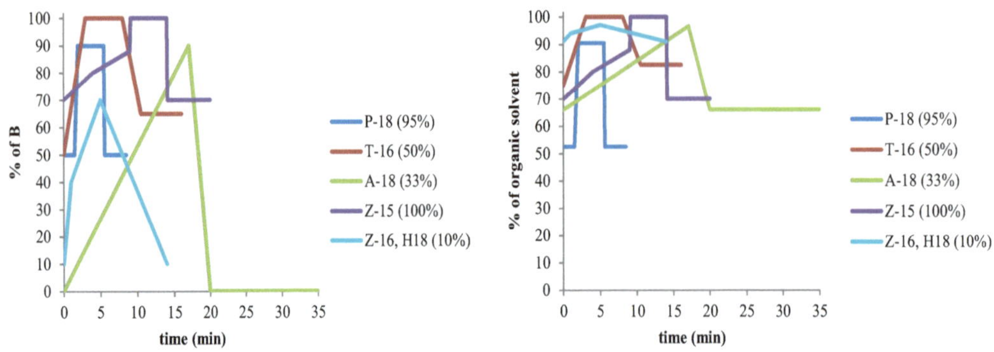

Fig. (6). Gradient of elution for the analysis of HBCD and TBBPA representing the percentage of mobile phase B, on the left, and the percentage of total organic solvent, on the right. The legend shows the method and the amount of water in mobile phase A between brackets.

The last group of compounds listed in Table **4** are **OPFRs**. Methods X-15 and X-17 (by the same authors) perform the same GC-EI-MS instrumental analysis. S-13 analyses OPFRs together with HFRs by low-pressure GC-EI-MS/MS. A-18 and V-18 use LC-ESI-MS/MS. Method S-13 performs a QuEChERS extraction of all FRs and analyses them in a single instrumental run, which saves time, efforts and resources. On the other hand, this means that the amount of OPFRs included in the analysis is reduced to 4 of the most commonly targeted. Table **5** shows what OPFRs are included in each method. X-15 and X-17, which separate OPFRs in the clean-up step after extracting all FRs together from the food samples, take advantage of the fact that isomers of OPFRs can be separated in GC. These methods are able to differentiate 2 isomers of TCIPP and 4 isomers of TCP. Although it is not possible to separate these isomers by LC, LC allows the analysis of more OPFRs. Methods A-18 and V-18 are able to analyze 14 and 17 OPFRs by LC, respectively. It is noteworthy that by performing the clean-up on-line using TFC (as explained in the clean-up section), A-18 requires an extraction as simple as the QuEChERS of S-13 yet it is one of the methods that include more OPFRs. See Fig. **7** for a chromatogram of OPFRs detected in a fish sample by TFC-LC-ESI-MS/MS.

Quality Parameters

Apart from a systematic analysis of blanks, a few methods use a standard reference material (SRM) for PBDEs. M-18, W-16 and V-18 use SRM NIST 1947, Lake Michigan fish tissue. M-18 also uses SRM NIST 1588b, cod liver oil. For V-18, an inter-laboratory comparison for eight PBDEs and the NIST 1947 showed between 2 and 35% variation in PBDE concentrations. Method S-16 carries out the extraction and clean-up in a clean room (class 10,000).

Table 5. OPFRs targeted by the methods.

	S-13	X-15	X-17	A-18	V-18
	LP-GC-EI-MS/MS	GC-EI-MS	GC-EI-MS	TFC-LC-ESI-MS/MS	HPLC-ESI-MS/MS
DCP				x	
EHDPP		x	x	x	x
IDPP				x	
IPPP				x	
T2B4MP					x
T3B4MP					x
T4B3MP					x
TBOEP			x	x	x
TBP		x	x	x	x
TCEP	x	x	x	x	x
TCIPP	x	2 isomers	2 isomers	x	x
TCP		4 isomers	4 isomers	x	
TDBPP					x
TDCPP	x	x	x	x	x
TEHP		x		x	x
TEP					x
THP				x	
TMPP					x
TPHP	x	x	x	x	x
TPPO				x	
TPrP		x			x
TTBP					x
V6					x

When working with OPFRs, blank contamination is an important issue. Indoor is contaminated with OPFRs and thus the contamination during sample preparation is an important factor. To solve this problem, different precautions are taken such as heating the non-volumetric material above 300 °C wrapped in aluminum foil and lastly rinse it with an appropriate solvent right before use. Even taking precautions, the blank signal is inevitable and uncontrollable, and different from day to day. The blank signal can come from different sources that cannot be controlled, like ambient air or nitrogen from the evaporator. A realistic goal is to minimize as much as possible the blank signal. In the first worldwide interlaboratory study of OPFRs, there is a more in depth explanation about the blank signal and how to minimize it [124].

Table **6** shows the limits of detection (LODs), limits of quantification (LOQs) and recoveries for the different compounds using the methods in Table **3**. The method column provides a condensed description of the sample type, extraction, clean-up and instrumental analysis. When some compounds are explicitly listed in the LODs or LOQs boxes (PBDEs, DP), they are not included in the corresponding generic group in the same box (BFRs, HFRs).

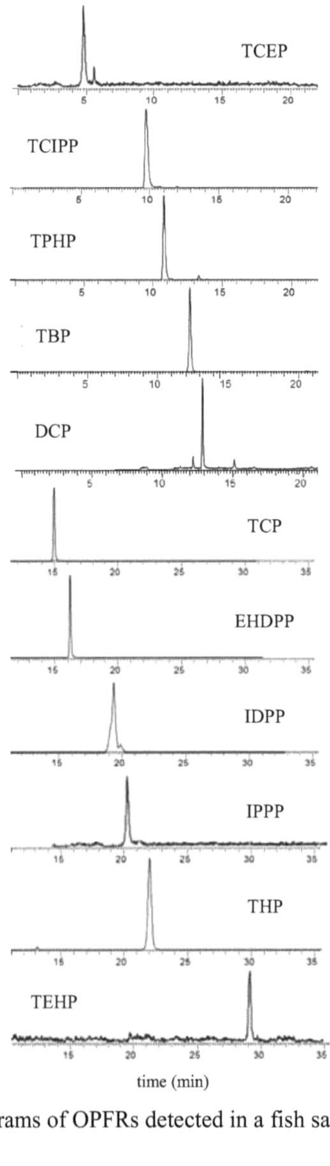

Fig. (7). TFC-LC-ESI-MS/MS chromatograms of OPFRs detected in a fish sample.

It can be seen in Table 6 that **LODs and LOQs** are reported either in wet weight (ww), dry weight (dw) or lipid weight (lw). For this reason, before comparing the differences amongst LODs and LOQs, a homogenization of the values is required. The water and lipid contents of samples from previous seafood ($n = 75$) [125], dolphin muscle ($n = 42$) and bird eggs ($n = 9$) (the two latter still unpublished) were added into a water-content set of data and a lipid-content set of data. The first quartile and the third quartile of each data set were taken as the minimum and maximum value, respectively. With these values LODs and LOQs in ww and in lw were transformed to a range of LODs and LOQs in dw. In better detail, the minimum value of a LOD or LOQ range in ww or lw was converted to dw using the first quartile value of the water or lipid content, respectively, and the maximum value of a LOD or LOQ range was converted using the third quartile value. This produced new ranges of LODs and LOQs in dw that allowed comparison among methods. On the other hand, this comparison has to be cautious bearing in mind the new homogenized ranges are approximations.

Probably the safest comparison to make is to confront LOD ranges with LOQ ranges. In that respect, the minimum converted LOD for **PBDEs** for the QuEChERS method S-13 is 1.6 ng g^{-1} dw, while LOQs for other methods are mostly below 1 ng g^{-1} dw. Converted LODs for PBDEs for S-15 (0.02-11.5 ng g^{-1} dw) are also relatively high compared to general LOQs. Finally, methods like M-18, V-17 and A-18 have a relatively high maximum LOD because of some higher brominated PBDE, usually BDE-209; even if the lower brominated PBDEs have LODs similar to the rest of the methods. It has been mentioned that PBDEs with a higher level of bromination are eluted at the end of the run with more challenging quality parameters. Methods Z-16 and H-18 (Soxhlet, ECNI-MS), P-17 (PLE, ECNI-MS) and Z-15 (UAE, EI-MS/MS) have the lowest LODs (0.001-0.19 ng g^{-1} dw). X-17 (UAE, ECNI-MS) and P-17 (PLE, ECNI-MS) have the lowest LOQs (0.02-0.22 ng g^{-1} dw). Out of these methods, P-17 was the only to report both LODs and LOQs —albeit in ww—, so the addition of these two groups of methods could be considered the ones with better LODs and LOQs for PBDEs in general.

Regarding **other BFRs** (other than PBDEs, HBCD or TBBPA), comparison is even more complex as some methods target as few as three compounds (A-18, S-15, H-18) while others analyze five to ten of them or even over twenty (S-16). Again, method S-13 (QuEChERS) has a minimum converted LOD of 3.2 ng g^{-1} dw, whereas most methods have LOQ ranges below 1.2 ng g^{-1} dw. Method S-16 (reporting in dw) includes so many compounds that its LODs range from 0.001 ng g^{-1} dw for some compound to 4.8 ng g^{-1} dw for 2,4,6-tribromophenol (2,4,6-TBP) and its LOQs range from 0.003 ng g^{-1} dw for some compound to 16 ng g^{-1} dw for bis(2-ethylhexyl)-3,4,5,6-tetrabromobenzoate (BEH-TEBP).

2,4,6-TBP and BEH-TEBP are rarely included in other methods. M-18, P-17 and V-18 (PLE, ECNI-MS), H-18 (Soxhlet, ECNI-MS) and Z-15 (UAE, EI-MS/MS) report the best LODs. X-15 and X-17 (UAE, ECNI-MS) and P-17 (PLE, ECNI-MS) report the best LOQs.

Looking at **dechloranes**, method S-13 (QuEChERS) has a minimum converted LOD of 3.2 ng g^{-1} dw for HFRs (not specifying the LOD for DP), whereas most methods report LOQ ranges below 0.2 ng g^{-1} dw. Converted LODs for HFRs for S-15 (0.02-11.5 ng g^{-1} dw) are also relatively high compared to general LOQs. Divergence in LODs and LOQs for analysis of dechloranes is not as great as for BFRs, in part probably because they are restricted to the same few compounds. The lowest LOD ranges are reported by Z-15 (<0.0002 ng g^{-1} dw), A-18 (<0.004 ng g^{-1} dw) and M-18 (<0.007 ng g^{-1} dw), and the lowest LOQ ranges are reported by A-18 (<0.03 ng g^{-1} dw), X-15/X-17 (<0.07 ng g^{-1} dw) and P-16/P-18 (<0.11 ng g^{-1} dw). Out of these methods, A-18 was the only to report both LODs and LOQs —albeit in lw—, so the addition of these two groups of methods could be considered the ones with better LODs and LOQs for at least DP, as this is the only dechlorane included in all of them.

Table 6. LODs, LOQs and recoveries of the methods.

Code	Method	LODs (ng g^{-1})	LOQs (ng g^{-1})	Recoveries
P-16, P-18	All foodstuffs. UAE, 2 × SPE, GC-ECNI-MS		PBDEs 0.005 ww, BDE-209 0.1 ww, BFRs 0.02-0.2 ww, DP 0.02 ww	75-121% BDE-99 204% TBA 59%
X-15	All foodstuffs. UAE, 4 × SPE, GC-ECNI-MS		PBDEs 0.008-0.02 dw, BDE-209 0.4 dw, BFRs 0.03-0.2 dw, DP 0.04-0.06 dw	72-144%
X-17	All foodstuffs. UAE, 3 × SPE, GC-ECNI-MS		PBDEs 0.02-0.14 dw, BFRs, 0.025-0.19 dw, DP 0.04-0.07 dw	86-125%, DBDPE 143%
M-18	Meat and greens. PLE, GPC, 2 × SPE, GC-ECNI-LRMS	PBDEs 0.004-0.08 ww, BDE-209 0.53 ww, BFRs 0.004-0.05 ww, DP 0.006 ww		47-105%, PBBB 147%
S-15	Seafood. Soxhlet, acid, 1 × SPE, GC-ECNI-MS	HFRs 0.005-2.1 lw		88-106%
Z-16	Eggs. Soxhlet, 1 × SPE, GC-ECNI-MS	BFRs 0.2-0.4 lw		internal standards 72-117%

(Table 6) cont.....

Code	Method	LODs (ng g^{-1})	LOQs (ng g^{-1})	Recoveries
H-18	Eggs. Soxhlet, acid, 1 × SPE, GC-ECNI-MS	PBDEs 0.04-0.50 lw HBB 0.10 lw DBDPE 1.50 lw		IS 92-103%
P-17	Seafood. PLE, GPC, 1 × SPE, HRGC-ECNI-MS	HFRs 0.003-0.03 ww	HFRs 0.006-0.04 ww	internal standard 99 ± 10%
W-16	Fish. PLE, GPC, 1 × SPE, GC-ECNI-MS		PBDEs 0.5-2.5 lw, BFRs 0.6-3.5 lw, Decs 0.8-1.8 lw	78-104%
V-18	Meat. PLE, 2 × SPE, GC-ECNI-MS	PBDEs 0.01-0.09 ww, BDE-206 0.23 ww, BFRs 0.01-0.07 ww, DP 0.01-0.03 ww	PBDEs 0.01-0.29 ww, BDE-206 0.76 ww, BFRs 0.01-0.21 ww, DP 0.03-0.09 ww	internal standards 61-99%
V-17	Eggs. Hot, 2 × SPE, GC-ECNI-MS	PBDEs 0.61-6.3 lw, BFRs 0.06-3.4 lw	PBDEs 2-21 lw, BFRs 0.2-11 lw	internal standards 70-100%
S-16	Fish. PLE, GPC, 1 × SPE, GC-ECNI-MS, GC-ECNI-MS/MS	PBDEs 0.003-0.012 dw, BDE-209 0.47 dw, BFRs 0.001-4.8 dw, Decs 0.001-0.2 dw	PBDEs 0.009-0.05 dw, BDE-209 1.6 dw, BFRs 0.003-16 dw, Decs 0.004-0.65 dw	internal standards 45-139%
Z-15	Fish. UAE, GPC, 1 × SPE, GC-EI-MS/MS	PBDEs 0.0004-0.004 ww, BFRs 0.0001-0.005 ww, DP 0.0002 ww		78-117%
T-16	All foodstuffs. Rapid, GPC, 2 × SPE, GC-EI-MS		BFRs 0.54-1.21 lw	56-98%, internal standards 90-117%
A-18	All foodstuffs. PLE, acid, 1 × SPE, GC-EI-MS/MS, GC-ECNI-MS, GC-ECNI-MS/MS	PBDEs 0.04-6.2 lw, BFRs 0.2-6.2 lw, Decs 0.002-0.02 lw	PBDEs 0.12-20.8 lw, BFRs 0.61-20.8 lw, Decs 0.008-0.07 lw	51-109%
S-13	Fish. QuEChERS, LP-GC-EI-MS/MS	PBDEs 0.5-10 ww, HFRs 1-10 ww		70-120%
V-17	Eggs. Hot, 2 × SPE, GC-ECNI-MS	HBCD, TBBPA 3.2 lw	HBCD, TBBPA 11 lw	internal standards 70-100%
P-18	All foodstuffs. UAE, 1 × SPE, UPLC-ESI-MS/MS		HBCD 0,004-1 ww, TBBPA 0.01-0.25 ww, bromophenols 0.01-8 ww	70-120%

(Table 6) cont.....

Code	Method	LODs (ng g^{-1})	LOQs (ng g^{-1})	Recoveries
T-16	All foodstuffs. Rapid, GPC, acid, LC-ESI-MS/MS		HBCD 0.02 lw	56-98%, internal standards 90-117%
A-18	All foodstuffs. PLE, acid, 1 × SPE, LC-ESI-MS/MS	HBCD 0.2 lw	HBCD 0.4 lw	
Z-15	Fish. UAE, GPC, filter, LC-ESI-MS/MS	HBCD 0.001-0.002 ww, TBBPA 0.0007 ww		88-106%
Z-16	Eggs. Soxhlet, acid, 1 × SPE, LC-ES-MS/MS	HBCD, TBBPA 0.7-1.7 lw		internal standards 50-108%
H-18	Eggs. Soxhlet, acid, 1 × SPE, LC-APCI-MS/MS	HBCD 0.98-3.42 lw		internal standards 86-97%
S-13	Fish. QuEChERS, LP-GC-MS/MS	OPFRs 1-10 ww		73-121%
X-15	All foodstuffs. UAE, 4 × SPE, GC-EI-MS		OPFRs 1.4-3.7 dw	58-121%
X-17	All foodstuffs. UAE, 3 × SPE, GC-EI-MS		OPFRs 1.0-2.5 dw	89-100%
A-18	All foodstuffs. UAE, TFC-LC-ESI-MS/MS	OPFRs 0.19-19.3 lw	OPFRs 0.97-24.8 lw	47-98%
V-18	Meat. PLE, 1 × SPE, HPLC-ESI-MS/MS	OPFRs 0.03-0.26 ww	OPFRs 0.09-0.78 ww	15-142% (mean 81%)

While TBBPA is analyzed with **HBCD**, it is not always included in the already few methods so it has to be left out of the comparison. The lowest converted LOD ranges are reported by Z-15 (<0.01 ng g^{-1} dw) and A-18 (<0.08 ng g^{-1} dw), the other reporting 0.04-1.3 ng g^{-1} dw. The lowest LOQ ranges are reported by T-16 (<0.008 ng g^{-1} dw) and A-18 (<0.15 ng g^{-1} dw), the others reporting up to 5.5 ng g^{-1} dw. A-18 was the only of these methods that reported both LODs and LOQs, albeit in lw. Interestingly enough, the three methods applied completely different extraction strategies as seen in Tables **3** and **6**.

Comparison of LODs and LOQs for **OPFRs** show overlapping ranges for all methods, except for S-13. Method S-13 (QuEChERS) has a converted LOD range of 3.2-55 ng g^{-1} dw, whereas most methods report LOQ ranges below 4.2 ng g^{-1} dw and A-18 below 9.4 ng g^{-1} dw.

In Table **6**, it is important to notice that some studies reference their **recovery values** to internal standards and not the native compounds. In the cases when all native compounds have their own internal standard and it is the same compound as the analyte but labelled isotopically, these should still be reliable data in the sense of comparable to the values reported by other methods. When not all analytes have their analogue internal standard, recovery ranges might turn out slightly favorable as fewer compounds may translate into smaller dispersion. For example, methods X-17 and M-18 assessed the individual recoveries for the native compounds and while their recoveries ranged 86-125% and 47-105%, respectively, X-17 obtained a recovery of 143% for DBDPE and M-18 obtained 147% for pentabromobenzyl bromide (PBBB). At the other side of the spectrum, P-17 only reports a 99 ± 10% recovery for BDE-71, which is used as an internal standard. Therefore, a reduced number of internal standards might not represent the real range of recovery of analytes. For the same reason, methods targeting more analytes may have a greater dispersion than methods targeting fewer compounds, increasing the chance of values further away from 100%.

Most methods have recoveries fitting between 60 and 110%. Only V-18 goes below 40%, which is the ideal lower limit for methods using surrogate standards, with its lowest recovery value at 15% for T3B4MP. However, V-18 is the only method to target this OPFR. Although all methods have acceptable recoveries, S-15, H-18 and P-17 show the smallest dispersion (88-109%) for HFRs, Z-15 and H-18 show the smallest dispersion for HBCD (86-106%) and X-17 shows the smallest dispersion for OPFRs (89-100%).

Taking into account the discussed quality parameters, method Z-15 for fish is reported to be one of the best performing methods for PBDEs, HBCD, other BFRs and dechloranes, that is, all **HFRs**. The extraction method is quite simple, using UAE, GPC for lipid removal and one SPE for clean-up, with DCM:hexane (1:1) as extracting or eluting solvent in all steps. The instrumental analysis is performed by GC-EI-MS/MS. Other well-performing methods for HFRs use GC-ECNI-MS: X-15/X-17, P-17, M-18 and H-18. Of all these methods, only H-18 targets HBCD. X-17 is an update of X-15 for all foodstuffs extracting by UAE and reducing clean-up from four SPEs to three, separating OPFRs. P-17 for seafood performs the same extraction and clean-up steps as Z-15 but changing the solvents mixture. M-18 for seafood performs adds a second SPE and has higher LODs for high-brominated PBDEs. Although H-18 has good quality parameters for PBDEs, and 2 other BFRs, it targets a limited amount of HFRs in eggs, it performs a more elaborate Soxhlet extraction and it requires a different chromatographic run for

HBCD, with LODs that are improved by other methods. A-18 also has decent LODs and LOQs for HFRs, including HBCD; but, like M-18, has higher LODs for high-brominated PBDEs. On a positive note, it extracts all compounds together by PLE, an acid attack for lipid removal and an SPE. On the other hand, it requires four chromatographic runs (see Table **6**). To sum up, X-17 (all foodstuffs, excluding HBCD) and Z-15 (fish, including HBCD) seem to be simpler methods with better performance. S-13 performs a simple QuEChERS extraction and includes four OPFRs, but performs very poorly compared to any of the other methods.

Regarding OPFRs, although all methods perform similarly (except for S-13), X-17 and A-18 seem to be preferable to X-15 and V-18. X-17 is an update of X-15 that introduces a change in the clean-up to successfully improve LOQs and recoveries. Additionally, the same extract can be divided during the clean-up into a fraction for the analysis of 9 OPFRs (including some isomers) and a fraction for the analysis of HFRs. A-18 targets 14 OPFRs with the simplest sample treatment consisting of an UAE and the extract is ready for instrumental analysis as clean-up is performed online by TFC as explained in the clean-up section of this chapter. While X-17 and A-18 are applied to analyze all foodstuffs, V-18 has only been applied to liver. V-18 also requires a separate extraction from HFRs but this extraction is just as complex as the other, which are disadvantages in front of X-17 and A-18.

LEVELS IN FOOD AND INTAKE

Many of the studies cited in this text to compare their methods analyzed food samples. Some even calculated FR intake through food consumption or performed a risk assessment. Here are a few results.

P-18 analyzed all kinds of foodstuffs. P-18 detected HFRs to some extent in 72% of the samples. The groups of potatoes, fish and meat had the highest number of contaminated samples, whereas food for infants was the least contaminated group. Seafood and Emmental cheese sowed the highest HFR levels. HFR content went up to 35.4 ng g^{-1} ww, with an average content of 1.2 ng g^{-1} ww. The most frequently found contaminants were BDE-47 (53%), BDE-209 (46%) and 2,4,6-TBP (40%). TBBPA, DP, and γ-HBCD occurred in less than 5% of the samples. HBB and BTBPE were not detected.

Overall, X-17 detected most of the target compounds in the food samples, with concentrations of total PBDEs, total HFRs —excluding PBDEs— and total OPFRs ranging from not-quantifiable to 0.93 ng g^{-1} ww, 0.14 ng g^{-1} ww and 150 ng g^{-1} ww, respectively. EHDPP was the main contributor to the OPFR contamination in food with a median of 3.0 ng g^{-1} ww (present in 80% of the samples), followed by TCIPP (up to 6.6 ng g^{-1} ww, in 40% of the samples). TCEP and TPHP were detected less than 20% of the samples, while other OPFRs were below their LOQs.

A-18 detected PBDEs in 95% of their farmed mussel samples with mean levels of 0.07 ± 0.18 ng g^{-1} dw. DBDPE and DP were found in about half the samples, while PBEB and HBB were not detected. OPFR levels ranged from 2.74-1380 ng g^{-1} lw. The same method had been applied in a study monitoring commercial seafood from European markets to find HFRs in 90.5% of the samples ranging from not detected to 456 ng g^{-1} lw or 41.1 ng g^{-1} ww [125]. In that study, DBDPE was not detected and 2,4,6-TBP was detected only in mussels, but at levels comparable to PBDEs.

X-17 estimated median dietary intakes (95th percentile) of total PBDEs and HFRs and OPFRs of 1.3 (14), < 0.01 (3.4) and 87 (340) ng kg^{-1} bw day^{-1}, respectively. The intakes for BDE47 and BDE209 were 0.20 and 0.86 ng kg^{-1} bw day^{-1}. The dietary intake of BEH-TEBP and *anti*-DP were 1.2 and 1.1 ng kg^{-1} bw day^{-1}, in the worst cases. The results of X-17 suggested that the dietary exposure was safely below the oral reference dose proposed by the United States EPA for PBDEs and TCEP. S-15 estimated daily intakes of PBDEs and other HFRs to be 1.0, and 0.18 ng kg^{-1} bw day^{-1}, respectively, via seafood consumption. These concentrations are not expected to pose health risks to humans.

H-18 estimated high dietary intake of HFRs via home-produced eggs. Considering the most polluted eggs, the margin of exposure of BDE-99 for adults and children in 2016 were 1.1 and 0.2, respectively. These values were below the 2.5 threshold set by the CONTAM panel, thus representing a significant potential health concern due to adverse effects of PBDEs on neurodevelopment and fertility. Z-16 also reported a potential health concern for BDE-99 due to dietary exposure via eggs. The study on European seafood performed using method A-18 carried out a risk assessment that revealed no health risk related to the exposure to BFRs *via* seafood consumption, except regarding BDE-99 for a 0.05% of the adult Portuguese consumers [125].

To sum up, studies report that PBDEs are still found in many or most of the samples despite the almost two-decade ban. While other HFRs are usually found at lower levels and with lower frequencies of detection, OPFRs can present concentrations some orders of magnitude greater than PBDEs. As for food safety, findings are that health risks are generally not expected, although BDE-99 does show concerning results in some occasions.

CONCLUDING REMARKS

FRs are necessary to reduce fire hazards related to the flammability of the many materials used nowadays. On the other hand, they could turn into a health hazard if not controlled since they leak from materials into the environment, they are persistent and bioaccumulate in organisms with eventual toxic effects. Food ingestion is the major route of exposition to FRs for humans, thus making the monitoring of FR levels in food one of the main public health interests. Several methods of analysis of FRs are currently in use targeting different groups of compounds. PBDEs and most HFRs are commonly analyzed together by GC. HBCD tends to be extracted separately and analyzed by LC. OPFRs are also extracted and analyzed independently, but few methods target them currently. The eight PBDEs included in almost all methods are BDE-28, -47, -99, -100, -153, -154, -183 and -209. HBB, BTBPE, PBEB and DBDPE are other BFRs that are usually targeted. *Syn*-DP and *anti*-DP are monitored in all studies including chlorinated FRs. The eight OPFRs usually analyzed are EHDPP, TBOEP, TBP, TCEP, TCIPP, TDCPP, TEHP and TPHP. PLE and UAE are the most used extraction techniques with the best results. Soxhlet extractions are rare and tend to perform worse and are far more complex. One method using QuEChERS reported poor results compared to any other method and included a limited amount of FRs. Clean-up generally includes one or two SPEs, sometimes replacing the first SPE with GPC or an acid attack to remove lipids. Florisil and silica are recurring sorbents and DCM:hexane is a recurring solvent mixture with an increased percentage of DCM for elution. One of the methods with lower LODs and LOQs extracted al HFRs with UAE and removes lipids by GPC. The extract is divided into a fraction for HBCD and TBBPA (LC-ESI-MS/MS), purified through a syringe filter, and a fraction for the other HFRs (GC-MS/MS), which undergoes an SPE. The other best performing method extracted HFRs (not HBCD or TBBPA) and OPFRs together by UAE and obtained to separate extracts after four SPEs to be analyzed by GC-MS. Finally, a method for OPFRs reduced sample treatment to only UAE, performing on-line clean-up during the instrumental analysis by TFC-LC-MS/MS, with satisfactory LODs and LOQs. Food samples analyzed using various methods contained OPFRs in concentrations some orders of magnitude higher than PBDEs. These two groups of FRs were detected in most samples. Other HFRs showed in fewer samples and at lower concentrations. Estimated intakes and risk assessments suggest that levels of FRs in food are safe, although BDE-99 might cause concern in some cases.

LIST OF ABBREVIATIONS

2,4,6-TBP	2,4,6-tribromophenol
ACN	acetonitrile

APCI	atmospheric pressure chemical ionization
BAF	bioaccumulation factor
BEH-TEBP	bis(2-ethylhexyl)-3,4,5,6-tetrabromobenzoate
BFR	brominated flame retardant
BMD	benchmark dose
BMF	biomagnification factor
BSAF	biota-sediment accumulation factor
BTBPE	1,2-bis-(2,4,6-tribromophenoxy)ethane
DBDPE	decabromodiphenyl
DCM	dichloromethane
DCP	diphenil cresyl phosphate
Dec	Dechlorane
decaBDE	decabrominated diphenyl ether
DP	Dechlorane Plus
dSPE	dispersive solid phase extraction
dw	dry weight
ECNI	electron capture negative ionization
EFR	emerging flame retardant
EFSA	European Food Safety Authority
EHDPP	2-ethylhexyl diphenyl phosphate
EI	electron ionization
EPA	Environmental Protection Agency
ESI	electrospray ionization
EU	European Union
FR	flame retardant
GC	gas chromatography
GPC	gel permeation chromatography
HBB	hexabromobenzene
HBCD	hexabromocyclododecane
HFR	halogenated flame retardant
HN	halogenated norbornene
IDPP	isodecyl diphenyl phosphate
IPPP	isopropyl phenyl phosphate
K_{ow}	octanol-water partition coefficient
LC	liquid chromatography

LOD	limit of detection
LOQ	limit of quantification
LP	low-pressure
lw	lipid weight
m/z	mass-to-charge ratio
MRM	multiple reaction monitoring
MS	mass spectrometry
NICI	negative ion chemical ionization
NOAEL	no observed adverse effect level
NOEC	no observed effect concentration
octaBDE	octabrominated diphenyl ether
OPFR	organophosphorus flame retardant
PBBB	pentabromobenzyl bromide
PBDE	polybrominated diphenyl ether
PBEB	pentabromoethylbenzene
PCB	polychlorinated biphenyl
pentaBDE	pentabrominated diphenyl ether
PLE	pressurised liquid extraction
POP	persistent organic pollutant
PTFE	polytetrafluoroethylene
REACH	Registration, Evaluation, Authorisation and Restriction of Chemicals
SIM	single ion monitoring
SPE	solid phase extraction
SRM	standard reference material
$t_{1/2}$	half-life
T2B4MP	tris(2-bromo-4-methylphenyl) phosphate
T3B4MP	tris(3-bromo-4-methylphenyl) phosphate
T4B3MP	tris(4-bromo-3-methylphenyl) phosphate
TBBPA	tetrabromobisphenol A
TBOEP	tris(2-butoxyethyl) phosphate
TBP	tributyl phosphate
TCEP	tris(2-chloroethyl) phosphate
TCIPP	tris(chloroisopropyl) phosphate
TCP	tricresyl phosphate
TDBPP	tris(2,3-dibromopropyl) phosphate

TDCPP	tris(1,3-dichloroisopropyl) phosphate
TEHP	tris(2-ethylhexyl) phosphate
TEP	triethyl phosphate
TFC	turbulent flow chromatography
THP	trihexyl phosphate
TMPP	tris(methylphenyl) phosphate
TPHP	triphenyl phosphate
TPPO	triphenylphosphine oxide
TPrP	tripropyl phosphate
Tris	tris(2,3-dibromopropyl) phosphate
TTBP	tris(tribromoneopentyl) phosphate
UAE	ultrasound assisted extraction
V6	2,2-bis(chloromethyl)trimethylene bis[bis(2-chloroethyl)phosphate]
ww	wet weight

CONSENT FOR PUBLICATION

Not applicable.

CONFLICT OF INTEREST

The authors confirm that this chapter contents have no conflict of interest.

ACKNOWLEDGEMENTS

Declared none.

REFERENCES

[1] Kaur Bindra S, Narang RS. Combustion of flame retardants. Chemosphere 1995; 31(11): 4413-25. [http://dx.doi.org/10.1016/0045-6535(95)00310-5]

[2] Jensen S. PCB Conference. National Swedish Environmental Protection Board; Solna, Sweden. 1970.

[3] Hussain S. A history of halogenated flame retardants.History of polymeric composites. Utrecht: VNU Science Press BV 1987; pp. 271-98.

[4] Strauss F, Kollek L, Heyn W. Über den Ersatz positiven Wasserstoffs durch Halogen (I. Mitteil.). Ber Dtsch Chem Ges 1930; 63: 1868-85. [http://dx.doi.org/10.1002/cber.19300630739]

[5] van der Veen I, de Boer J. Phosphorus flame retardants: properties, production, environmental occurrence, toxicity and analysis. Chemosphere 2012; 88(10): 1119-53. [http://dx.doi.org/10.1016/j.chemosphere.2012.03.067] [PMID: 22537891]

[6] Anh HQ, Nam VD, Tri TM, *et al.* Polybrominated diphenyl ethers in plastic products, indoor dust, sediment and fish from informal e-waste recycling sites in Vietnam: a comprehensive assessment of contamination, accumulation pattern, emissions, and human exposure. Environ Geochem Health 2017;

39(4): 935-54.
[http://dx.doi.org/10.1007/s10653-016-9865-6] [PMID: 27542012]

[7] FAO/WHO, Joint FAO/WHO expert committee on food additives.. Sixty-fourth meeting. Rome 2005 February; 8-17. Summary and conclusions. 2005. ftp://ftp.fao.org/es/esn/jecfa/jecfa64_summary.pdf (accessed September 2017).

[8] Darnerud PO, Atuma S, Aune M, *et al.* Dietary intake estimations of organohalogen contaminants (dioxins, PCB, PBDE and chlorinated pesticides, *e.g.* DDT) based on Swedish market basket data. Food Chem Toxicol 2006; 44(9): 1597-606.
[http://dx.doi.org/10.1016/j.fct.2006.03.011] [PMID: 16730400]

[9] FAO, The State of World Fisheries and Aquaculture 2016. Contributing to food security and nutrition for all , 2016http://reliefweb.int/sites/reliefweb.int/files/resources/a-i5555e.pdf [accessed September 2017];

[10] Alaee M, Arias P, Sjödin A, Bergman A. An overview of commercially used brominated flame retardants, their applications, their use patterns in different countries/regions and possible modes of release. Environ Int 2003; 29(6): 683-9.
[http://dx.doi.org/10.1016/S0160-4120(03)00121-1] [PMID: 12850087]

[11] Ballschmiter K, Zell M. Analysis of polychlorinated biphenyls (PCB) by glass capillary gas chromatography - Composition of technical aroclor- and Clophen-PCB mixtures. Z Anal Chem 1980; 302(1): 20-31.
[http://dx.doi.org/10.1007/BF00469758]

[12] Brominated diphenylethers. Environmental Health Criteria. 1994; 162.

[13] Koch C, Schmidt-Kötters T, Rupp R, Sures B. Review of hexabromocyclododecane (HBCD) with a focus on legislation and recent publications concerning toxicokinetics and -dynamics. Environ Pollut 2015; 199: 26-34.
[http://dx.doi.org/10.1016/j.envpol.2015.01.011] [PMID: 25618363]

[14] Harrad S, Abdallah MAE, Rose NL, Turner SD, Davidson TA. Current-use brominated flame retardants in water, sediment, and fish from English lakes. Environ Sci Technol 2009; 43(24): 9077-83.
[http://dx.doi.org/10.1021/es902185u] [PMID: 19921842]

[15] Abdallah MAE, Uchea C, Chipman JK, Harrad S. Enantioselective biotransformation of hexabromocyclododecane by *in vitro* rat and trout hepatic sub-cellular fractions. Environ Sci Technol 2014; 48(5): 2732-40.
[http://dx.doi.org/10.1021/es404644s] [PMID: 24483621]

[16] Janák K, Covaci A, Voorspoels S, Becher G. Hexabromocyclododecane in marine species from the Western Scheldt Estuary: diastereoisomer- and enantiomer-specific accumulation. Environ Sci Technol 2005; 39(7): 1987-94.
[http://dx.doi.org/10.1021/es0484909] [PMID: 15871228]

[17] de Wit CA, Alaee M, Muir DCG. Levels and trends of brominated flame retardants in the Arctic. Chemosphere 2006; 64(2): 209-33.
[http://dx.doi.org/10.1016/j.chemosphere.2005.12.029] [PMID: 16458344]

[18] Covaci A, Harrad S, Abdallah MAE, *et al.* Novel brominated flame retardants: a review of their analysis, environmental fate and behaviour. Environ Int 2011; 37(2): 532-56.
[http://dx.doi.org/10.1016/j.envint.2010.11.007] [PMID: 21168217]

[19] Yamaguchi Y, Kawano M, Tatsukawa R, Moriwaki S. Hexabromobenzene and its debrominated compounds in human adipose tissues of Japan. Chemosphere 1988; 17(4): 703-7.
[http://dx.doi.org/10.1016/0045-6535(88)90250-0]

[20] Watanabe I, Kashimoto T, Tatsukawa R. Hexabromobenzene and its debrominated compounds in river and estuary sediments in Japan. Bull Environ Contam Toxicol; (United States) 1986. Medium: X;

Size: Pages: 778-784.

[21] Watanabe I, Sakai S. Environmental release and behavior of brominated flame retardants. Environ Int 2003; 29(6): 665-82.
[http://dx.doi.org/10.1016/S0160-4120(03)00123-5] [PMID: 12850086]

[22] Buser HR. Polybrominated dibenzofurans and dibenzo-p-dioxins: thermal reaction products of polybrominated diphenyl ether flame retardants. Environ Sci Technol 1986; 20(4): 404-8.
[http://dx.doi.org/10.1021/es00146a015] [PMID: 22300214]

[23] Hoh E, Zhu L, Hites RA. Novel flame retardants, 1,2-bis(2,4,6-tribromophenoxy)ethane and 2,3,4,5,6-pentabromoethylbenzene, in United States' environmental samples. Environ Sci Technol 2005; 39(8): 2472-7.
[http://dx.doi.org/10.1021/es048508f] [PMID: 15884337]

[24] Kajiwara N, Noma Y, Takigami H. Photolysis studies of technical decabromodiphenyl ether (DecaBDE) and ethane (DeBDethane) in plastics under natural sunlight. Environ Sci Technol 2008; 42(12): 4404-9.
[http://dx.doi.org/10.1021/es800060j] [PMID: 18605562]

[25] Hoh E, Zhu L, Hites RA. Dechlorane plus, a chlorinated flame retardant, in the Great Lakes. Environ Sci Technol 2006; 40(4): 1184-9.
[http://dx.doi.org/10.1021/es051911h] [PMID: 16572773]

[26] Sverko E, Reiner EJ, Tomy GT, *et al.* Compounds structurally related to Dechlorane Plus in sediment and biota from Lake Ontario (Canada). Environ Sci Technol 2010; 44(2): 574-9.
[http://dx.doi.org/10.1021/es9025535] [PMID: 20025285]

[27] Wang D-G, Yang M, Qi H, *et al.* An Asia-specific source of dechlorane plus: concentration, isomer profiles, and other related compounds. Environ Sci Technol 2010; 44(17): 6608-13.
[http://dx.doi.org/10.1021/es101224y] [PMID: 20681512]

[28] Tomy GT, Pleskach K, Ismail N, *et al.* Isomers of dechlorane plus in Lake Winnipeg and Lake Ontario food webs. Environ Sci Technol 2007; 41(7): 2249-54.
[http://dx.doi.org/10.1021/es062781v] [PMID: 17438771]

[29] Tomy GT, Thomas CR, Zidane TM, *et al.* Examination of isomer specific bioaccumulation parameters and potential *in vivo* hepatic metabolites of *syn*- and *anti*-Dechlorane Plus isomers in juvenile rainbow trout (*Oncorhynchus mykiss*). Environ Sci Technol 2008; 42(15): 5562-7.
[http://dx.doi.org/10.1021/es800220y] [PMID: 18754476]

[30] Sverko E, Tomy GT, Reiner EJ, *et al.* Dechlorane plus and related compounds in the environment: a review. Environ Sci Technol 2011; 45(12): 5088-98.
[http://dx.doi.org/10.1021/es2003028] [PMID: 21574656]

[31] Krackeler JJ, Biddell WG. Flame retardant polyethylene composition and coating process. US 3900533 1975.

[32] Schreder ED, La Guardia MJ. Flame retardant transfers from U.S. households (dust and laundry wastewater) to the aquatic environment. Environ Sci Technol 2014; 48(19): 11575-83.
[http://dx.doi.org/10.1021/es502227h] [PMID: 25288150]

[33] Andresen JA, Grundmann A, Bester K. Organophosphorus flame retardants and plasticisers in surface waters. Sci Total Environ 2004; 332(1-3): 155-66.
[http://dx.doi.org/10.1016/j.scitotenv.2004.04.021] [PMID: 15336899]

[34] Chung H-W, Ding W-H. Determination of organophosphate flame retardants in sediments by microwave-assisted extraction and gas chromatography-mass spectrometry with electron impact and chemical ionization. Anal Bioanal Chem 2009; 395(7): 2325-34.
[http://dx.doi.org/10.1007/s00216-009-3139-4] [PMID: 19784832]

[35] Gao Z, Deng Y, Yuan W, He H, Yang S, Sun C. Determination of organophosphorus flame retardants in fish by pressurized liquid extraction using aqueous solutions and solid-phase microextraction

coupled with gas chromatography-flame photometric detector. J Chromatogr A 2014; 1366: 31-7. [http://dx.doi.org/10.1016/j.chroma.2014.09.028] [PMID: 25282313]

[36] Giulivo M, Capri E, Eljarrat E, Barceló D. Analysis of organophosphorus flame retardants in environmental and biotic matrices using on-line turbulent flow chromatography-liquid chromatography-tandem mass spectrometry. J Chromatogr A 2016; 1474: 71-8. [http://dx.doi.org/10.1016/j.chroma.2016.10.042] [PMID: 27817833]

[37] Hou R, Xu Y, Wang Z. Review of OPFRs in animals and humans: Absorption, bioaccumulation, metabolism, and internal exposure research. Chemosphere 2016; 153: 78-90. [http://dx.doi.org/10.1016/j.chemosphere.2016.03.003] [PMID: 27010170]

[38] Covaci A, Voorspoels S, Ramos L, Neels H, Blust R. Recent developments in the analysis of brominated flame retardants and brominated natural compounds. J Chromatogr A 2007; 1153(1-2): 145-71. [http://dx.doi.org/10.1016/j.chroma.2006.11.060] [PMID: 17156788]

[39] Kuramochi H, Takigami H, Scheringer M, Sakai S. Estimation of physicochemical properties of 52 non-PBDE brominated flame retardants and evaluation of their overall persistence and long-range transport potential. Sci Total Environ 2014; 491-492: 108-17. [http://dx.doi.org/10.1016/j.scitotenv.2014.04.004] [PMID: 24802073]

[40] Tokarz JA III, Ahn M-Y, Leng J, Filley TR, Nies L. Reductive debromination of polybrominated diphenyl ethers in anaerobic sediment and a biomimetic system. Environ Sci Technol 2008; 42(4): 1157-64. [http://dx.doi.org/10.1021/es071989t] [PMID: 18351087]

[41] Meylan WM, Howard PH. Computer estimation of the atmospheric gas-phase reaction rate of organic compounds with hydroxyl radicals and ozone. Chemosphere 1993; 26(12): 2293-9. [http://dx.doi.org/10.1016/0045-6535(93)90355-9]

[42] Shen L, Reiner EJ, Helm PA, *et al.* Historic trends of dechloranes 602, 603, 604, dechlorane plus and other norbornene derivatives and their bioaccumulation potential in lake ontario. Environ Sci Technol 2011; 45(8): 3333-40. [http://dx.doi.org/10.1021/es104328r] [PMID: 21434636]

[43] Bergman A, Rydén A, Law RJ, *et al.* A novel abbreviation standard for organobromine, organochlorine and organophosphorus flame retardants and some characteristics of the chemicals. Environ Int 2012; 49: 57-82. [http://dx.doi.org/10.1016/j.envint.2012.08.003] [PMID: 22982223]

[44] Wei G-L, Li D-Q, Zhuo M-N, *et al.* Organophosphorus flame retardants and plasticizers: sources, occurrence, toxicity and human exposure. Environ Pollut 2015; 196: 29-46. [http://dx.doi.org/10.1016/j.envpol.2014.09.012] [PMID: 25290907]

[45] Gerecke AC, Hartmann PC, Heeb NV, *et al.* Anaerobic degradation of decabromodiphenyl ether. Environ Sci Technol 2005; 39(4): 1078-83. [http://dx.doi.org/10.1021/es048634j] [PMID: 15773480]

[46] Gerecke AC, Giger W, Hartmann PC, *et al.* Anaerobic degradation of brominated flame retardants in sewage sludge. Chemosphere 2006; 64(2): 311-7. [http://dx.doi.org/10.1016/j.chemosphere.2005.12.016] [PMID: 16442150]

[47] Hunziker RW, Gonsior S, MacGregor JA, Desjardins D, Ariano J, Friederich U. Fate and effect of hexabromocyclododecane in the environment. Organohalogen Compd 2004; 66: 2300-5.

[48] Davis JW, Gonisor S, West B, Marty G. Evaluation of the aerobic and anaerobic transformation of hexabromocyclododecane in soil and sediments. The Fifth Annual Workshop on Brominated Flame Retardants in the Environment Book of Abstracts. 31.

[49] EFSA, Scientific opinion on emerging and novel brominated flame retardants (BFRs) in food. EFSA Journal [Online] 2012, http://www.efsa.europa.eu/en/efsajournal/pub/2908.htm# [accessed August

2015];

[50] DeCarlo VJ. Studies on brominated chemicals in the environment. Ann N Y Acad Sci 1979; 320(1): 678-81. [http://dx.doi.org/10.1111/j.1749-6632.1979.tb56642.x] [PMID: 287406]

[51] Norén K, Meironyté D. Certain organochlorine and organobromine contaminants in Swedish human milk in perspective of past 20-30 years. Chemosphere 2000; 40(9-11): 1111-23. [http://dx.doi.org/10.1016/S0045-6535(99)00360-4] [PMID: 10739053]

[52] Macdonald RW, Bewers JM. Contaminants in the arctic marine environment: priorities for protection. ICES J Mar Sci 1996; 53(3): 537-63. [http://dx.doi.org/10.1006/jmsc.1996.0077]

[53] Lee HJ, Kim GB. An overview of polybrominated diphenyl ethers (PBDEs) in the marine environment. Ocean Sci J 2015; 50(2): 119-42. [http://dx.doi.org/10.1007/s12601-015-0010-8]

[54] Zhang Y, Wang W, Song J, *et al.* Environmental characteristics of polybrominated diphenyl ethers in marine system, with emphasis on marine organisms and sediments. BioMed Res Int 2016; 2016 [http://dx.doi.org/10.1155/2016/1317232]

[55] Mikula P, Svobodova Z. Brominated flame retardants in the environment: Their sources and effects (a review). Acta Vet Brno 2006; 75(4): 587-99. [http://dx.doi.org/10.2754/avb200675040587]

[56] Costa LG, Giordano G. Is decabromodiphenyl ether (BDE-209) a developmental neurotoxicant? Neurotoxicology 2011; 32(1): 9-24. [http://dx.doi.org/10.1016/j.neuro.2010.12.010] [PMID: 21182867]

[57] Branchi I, Capone F, Alleva E, Costa LG. Polybrominated diphenyl ethers: neurobehavioral effects following developmental exposure. Neurotoxicology 2003; 24(3): 449-62. [http://dx.doi.org/10.1016/S0161-813X(03)00020-2] [PMID: 12782110]

[58] Law K, Halldorson T, Danell R, *et al.* Bioaccumulation and trophic transfer of some brominated flame retardants in a Lake Winnipeg (Canada) food web. Environ Toxicol Chem 2006; 25(8): 2177-86. [http://dx.doi.org/10.1897/05-500R.1] [PMID: 16916037]

[59] Sühring R, Busch F, Fricke N, Kötke D, Wolschke H, Ebinghaus R. Distribution of brominated flame retardants and dechloranes between sediments and benthic fish — A comparison of a freshwater and marine habitat. Sci Total Environ 2016; 542, Part A: 578-85.

[60] Tian S, Zhu L, Bian J, Fang S. Bioaccumulation and metabolism of polybrominated diphenyl ethers in carp (*Cyprinus carpio*) in a water/sediment microcosm: important role of particulate matter exposure. Environ Sci Technol 2012; 46(5): 2951-8. [http://dx.doi.org/10.1021/es204011k] [PMID: 22313265]

[61] Sellström U, de Wit CA, Lundgren N, Tysklind M. Effect of sewage-sludge application on concentrations of higher-brominated diphenyl ethers in soils and earthworms. Environ Sci Technol 2005; 39(23): 9064-70. [http://dx.doi.org/10.1021/es051190m] [PMID: 16382926]

[62] Gu S-Y, Ekpeghere KI, Kim H-Y, *et al.* Brominated flame retardants in marine environment focused on aquaculture area: Occurrence, source and bioaccumulation. Sci Total Environ 2017; 601-602: 1182-91. [http://dx.doi.org/10.1016/j.scitotenv.2017.05.209] [PMID: 28605836]

[63] Nyholm JR, Asamoah RK, van der Wal L, Danielsson C, Andersson PL. Accumulation of polybrominated diphenyl ethers, hexabromobenzene, and 1,2-dibromo-4-(-,2-dibromoethyl)cyclohexane in earthworm (*Eisenia fetida*). Effects of soil type and aging. Environ Sci Technol 2010; 44(23): 9189-94. [http://dx.doi.org/10.1021/es1023288] [PMID: 21028802]

[64] Zhou T, Ross DG, DeVito MJ, Crofton KM. Effects of short-term *in vivo* exposure to polybrominated diphenyl ethers on thyroid hormones and hepatic enzyme activities in weanling rats. Toxicol Sci 2001; 61(1): 76-82.
[http://dx.doi.org/10.1093/toxsci/61.1.76] [PMID: 11294977]

[65] Hall AJ, Kalantzi OI, Thomas GO. Polybrominated diphenyl ethers (PBDEs) in grey seals during their first year of life--are they thyroid hormone endocrine disrupters? Environ Pollut 2003; 126(1): 29-37.
[http://dx.doi.org/10.1016/S0269-7491(03)00149-0] [PMID: 12860100]

[66] Meerts IA, Letcher RJ, Hoving S, *et al. In vitro* estrogenicity of polybrominated diphenyl ethers, hydroxylated PDBEs, and polybrominated bisphenol A compounds. Environ Health Perspect 2001; 109(4): 399-407.
[http://dx.doi.org/10.1289/ehp.01109399] [PMID: 11335189]

[67] Nakari T, Pessala P. *In vitro* estrogenicity of polybrominated flame retardants. Aquat Toxicol 2005; 74(3): 272-9.
[http://dx.doi.org/10.1016/j.aquatox.2005.06.004] [PMID: 16024102]

[68] Stoker TE, Cooper RL, Lambright CS, Wilson VS, Furr J, Gray LE. *In vivo* and *in vitro* anti-androgenic effects of DE-71, a commercial polybrominated diphenyl ether (PBDE) mixture. Toxicol Appl Pharmacol 2005; 207(1): 78-88.
[http://dx.doi.org/10.1016/j.taap.2005.05.010] [PMID: 16005038]

[69] Eriksson P, Viberg H, Jakobsson E, Örn U, Fredriksson A. A brominated flame retardant, 2,2′,4,4′,5-pentabromodiphenyl ether: uptake, retention, and induction of neurobehavioral alterations in mice during a critical phase of neonatal brain development. Toxicol Sci 2002; 67(1): 98-103.
[http://dx.doi.org/10.1093/toxsci/67.1.98] [PMID: 11961221]

[70] Viberg H, Fredriksson A, Eriksson P. Neonatal exposure to the brominated flame retardant 2,2′,4,4′,5-pentabromodiphenyl ether causes altered susceptibility in the cholinergic transmitter system in the adult mouse. Toxicol Sci 2002; 67(1): 104-7.
[http://dx.doi.org/10.1093/toxsci/67.1.104] [PMID: 11961222]

[71] Viberg H, Fredriksson A, Eriksson P. Neonatal exposure to polybrominated diphenyl ether (PBDE 153) disrupts spontaneous behaviour, impairs learning and memory, and decreases hippocampal cholinergic receptors in adult mice. Toxicol Appl Pharmacol 2003; 192(2): 95-106.
[http://dx.doi.org/10.1016/S0041-008X(03)00217-5] [PMID: 14550744]

[72] Viberg H, Fredriksson A, Jakobsson E, Örn U, Eriksson P. Neurobehavioral derangements in adult mice receiving decabrominated diphenyl ether (PBDE 209) during a defined period of neonatal brain development. Toxicol Sci 2003; 76(1): 112-20.
[http://dx.doi.org/10.1093/toxsci/kfg210] [PMID: 12915714]

[73] Hardy ML. The toxicology of the three commercial polybrominated diphenyl oxide (ether) flame retardants. Chemosphere 2002; 46(5): 757-77.
[http://dx.doi.org/10.1016/S0045-6535(01)00240-5] [PMID: 11999799]

[74] EFSA, Scientific opinion on hexabromocyclododecanes (HBCDDs) in food. EFSA Journal [Online] 2011, .http://www.efsa.europa.eu/en/efsajournal/doc/2296.pdf [Accessed August 2015].

[75] Feng M, Qu R, Li Y, Wei Z, Wang Z. Biochemical biomarkers in liver and gill tissues of freshwater fish *Carassius auratus* following *in vivo* exposure to hexabromobenzene. Environ Toxicol 2014; 29(12): 1460-70.
[http://dx.doi.org/10.1002/tox.21876] [PMID: 23804377]

[76] Frydrych B, Szymańska JA, Sala M, Bruchajzer E. The effect of selected aromatic bromine derivatives on the activity of glutathione peroxidase and transferase. Int J Occup Med Environ Health 2005; 18(2): 127-32.
[PMID: 16201203]

[77] Feng M, Qu R, Wang C, Wang L, Wang Z. Comparative antioxidant status in freshwater fish

Carassius auratus exposed to six current-use brominated flame retardants: a combined experimental and theoretical study. Aquat Toxicol 2013; 140-141: 314-23. [http://dx.doi.org/10.1016/j.aquatox.2013.07.001] [PMID: 23880106]

[78] Sun RB, Xi ZG, Zhang HS, Zhang W. Subacute effect of decabromodiphenyl ethane on hepatotoxicity and hepatic enzyme activity in rats. Biomed Environ Sci 2014; 27(2): 122-5. [PMID: 24625403]

[79] Liang X, Li W, Martyniuk CJ, *et al.* Effects of dechlorane plus on the hepatic proteome of juvenile Chinese sturgeon (*Acipenser sinensis*). Aquat Toxicol 2014; 148: 83-91. [http://dx.doi.org/10.1016/j.aquatox.2014.01.003] [PMID: 24463492]

[80] Wu ZY, Huang L, Gao YL, Guo QH, Sun YQ. The effects of Dechlorane Plus on toxicity. Appl Mech Mater 2013; 380-384: 4163-6. [http://dx.doi.org/10.4028/www.scientific.net/AMM.380-384.4163]

[81] Liu X, Ji K, Choi K. Endocrine disruption potentials of organophosphate flame retardants and related mechanisms in H295R and MVLN cell lines and in zebrafish. Aquat Toxicol 2012; 114-115: 173-81. [http://dx.doi.org/10.1016/j.aquatox.2012.02.019] [PMID: 22446829]

[82] WHO Environmental Health Criteria 209: Flame retardants: tris-(chloropropyl)phosphate and tris-(2-chloroethyl)phosphate. http://wwwinchemorg/documents/ehc/ehc/ehc44htm , [accessed October 2018];

[83] Betts KS. New thinking on flame retardants. Environ Health Perspect 2008; 116(5): A210-3. [http://dx.doi.org/10.1289/ehp.116-a210] [PMID: 18470294]

[84] REACH, Regulation (EC) No 1907/2006 of the European Parliament and of the Council of 18 December 2006 concerning the Registration, Evaluation, Authorisation and Restriction of Chemicals (REACH), establishing a European Chemicals Agency, amending Directive 1999/45/EC and repealing Council Regulation (EEC) No 793/93 and Commission Regulation (EC) No 1488/94 as well as Council Directive 76/769/EEC and Commission Directives 91/155/EEC, 93/67/EEC, 93/105/EC and 2000/21/EC. 2006http://eur-lex.europa.eu/legal-content/EN/TXT/?uri=CELEX:32006R1907 [accessed August 2015];

[85] SC Stockholm. Convention on Persistent Organic Pollutants. http://chm.pops.int/ [accessed August 2017];

[86] 2014 Directive, Commission Directive 2014/79/EU of 20 June 2014 amending Appendix C of Annex II to Directive 2009/48/EC of the European Parliament and of the Council on the safety of toys, as regards TCEP, TCPP and TDCP Text with EEA relevance , 2014https://eur-lex.europa.eu/legal-content/EN/ALL/?uri=celex%3A32014L0079 [accessed October 2018];

[87] EPA, Petition to Order Testing of the Chlorinated Phosphate Ester Cluster Flame Retardants (TCEP, TCPP AND TDCPP) Under Section 4(a) of the Toxic Substances Control Act , 2017 https://www.epa.gov/sites/production/files/2017-01/documents/cpe_test_ petition_appx_final_0.pdf [accessed October 2018];

[88] Poma G, Malarvannan G, Voorspoels S, *et al.* Determination of halogenated flame retardants in food: Optimization and validation of a method based on a two-step clean-up and gas chromatography–mass spectrometry. Food Control 2016; 65: 168-76. [http://dx.doi.org/10.1016/j.foodcont.2016.01.027]

[89] Poma G, Malysheva SV, Goscinny S, *et al.* Occurrence of selected halogenated flame retardants in Belgian foodstuff. Chemosphere 2018; 194: 256-65. [http://dx.doi.org/10.1016/j.chemosphere.2017.11.179] [PMID: 29216545]

[90] Xu F, García-Bermejo Á, Malarvannan G, Gómara B, Neels H, Covaci A. Multi-contaminant analysis of organophosphate and halogenated flame retardants in food matrices using ultrasonication and vacuum assisted extraction, multi-stage cleanup and gas chromatography-mass spectrometry. J Chromatogr A 2015; 1401: 33-41. [http://dx.doi.org/10.1016/j.chroma.2015.05.001] [PMID: 25997844]

[91] Xu F, Tay J-H, Covaci A, *et al.* Assessment of dietary exposure to organohalogen contaminants, legacy and emerging flame retardants in a Norwegian cohort. Environ Int 2017; 102: 236-43. [http://dx.doi.org/10.1016/j.envint.2017.03.009] [PMID: 28335995]

[92] Tao F, Matsukami H, Suzuki G, *et al.* Emerging halogenated flame retardants and hexabromocyclododecanes in food samples from an e-waste processing area in Vietnam. Environ Sci Process Impacts 2016; 18(3): 361-70. [http://dx.doi.org/10.1039/C5EM00593K] [PMID: 26843139]

[93] Aznar-Alemany Ò, Aminot Y, Vilà-Cano J, *et al.* Halogenated and organophosphorus flame retardants in European aquaculture samples. Sci Total Environ 2018; 612: 492-500. [http://dx.doi.org/10.1016/j.scitotenv.2017.08.199] [PMID: 28865267]

[94] Morris AD, Muir DCG, Solomon KR, Teixeira CF, Duric MD, Wang X. Bioaccumulation of polybrominated diphenyl ethers and alternative halogenated flame retardants in a vegetation–caribou–wolf food chain of the Canadian Arctic. Environ Sci Technol 2018; 52(5): 3136-45. [http://dx.doi.org/10.1021/acs.est.7b04890] [PMID: 29320633]

[95] Sun RX, Luo XJ, Tan XX, Tang B, Li ZR, Mai BX. Legacy and emerging halogenated organic pollutants in marine organisms from the Pearl River Estuary, South China. Chemosphere 2015; 139: 565-71. [http://dx.doi.org/10.1016/j.chemosphere.2015.07.044] [PMID: 26318119]

[96] Pedro S, Fisk AT, Tomy GT, *et al.* Mercury and persistent organic pollutants in native and invading forage species of the Canadian Arctic: Consequences for food web dynamics. Environ Pollut 2017; 229: 229-40. [http://dx.doi.org/10.1016/j.envpol.2017.05.085] [PMID: 28599207]

[97] Zhang H, Bayen S, Kelly BC. Co-extraction and simultaneous determination of multi-class hydrophobic organic contaminants in marine sediments and biota using GC-EI-MS/MS and LC-ES--MS/MS. Talanta 2015; 143: 7-18. [http://dx.doi.org/10.1016/j.talanta.2015.04.084] [PMID: 26078122]

[98] Sapozhnikova Y, Lehotay SJ. Multi-class, multi-residue analysis of pesticides, polychlorinated biphenyls, polycyclic aromatic hydrocarbons, polybrominated diphenyl ethers and novel flame retardants in fish using fast, low-pressure gas chromatography-tandem mass spectrometry. Anal Chim Acta 2013; 758: 80-92. [http://dx.doi.org/10.1016/j.aca.2012.10.034] [PMID: 23245899]

[99] Widelka M, Lydy MJ, Wu Y, Chen D. Statewide surveillance of halogenated flame retardants in fish in Illinois, USA. Environ Pollut 2016; 214: 627-34. [http://dx.doi.org/10.1016/j.envpol.2016.04.063] [PMID: 27131823]

[100] Verreault J, Letcher RJ, Gentes M-L, Braune BM. Unusually high Deca-BDE concentrations and new flame retardants in a Canadian Arctic top predator, the glaucous gull. Sci Total Environ 2018; 639: 977-87. [http://dx.doi.org/10.1016/j.scitotenv.2018.05.222] [PMID: 29929336]

[101] Zeng YH, Luo XJ, Tang B, Mai BX. Habitat- and species-dependent accumulation of organohalogen pollutants in home-produced eggs from an electronic waste recycling site in South China: Levels, profiles, and human dietary exposure. Environ Pollut 2016; 216: 64-70. [http://dx.doi.org/10.1016/j.envpol.2016.05.039] [PMID: 27239689]

[102] Huang CC, Zeng YH, Luo XJ, *et al.* Level changes and human dietary exposure assessment of halogenated flame retardant levels in free-range chicken eggs: A case study of a former e-waste recycling site, South China. Sci Total Environ 2018; 634: 509-15. [http://dx.doi.org/10.1016/j.scitotenv.2018.03.386] [PMID: 29631140]

[103] Vetter W, Gallistl C, Schlienz A, Preston T, Müller J, von der Trenck KT. Brominated flame retardants (BFRs) in eggs from birds of prey from Southern Germany, 2014. Environ Pollut 2017;

231(Pt 1): 569-77.
[http://dx.doi.org/10.1016/j.envpol.2017.08.047] [PMID: 28843896]

[104] Stolker AAM, Peters RJB, Zuiderent R, DiBussolo JM, Martins CPB. Fully automated screening of veterinary drugs in milk by turbulent flow chromatography and tandem mass spectrometry. Anal Bioanal Chem 2010; 397(7): 2841-9.
[http://dx.doi.org/10.1007/s00216-010-3660-5] [PMID: 20379812]

[105] Llorca M, Pérez F, Farré M, Agramunt S, Kogevinas M, Barceló D. Analysis of perfluoroalkyl substances in cord blood by turbulent flow chromatography coupled to tandem mass spectrometry. Sci Total Environ 2012; 433: 151-60.
[http://dx.doi.org/10.1016/j.scitotenv.2012.05.080] [PMID: 22789815]

[106] Barón E, Eljarrat E, Barceló D. Gas chromatography/tandem mass spectrometry method for the simultaneous analysis of 19 brominated compounds in environmental and biological samples. Anal Bioanal Chem 2014; 406(29): 7667-76.
[http://dx.doi.org/10.1007/s00216-014-8196-7] [PMID: 25260412]

[107] Björklund J, Tollbäck P, Hiärne C, Dyremark E, Östman C. Influence of the injection technique and the column system on gas chromatographic determination of polybrominated diphenyl ethers. J Chromatogr A 2004; 1041(1-2): 201-10.
[http://dx.doi.org/10.1016/j.chroma.2004.04.025] [PMID: 15281270]

[108] Van den Eede N, Dirtu AC, Ali N, Neels H, Covaci A. Multi-residue method for the determination of brominated and organophosphate flame retardants in indoor dust. Talanta 2012; 89: 292-300.
[http://dx.doi.org/10.1016/j.talanta.2011.12.031] [PMID: 22284495]

[109] Tomy GT, Pleskach K, Oswald T, *et al.* Enantioselective bioaccumulation of hexabromocyclododecane and congener-specific accumulation of brominated diphenyl ethers in an eastern Canadian Arctic marine food web. Environ Sci Technol 2008; 42(10): 3634-9.
[http://dx.doi.org/10.1021/es703083z] [PMID: 18546701]

[110] Gentes M-L, Letcher RJ, Caron-Beaudoin E, Verreault J. Novel flame retardants in urban-feeding ring-billed gulls from the St. Lawrence River, Canada. Environ Sci Technol 2012; 46(17): 9735-44.
[http://dx.doi.org/10.1021/es302099f] [PMID: 22845168]

[111] Zhang Y, Luo X-J, Wu J-P, *et al.* Contaminant pattern and bioaccumulation of legacy and emerging organhalogen pollutants in the aquatic biota from an e-waste recycling region in South China. Environ Toxicol Chem 2010; 29(4): 852-9.
[http://dx.doi.org/10.1002/etc.122] [PMID: 20821514]

[112] Bendig P, Hägele F, Vetter W. Widespread occurrence of polyhalogenated compounds in fat from kitchen hoods. Anal Bioanal Chem 2013; 405(23): 7485-96.
[http://dx.doi.org/10.1007/s00216-013-7194-5] [PMID: 23877177]

[113] Möller A, Xie Z, Sturm R, Ebinghaus R. Large-scale distribution of dechlorane plus in air and seawater from the Arctic to Antarctica. Environ Sci Technol 2010; 44(23): 8977-82.
[http://dx.doi.org/10.1021/es103047n] [PMID: 21047104]

[114] Barón E, Eljarrat E, Barceló D. Analytical method for the determination of halogenated norbornene flame retardants in environmental and biota matrices by gas chromatography coupled to tandem mass spectrometry. J Chromatogr A 2012; 1248: 154-60.
[http://dx.doi.org/10.1016/j.chroma.2012.05.079] [PMID: 22695692]

[115] Eljarrat E, de la Cal A, Barceló D. Determination of decabromodiphenyl ether in sediments using selective pressurized liquid extraction followed by GC-NCI-MS. Anal Bioanal Chem 2004; 378(3): 610-4.
[http://dx.doi.org/10.1007/s00216-003-2220-7] [PMID: 14513196]

[116] Labunska I, Abdallah MA-E, Eulaers I, *et al.* Human dietary intake of organohalogen contaminants at e-waste recycling sites in Eastern China. Environ Int 2015; 74: 209-20.
[http://dx.doi.org/10.1016/j.envint.2014.10.020] [PMID: 25454238]

[117] Koesukwiwat U, Lehotay SJ, Miao S, Leepipatpiboon N. High throughput analysis of 150 pesticides in fruits and vegetables using QuEChERS and low-pressure gas chromatography-time-of-flight mass spectrometry. J Chromatogr A 2010; 1217(43): 6692-703. [http://dx.doi.org/10.1016/j.chroma.2010.05.012] [PMID: 20627307]

[118] Malysheva SV, Goscinny S, Malarvannan G, *et al.* Development and validation of a quantitative UHPLC-MS/MS method for selected brominated flame retardants in food. Food Addit Contam Part A Chem Anal Control Expo Risk Assess 2018; 35(2): 292-304. [http://dx.doi.org/10.1080/19440049.2017.1393110] [PMID: 29087783]

[119] Abdallah MA-E, Harrad S. Tetrabromobisphenol-A, hexabromocyclododecane and its degradation products in UK human milk: relationship to external exposure. Environ Int 2011; 37(2): 443-8. [http://dx.doi.org/10.1016/j.envint.2010.11.008] [PMID: 21167604]

[120] Guerra P, de la Torre A, Martínez MA, Eljarrat E, Barceló D. Identification and trace level determination of brominated flame retardants by liquid chromatography/quadrupole linear ion trap mass spectrometry. Rapid Commun Mass Spectrom 2008; 22(7): 916-24. [http://dx.doi.org/10.1002/rcm.3443] [PMID: 18302167]

[121] Feng AH, Chen SJ, Chen MY, He MJ, Luo XJ, Mai BX. Hexabromocyclododecane (HBCD) and tetrabromobisphenol A (TBBPA) in riverine and estuarine sediments of the Pearl River Delta in southern China, with emphasis on spatial variability in diastereoisomer- and enantiomer-specific distribution of HBCD. Mar Pollut Bull 2012; 64(5): 919-25. [http://dx.doi.org/10.1016/j.marpolbul.2012.03.008] [PMID: 22475420]

[122] Chen D, Letcher RJ, Chu S. Determination of non-halogenated, chlorinated and brominated organophosphate flame retardants in herring gull eggs based on liquid chromatography-tandem quadrupole mass spectrometry. J Chromatogr A 2012; 1220: 169-74. [http://dx.doi.org/10.1016/j.chroma.2011.11.046] [PMID: 22186491]

[123] Chen D, Martin P, Burgess NM, *et al.* European starlings (*Sturnus vulgaris*) suggest that landfills are an important source of bioaccumulative flame retardants to Canadian terrestrial ecosystems. Environ Sci Technol 2013; 47(21): 12238-47. [http://dx.doi.org/10.1021/es403383e] [PMID: 24059974]

[124] Brandsma SH, de Boer J, Leonards PEG, Cofino WP, Covaci A, Leonards PEG. Organophosphorus flame-retardant and plasticizer analysis, including recommendations from the first worldwide interlaboratory study. *TrAC-*. Trends Analyt Chem 2013; 43: 217-28. [http://dx.doi.org/10.1016/j.trac.2012.12.004]

[125] Aznar-Alemany Ò, Trabalón L, Jacobs S, *et al.* Occurrence of halogenated flame retardants in commercial seafood species available in European markets. Food Chem Toxicol 2017; 104: 35-47. [http://dx.doi.org/10.1016/j.fct.2016.12.034] [PMID: 28027981]

CHAPTER 5

Dioxins and PCBs in Food and Feed Matrices: Advances in Physico-Chemical Methods and EU Regulatory Framework

Jordi Parera, Manuela Ábalos and **Esteban Abad***

Laboratorio de Dioxinas, Departamento de Química Ambiental, IDAEA-CSIC, C/ Jordi Girona, 18-26, 08034 Barcelona, Spain

Abstract: Polychlorinated dibenzo-p-dioxins (PCDDs), polychlorinated dibenzofurans (PCDFs) and polychlorinated biphenyls (PCBs) are major representatives of persistent organic pollutants. While PCDD/Fs are unwanted by-products, mainly from waste incineration and industrial processes, PCBs were manufactured and widely used as transformer oils until bans enter in force at the late '70s. These compounds are highly toxic and can easily bioaccumulate and biomagnify throughout the food chain reaching the top living organisms, including human beings. Food is the main route of human exposure to PCDD/Fs and PCBs, with products from animal origin contributing largely to the dietary intake. In this sense, several contamination episodes involving feed and food products that occurred at the late '90s led to the establishment of a European regulatory framework that aims to both, set maximum levels for these compounds in different food/feed categories and to lay down analytical methods for the determination of these compounds. In this work, an overview of the different chemical methodologies that have been applied during the last decades to the determination of PCDD/Fs and PCBs, more in particular dioxin-like PCBs, in food and feed samples is presented. Advances in extraction and purification steps are described, but special attention is given to the evaluation of several mass spectrometric techniques in comparison to gas chromatography coupled to high-resolution mass spectrometry (GC-HRMS), which has traditionally been the unique confirmatory technique until recently.

Keywords: Clean-up, Dioxin-like PCBs, EU regulations, Extraction, Feed, Food, GC-HRMS, GC-MS/MS, Ion trap, Mass spectrometry, PCBs, PCDDs, PCDFs, Triple quadrupole.

---

* **Corresponding author Esteban Abad:** Laboratorio de Dioxinas, Departamento de Química Ambiental, IDAEA-CSIC, C/ Jordi Girona, 18-26, 08034 Barcelona, Spain; Tel: +34 934006185; E-mail: esteban.abad@idaea.csic.es

Belén Gómara & María Luisa Marina (Eds.)
Current and Future Developments in Food Science (Vol. 1)

INTRODUCTION

Polychlorinated dibenzo-*p*-dioxins (PCDDs) and polychlorinated dibenzofurans (PCDFs) are two families of compounds that comprise a total of 210 different congeners (*i.e.*, 75 PCDDs and 135 PCDFs).

They are also commonly known as dioxins and furans, respectively, or only as dioxins. Their molecule is characterized by two benzene rings connected by one oxygen atom, in the case of furans, or two oxygen atoms, in the case of dioxins. Each one of the benzene rings can have several chlorine atoms; therefore a total of eight homologues groups can be defined taking into account the degree of chlorination, from one chlorine atom present (mono-) to a maximum of eight chlorine atoms (octa-). This particular molecular structure presents high stability; in consequence, dioxins are characterized by their resistance to chemical and biological degradation.

Dioxins were first identified in fly ash and gas emissions from incinerator facilities at the end of the '70s [1]. Nowadays, it is well-known that these compounds are unwanted by-products mainly originated from anthropogenic activities, being the major source of waste incineration and some specific industrial processes (*e.g.*, cement kilns, pulp and paper mills, sintering plants or pesticide manufacturing). There are also diffuse sources, such as vehicle exhaust [2, 3].

The toxic effects of PCDD/Fs to the living organisms, and in particular to humans, have been widely studied. It has been demonstrated that they act as endocrine disruptors and are associated with cancer risk [4, 5]. However, it has to be remarked that among the 210 PCDD/F congeners, only those with chlorine atoms at least in the 2,3,7,8 positions of the molecule have been found to show toxicological activity. This way, the number of target congeners to be considered for analysis is reduced to 17 (*i.e.*, 7 PCDDs and 10 PCDFs).

On the other hand, polychlorinated biphenyls (PCBs) are another family of compounds, which includes a total of 209 congeners. For these compounds, the degree of chlorination varies from one (mono-) to ten (deca-) chlorine atoms present at the molecule. Contrary to what it has been mentioned about PCDD/F sources, PCBs were manufactured and commercialized in 1929, mainly as electrical insulating fluids, and they were widely used until restrictions to their production and application came into force between the end of the '70s and the beginning of the '80s [6]. Among the 209 PCB congeners, those without chlorine atoms in the *ortho* position of the molecule (*i.e.*, 4 non-*ortho* PCBs) and those with only one chlorine atom in this position (*i.e.*, 8 mono-*ortho* PCBs) show similar physico-chemical characteristics and toxicological activity to dioxins.

These 12 PCBs are the so-called dioxin-like PCBs (dl-PCBs).

Due to the extreme stability of dioxins and PCBs, they are highly persistent once they have been released in the environment and can be transported over long distances. Besides, the lipophilic character of these compounds leads them to bioaccumulate and to biomagnify throughout the food chain reaching the top living organisms, including human beings. All these particularities, together with their proved high toxicity even at very low concentrations (trace levels), have placed dioxins and PCBs among the most representative persistent organic pollutants (POPs).

The toxic effects of dioxin-like compounds are related to the interaction with the aryl hydrocarbon receptor (AhR) on the cells [7]. Although all toxic congeners present a similar mechanism of interaction, the toxicological potential varies depending on the degree of chlorination and the distribution of the chlorine atoms at the molecule. In general, for PCDD/Fs, toxicity decreases when the number of chlorine atoms increases, being 2,3,7,8-tetrachlorodibenzo-*p*-dioxin (2,3,7,8-TCDD) the most toxic dioxin congener. Taking this into account, a methodology based on the assignment of relative value, defined as the toxic equivalency factor (TEF), to each dioxin-like compound has been proposed. The individual TEFs are related to the maximum value assigned to the 2,3,7,8-TCDD (TEF=1), this way the toxicity of a sample due to the presence of PCDD/Fs and dl-PCBs can be calculated as the sum of the products obtained by multiplying the concentration of each congener by its corresponding TEF. This final result (sum) is known as the total toxic equivalent (TEQ) and allows the comparison of samples in terms of their toxicological potential.

Several TEF schemes have been established with slight differences in the assigned TEF values. In particular, panel of expert of the World Health Organization (WHO) established consensual TEFs for dioxins and dl-PCBs for human, fish and wildlife risk assessment in 1997 [8]; later on, in 2005, these values were updated after a first re-evaluation [9]. Table **1** shows a comparison between the first WHO-TEF values assigned (1998) and the revised ones (2005). The WHO-TEF scheme is important since it was the one adopted by the end of the '90s to set tolerable weekly/daily/monthly intakes. Moreover, this scheme has later been used to establish maximum levels, expressed in TEQ, for PCDD/Fs and dl-PCBs in several food and feed products.

Levels of dioxins and PCBs increased dramatically until the late '70s or early '80s, due to the lack of knowledge about their presence in the environment and the adverse effects of these compounds for the living organisms. Since then, significant efforts have been made to adopt strategies to reduce unintentional releases and to restrict or ban the production of those compounds that have commercial use. At the '80s, countries such as Canada started to identify main sources of PCDD/F emissions to the environment and to carry out quantitative estimations of the levels that were released [10]. Moreover, measures were undertaken, for instance in facilities like waste incineration plants, to implement appropriate air cleaning systems with the aim to reduce PCDD/Fs in gas emissions [11]. In 2001, dioxins and PCBs were included in the first list of substances of the Stockholm Convention on POPs [12], which reflects the great concern that these pollutants raise in terms of protection of the environment and human health.

Dietary intake is the main route of human exposure to PCDD/Fs and PCBs. Due to the lipophilic character of these compounds, they tend to accumulate in the fatty tissues of the organisms and to biomagnify over the trophic chain. This way, food of animal origin contributes the most to the overall human exposure [13].

Table 1. Comparative WHO-TEF values from 2005, re-evaluated, and 1998 [8, 9].

PCDFs	**TEF_{2005}(TEF_{1998})**	**PCDDs**	**TEF_{2005}(TEF_{1998})**
2,3,7,8-TCDF	0.1 (0.1)	**2,3,7,8-TCDD**	1 (1)
1,2,3,7,8-PeCDF	**0.03** (0.05)	**1,2,3,7,8-PeCDD**	1 (1)
2,3,4,7,8-PeCDF	**0.3** (0.5)		
1,2,3,4,7,8-HxCDF	0.1 (0.1)	**1,2,3,4,7,8-HxCDD**	0.1 (0.1)
1,2,3,6,7,8-HxCDF	0.1 (0.1)	**1,2,3,6,7,8-HxCDD**	0.1 (0.1)
2,3,4,6,7,8-HxCDF	0.1 (0.1)	**1,2,3,7,8,9-HxCDD**	0.1 (0.1)
1,2,3,7,8,9-HxCDF	0.1 (0.1)		
1,2,3,4,6,7,8-HpCDF	0.01 (0.01)	**1,2,3,4,6,7,8-HpCDD**	0.01 (0.01)
1,2,3,4,7,8,9-HpCDF	0.01 (0.01)		
OCDF	**0.0003** (0.0001)	**OCDD**	**0.0003** (0.0001)
non-*ortho* PCBs	**TEF_{2005}(TEF_{1998})**	**mono-*ortho* PCBs**	**TEF_{2005}(TEF_{1998})**
PCB-77	0.0001 (0.0001)	**PCB-105**	**0.00003** (0.0001)
PCB-81	**0.0003** (0.0001)	**PCB-114**	**0.00003** (0.0005)
PCB-126	0.1 (0.1)	**PCB-118**	**0.00003** (0.0001)
PCB-169	**0.03** (0.01)	**PCB-123**	**0.00003** (0.0001)
		PCB-156	**0.00003** (0.0005)
		PCB-157	**0.00003** (0.0005)
		PCB-167	**0.00003** (0.00001)
		PCB-189	**0.00003** (0.0001)

During the '90s, several food contamination incidents generated a great concern, not only among scientists but to the general population as well. At the end of 1997, an increase of PCDD/Fs in milk was observed in Germany, which was finally related to the use of a contaminated citrus pulp from Brazil as a raw material in feed for cows [14]. However, the most significant contamination episode took place in Belgium, starting in January 1999. The origin was animal feed that appeared to contain very high concentrations of dioxins and PCBs as a result of the accidental addition of PCB oil to a recycled fat intended for animal feeding purposes [15]. The source of contamination was finally discovered after several months, but the consequences were already dramatic. This incident led to a major food crisis that made a difference in the strategy adopted by the EU to deal with the presence of PCDD/Fs and PCBs in food and feed products.

The need to undertake actions with the aim to protect human health, as well as the environment, from the toxic effects of these compounds was clearly stated. In this regard, a legal framework was established in 2001 setting maximum levels and action levels for PCDD/Fs, expressed in WHO-TEQ, in different food and feed category products. These levels have been revised over the last two decades. In a first revision, maximum levels were also established for PCDD/Fs and dl-PCBs together. Moreover, since 2011, with the publication of the latest updates related to this legal framework, limit values have also been set for 6 non-dioxin-like PCBs (NDL-PCBs). Since these compounds do not have associated TEFs, maximum levels are expressed in concentration basis for the sum of the 6 NDL-PCBs. On the other hand, the re-evaluated TEF values from 2005 were already considered for the revision of the WHO-TEQ maximum levels in recent Regulations for food [16] and feed [17].

The EU strategy for protecting human health and the environment from PCDD/F and PCB toxic effects is based on two main intervention lines: 1) reduction of their presence in the environment; and 2) reduction of their presence in feed and food. Considering that food products of animal origin are the predominant source of human exposure to these pollutants, monitoring on feed is imperative for proper control of the presence of PCDD/Fs and PCBs along the food chain. However, despite the fact that great efforts have been carried out in this regard, several incidents have still occurred in the last years related to high levels of dioxins and PCBs in feed. In almost all cases the problem originates from the use of contaminated raw materials, either presenting unexpected high levels themselves (*e.g.*, potato peels impregnated with kaolinitic clays) or becoming unintentionally contaminated during their production process (*e.g.*, recycled fat from gelatin production, dried breadcrumb from waste bread) [18]. In the light of these incidents, a new Regulation entered in force in 2012 setting requirements on the production and control of the presence of PCDD/Fs and PCBs in fats and oils used as feed ingredients, with the aim to achieve further improvements on the reduction of the levels of these compounds in the final food products [19].

The EU regulatory framework for PCDD/Fs and PCBs is not just limited to the establishment of maximum/action levels and to implement monitoring programs. Moreover, different Directives and Regulations have been issued laying down methods of sampling and analysis. These documents do not focus on detailed procedural steps, but instead specific requirements are set out that a certain procedure has to fulfil. The requirements refer to all the analytical steps (*i.e.*, extraction, purification and instrumental determination), for instance, labelled standards must be added prior to the extraction to allow isotope dilution quantification, as well as to control analytical recoveries. Among all the requirements described those more directly related with instrumental techniques are of particular attention. In this sense, it has to be pointed out that, while concentrations of most PCBs (*e.g.*, the 6 regulated NDL-PCBs) present in food and feed can be determined by different techniques, including GC-ECD, GC-MS, *etc*; comparatively, much lower levels of dioxins and dl-PCBs are usually found in the same kind of matrices. Therefore, in order to properly identify and quantify these compounds at trace levels the following basic requirements have to be fulfilled:

- High sensitivity and low detection limits. Detection must be at the lower picogram (pg) or upper femtogram (fg) level, and the limit of quantification shall be about one-fifth of the maximum level.
- High selectivity and specificity. The analytical methodology must allow clear differentiation between PCDD/Fs and dl-PCBs and other interfering compounds present at higher concentrations. Besides, a differentiation between congeners is mandatory with the aim to discriminate the toxic ones. For MS techniques two specific ions from the molecular cluster can be monitored allowing calculating the isotope ratio between them, this way the comparison of this value with the theoretical one serves as quality criteria for selectivity.
- High accuracy (trueness and precision). The result obtained should be close to the real value with as low dispersion as possible.

In 2001, when the legislative framework was established, gas chromatography coupled to high-resolution mass spectrometry (GC-HRMS) proved to be the unique technique able to fulfil all these requirements and therefore that could be considered a confirmatory technique. Only recently, in 2014, gas chromatography coupled to tandem mass spectrometry (GC-MS/MS) has also been included as a confirmatory technique. Confirmatory methods enable the unequivocal identification and quantification of PCDD/Fs and dl-PCBs at the maximum level or even at the action threshold. Conversely, screening methods allow to select those samples with levels that exceed the maximum levels or the action thresholds and can be based either on bioanalytical or physico-chemical (*i.e.*, GC-MS) methods.

Bioanalytical screening methods have not been included for discussion within the present work. On the contrary, attention is focused on physico-chemical methods. Therefore, the aim is to give an overview of the different chemical methodologies that have been applied to the analysis of PCDD/Fs and PCBs, more in particular dl-PCBs, in food and feed samples during the last decades. Advances in extraction and purification steps are described, but special attention has been paid to the assessment of the performance of several mass spectrometric techniques. Moreover, discussion is given about how these advances allow meeting the criteria included in the EU regulatory framework.

ADVANCES IN EXTRACTION AND PURIFICATION TECHNIQUES

Among the requirements described at the EU Regulations for the analysis of PCDD/Fs, dl-PCBs and NDL-PCBs in food and feed there is a specific mention about the control of recoveries by the addition of labelled standards at the beginning of the analysis. In general terms, recoveries of the individual labelled compounds must be between 60% and 120% [20, 21]. In this sense, the classical Soxhlet extraction is an easy and cost-effective technique able to fulfil these recovery criteria. PCBs can be extracted from food and feed samples by non-polar/medium-polar solvents (*e.g.*, *n*-hexane) or solvent mixtures (*e.g.*, *n*-hexane:dichloromethane); while higher polarity is usually needed to release PCDD/Fs from the matrices (*e.g.*, acetone, toluene, cyclohexane:toluene).

Taking into account the previous know-how about the performance of polar and non-polar solvent extraction mixtures, new methodologies have been developed based on the application of automated pressurised-liquid extraction (PLE) systems. Samples are subject to high temperatures and pressures, this way extraction efficiency (recoveries) can be improved and extraction time is dramatically reduced compared to classic Sohxlet procedures. Non-polar organic solvents, such as *n*-hexane [22], as well as mixtures of solvents, such as *n*-hexane:acetone (4:1, *v/v*) and *n*-hexane:dichloromethane (1:1, *v/v*) [23] have been used for the analysis of PCDD/Fs and PCBs in several meat samples and for PCBs in fish, respectively. Alternatively, Kitamura *et al.* proposed the use of dimethyl sulfoxide (DMSO):acetonitrile (1:9, *v/v*) as a solvent extraction mixture, combined with DMSO/acetonitrile/*n*-hexane partitioning, for PCDD/Fs and non-*ortho* PCBs in lipid-rich matrices (*i.e.*, meat and fish). This solvent mixture allows about 99% reduction of the co-extracted fat compared with other mixtures such as acetone:*n*-hexane (1:1, *v/v*) and consequently, no further sulfuric acid clean-up treatment has to be done to remove lipids from the extracts [24]. The main disadvantage of PLE systems is the need for an exhaustively cleaning step of the cells to prevent cross-contamination. In addition, limitations on the volume of commercial extraction cells can also be a drawback, particularly when PCDD/F analysis is intended on low-lipid content matrices (1-5% of lipids), since a larger amount of sample has to be extracted to reach enough sensitivity.

Although other extraction techniques, such as microwave-assisted extraction (MAE) or matrix solid-phase dispersion (MSPD), have also been reported in the literature for the analysis of PCDD/Fs and PCBs in food and feed samples, only very few works are related to their use. Weichbrodt *et al.* applied a MAE procedure for the determination of PCBs and other organochlorine compounds in fish fillets and cod liver [25]. On the other hand, Carro *et al.* studied the applicability of MSPD to the analysis of PCBs and brominated compounds in aquaculture feed and marine species [26]. Finally, solid phase extraction (SPE) could be applied to liquid food samples, but has mostly been used for the analysis of biological samples, such as human milk [27, 28].

Once the analytes have been extracted from a sample, the next step in most analytical methods is a clean-up process to remove co-extracted interfering compounds. Generally, clean-up is a crucial step in the analysis of dioxins and dl-PCBs because, even though GC-HRMS provides accurate selectivity and sensitivity, good quality of the final extracts is mandatory to fulfil the requirements indicated at the EU Regulations. Normally, removal of lipids and other oxidizable compounds present in the extract is the first step in the clean-up process of food and feed samples. To this end, the use of sulfuric acid digestion was already proposed in the '70s [29]. Nowadays, this approach is still commonly applied through the use of acid silica columns. Alternative lipid removal methods, based on size exclusion processes, such as gel permeation chromatography (GPC), have also been developed for dioxin analysis [22]. GPC systems have the advantage that they can be automated and columns can be re-used; however, on the other hand, one of the main drawbacks is that repeated GPC series may be required for complete purification of the extracts depending on the lipid content of the samples.

After the first purification step, fractionation is carried out to separate dioxins from PCBs and from other remaining compounds in the extracts. A classical manual solid-liquid adsorption chromatography approach based on the sequential use of a multilayer silica column followed by a Florisil or an alumina column is widely accepted. In addition, a complementary fractionation step involving activated carbon is commonly applied to improve the quality of the final extracts.

Other adsorbents, for instance, one based on a 2-(1pyrenyl)ethyl phase have also been evaluated for PCDD/Fs and PCBs [30].

Manual fractionation schemes are time-consuming, increase the risk of cross-contamination and, because of the use of open columns, there is a risk of human exposure to the solvents. Thus, automation of this process has been proposed, based on pressurized pre-packaged columns, with the aim to reduce sample manipulation and analysis time. The first fully automated clean-up system specially designed for the analysis of PCDD/Fs and PCBs was released on the market in the late 1990s. The instrument allowed unattended pre-conditioning of a set of columns, loading of the sample and fractionation to obtain a final extract ready for instrumental analysis in about 2h [31].

Within the development of PLE methodologies for PCDD/F and PCB analysis, some attempts about on-cell purification have been proposed to retain the co-extracted lipids. Björklund *et al.* studied the use of several adsorbents, such as Florisil, basic alumina, neutral alumina, acidic alumina and sulfuric acid-impregnated silica, as fat retainers in the selective extraction of PCBs from fatty samples. The best results were achieved with the latest adsorbent [32]. Later on, different fat/acid silica fat retainer ratios were checked using *n*-heptane as extraction solvent [33]. The optimised developed on-cell purification PLE method, followed by off-line carbon fractionation, was applied to the determination of PCDD/Fs and dl-PCBs in several food and feed matrices [34]. Other authors described the use of Florisil as an on-cell fat retainer in the analysis of PCBs in biota samples [35].

On the other hand, Haglund *et al.* evaluated a more complex approach aiming to carry out on-cell fractionation with carbon. To this end, a specially designed cell was filled with the sample and a carbon-celite mixture and three different fractions were obtained containing compounds separated in function of their chemical characteristics [36].

Automation of dioxin and PCB analysis was further improved by upgrading the fully automated clean-up system combining it with a PLE instrument. This way, extraction and purification/fractionation can be done unattended [37]. Automated systems proved to be less time consuming and allowed to reduce sample manipulation, decreasing cross contamination and the risk of human exposure to solvents. Therefore, these systems have shown to be suitable in routine analysis to obtain highly purified extracts in a short period. However, a high initial investment is required, and consumables (*i.e.*, pre-packaged columns) are expensive. Moreover, large volumes of solvents are used. In addition, when matrices with high lipid content are processed under these systems, a previous off-line purification is recommended in order to preserve the system from clogging, overpressure and leaks.

Recently, last advances in fully automated clean-up systems have focused on improving the valve design, allowing to dramatically reduce the solvent volume needed while maintaining adequate efficiency [38, 39]. Semi-automated systems are also commercially available that carry out unattended clean-up after a manual pre-conditioning of the columns and load of the sample extract [40]. However, for both automated and semi-automated clean-up systems, the main drawback is still the high investment needed to start working with this kind of instruments. In this sense, to satisfy the demand of most cost-effective clean-up systems, simply and more economic manual commercial kits have also been developed. Based on the features of the automated methods, such as labor reduction, short analysis time or the use of disposable columns, systems involving the use of a vacuum pump [41] or pressurized nitrogen gas [42] are available in the market.

ADVANCES IN MASS SPECTROMETRY TECHNIQUES

GC-HRMS as the Reference Technique

The reference method for the analysis of PCDD/Fs and dl-PCBs has traditionally been gas chromatography coupled to high-resolution mass spectrometry (GC-HRMS). More concretely, mass spectrometers with magnetic sector type analyzers are the instruments that proved to fulfil the specific technical requirements to properly identify and quantify the individual toxic congeners at trace levels. High sensitivity is obtained from the optimization of the production of molecular ions in the electron impact (EI) source and by operating in the selected ion monitoring (SIM) mode. The particular geometrical characteristics of this type of analyzers allow a good performance of the process to carry the selected ions towards the detector, this way limits of detection (LODs) down to the low femtogram (fg) level can be achieved. Moreover, high selectivity results from the elevated mass resolution (>10,000; 10% valley) that magnetic sector instruments can reach. High mass resolution is needed to avoid interferences from ions of similar *m/z* ratio coming from other potentially interfering compounds present in the final extracts, generally at higher concentrations than the target analytes (*i.e.*, PCDD/Fs and dl-PCBs) [43].

Taking these into account, GC-HRMS became the unique accepted approach in most official international standard methods referred to PCDD/Fs and dl-PCBs in environmental samples [44 - 46]. In a similar way, until only a few years ago EU Regulations and Directives dealing with the analysis of these compounds in food and feed considered GC-HRMS as the only confirmatory technique [47 - 52]. In contrast to screening methods, GC-HRMS enabled to accurately determine

PCDD/Fs and dl-PCBs at the very low levels already present in most food/feed samples as a consequence of the implementation of different actions to reduce their concentration in the environment.

On account of the decreasing levels observed, during the last decades, some advanced complex developments have been proposed to reduce even more the LODs. This is particularly interesting for the analysis of biological samples, such as human serum, for which concentrations can be at the attogram (ag) level. In that regard, Patterson *et al.* adapted a liquid nitrogen jet-cooled thermal modulator, typically used in comprehensive two-dimensional gas chromatography, to a GC oven coupled to a HRMS magnetic sector instrument, allowing narrower chromatographic peaks. This particular system to enhance sensitivity is known as cryogenic zone compression (CZC) [53].

However, even though GC-HRMS with magnetic sector analyzers provides adequate sensitivity, selectivity and robustness, it is a technique that requires an important investment to start-up and also has a high running cost. In addition, an ample space is needed to house the mass spectrometer as well as the necessary complementary accessories (*i.e.*, water cooling systems, air conditioning or uninterrupted power systems) and highly skilled operators together with an adequate maintenance are mandatory for an optimum performance of this type of instruments. For these reasons, the applicability of several MS techniques on the analysis of PCDD/Fs and dl-PCBs has been evaluated during the last two decades as alternatives to GC-HRMS. Particularly after the "Belgium crisis" when the EU made great efforts on the implementation of a strategy which included efficient monitoring programs to ensure the quality of food and feed concerning the presence of these compounds. The main goal was to reduce both the complexity of the instrumental analysis and the economic cost but keeping the required sensitivity, selectivity and robustness that is mandatory in routine analysis.

Time-of-Flight Mass Spectrometry (ToF-MS)

Several authors have tested time-of-flight mass spectrometry (ToF-MS) for the determination of dioxins and PCBs. Focant *et al.* described the use of fast GC-high-speed (HS)-ToF-MS for the analysis of PCBs in human serum, even though the LODs where higher than those achieved with GC-HRMS it was possible to quantify the most abundant PCBs present in this kind of samples. However, due to this limited sensitivity, the technique seemed not to be sufficiently suitable for the analysis of PCDD/Fs and non-*ortho* PCBs. Therefore, comprehensive two-dimensional chromatography (GCxGC) coupled to (HS)-ToF-MS was proposed by the same authors as a way to increase both sensitivity and selectivity [54].

GCxGC coupled to micro electron capture detection (µECD) was also evaluated for the analysis of PCBs [55] and PCDD/Fs together with dl-PCBs [56] with LODs between 30 and 150 fg (injected; S/N=3:1) for a standard solution, that means in the same order of magnitude than GC-HRMS and better than those obtained with GCxGC-(HS)-ToF-MS. However, a lack of selectivity is observed when real samples are analyzed since it is not possible to confirm the identity of the signals [57], on the contrary, ToF-MS mass spectra can be used to identify the presence of the target analytes properly, then GCxGC-(HS)-ToF-MS seemed to be more appropriate for the analysis of complex mixtures.

In this sense, during the early years of the 2000s, several studies showed the relevance of using this comprehensive technique for dioxin compounds analysis in food and feed matrices. Hoh *et al.* were able to identify 0.25 pg of TCDD in a standard mixture [58] and applied the developed GCxGC-(HS)-ToF-MS method to the analysis of PCDD/Fs and non-*ortho* PCBs in fish oil [59].

Focant *et al.* also reported a similar LOD for TCDD and analyzed fish, pork and milk samples. These authors concluded that even in general comparable results were obtained in terms of TEQ among different MS techniques; sensitivity has still to be improved for GCxGC-(HS)-ToF-MS in order to be competitive with GC-HRMS. Moreover, larger relative standard deviations (RSDs) were observed, particularly for the case of the lowest level samples [60]. Additionally, data handling and evaluation was laborious due to limited software development for these particular comprehensive techniques, which made the overall analysis costs similar or even higher than those of GC-HRMS. Nowadays, this is still possibly the most important drawback of GCxGC-(HS)-ToF-MS for its application in routine analysis at trace levels. Therefore, further research in the use of this MS technique for the analysis of PCDD/Fs and dl-PCBs has not been significant during the last years, with only a few articles dealing with this topic [61].

In contrast to high-acquisition speed unit-resolution ToF spectrometers, GC-ToF-MS instruments with high-resolution (HR)/accurate mass analyzers were also available, although with a lower speed acquisition. Despite the gain in selectivity that can be obtained, the limited dynamic range of HR-ToF-MS instruments reduces their potential for quantitative analysis. Among the scarce literature found on the application of this technique in dioxin and PCB analysis it should be mentioned the development of an experimental GC-HR-ToF-MS with multiphoton ionization that provided a resolution of 1,600 from the full width at half maximum (FWHM) and a LOD of 19 fg for 1,3,6,8-TCDF [62].

Nowadays, high-speed high-resolution/accurate ToF analyzers have been placed in the market; however, the assessment of this new instrumentation in the area of dioxin and dioxin-like compounds has not yet been fully explored. A reason for that may be the preference in development of tandem mass spectrometry over other MS techniques for PCDD/Fs and dl-PCBs analysis, notably since GC-MS/MS was also included as a confirmatory technique in EU food and feed Regulations in 2014 [63, 64].

Tandem Mass Spectrometry with Ion Trap Analyzer (IT-MS/MS)

The development of gas chromatography coupled to ion trap mass spectrometry operating in the tandem mode (GC-IT-MS/MS) for the analysis of dioxins and PCBs was parallel to the previously described research on ToF-MS on the same topic. At the beginning of the 2000s, these were the two approaches considered to be the main promising competitors to the reference GC-HRMS method [54].

In IT-MS/MS selectivity is based on the selection of precursor ions that are produced, isolated and stored in the ion trap and afterwards dissociated by collision-induced dissociation (CID), this way several product ions are obtained from each precursor ion and further detected [65]. Therefore, while usually the two most intense ions from the molecular cluster of each PCDD/F or PCB are detected in HRMS; in the case of IT-MS/MS, two different product ions from a single precursor ion are monitored for each compound. For the particular case of PCDD/Fs the production of product ions is based on the well-known specific loss of a $COCl^{\bullet}$ fragment [66]; on the other hand, for dl-PCBs, the loss of two chlorine atoms from the precursor ion generates the product ions, being those corresponding to the loss of two ^{35}Cl and from the loss of $^{35}Cl^{37}Cl$ the most intense ones in the MS/MS spectra [67].

GC-IT-MS/MS was first applied with good results to the analysis of PCDD/Fs and PCBs in environmental samples such as sewage effluents [68, 69]; air samples [70]; soils and sediments [69]; or fly ashes [69, 71, 72]. However, to consider the technique in routine analysis for food and feed samples, both sensitivity and selectivity had to be optimized, and further assessment had to be done by analyzing different kinds of matrices (*e.g.*, meat, fish, vegetable oil...).

In order to have the following product ions from a single precursor ion: $[M-CO^{35}Cl]^+$ and $[M-CO^{37}Cl]^+$, for PCDD/Fs; and $[M-^{35}Cl^{35}Cl]^+$ and $[M-^{35}Cl^{37}Cl]^+$, for PCBs, respectively, it is mandatory to select species containing at least one ^{37}Cl atom ($[M + 2]^{\bullet +}$) from the molecular cluster.

Maximum sensitivity can be reached by isolating as much precursor ions as possible; therefore the preferred precursor ion for each analyte is the one with the highest relative isotope abundance. In this sense, Malavia *et al.* (2007a) considered $[M + 2]^{\bullet +}$ as the selected precursor ion for all dl-PCBs and most of the

PCDD/Fs, except for the octachlorinated ones for which $[M + 4]^{\bullet+}$ is more abundant. The corresponding transitions selected for each homologue group of PCDD/Fs and dl-PCBs are shown in Table **2** [73]. On the other hand, other authors pointed out that sensitivity can also be enhanced by having similar abundances of the two product ions, even though they could not be obtained from the most abundant precursor ion. The explanation to this fact is that this way the measurement of the two products ions is more accurate, particularly at low concentrations, compensating for the loss in abundance of the precursor ion [54]. In addition, a good measurement of the product ions is essential in order to have a proper check of the isotope ratio, which is one of the main quality criteria for selectivity assessment.

Sensitivity is also highly dependent on the configuration of each ion trap mass spectrometer instrument [74]. Moreover, maximum trapping efficiency of the precursor ions as well as the optimization of the parameters related to the CID process (*i.e.*, precursor ion isolation window, excitation time, excitation voltage and the excitation energy-*q*z parameter) are necessary to improve the LODs and get maximum selectivity.

Hayward *et al.* already applied GC-IT-MS/MS to the analysis of PCDD/Fs in a substantial number of food samples in US, including several fish and shellfish species, eggs and milk products, and compared the results with those obtained with GC-HRMS. Despite that, the LODs were not as good as the ones reported for GC-HRMS, the authors concluded that the technique was an interesting alternative to the reference method allowing the screening of PCDD/Fs in the samples at a lower cost and in a shorter time [75]. The differences in sensitivity compared to GC-HRMS have also been described by other authors, as an example, the instrumental LOD for TCDD has been estimated as 5-10 times higher than the LOD obtained with GC-HRMS. Eppe *et al.* reported a LOD of 200 fg (injected; S/N=5:1) for TCDD and proposed the use of a programmable temperature vaporizer-large volume (PTV-LV) injector coupled to the GC-I--MS/MS system to improve sensitivity. Different injection volumes were checked and finally, 10μl was found to be the best compromise for getting good sensitivity but also for keeping a high-performance capacity after several injections of real samples [76].

Malavia *et al.* optimized a GC-IT-MS/MS method for PCDD/Fs and dl-PCBs and assessed its suitability by analyzing different food/feed matrices (*i.e.*, vegetable oil, milk, fish oil, chicken, pork, fish, eggs and a chicken compound feed) [73, 77]. The instrumental LOD for TCDD was similar to that reported by Eppe *et al.*

Table 2. Precursor ion (most abundant from the molecular cluster) and two corresponding product ions selected for each PCDD/F and dl-PCB homologue group in a GC-IT-MS/MS method. Reprinted with permission from [73].

MS-MS Transitions		
Compound	**Precursor Ions (*m/z*)**	**Product Ions (*m/z*)**
TCDFs	305.9 (M+2)	240.9 + 242.9
^{13}C-TCDFs	317.9 (M+2)	251.9 + 253.9
TCDDs	321.9 (M+2)	256.9 + 258.9
^{13}C-TCDDs	333.9 (M+2)	267.9 + 269.9
PeCDFs	339.9 (M+2)	274.9 + 276.9
^{13}C-PeCDFs	351.9 (M+2)	285.9 + 287.9
PeCDDs	355.9 (M+2)	290.9 + 292.9
^{13}C-PeCDDs	367.9 (M+2)	301.9 + 303.9
HxCDFs	373.8 (M+2)	308.8 + 310.8
^{13}C-HxCDFs	385.9 (M+2)	319.9 + 321.9
HxCDDs	389.8 (M+2)	324.8 + 326.8
^{13}C-HxCDDs	401.9 (M+2)	335.9 + 337.9
HpCDFs	407.8 (M+2)	342.8 + 344.8
^{13}C-HpCDFs	419.8 (M+2)	353.8 + 355.8
HpCDDs	423.8 (M+2)	358.8 + 360.8
^{13}C-HpCDDs	435.8 (M+2)	369.8 + 371.8
OCDFs	443.7 (M+4)	378.7 + 380.7
OCDDs	459.7 (M+4)	394.7 + 396.7
^{13}C-OCDDs	471.8 (M+4)	405.8 + 407.8
TeCBs	291.9 (M+2)	219.9 + 221.9
^{13}C-TeCBs	304.0 (M+2)	232.0 + 234.0
PeCBs	325.9 (M+2)	253.9 + 255.9
^{13}C-PeCBs	337.9 (M+2)	265.9 + 267.9
HxCBs	359.8 (M+2)	287.8 + 289.8
^{13}C-HxCBs	371.9 (M+2)	299.9 + 301.9
HpCBs	393.8 (M+2)	321.8 + 323.8
^{13}C-HpCBs	405.8 (M+2)	333.8 + 335.8

Moreover, long-term precision values established from the analysis of several vegetable oil samples (*i.e.*, one non-spiked + five spiked at different concentration levels) were even better than those from GC-HRMS, with RSD% ranging from

3% to 19% for all the individual toxic congeners. The determined concentrations varied from 0.35 to 5.63 pg WHO-TEQ g^{-1} oil and from 0.02 to 5.37 pg WHO-TEQ g^{-1} oil, for PCDD/Fs and dl-PCBs, respectively; with LOD values of 0.28 pg WHO-TEQ g^{-1} oil (PCDD/Fs) and 0.009 pg WHO-TEQ g^{-1} oil (dl-PCBs) for a sample amount of 5 g oil. Fig. **(1)** shows a comparison between the concentrations found for all the individual PCDD/Fs and dl-PCBs in one of the spiked oil samples by applying the GC-IT-MS/MS and the GC-HRMS methods.

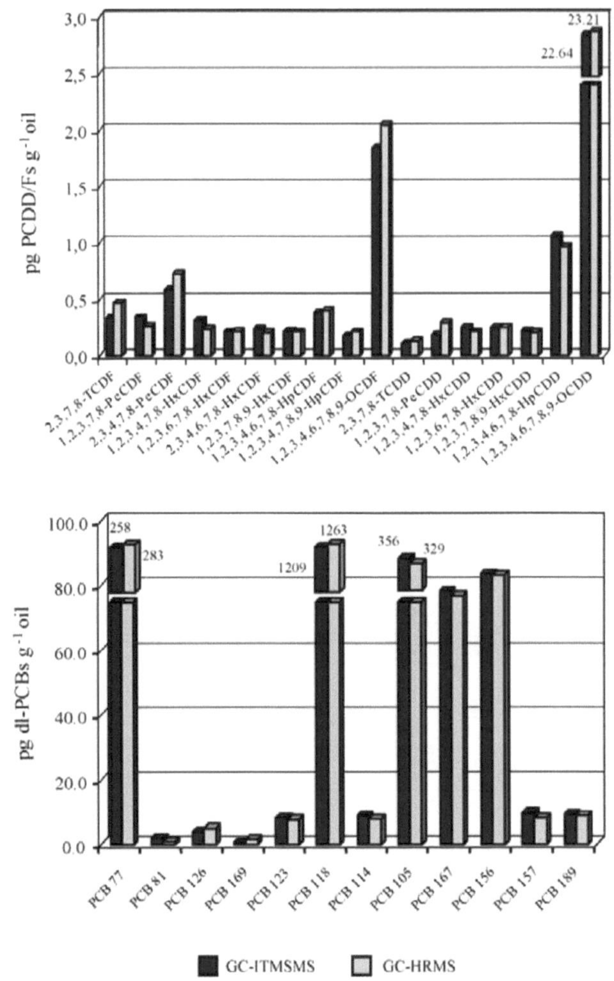

Fig. (1). Individual concentrations of PCDD/Fs and dl-PCB congeners in a spiked vegetable oil obtained by GC-IT-MS/MS and GC-HRMS methods. Reprinted with permission from [73].

In order to assess the selectivity of the GC-IT-MS/MS method, some vegetable oil samples containing different interfering compounds were also analyzed. The results indicated that previous efficient clean-up processes are needed before instrumental analysis, since the presence of interferences, such as lipids, in the final extracts dramatically affected the instrument performance (Fig. **2**).

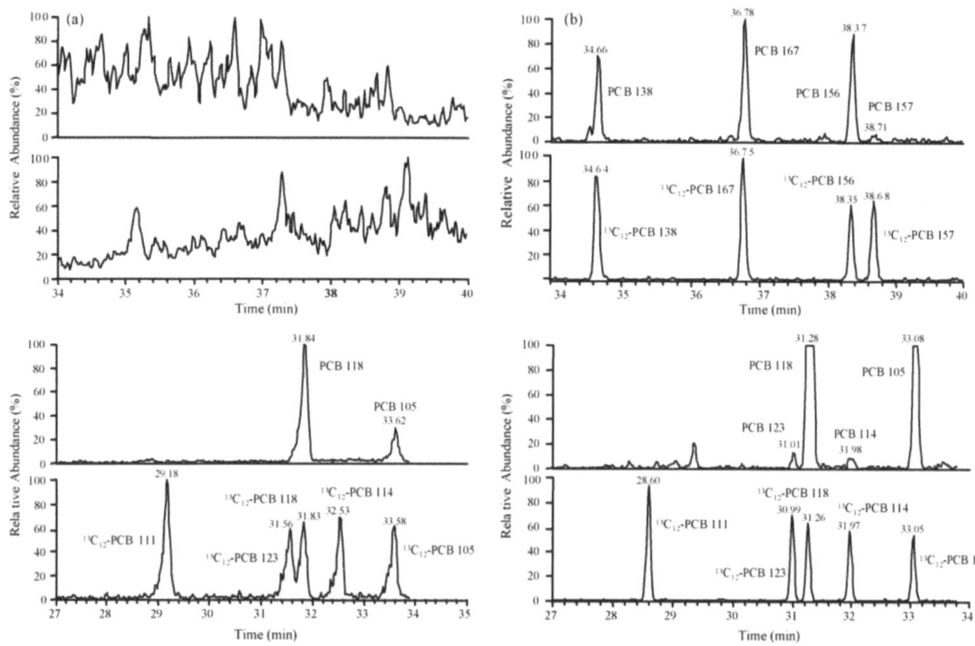

Fig. (2). GC-IT-MS/MS chromatograms of penta-mono-*ortho* PCBs (PCB 105, PCB 114, PCB 118 and PCB 123) and hexa-mono-*ortho* PCBs (PCB 156, PCB 157 and PCB 167) for a vegetable oil sample with: (a) problems in the clean-up process and (b) suitable clean-up process. Reprinted with permission from [73].

This is related to the particular operational characteristics of the ion trap analyzers, since signal suppression can be observed for the target analytes due to the presence of ions from other co-extracted compounds, with the consequent decrease in sensitivity and the possibility to give false negatives. In this sense, Focant *et al.* compared the robustness of IT-MS/MS vs. ToF-MS systems for dioxin compounds and also concluded that the first can be easier contaminated by interfering ions in routine analysis [54, 60].

In summary, although IT-MS/MS and ToF-MS are suitable techniques for the analysis of PCDD/Fs and dl-PCBs, in general, they show higher detection limits and lower selectivity than HRMS which represents an important drawback for the determination of these compounds in low-level food/feed samples. More important, they seem not to be as robust as HRMS when dealing with a large number of analyses. For all these reasons, these techniques do not fulfil the minimum requirements for confirmatory purposes described in the EU Directives and Regulations for the determination of dioxins and related compounds in food and feed; on the contrary, they were classified as screening techniques able to discriminate samples with concentrations exceeding the maximum levels or the action levels.

Tandem Mass Spectrometry with Triple Quadrupole Configuration (QqQ-MS/MS)

In mass spectrometers with ion trap analyzers, the production of precursor and product ions takes place in the same physical space but in sequential steps. Conversely, for instruments with a triple quadrupole configuration (QqQ) tandem mass spectrometry is done in space. A first quadrupole allows selecting the target precursor ions among all those that are obtained in the ion source, and the second quadrupole acts as a collision cell where these ions are fragmented to give the products ions, which are separated in a third (last) quadrupole.

Triple quadrupole instruments for the analysis of dioxins were already investigated in the past [78]. In 2000, a comparative study about the performance on the determination of PCDD/Fs of three different mass spectrometers, including a QqQ instrument, an ion trap instrument and a magnetic sector high-resolution mass spectrometer, was carried out [79]. However, it was at the beginning of the present decade, when a new generation of triple quadrupole GC-MS/MS instruments of higher sensitivity became commercially available, that this technique was considered as a potential reliable alternative to GC-HRMS for the analysis of dioxin-like compounds.

The triple quadrupole systems have a significant advantage over the ion trap analyzers since more than one precursor ion can be selected for each target analyte. This way, several authors have reported the use of two different precursor ions from the corresponding molecular cluster for the analysis of PCDD/Fs and PCBs in food and feed samples [80 - 83]. Taking into account the system of identification points established at the Directive 96/23/EC for the confirmation of substances, a QqQ-MS/MS method based on two precursor ions and two different product ions (*i.e.*, a transition from each precursor ion) reaches five identification points, while in HRMS with two ions or IT-MS/MS with one precursor ion and two product ions the maximum number of identification points is four [84].

However, even though the use of more than one precursor ion is preferred in terms of selectivity, it is also important to check the most intense transitions in order to have maximum sensitivity. In this sense, García-Bermejo *et al.* investigated different transitions for PCDD/Fs other than the most studied loss of the $COCl^{\bullet}$ fragment; while this transition was favoured at low CID voltages (*i.e.*, 20-30 V), the loss of a $COCl_3^{\bullet}$ group and two $COCl^{\bullet}$ groups also showed a good signal intensity for PCDFs and PCDDs, respectively, at higher CID voltages (*i.e.*, 30-50 V). The relative abundance of these secondary transitions depends on the chlorination degree, being more intense for the lowest chlorinated congeners and even more significant than the loss of the COCl˙ fragment from the second precursor ion. For this reason, contrary to what was previously proposed by other authors, García-Bermejo *et al.* selected a single precursor ion for the pentachlorinated dioxin and pentachlorinated furans in order to get maximum sensitivity for these particular congeners (Table **3**) [85]. A comparison of the intensity of several studied transitions, as a function of the CID voltage, is shown in Fig. (**3a and b**).

Table 3. Precursor and product ions and CID voltages selected for native PCDD/Fs in a QqQ-MS/MS system (a single precursor ion is selected for penta-chlorinated compounds). Reprinted with permission from [85].

Segment No.	Time (min)	Congener	Precursor Ion	Product Ion	CID (V)
1	24.0-27.0	TeCDF	304	241	27
			306	243	27
		TeCDD	320	257	21
			322	259	20
2	27.0-30.0	PeCDFs	340	277	27
			340	205	50
		PeCDD	356	293	19
			356	230	38
3	30.0-32.4	HxCDFs	374	311	28
			376	313	28
		HxCDDs	390	327	20
			392	329	21
4	32.4-35.2	HpCDFs	408	345	27
			410	347	28
		HpCDD	424	361	20
			426	363	20
5	35.2-37.0	OCDF	442	379	27
			444	381	27
		OCDD	458	395	19
			460	397	18

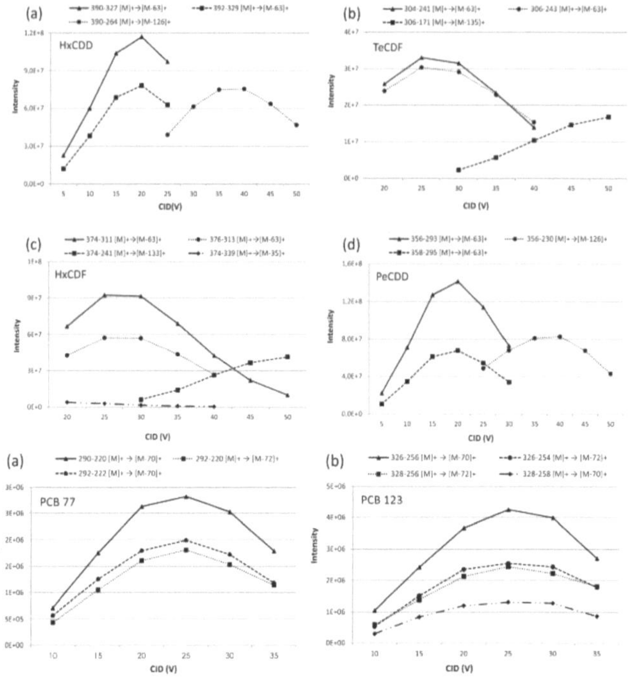

Fig. (3). (A): Intensity of several studied transitions for (a) HxCDD, (b) TeCDF, (c) HxCDF and (d) PeCDD and **(B):** Intensity of the studied transitions for PCBs (a) 77 and (b) 123; depending on the voltage of collision induced dissociation (CID) applied in a QqQ-MS/MS system. Reprinted with permission from [85].

In 2014, the good performance exhibit by the new generation of GC-QqQ-MS/MS instruments eventually led to amend EU Regulations related to the analysis of PCDD/Fs and PCBs in food and feed matrices by including GC-MS/MS as a confirmatory technique, in addition to GC-HRMS [63, 64]. Despite the use of the triple quadrupole configuration is not explicitly mentioned, GC-MS/MS methods intended for confirmatory purposes have to fulfil some specific requirements. Among them, the first one indicates that at least 2 precursor ions, each with one specific corresponding transition product ion, have to be monitored for each congener; this way excluding other MS/MS systems, such as those based on ion trap analyzers.

Ábalos *et al.* and L'Homme *et al.* evaluated the performance of two analytical methods for PCDD/Fs and dl-PCBs, based on two different GC-QqQ-MS/MS instruments, against the criteria indicated at the new EU Regulations [81, 82]. For instance, the ion ratio (*i.e.*, ratio between the intensities of the two selected transition product ions) of the different congeners was checked in several food/feed samples. For each congener, a ± 15% maximum permitted tolerance for the ion ratio compared to the calculated/measured values is set as a quality requirement. For both methods, the ion ratio values for all compounds were within this percentage of tolerance for the different samples analysed (*i.e.*, fish, milk powder and feed [81] (Fig. **4**) and vegetable oil [82]).

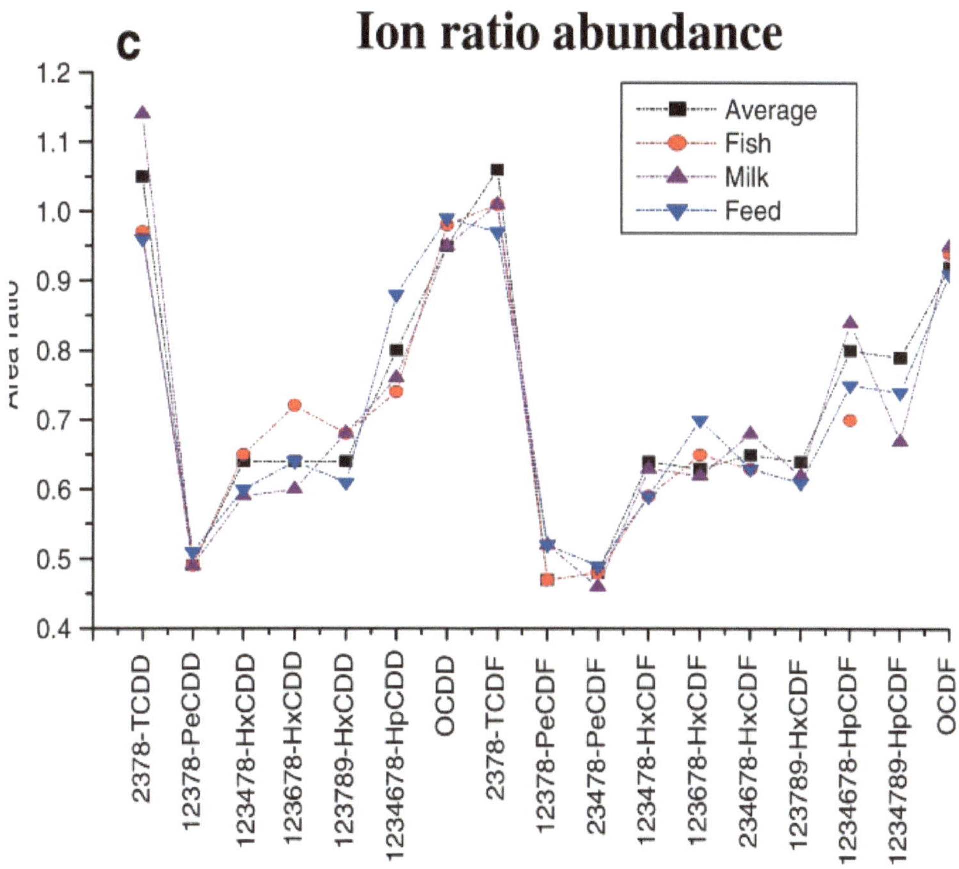

Fig. (4). Comparison of the ion ratio between the intensities of two selected product ions for PCDD/F congeners in food/feed sample extracts vs. the average value from calibration standards. Reprinted with permission from [81].

The assessment of sensitivity and more particularly an accurate determination of the limits of detection and/or limits of quantification (LOQs) was an important issue in the two studies above mentioned. In 2004 and 2005, a specific amendment was added to Directives 2002/69/EC (food) and 2002/70/EC (feed), respectively, related to the way to calculate the LOQ of the individual congeners [86, 87]. For PCDD/F analysis in food and feed the LOQ for a certain congener was established as 'the concentration of an analyte in the extract of a sample which produces an instrumental response at two different ions, to be monitored with an S/N ratio of 3:1 for the less sensitive signal'. This calculation is possible for GC-HRMS but is not reliable for GC-MS/MS systems since they usually show a very low noise level. An example of the unrealistic low LODs/LOQs that can be obtained by using this approach was given by García-Bermejo *et al.* (Fig. **5**) [85].

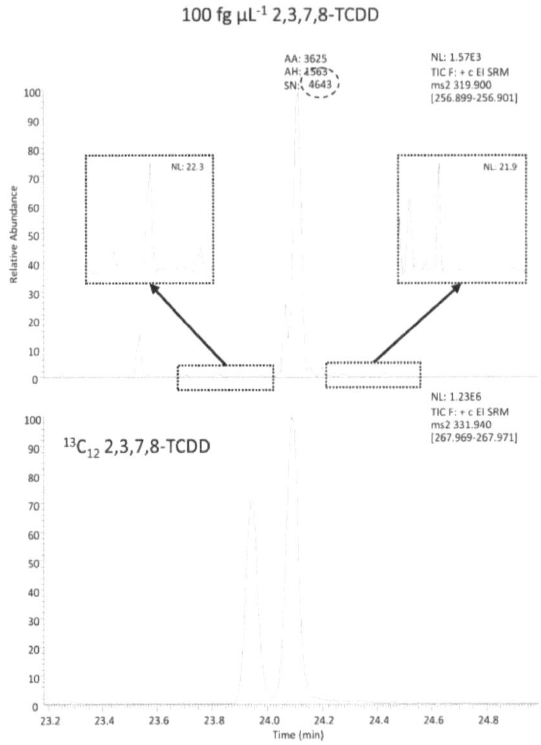

S/N	iLOD (fg µL^{-1}) (S/N=3)
4643 (provided by the software)	0.06
71 (manual calculation)	4

Fig. (5). Estimation of the LOD for 2,3,7,8-TCDD, using the S/N ratio approach from a low concentration standard (100 fg injected), for GC-QqQ-MS/MS. Reprinted with permission from [85].

Alternatively, in the new legislation from 2014 another way to calculate LOQ is suggested for this kind of instruments, the LOQ is then considered as 'the lowest concentration point on a calibration curve that gives an acceptable (≤ 30%) and consistent deviation to the average relative response factor calculated for all points on the calibration curve in each series of samples'. Based on this approach, Ábalos *et al.* found individual LOQs between 10 and 70 fg (injected) for PCDD/Fs, being tetrachlorinated congeners the ones with the lowest values (Fig.

6) [81]. On the other hand, other authors have questioned this alternative way to estimate/calculate the LOQs and proposed to define LOQs as 10 times the standard deviation (SD) associated to several replicate injections of the lowest acceptable calibration point [82]. However, LOQs were comparable to those reported by Ábalos *et al.* for most of the PCDD/Fs congeners.

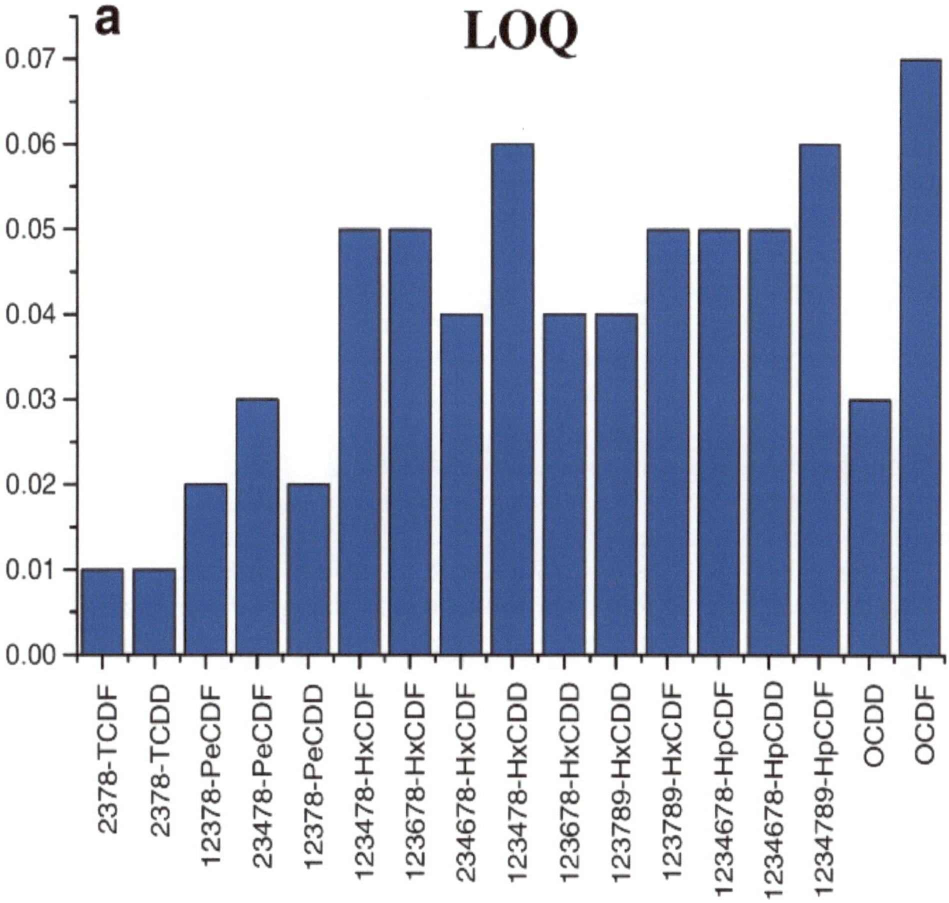

Fig. (6). Calculated LOQs for PCDD/Fs following the recommendation indicated at the EU Regulations for GC-QqQ-MS/MS. Reprinted with permission from [81].

The developed GC-QqQ-MS/MS methods have been applied to the analysis of real samples and results have been compared to GC-HRMS. In general, good levels of agreement were found, either when individual congener results were compared [81, 85] and when WHO-TEQ values were considered. In particular, deviations were usually below 20% for different food samples in a concentration range between 1 and 10 pg WHO-PCDD/F-TEQ g^{-1} fat, while higher deviations were observed when the levels were below 0.5 pg WHO-PCDD/F-TEQ g^{-1} fat [83].

Regardless of the MS technique applied for the analysis of dioxin-like compounds (*i.e.*, HRMS or MS/MS) molecular or precursor ions are normally obtained as a result of electron impact ionization (EI). This ionization mode generates high fragmentation; in this sense, even though lower ionization energies are commonly used in the determination of PCDD/Fs and dl-PCBs (*i.e.*, 30-40 eV instead of 70 eV), a significant loss in sensitivity already takes place in the source.

A few years ago a new atmospheric-pressure chemical ionization (APCI) source able to be coupled to GC was developed. An MS system based on a GC device equipped with this new source (APGC) coupled to a triple quadrupole mass spectrometer became commercially available, and its applicability for PCDD/F analysis has been evaluated [88, 89]. Since APCI produces softer ionization, the spectra are dominated by the majority presence of molecular ions, which dramatically increased the sensitivity. This way, for instance, a LOD of 2 fg for TCDD can be reached (Fig. 7), with 10 fg (injected) being detected at an S/N>50. The gain in sensitivity compared to GC-HRMS is clear, and the use of triple quadrupole configuration also seemed to guarantee the fulfilment of the criteria for the purpose of GC-MS/MS as a confirmatory method. PCDD/Fs, as well as dl-PCBs, have been determined by using this MS system in several matrices, including food, showing good correlation with the results obtained by GC-HRMS [88 - 90]. However, the performance/robustness of the APGC source has still to be properly assessed after the injection of a large number of samples in routine analyses.

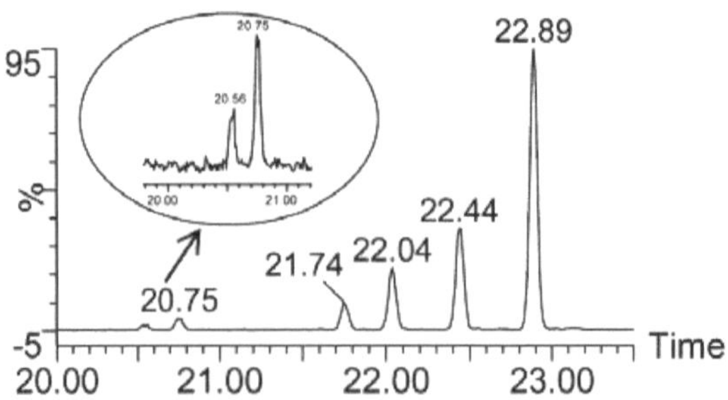

Fig. (7). APGC-QqQ-MS/MS sensitivity assessment, using a standard solution containing from 2 to 100 pg μL^{-1} of different TCDD congeners. Reprinted with permission from [88].

New HRMS Techniques

As it has been stated, in the field of dioxin analysis the term HRMS usually refers to magnetic sector analyzers. However, a few studies have also focused on the use of other new ultra-high-resolution MS techniques for the determination of PCDD/Fs. In this sense, it can be mentioned the research done by Taguchi *et al.* on the performance of a gas chromatograph-tandem quadrupole-Fourier transform ion cyclotron resonance mass spectrometer (GC-MS/MS-FTICRMS) operating in the GC-FTICRMS mode. The authors found that, at a resolving power between 50,000 and 100,000, the LOD for TCDD was about 1 pg. Therefore, the technique can be useful for the analysis of environmental complex matrices with medium-high PCDD/F concentrations but seems not suitable for low-level samples, such as food [91].

More recently, important developments have been carried out on Orbitrap technology allowing this kind of mass analyzer to be coupled with GC. The first commercially available GC-Orbitrap was placed in the market in 2015, and some work has already been done to check the applicability of this ultra-high-resolution MS technique to the analysis of PCDD/Fs. For instance, Cojocariu *et al.* reported the use of GC-Orbitrap MS for the study of persistent organic pollutants, including PCDD/Fs and PCBs. Using the GC-Orbitrap at a resolving power of 60,000, halogenated POPs were detected and identified in several complex matrix samples (*i.e.*, soil, sewage sludge, fly ash and fish) [92].

CONCLUDING REMARKS

The determination of PCDD/Fs and dl-PCBs in food and feed samples is subject to specific requirements that can only be fulfilled by using certain analytical techniques. In this sense, advances in extraction and clean-up steps during the last decade have been limited, with not many significant improvements since automated PLE and purification/fractionation systems were introduced in the market. On the other hand, research on alternative mass spectrometric techniques has been successful and; accordingly, at the present moment GC-MS/MS with a triple quadrupole configuration is in a position to compete with the reference GC-HRMS. Thus, it was already included as a confirmatory technique within the EU regulatory framework. Very recently, the potential of new ultra-high-resolution instruments (*i.e.*, GC-Orbitrap) for dioxin-like compound analysis has also been assessed, with promising preliminary results; however, further research on that topic is needed.

CONSENT FOR PUBLICATION

Not applicable.

CONFLICT OF INTEREST

The authors confirm that this chapter contents have no conflict of interest.

ACKNOWLEDGEMENT

J. Parera is grateful to the Spanish Ministry of Economy and Competitiveness for a Juan de la Cierva – Formación research fellowship (FJCI-2015-26722).

REFERENCES

[1] Olie K, Vermeulen PL, Hutzinger O. Chlorodibenzo-p-dioxins and chlorodibenzofurans are trace components of fly ash and flue gas of some municipal incinerators in The Netherlands. Chemosphere 1977; 8: 455-9. [http://dx.doi.org/10.1016/0045-6535(77)90035-2]

[2] Fiedler H. Sources of PCDD/PCDF and impact on the environment. Chemosphere 1996; 32(1): 55-64. [http://dx.doi.org/10.1016/0045-6535(95)00228-6] [PMID: 8564435]

[3] Abad E, Martínez K, Caixach J, Rivera J. Polychlorinated dibenzo-p-dioxin/polychlorinated dibenzofuran releases into the atmosphere from the use of secondary fuels in cement kilns during clinker formation. Environ Sci Technol 2004; 38(18): 4734-8. [http://dx.doi.org/10.1021/es049641a] [PMID: 15487780]

[4] IARC, International Agency for Research on Cancer. IARC Monographs on the evaluation of carcinogenic risks to humans. Polychlorinated dibenzo-p-dioxins and polychlorinated dibenzofurans Lyon. 1997; 69.

[5] Kavlock RJ. Overview of endocrine disruptor research activity in the United States. Chemosphere 1999; 39(8): 1227-36. [http://dx.doi.org/10.1016/S0045-6535(99)00190-3] [PMID: 10467717]

[6] De Voogt P, Brinkman UATh. Production, properties and usage of polychlorinated biphenyls In: Kimbrough, RD, Jensen, AA, Eds. Halogened biphenyls, terphenyls, naphthalene, dibenzodioxins and related products. 2nd. Elsevier Science B.V: Amsterdam 1989; pp. 1. 3-45 . [http://dx.doi.org/10.1016/B978-0-444-81029-8.50005-9]

[7] Safe S. Polychlorinated biphenyls (PCBs) and polybrominated biphenyls (PBBs): biochemistry, toxicology, and mechanism of action. Crit Rev Toxicol 1984; 13(4): 319-95. [http://dx.doi.org/10.3109/10408448409023762] [PMID: 6091997]

[8] Van den Berg M, Birnbaum L, Bosveld ATC, *et al.* Toxic equivalency factors (TEFs) for PCBs, PCDDs, PCDFs for humans and wildlife. Environ Health Perspect 1998; 106(12): 775-92. [http://dx.doi.org/10.1289/ehp.98106775] [PMID: 9831538]

[9] Van den Berg M, Birnbaum LS, Denison M, *et al.* The 2005 World Health Organization reevaluation of human and Mammalian toxic equivalency factors for dioxins and dioxin-like compounds. Toxicol Sci 2006; 93(2): 223-41. [http://dx.doi.org/10.1093/toxsci/kfl055] [PMID: 16829543]

[10] Sheffield A. Sources and releases of PCDD's and PCDF's to the Canadian environment. Chemosphere 1985; 14: 811-4. [http://dx.doi.org/10.1016/0045-6535(85)90193-6]

[11] Abad E, Caixach J, Rivera J. Improvements in dioxin abatement strategies at a municipal waste management plant in Barcelona. Chemosphere 2003; 50(9): 1175-82. [http://dx.doi.org/10.1016/S0045-6535(02)00483-6] [PMID: 12547331]

[12] UNEP (United Nations Environment Programme). Stockholm Convention on Persistent Organic Pollutants.Geneva, Switzerland 2001.

[13] Malisch R, Kotz A. Dioxins and PCBs in feed and food--review from European perspective. Sci Total Environ 2014; 491-492: 2-10. [http://dx.doi.org/10.1016/j.scitotenv.2014.03.022] [PMID: 24804623]

[14] Malisch R. Increase of the PCDD/F-contamination of milk, butter and meat samples by use of contaminated citrus pulp. Chemosphere 2000; 40(9-11): 1041-53. [http://dx.doi.org/10.1016/S0045-6535(99)00352-5] [PMID: 10739045]

[15] Bernard A, Broeckaert F, De Poorter G, *et al.* The Belgian PCB/dioxin incident: analysis of the food chain contamination and health risk evaluation. Environ Res 2002; 88(1): 1-18. [http://dx.doi.org/10.1006/enrs.2001.4274] [PMID: 11896663]

[16] European Commission. COMMISSION REGULATION (EU) No 1259/2011 of 2 December 2011 amending Regulation (EC) No 1881/2006 as regards maximum levels for dioxins, dioxin-like PCBs and non dioxin-like PCBs in foodstuffs Off J Eur Union 2011; (320): 18-23.

[17] European Commission.. COMMISSION REGULATION (EU) No 277/2012 of 28 March 2012 amending Annexes I and II to Directive 2002/32/EC of the European Parliament and of the Council as regards maximum levels and action thresholds for dioxins and polychlorinated biphenyls Off J Eur Union 2012; (91): 1-7.

[18] Malisch R. Incidents with dioxins and PCBs in food and feed-Investigative work, risk management and economic consequences. J Environ Prot 2017; 8: 744-85. [http://dx.doi.org/10.4236/jep.2017.86048]

[19] European Commission. COMMISSION REGULATION (EU) No 225/2012 of 15 March 2012 amending Annex II to Regulation (EC) No 183/2005 of the European Parliament and of the Council as regards the approval of establishments placing on the market, for feed use, products derived from vegetable oils and blended fats and as regards the specific requirements for production, storage, transport and dioxin testing of oils, fats and products derived thereof. Off J Eur Union 2012; (77): 1-5.

[20] European Commission. COMMISSION REGULATION (EU) 2017/644 of 5 April 2017 laying down methods of sampling and analysis for the control of levels of dioxins, dioxin-like PCBs and non-dioxin-like PCBs in certain foodstuffs and repealing Regulation (EU) No 589/2014. Off J Eur Union 2017; (92): 9-34.

[21] European Commission. COMMISSION REGULATION (EU) 2017/771 of 3 May 2017 amending Regulation (EC) No 152/2009 as regards the methods for the determination of the levels of dioxins and polychlorinated biphenyls. Off J Eur Union 2017; (115): 22-42.

[22] Pirard C, Focant JF, De PE. An improved clean-up strategy for simultaneous analysis of polychlorinated dibenzo-p-dioxins (PCDD), polychlorinated dibenzofurans (PCDF), and polychlorinated biphenyls (PCB) in fatty food samples. Anal Bioanal Chem 2002; 372(2): 373-81. [http://dx.doi.org/10.1007/s00216-001-1181-y] [PMID: 11936114]

[23] Suchan P, Pulkrabová J, Hajšlová J, Kocourek V. Pressurized liquid extraction in determination of polychlorinated biphenyls and organochlorine pesticides in fish samples. Anal Chim Acta 2004; 520: 193-200. [http://dx.doi.org/10.1016/j.aca.2004.02.061]

[24] Kitamura K, Takazawa Y, Hashimoto S, Choi JW, Ito H, Morita M. Effective extraction method for dioxin analysis from lipid-rich biological matrices using a combination of pressurized liquid extraction and dimethyl sulfoxide/acetonitrile/hexane partitioning. Anal Chim Acta 2004; 512: 27-37. [http://dx.doi.org/10.1016/j.aca.2004.02.040]

[25] Weichbrodt M, Vetter W, Luckas B. Microwave-assisted extraction and accelerated solvent extraction with ethyl acetate-cyclohexane before determination of organochlorines in fish tissue by gas chromatography with electron-capture detection. J AOAC Int 2000; 83(6): 1334-43. [PMID: 11128135]

[26] Carro AM, Lorenzo RA, Fernández F, Rodil R, Cela R. Multi-residue screening of chlorinated and brominated compounds from aquaculture samples using matrix solid-phase dispersion--gas chromatography-mass spectrometry. J Chromatogr A 2005; 1071(1-2): 93-8. [http://dx.doi.org/10.1016/j.chroma.2004.12.005] [PMID: 15865179]

[27] Focant JF, Pirard C, De Pauw E. Automated sample preparation-fractionation for the measurement of dioxins and related compounds in biological matrices: a review. Talanta 2004; 63(5): 1101-13. [http://dx.doi.org/10.1016/j.talanta.2004.05.025] [PMID: 18969540]

[28] Lin Y, Feng C, Xu Q, *et al.* A validated method for rapid determination of dibenzo-p-dioxins/furans (PCDD/Fs), polybrominated diphenyl ethers (PBDEs) and polychlorinated biphenyls (PCBs) in human milk: focus on utility of tandem solid phase extraction (SPE) cleanup. Anal Bioanal Chem 2016; 408(18): 4897-906. [http://dx.doi.org/10.1007/s00216-016-9576-y] [PMID: 27178547]

[29] Hummel RA. Clean-up techniques for the determination of parts per trillion residue levels of 2,3,7,8-tetrachlorodibenzo-p-dioxin (TCDD). J Agric Food Chem 1977; 25(5): 1049-53. [http://dx.doi.org/10.1021/jf60213a038] [PMID: 893826]

[30] Concejero M, Ramos L, Jiménez B, *et al.* Suitability of several carbon sorbents for the fractionation of various sub-groups of toxic polychlorinated biphenyls, polychlorinated dibenzo-p-dioxins and polychlorinated dibenzofurans. J Chromatogr A 2001; 917(1-2): 227-37. [http://dx.doi.org/10.1016/S0021-9673(01)00667-7] [PMID: 11403473]

[31] Abad E, Sauló J, Caixach J, Rivera J. Evaluation of a new automated cleanup system for the analysis of polychlorinated dibenzo-p-dioxins and dibenzofurans in environmental samples. J Chromatogr A 2000; 893(2): 383-91. [http://dx.doi.org/10.1016/S0021-9673(00)00760-3] [PMID: 11073306]

[32] Björklund E, Müller A, von Holst C. Comparison of fat retainers in accelerated solvent extraction for the selective extraction of PCBs from fat-containing samples. Anal Chem 2001; 73(16): 4050-3. [http://dx.doi.org/10.1021/ac010178j] [PMID: 11534735]

[33] Sporring S, von Holst C, Björklund E. Selective pressurized liquid extraction of PCB's from food and feed samples: Effects of high lipid amounts and lipid type on fat retention. Chromatographia 2006; 64: 553-7. [http://dx.doi.org/10.1365/s10337-006-0059-8]

[34] Wiberg K, Sporring S, Haglund P, Björklund E. Selective pressurized liquid extraction of polychlorinated dibenzo-p-dioxins, dibenzofurans and dioxin-like polychlorinated biphenyls from food and feed samples. J Chromatogr A 2007; 1138(1-2): 55-64. [http://dx.doi.org/10.1016/j.chroma.2006.10.074] [PMID: 17109872]

[35] Gómez-Ariza JL, Bujalance M, Giráldez I, Velasco A, Morales E. Determination of polychlorinated biphenyls in biota samples using simultaneous pressurized liquid extraction and purification. J Chromatogr A 2002; 946(1-2): 209-19. [http://dx.doi.org/10.1016/S0021-9673(01)01534-5] [PMID: 11873970]

[36] Haglund P, Sporring S, Wiberg K, Björklund E. Shape-selective extraction of PCBs and dioxins from fish and fish oil using in-cell carbon fractionation pressurized liquid extraction. Anal Chem 2007; 79(7): 2945-51. [http://dx.doi.org/10.1021/ac0624501] [PMID: 17343366]

[37] Focant JF, Shirkhan H, De Pauw E. On-line automated PLE and multi-column clean-up system for PCDD/F and PCB analysis in food samples. Organohalogen Compd 2002; 55: 33-6.

[38] Bernsmann T, Albrecht M, Fürst P. Determination of dioxin/PCB and BDE in one automatic system with different sample clean-up column settings. Organohalogen Compd 2016; 78: 797-801.

[39] Calaprice C, Focant JF. Revisited sample preparation approach for dioxin measurements in human serum samples. Organohalogen Compd 2016; 78: 773-6.

[40] Marchand P, Lesquin E, Brosseaud A, Vaccher V, Vénisseau A, Le Bizec B. A new and highly innovative automatic purification system evaluated for dioxins and PCBs. Organohalogen Compd 2014; 76: 546-9.

[41] Aries E, Anderson DR, Ordsmith N, Fitzpatrick L. The SUPELCO dioxin prep system – Florisil

version: A new sample preparation method for fast analyses of PCDD/Fs and WHO-12 PCBs in environmental samples. Organohalogen Compd 2006; 68: 911-5.

[42] Yang S, Reiner EJ, Kolic TM. Lega. R.; Harrison, R.O. A rapid and simple cleanup method for analysis of dioxins/furans, dioxin-like PCBs, PBDEs and PCNs. Organohalogen Compd 2010; 72: 1788-91.

[43] Tondeur Y, Niederhut WN, Campana JE, Missler SR. A hybrid HRGC/MS/MS method for the characterization of tetrachlorinated-p-dioxins in environmental samples. Biomed Environ Mass Spectrom 1987; 14(8): 449-56.
[http://dx.doi.org/10.1002/bms.1200140812] [PMID: 2957006]

[44] US EPA Method 1613 Revision B. Tetra- through octa- chlorinated dioxins and furans by isotope dilution HRGC/HRMS Washington, DC: U.S. Environmental Protection Agency Office of Water Engineering and Analysis Division 1994.

[45] US EPA Method 1668 Revision A. Chlorinated biphenyl congeners in water, soil, sediment, biosolids and tissue by HRGC/ HRMS Washington, DC: Environmental Protection Agency Office of Water Engineering and Analysis Division 1999.

[46] EN 1948. Stationary source emissions. Determination of the mass concentration of PCDDs/PCDFs and dioxin-like PCBs 2006.
[http://dx.doi.org/10.1006/enrs.2001.4274] [PMID: 11896663]

[47] European Commission. COMMISSION DIRECTIVE 2002/69/EC of 26 July 2002 laying down the sampling methods and the methods of analysis for the official control of dioxins and the determination of dioxin-like PCBsin foodstuffs Off J Eur Union 2002; (209): 5-14.

[48] European Commission. COMMISSION DIRECTIVE 2002/70/EC of 26 July 2002 establishing requirements for the determination of levels of dioxins and dioxin-like PCBs in feedingstuffs Off J Eur Union 2002; (209): 15-21.

[49] European Commission. COMMISSION REGULATION (EC) No 1883/2006 of 19 December 2006 laying down methods of sampling and analysis for the official control of levels of dioxins and dioxin-like PCBs in certain foodstuffs Off J Eur Union 2006; (364): 32-43.

[50] European Commission. COMMISSION REGULATION (EC) No 152/2009 of 27 January 2009 laying down the methods of sampling and analysis for the official control of feed COMMISSION REGULATION (EU) No 252/2012 of 21 March 2012 laying down methods of sampling and analysis for the official control of levels of dioxins, dioxin- like PCBs and non-dioxin-like PCBs in certain foodstuffs and repealing Regulation (EC) No 1883/2006 Off J Eur Union, 2009; (54): 1-130.

[51] European Commission. COMMISSION REGULATION (EU) No 252/2012 of 21 March 2012 laying down methods of sampling and analysis for the official control of levels of dioxins, dioxin- like PCBs and non-dioxin-like PCBs in certain foodstuffs and repealing Regulation (EC) No 1883/2006 Off J Eur Union, 2012; (84): 1-22.

[52] European Commission. COMMISSION REGULATION (EU) No 278/2012 of 28 March 2012 amending Regulation (EC) No 152/2009 as regards the determination of the levels of dioxins and polychlorinated biphenyls Off J Eur Union, 2012; (91): 8-22.

[53] Patterson DG Jr, Welch SM, Turner WE, Sjödin A, Focant JF. Cryogenic zone compression for the measurement of dioxins in human serum by isotope dilution at the attogram level using modulated gas chromatography coupled to high resolution magnetic sector mass spectrometry. J Chromatogr A 2011; 1218(21): 3274-81.
[http://dx.doi.org/10.1016/j.chroma.2010.10.084] [PMID: 21112056]

[54] Focant JF, Pirard C, Eppe G, De Pauw E. Recent advances in mass spectrometric measurement of dioxins. J Chromatogr A 2005; 1067(1-2): 265-75.

[55] Korytár P, Leonards PEG, de Boer J, Brinkman UATh. High-resolution separation of polychlorinated biphenyls by comprehensive two-dimensional gas chromatography. J Chromatogr A 2002; 958(1-2):

203-18.
[http://dx.doi.org/10.1016/S0021-9673(02)00327-8] [PMID: 12134818]

[56] Korytár P, Danielsson C, Leonards PEG, Haglund P, de Boer J, Brinkman UATh. Separation of seventeen 2,3,7,8-substituted polychlorinated dibenzo-p-dioxins and dibenzofurans and 12 dioxin-like polychlorinated biphenyls by comprehensive two-dimensional gas chromatography with electron-capture detection. J Chromatogr A 2004; 1038(1-2): 189-99.
[http://dx.doi.org/10.1016/j.chroma.2004.03.026] [PMID: 15233534]

[57] Danielsson C, Wiberg K, Korytár P, Bergek S, Brinkman UATh, Haglund P. Trace analysis of polychlorinated dibenzo-p-dioxins, dibenzofurans and WHO polychlorinated biphenyls in food using comprehensive two-dimensional gas chromatography with electron-capture detection. J Chromatogr A 2005; 1086(1-2): 61-70.
[http://dx.doi.org/10.1016/j.chroma.2004.11.071] [PMID: 16130656]

[58] Hoh E, Mastovska K, Lehotay SJ. Optimization of separation and detection conditions for comprehensive two-dimensional gas chromatography--time-of-flight mass spectrometry analysis of polychlorinated dibenzo-p-dioxins and dibenzofurans. J Chromatogr A 2007; 1145(1-2): 210-21.
[http://dx.doi.org/10.1016/j.chroma.2007.01.064] [PMID: 17306276]

[59] Hoh E, Lehotay SJ, Mastovska K, Huwe JK. Evaluation of automated direct sample introduction with comprehensive two-dimensional gas chromatography/time-of-flight mass spectrometry for the screening analysis of dioxins in fish oil. J Chromatogr A 2008; 1201(1): 69-77.
[http://dx.doi.org/10.1016/j.chroma.2008.05.089] [PMID: 18556010]

[60] Focant JF, Eppe G, Scippo ML, *et al.* Comprehensive two-dimensional gas chromatography with isotope dilution time-of-flight mass spectrometry for the measurement of dioxins and polychlorinated biphenyls in foodstuffs. Comparison with other methods. J Chromatogr A 2005; 1086(1-2): 45-60.
[http://dx.doi.org/10.1016/j.chroma.2005.05.090] [PMID: 16130655]

[61] Planche C, Ratel J, Mercier F, Blinet P, Debrauwer L, Engel E. Assessment of comprehensive two-dimensional gas chromatography-time-of-flight mass spectrometry based methods for investigating 206 dioxin-like micropollutants in animal-derived food matrices. J Chromatogr A 2015; 1392: 74-81.
[http://dx.doi.org/10.1016/j.chroma.2015.02.054] [PMID: 25814331]

[62] Watanabe-Ezoe Y, Li X, Imasaka T, Uchimura T, Imasaka T. Gas chromatography/femtosecond multiphoton ionization/time-of-flight mass spectrometry of dioxins. Anal Chem 2010; 82(15): 6519-25.
[http://dx.doi.org/10.1021/ac1009063] [PMID: 20593780]

[63] European Commission. COMMISSION REGULATION (EU) No 589/2014 of 2 June 2014 laying down methods of sampling and analysis for the control of levels of dioxins, dioxin-like PCBs and non-dioxin-like PCBs in certain foodstuffs and repealing Regulation (EU) No 252/2012 Off J Eur Union, 2014; (164): 18-40.

[64] European Commission. COMMISSION REGULATION (EU) No 709/2014 of 20 June 2014 amending Regulation (EC) No 152/2009 as regards the determination of the levels of dioxins and polychlorinated biphenyls Off J Eur Union, 2014; (188): 1-18.

[65] Brunnée C. The ideal mass analyzer: fact or fiction? Int J Mass Spectrom Ion Process 1987; 76: 125-237.
[http://dx.doi.org/10.1016/0168-1176(87)80030-7]

[66] Chess EK, Gross ML. Determination of tetrachlorodibenzo-p-dioxins by mass spectrometric metastable decomposition monitoring. Anal Chem 1980; 52: 2057-61.
[http://dx.doi.org/10.1021/ac50063a015]

[67] Malavia J, Santos FJ, Galceran MT. Gas chromatography-ion trap tandem mass spectrometry versus GC-high-resolution mass spectrometry for the determination of non-*ortho*-polychlorinated biphenyls in fish. J Chromatogr A 2004; 1056(1-2): 171-8.
[http://dx.doi.org/10.1016/j.chroma.2004.05.066] [PMID: 15595547]

[68] Küchler T, Brzezinski H. Application of GC-MS/MS for the analysis of PCDD/Fs in sewage effluents. Chemosphere 2000; 40(2): 213-20. [http://dx.doi.org/10.1016/S0045-6535(99)00301-X] [PMID: 10665434]

[69] Fabrellas B, Sanz P, Abad E, Rivera J, Larrazábal D. Analysis of dioxins and furans in environmental samples by GC-ion-trap MS/MS. Chemosphere 2004; 55(11): 1469-75. [http://dx.doi.org/10.1016/j.chemosphere.2004.01.039] [PMID: 15099726]

[70] Mandalakis M, Tsapakis M, Stephanou EG. Optimization and application of high-resolution gas chromatography with ion trap tandem mass spectrometry to the determination of polychlorinated biphenyls in atmospheric aerosols. J Chromatogr A 2001; 925(1-2): 183-96. [http://dx.doi.org/10.1016/S0021-9673(01)01033-0] [PMID: 11519804]

[71] Focant JF, Eppe G, De Pauw E. Optimisation and use of tandem-in-time mass spectrometry in comparison with immunoassay and HRGC/HRMS for PCDD/F screening. Chemosphere 2001; 43(4-7): 417-24. [http://dx.doi.org/10.1016/S0045-6535(00)00389-1] [PMID: 11372821]

[72] Kemmochi Y, Tsutsumi K. Rapid PCDD/PCDF screening method for fly ash with ion trap MS/MS. Chemosphere 2001; 43(4-7): 433-7. [http://dx.doi.org/10.1016/S0045-6535(00)00391-X] [PMID: 11372823]

[73] Malavia J, Abalos M, Santos FJ, Abad E, Rivera J, Galceran MT. Analysis of polychlorinated dibenzo-p-dioxins, dibenzofurans and dioxin-like polychlorinated biphenyls in vegetable oil samples by gas chromatography-ion trap tandem mass spectrometry. J Chromatogr A 2007; 1149(2): 321-32. [http://dx.doi.org/10.1016/j.chroma.2007.03.052] [PMID: 17399728]

[74] Malavia J, Santos FJ, Galceran MT. Comparison of gas chromatography-ion-trap tandem mass spectrometry systems for the determination of polychlorinated dibenzo-p-dioxins, dibenzofurans and dioxin-like polychlorinated biphenyls. J Chromatogr A 2008; 1186(1-2): 302-11. [http://dx.doi.org/10.1016/j.chroma.2007.11.089] [PMID: 18096174]

[75] Hayward DG, Holcomb J, Glidden R, *et al.* Quadrupole ion storage tandem mass spectrometry and high-resolution mass spectrometry: complementary application in the measurement of 2,3,7,8-chlorine substituted dibenzo-p-dioxins and dibenzofurans in US foods. Chemosphere 2001; 43(4-7): 407-15. [http://dx.doi.org/10.1016/S0045-6535(00)00388-X] [PMID: 11372820]

[76] Eppe G, Focant JF, Pirard C, Pauw ED. PTV-LV-GC/MS/MS as screening and complementary method to HRMS for the monitoring of dioxin levels in food and feed. Talanta 2004; 63(5): 1135-46. [http://dx.doi.org/10.1016/j.talanta.2004.05.056] [PMID: 18969543]

[77] Malavia J, Abalos M, Santos FJ, Abad E, Rivera J, Galceran MT. Ion-trap tandem mass spectrometry for the analysis of polychlorinated dibenzo-p-dioxins, dibenzofurans, and dioxin-like polychlorinated biphenyls in food. J Agric Food Chem 2007; 55(26): 10531-9. [http://dx.doi.org/10.1021/jf0719858] [PMID: 18052095]

[78] Clement RE, Bobbie B, Taguchi V. Comparison of instrumental methods for chlorinated dibenzo--dioxin (CDD) Determination-Interim results of a Round-Robin study involving GC-MS, MS-MS, and high resolution MS. Chemosphere 1986; 15: 1147-56. [http://dx.doi.org/10.1016/0045-6535(86)90385-1]

[79] March RE, Splendore M, Reiner EJ, *et al.* A comparison of three mass spectrometric methods for the determination of dioxins/furans. Int J Mass Spectrom 2000; 194: 235-46. [http://dx.doi.org/10.1016/S1387-3806(99)00228-6]

[80] Sandy C, Fürst P, Bernsmann T, Baumeister D. Determination of polychlorinated biphenyl congeners in foodstuffs and animal feed using a triple quadrupole GC-MS/MS instrument. Organohalogen Compd 2011; 73: 1370-1.

[81] Ábalos M, Cojocariu CI, Silcock P, *et al.* Meeting the European Commission performance criteria for the use of triple quadrupole GC-MS/MS as a confirmatory method for PCDD/Fs and dl-PCBs in food

and feed samples. Anal Bioanal Chem 2016; 408(13): 3511-25.
[http://dx.doi.org/10.1007/s00216-016-9428-9] [PMID: 26922340]

[82] L'Homme B, Scholl G, Eppe G, Focant JF. Validation of a gas chromatography-triple quadrupole mass spectrometry method for confirmatory analysis of dioxins and dioxin-like polychlorobiphenyls in feed following new EU Regulation 709/2014. J Chromatogr A 2015; 1376: 149-58.
[http://dx.doi.org/10.1016/j.chroma.2014.12.013] [PMID: 25533395]

[83] Kotz A, Malisch R, Wahl K, *et al.* GC-MS/MS determination of PCDD/Fs and PCBs in feed and food-comparison with GC-HRMS. Organohalogen Compd 2011; 73: 688-91.

[84] European Commission. COMMISSION DECISION of 12 August 2002 implementing Council Directive 96/23/EC concerning the performance of analytical methods and the interpretation of results. Off J Eur Union 2002; (221): 8-36.

[85] García-Bermejo Á, Ábalos M, Sauló J, Abad E, González MJ, Gómara B. Triple quadrupole tandem mass spectrometry: A real alternative to high resolution magnetic sector instrument for the analysis of polychlorinated dibenzo-p-dioxins, furans and dioxin-like polychlorinated biphenyls. Anal Chim Acta 2015; 889: 156-65.
[http://dx.doi.org/10.1016/j.aca.2015.07.039] [PMID: 26343438]

[86] European Commission. COMMISSION DIRECTIVE 2004/44/EC of 13 April 2004 amending Directive 2002/69/EC laying down the sampling methods and the methods of analysis for the official control of dioxins and the determination of dioxin-like PCBs in foodstuffs Off J Eur Union, 2004; (113): 17-8.
[http://dx.doi.org/10.1006/enrs.2001.4274] [PMID: 11896663]

[87] European Commission. COMMISSION DIRECTIVE 2005/7/EC of 27 January 2005 amending Directive 2002/70/EC establishing requirements for the determination of levels of dioxins and dioxin-like PCBs in feedingstuffs Off J Eur Union, 2005; (27): 41-3.
[http://dx.doi.org/10.1006/enrs.2001.4274] [PMID: 11896663]

[88] van Bavel B, Geng D, Cherta L, *et al.* Atmospheric-pressure chemical ionization tandem mass spectrometry (APGC/MS/MS) an alternative to high-resolution mass spectrometry (HRGC/HRMS) for the determination of dioxins. Anal Chem 2015; 87(17): 9047-53.
[http://dx.doi.org/10.1021/acs.analchem.5b02264] [PMID: 26267710]

[89] Organtini KL, Haimovici L, Jobst KJ, *et al.* Comparison of atmospheric pressure ionization gas chromatography-triple quadrupole mass spectrometry to traditional high-resolution mass spectrometry for the identification and quantification of halogenated dioxins and furans. Anal Chem 2015; 87(15): 7902-8.
[http://dx.doi.org/10.1021/acs.analchem.5b01705] [PMID: 26140516]

[90] Portolés T, Sales C, Abalos M, Sauló J, Abad E. Evaluation of the capabilities of atmospheric pressure chemical ionization source coupled to tandem mass spectrometry for the determination of dioxin-like polychlorobiphenyls in complex-matrix food samples. Anal Chim Acta 2016; 937: 96-105.
[http://dx.doi.org/10.1016/j.aca.2016.06.038] [PMID: 27590550]

[91] Taguchi VY, Nieckarz RJ, Clement RE, Krolik S, Williams R. Dioxin analysis by gas chromatography-Fourier transform ion cyclotron resonance mass spectrometry (GC-FTICRMS). J Am Soc Mass Spectrom 2010; 21(11): 1918-21.
[http://dx.doi.org/10.1016/j.jasms.2010.07.010] [PMID: 20727785]

[92] Cojocariu CI, Abalos M, Abad E, *et al.* Potential of gas chromatography coupled with Orbitrap-based mass spectrometry for the analysis of halogenated persistent organic pollutants in food and environmental samples. Organohalogen Compd 2015; 77: 154-7.

CHAPTER 6

Pesticides

Vicente Andreu and **Yolanda Picó***

Environmental and Food Safety Research Group of the University of Valencia (SAMA-UV), Desertification Research Center (CIDE), Joint Research Center CSIC-UV-GV, Moncada, Spain

Abstract: Analysis of pesticide residues is very important to enforce legislation and guarantee food safety. The correct use of pesticides is still crucial in agriculture because they provide spectacular increases in crop yields and ensure global demand for grain. However, the indiscriminate, incorrect and/or excessive use of pesticides in agriculture may have some serious adverse effects such as the accumulation of residues in food. Pesticide residues are controlled worldwide by maximal residues limits (MRLs), not the same in all countries but generally ranging from a few μg kg^{-1} (usually for pesticides that are banned) to a few tens of mg kg^{-1}. Determining pesticides at this concentration requires sensitive, accurate and robust instrumentation, and trained personnel as well. This chapter explores the latest advances to determine pesticide residues as accurately as possible in the shortest time. A description of aspects like improvement of high-throughput methods specificity and advances in the determination by gas chromatography-mass spectrometry (GC-MS), liquid chromatography-mass spectrometry (LC-MS) or (bio)sensors, are presented in this chapter. The focus is on multi-residue o multiplexed analysis that will offer rapidity and economy in order to achieve the required sensitivity (<0.01 mg kg^{-1}). The primary purpose of this chapter is to provide the reader with a state- of- the- art assessment and identification of gaps within this field, and to establish future trends in the extraction, purification, and determination of pesticide residues.

Keywords: Dispersive liquid-liquid extraction, Food of animal origin, Fruits and vegetables, Gas chromatography-mass spectrometry, Liquid chromatography-mass spectrometry, QuEChERS, QuPPe, Sensors.

INTRODUCTION

Pesticides are a group of just a few natural molecules and a majority of synthetic ones developed to eliminate pests in both agriculture and livestock. To give

* **Corresponding author Yolanda Picó:** Environmental and Food Safety Research Group of the University of Valencia (SAMA-UV), Desertification Research Centre (CIDE), Joint Research Centre CSIC-UV-GV, Moncada-Naquera Road Km 4.5, 46113 Moncada, Spain; Tel: +34 963424215; E-mail: Yolanda.Pico@uv.es

Belén Gómara & María Luisa Marina (Eds.)
Current and Future Developments in Food Science (Vol. 1)

an idea of the number of this type of compounds, the reare approximately 1000 active ingredients which have no chemical relationship among them and cover a wide range of physico-chemical properties.

Pesticides play a significant role in food production because they increase crop yields and the number of harvests that can be obtained per year. This is particularly important in countries with endemic food shortages that may expect an exponential growth of population in the coming years. However, pesticides are significantly toxic to biota and people. Toxicity depends on their function, action mechanism, doses, and exposure time. Several studies indicate that dietary exposure to pesticides (or long-term exposure) is associated with a broad spectrum of adverse effects on the development and reproduction of human beings, and on their nervous and immune systems. Pesticides can also cause oxidative stress, cell damage, endocrine disruption and different types of cancer.

The production, distribution and application of pesticides is strictly regulated and subject to a tight control due to their toxicity. Since pesticides are purposely applied on large surfaces, their exact destination is difficult to control. Monitoring their presence in food to ensure food safety is also crucial.

Analytical laboratories aim to detect, identify and quantitate many different pesticides with diverse physical-chemical properties in increasingly complex matrices at trace concentrations (ca. 10 $\mu g\ kg^{-1}$).

This chapter outlines the basic principles of advanced extraction and determination approaches, their advantages and drawbacks, the suitability of analytical validation parameters and their robustness and usefulness for pesticide residue determination.

LEGISLATION

The legislation regulating pesticide residues in food is very extensive. Only specifically authorized pesticides may be used, and pesticide levels in food must be below the established maximal residues limits (MRLs) to be apt for consumption. Additionally, the guidelines concerning the analytical methods used to determine these residues are also very strict.

The first guideline marks the number of substances to be determined. There are about 1000 active substances considered as pesticides but important differences exist in the authorized products among countries. The European Union (EU) reduced in 2008 the number of authorized substances. However, this is only relatively useful since markets are global and products banned in the European Union are authorized in other parts of the world. Therefore, at least in theory, the nearly 1000 active substances that exist need to be monitored.

The second guideline establishes the limits of detection to be reached by analytical methods. Residues are the rest of the pesticide formulations that remain on or in the plant after application. MRLs are defined by the EU [1] as "the upper legal levels of a concentration for pesticide residues (expressed in mg kg^{-1}) in or on food or feed based on good agricultural practices (GAP) and to ensure the lowest possible consumer exposure". MRLs for different crops and pesticides, as established in the EU, can be found in the MRL database on the Commission website [1]. If a pesticide is not regulated within the EU, the MRL applied is 0.01 mg kg^{-1}. It is assumed that this value is the limit of detection achievable using state-of-the-art instrumentation. "Pesticide residues" according to the EU include conversion and degradation products, metabolites, reaction products, and impurities that have toxicological or environmental significance [1]. This is what makes these compounds determination more complicated, because it is mandatory to identify the pesticide and a variable number of its metabolites.

The EU legislation does not establish official analytical methods for the determination of pesticide residues in food. However, although their use is not mandatory, standardization agencies propose a number of well-validated methods that could be used. Instead, the EU has approved guidelines establishing the minimum quality requirements of an analytical method to be applied. The main problem with these analyses is that pesticides are at a very low concentration in the samples, and therefore, determination has to be precise and accurate and reach a very high sensitivity. The most important guidelines applicable to pesticide residues can be accessed on the EU website.

ANALYTICAL METHODS

Modern scientific methods to measure pesticide residues in plants involved three different phases that can be summarized as: 1) Sampling, and sample preparation, 2) Extraction and clean-up of the extract, 3) Determination. Fig. (**1**) schematizes the workflow of these methods as well as the time spent in each step.

Sample and sample preparation involve a series of well-known procedures that are outside of the scope of this chapter. Sampling procedure to determine pesticides in food is well established in the EU guidelines. Sample preparation involves cutting, chopping and homogenization. Samples are commonly preserved frozen at -20 ºC.

The extraction and clean-up procedures are commonly determined by: (i) the characteristics of the matrix (% of proteins, lipids and carbohydrates, presence of salts, *etc*.), (ii) type of pesticides to be determined, and (iii) determination technique used. As a general rule, food requires a solvent extraction step; then, solid-phase extraction (SPE) is the most used clean-up. The EU guidelines establish the need to validate analytical methods at least for one matrix of each representative type (based on complexity) [1], commodities are distinguished into four groups: (i) dried samples characterized by high protein/high starch content (*i.e.* cereals and legumes); (ii) high water content (*i.e.* most fruits and vegetables); (iii) fatty products (*i.e.* avocados, linseed, nuts, olives, and rape seed), and (iv) high acidic content (mostly citrus fruits, such as grapefruits, oranges, tangerines, lemons). In addition to these four groups, there are several "non-classified" matrices (coffee beans, cocoa beans, herbal infusions, hops, spices, tea and tobacco) that are considered incapable of being grouped and require specific method validation.

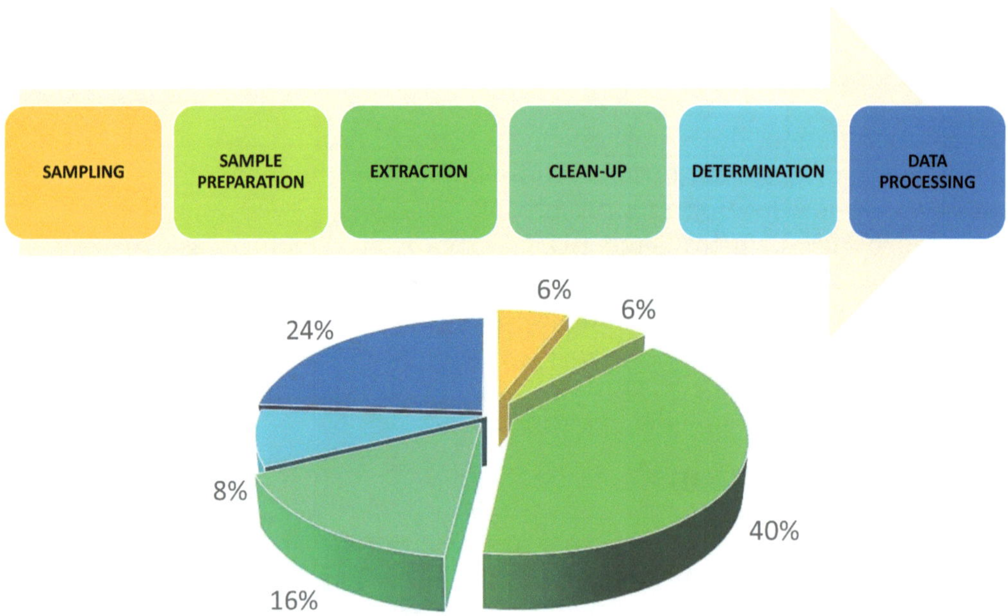

Fig. (1). Workflow for pesticides determination, and weight of each step.

Once extracted, phytosanitary residues on plants are determined by:

- Chromatographic methods (mostly coupled to mass spectrometry).
- Direct analysis by MS.
- Sensors.

The chromatographic methods most commonly applied for pesticide residue determination are gas chromatography-mass spectrometry (GC-MS) and liquid chromatography-mass spectrometry (LC-MS) [2 - 7]. Both techniques can provide confirmation of multiple classes and many pesticide residues simultaneously. However, recent advances, including high-resolution mass spectrometry (HRMS), have evolved until almost eliminating the previous LC or GC separation. Other emerging techniques, such as sensors and biosensors, are highly promising to speed up the analysis. These techniques are advancing quickly thanks to new materials, such as nanomaterials, molecularly imprinted polymers (MIPs), and aptamers [8 - 11].

In this regard, the most important chromatographic methods have been summarized in Tables **1** and **2.** To shed light on the different sections of this part describing methods, the number of pesticides covered, sample preparation method, and characteristics of the determination have been outlined.

Most of the developed methods tend to be universal, and this is applicable to any type of plant or animal material. These methods are characterized by the use of a generic extraction method capable of simply and economically extracting all possible pesticides without carrying endogenous matrix compounds, which are further identified using a very selective and specific determination method. These methods are called "multi-residue methods" and are commonly related to the determination by GC or LC. They could be able to easily cover up to 500 pesticides in one method. However, as observed in Tables **1** and **2,** most of the methods are able to cover between 100 and 200 pesticide residues. The main disadvantage of these multi-residue methods is that they cannot be applied to determine the whole range of selected pesticides with optimal accuracy and precision using only one approach. This is because even those that can be easily extracted by a generic extraction procedure present a high diversity in chemical composition requiring several strategies for their proper determination. It is frequent that at least one GC and one LC injection are required in order to enlarge the number of determinable pesticides. It does not mean, however, that all existing pesticides can be extracted by a multi-residue method.

Table 1. Recent relevant examples of papers published on pesticide residue determination in food using GC coupled to several detectors.

Pesticides	Extraction	Determination	Quantification	Ref
14 multiclass	HS-SPME	GC-MS (EI) Conventional	SIM	[12]
Organophosphorus pesticides	Solvent and SBSE (new coating)	GC-FID	Conventional	[13]

(Table 1) cont.....

Pesticides	Extraction	Determination	Quantification	Ref
Organophosphorus triazole, organochlorine	MWAE and SPME (new coating)	GC-MS(EI)	Time-scheduled SIM	[14]
Multi-residue	Ethyl acetate	GC-QToF-MS (NCI) Blackflusing	Automated accurate mass data processing	[15]
47 multiclass	QuEChERS (modified)	GC-MS/MS	SRM Two transitions	[16]
OPP	QuEChERS	GC-QToF-MS (API)	MS^{E}-Automated processing	[17]
Quinalfos	QuEChERS	GC-ECD	Conventional	[18]
5 multiclass	HS-SPME (new fiber)	GC-ECD	Conventional	[19]
36 multiclass	QuEChERS	GC-ECD GC-FPD	Conventional	[20]
385 multi-residue	QuEChERS	GC-MS (EI)	SIM	[21]
36 multiclass	QuEChERS	GC-ECD GC-FPD	Conventional	[22]
12 pesticides	SE-LLP-DLLME No filtration	GC-FID	Conventional	[23]
Organophosphorus	SE-DLLME	GC-NPD	Conventional	[24]
Pyrethroids	SE-LLP-DLLME	GC-NPD GC-MS (EI)	Conventional SIM	[25]
4 pesticides	LLE (continuous)	GC-FID	Conventional	[26]
8 pesticides	GA-DLLME (eutectic solvent)	GC-FID	Conventional	[27]
Organophosphorus	QuEChERS with MdSPE	GC-PFD GC-MS	Conventional SIM	[28]
75 priority	10-fold dilution	GC-QqQ-MS/MS (EI) Back flushing	MRM: 2-3 MS/MS transitions	[29]
5 OPP	MSPD	GC-ECD GC-MS	Conventional SIM (3 ions)	[30]
15 Pesticides	QuEChERS	GC-MS (EI)	SIM (3 ions)	[31]
Ethion	QuEChERS	GC-ECD	Conventional	[32]
7 EDCs	USEME	GC-MS (EI)	SIM	[33]
113 Pesticides	QuEChERS	GC-QqQ-MS/MS (EI)	SRM (2 transitions)	[34]
10 Pyrethroids	QuEChERS	GC-QqQ-MS/MS (EI) Back flushing	SRM (2 transitions)	[35]
184 multiclass	MSPD	(GC-μECD/NPD),	Conventional	[36]
35 pesticides	QuEChERS	GC-MS (EI)	SIM	[37]

(Table 1) cont.....

Pesticides	Extraction	Determination	Quantification	Ref
Organophosphorus	SE-MSPE	GC-NPD	Conventional	[38]
Fenitrothion	QuEChERS	GC-FPD	Conventional	[39]
Multi-residue Multiclass	Mini-Luke SE	GC-ToF-MS (EI) No HR	Full scan *m/z* 40 and 450	[40]
85 pesticides	Mini-Luke SE QuEChERS	GC-ToF-MS (EI) No HR	Full scan *m/z* 40 and 450	[41]
Trifloxystrobin and tebuconazole	QuEChERS	GC-MS (EI)	SIM	[42]
Propiconazole isomers and tebuconazole	Diluted SPME	GC-NPD GC-MS (EI)	Conventional SIM	[43]
Multi-residue Multiclass	Non-cleaned QuEChERS	LVI GC-MS/MS (EI) Back flushing	SRM (2 transitions)	[44]
Multi-residue Multiclass	Sonication-SE SDME	GC-MS	Quali-Full scan Quanti-SIM	[45]
Organophosphorus	SE-SPE	GC-FPD	Conventional	[46]
Multi-residue Multiclass	QuEChERS mPFC	GC-MS	SIM (≥ 2 ions)	[47]
36 Multiclass	QuEChERS DLLME	GC-MS	SIM	[48]
3 Pesticides	MSPE	GC-µECD	Conventional	[49]
Organophosphorus	H_2O dilution SPME (new coating)	GC-CD-IM-MS	Full scan	[50]
131 Multiclass	QuEChERS	GC-µECD/GC-NPD	Conventional	[51]
10 pesticides and their metabolites	QuEChERS	GC-ECD	Conventional	[52]
Chlorpyrifos Imidacloprid	QuEChERS	GC-MS (EI)	SIM	[53]
220 Multiclass	QuEChERS Ethyl acetate Mini-Luke NL-method	GC-MS/MS (EI)	MRM (2 transitions per pesticide)	[54]
3 OPPs	Dilution SPE	GC-MS (EI)	SIM	[55]
Pesticides and PCBs	QuEChERS	GC-QOrbitrap	Full scan (60000 FWHM) SIM	[56]
Multiclass Multi-residue	QuEChERS	GC-QOrbitrap	Full scan (60000 FWHM)	[57]

Table 2. Recent relevant examples of papers published on pesticide residue determination in plant materials using LC coupled to any detector.

Pesticides	Extraction	Determination	Quantification	Ref
Spinosad	SE LLP	HPLC-UV	Conventional	[58]
Multiclass Multi-residue	DSPE (new material)	HPLC-PDA	Conventional	[59]
Fungicides	QuEChERS	LC-MS/MS	MRM (2 transitions per compound	[60]
42 Pesticides	QuEChERS	LC-MS/MS	MRM (2 transitions per compound)	[61]
57 Pesticides	QuEChERS	LC-MS/MS	Dynamic MRM	[62]
Diazinon	H_2O dilution MIPs	HPLC-PDA	Conventional	[63]
Multiclass	QuEChERS (modified)	HPLC-PDA	Convectional	[64]
28 pesticides	Ethyl acetate	LC-HRMS ToF	Accurate mass extraction	[65]
20 Multiclass	QuEChERS	LC-MS/MS	MRM (2 transitions per compound)	[66]
Six acidic pesticides	H_2O dilution On-line SPE (new sorbent)	LC-MS/MS	MRM (2 transitions per compound)	[67]
42 multiclass	QuEChERS	LC-MS/MS	MRM (2 transitions per compound)	[61]
Polar pesticides	QuPPE	LC-MS/MS	MRM (2 transitions per compound)	[68]
253 multiclass	Different QuEChERS	LC-MS/MS	MRM (2 transitions per compound)	[69]
Multi-residue pesticides and others (28)	QuPPE	HILIC-MS/MS	MRM (2 transitions per compound)	[70]
385 multi-residue	QuEChERS	LC-MS/MS	MRM Two transitions	[21]
75 priority	10-fold dilution	LC-QToF-MS	MS^E Qualitative and quantitative	[29]
256 pesticides	dilution	LC-MS/MS QTRAP	MRM Two transitions	[71]
8 pesticides	QuEChERS	LC-MS/MS	MRM Two transitions	[72]

(Table 2) cont.....

Pesticides	Extraction	Determination	Quantification	Ref
Quaternary ammonium pesticides	QuEChERS	HILIC-MS/MS	MRM Two transitions	[73]
220 Multiclass	QuEChERS Ethyl acetate Mini-Luke NL-method	LC-MS/MS	MRM (2 transitions per pesticide)	[54]
63 pesticides	QuEChERS	LC-MS/MS	MRM (2 transitions)	[37]
3 pesticides	Mini-Luke SE QuEChERS	LC-MS/MS	MRM (2 transitions)	[41]
64 Multiclass	QuEChERS	nLC-QOrbitrap	AIF mode	[74]
68 pesticides	SE dSPE	UHPLC-QToF	MS^E Quali and Quanti	[75]
600 pesticides	QuEChERS	UHPLC-QToF	MS^E Quali and Quanti	[76]
Pyrisoxazole Enantiomers	QuEChERS (modified)	UHPLC-QToF	MS^E Quali and Quanti	[77]
60 pesticides	SE and SPE	UHPLC-ToF	Quali and Quanti	[78]
Glyphosate AMPA	H_2O dilution	UHPLC-QOrbitrap (anionic exchanger)	Quali and Quanti	[79]
184 multiclass	QuEChERS	UHPLC-QOrbitrap	AIF Quali and Quanti	[80]
15 LC amenable pesticides	QuEChERS	UHPLC-QOrbitrap	SWATH Fix window Quali and Quanti	[81]
8 acidic herbicides	SE	LC-MS/MS	MRM (2 transitions)	[4]
109 multiclass	QuEChERS	LC-MS/MS	MRM (2 transitions)	[82]
200 pesticides	SE	LC-MS/MS	MRM (2 transitions)	[83]
Carbendazim, Chlorpyrifos	QuEChERS	HPLC-FL	Conventional	[84]
10 Chiral compound	QuEChERS	LC-MS/MS QTRAP	MRM (2 transitions)	[85]
Ametoctradin residue	QuEChERS	LC-MS/MS	MRM (2 transitions)	[86]
Dithiocarbamate fungicides	QuEChERS (modified)	LC-MS/MS	MRM (2 transitions)	[87]
170 multiclass	QuEChERS (without clean-up)	LC-MS/MS	MRM (2 transitions)	[88]

(Table 2) cont.....

Pesticides	Extraction	Determination	Quantification	Ref
Piridaben residues	QuEChERS	LC-MS/MS	MRM (2 transitions)	[89]
49 multiclass	QuEChERS	LC-MS/MS	MRM (2 transitions)	[90]
Multiclass	QuEChERS ILs dSPE	UHPLC-MS/MS	MRM Multivariate Surface response	[91]
Multiclass	QuEChERS dSPE and DLLME	LC-MS/MS	MRM (2 transitions)	[92]
Trichlofon Monocrotofos	SE-MIP	HPLC-UV	Conventional	[93]
28 pesticides	QuEChERS	LC-MS/MS (small injection volume)	MRM (2 transitions)	[94]
Multiclass	QuEChERS (chemometric optimization)	LC-MS/MS	MRM (2 transitions)	[95]
Multiclass	SE MSPE	HPLC-UV	Conventional	[96]

A few particular cases of pesticides or pesticide families still require single-residue methods. This happens mostly with:

- Very polar pesticides (glyphosate, glufosinate ammonium).
- Ionic pesticides (fosetyl-Al).
- Unstable pesticides (cyromazine, etephon).

However, the trend is clearly to develop high coverage multiresidue methods to the extent that, recently, the EU reference laboratories for pesticide residues developed a method able to simultaneously extract those pesticides not well-extracted by the QuEChERS method due to their high polarity [97].

Pesticides included in the multi-residue schemes could also be analyzed using specific methods. This is not very frequent because it is not economically efficient, but in some specific cases, it could be useful. Most of these single or few similar classes of residues methods are based on a very selective extraction using MIPs or, less frequently, immunosorbents. Others are based on the development of specific immunoassays or sensor/biosensors. In this case, rapidity could be considered an advantage.

In the case of ionic or very polar pesticides, in addition to the general extraction method, two alternative LC strategies could be used [97]:

i. Decreasing the analytes polarity, either by transforming them by chemical reaction into another less polar substance (derivatization) or by adding an opposing ion charge counter-ion forming an ion pair, to the mobile phase.
ii. Applying alternative LC mechanisms more appropriate for the separation of polar compounds as hydrophilic interaction (HILIC), carbon or ion exchange chromatography instead of reversed phase.

Both strategies increase retention of analytes in the analytical column and improve chromatographic peak shape of the analytes.

Sample preparation step is necessary in order to eliminate co-extracted substances that are endogenous of the matrix. This step is long (commonly described as one of the bottle neck of the analysis), expensive, and important to ensure the appropriate accuracy and precision.

Extraction

Solvent Extraction (SE)

The extraction and purification of pesticide residues in food have made significant progress. The first multi-residue scheme applied was that proposed by Mills in the 1960s [98]. This method first involved solvent extraction, then partition and finally column separation requiring at least 1 or 2 days for execution per sample. Nowadays, a QuEChERS requires approximately 20 min of execution time. The interest in decreasing the analysis time has been mainly due to the fact that food products are highly perishable and must be commercialized as soon as possible [2].

Solvent extraction (SE) is a technique that requires energy, which is obtained through mechanical shaking, ultrasounds, pressurized liquid extraction or microwave [14]. Ultrasound and mechanical shaking are sufficient to ensure proper extraction. SE of pesticide residues from food is usually carried out with water miscible organic solvents (polar). The most common solvents include acetone, acetonitrile and methanol and their mixtures with water because they provide good recoveries for pesticides in a wide range of polarities. Ethyl acetate is also frequently used to extract pesticides. Some of these methods are referred to by their own name (*e.g.* mini-Luke procedure, QuEChERS, *etc.*) [99, 100].

QuEChERS is nowadays the extraction platform that has improved rapidity much more. The name is the abbreviation of "Quick, Easy, Cheap, Effective, Rugged, and Safe" method. This method involves an acetonitrile extraction (where the phases are separated by salt addition) followed by dispersive solid-phase extraction (dSPE) clean up. It is a method that reduces the volume of solvent used in extraction, allows a large number of samples to be analyzed in a short time and is capable of extracting a broad spectrum of pesticides with appropriate recoveries. The advantage of QuEChERS method is that many variants have been developed depending on the analyst's objectives. One of the most common is to buffer with a slightly acidic pH to stabilize pH-sensitive pesticides. Two official methods have been described: one using ammonium acetate (pH 4.8) proposed by the Association of Analytical Chemists (AOAC) (official method 2007.01) and another using a citrate buffer (pH 5.5) proposed by the European Committee for Standardization (CEN) (Standard EN method 15662) [101].

An alternative multi-residue method for the residue analysis of very polar, non-QuEChERS-amenable, pesticides in foods, was also developed pointing out the interest of generic procedures to extract pesticides, even if the pesticides cannot be determined under common conditions. In this method, named Quick Method for the Analysis of numerous Highly Polar Pesticides (QuPPE-method), pesticides residues are extracted after adjusting the water ratio and adding acidified methanol and separating them from the matrix by centrifugation without any further clean-up. The QuPPE method also proposes the use of different LC platforms to achieve more accurate separation and quantification of the pesticides.

Matrix Solid-Phase Dispersion (MSPD)

Extraction by MSPD is based on dispersing small quantities of the solid sample with a sorbent by mixing with a mortar pistil (this produces a disruption of the food structure). Once a homogeneous mixture is formed, it is placed in a SPE cartridge and the pesticides are eluted with a small amount of organic solvents [30, 36]. Although this technique has been in use since the 1990s, its use has not become widespread because, although based on a very elegant principle, solvent extraction works well and is cheaper since it does not require the use of sorbents whereas sample homogenization is not always easy and can be a lengthy process.

Head-Space Solid Phase Micro-Extraction (HS-SPME)

HS-SPME is the solvent-free version of SPME technique for food. In this technique, the sample is heated and the fiber is introduced in the headspace above the sample, volatile pesticides reach the headspace and are adsorbed on the fiber and then, the fiber is placed in the injector of the instrument where the analytes are desorbed on-line. If the instrumental analysis is carried out by GC, then thermal desorption could be applied and the method is completely solvent-free. This modification of the SPME methods shortens the time of the extraction and allows direct application to solid-samples.

One of the disadvantages of this procedure is the number of method’s variables that require optimization. This difficulty is lessened through the application of different multifactorial designs. The Plackett-Burman (P-B) design was one of the most commonly applied to establish the significance of various factors that could be further optimized using central composite design (CCD) to determine multiclass pesticide residues in fruit and vegetable [12] or fruit juices [19] by HS-SPME-GC-MS.

Furthermore, most of the commercial coatings for fibers are appropriate for non- or moderately polar compounds, whereas currently used pesticides tend to possess a more polar character. Research is on-going on new coating polymeric films such as 4-(2,5-di(thiophen-2-yl)-1H-pyrrol-1-yl) benzene [19], *etc.*

Clean-up

Liquid-Liquid Partitioning (LLP)

This procedure is the oldest one within pesticide residue determination. The disadvantages of LLP are that it uses high volumes of expensive and toxic organic solvents and it is long and not always suitable for determining compounds that could be at very low concentrations. There are some examples reported in the recent literature, such as pH-dependent continuous homogeneous liquid-liquid extraction that involves homemade designed extraction devices and two solvents. The method is based on the use of an amine (basic organic solvent) and acidified deionized water containing the analytes to get a homogeneous solution of both solvents. Both phases are separated by changing the pH [27].

The methods based on partition have evolved making possible to overcome the above-mentioned disadvantages. As a result, liquid-phase micro-extraction (LPME) that requires minimal amounts of solvents has been developed [102]. Many LPME techniques were implemented. The most used for pesticide residue determination in plants are:

- Ultrasound-assisted emulsification micro-extraction (USAEME) in which an organic water immiscible solvent is emulsified in the sample (diluted apple juice an extract dissolved in water) by ultrasounds. Both liquid phases are separated by centrifugation.
- Single drop micro-extraction (SDME) requires generating a stable drop of organic solvent at the tip of a syringe needle. Pesticides are isolated in the drop either by direct immersion of the drop in water or by headspace. After extraction, the micro drop with the analytes is retracted into the syringe and then introduced into the determination system [45].
- Dispersive liquid-liquid micro-extraction (DLLME) consists of the rapid

dispersion of an immiscible solvent in a phase composed of water and a dispersing agent (miscible in water and in the extraction solvent). Once the dispersion has been formed, the solvent agent in which the analytes have been dissolved is separated.

DLLME is the most commonly applied method (see Tables **1** and **2**). It uses high volumes of the dispersing agent and requires the use of a non-polar organic solvent (mostly hazardous halogenated hydrocarbons) but it is fast, simple and in most cases very effective. There are different proposals in the literature to overcome the disadvantages of DLLME which can be basically grouped into two approaches: (i) the elimination of the dispersant, and (ii) replacing halogenated extracting solvents with other less toxic ones. Within the first group, there are techniques such as ultrasound-assisted emulsification micro-extraction, vortex-assisted liquid-liquid micro-extraction, air-assisted liquid-liquid micro-extraction, and recently, gas-assisted dispersive liquid-phase micro-extraction (GA-DLPME). The second group includes those techniques based on the use of lighter-than-water solvents, for example, long-chain alcohols or hydrocarbons, ionic liquids (ILs), and deep eutectic solvents (DESs) [27].

Solid-Phase Extraction (SPE)

Solid-phase extraction (SPE) was developed in the early 1980s. It has replaced, with very few exceptions, solvent partition (LLP) because it enables to decrease the volume of solvents and provides cleaner extracts. The typical SPE columns used for the clean-up of food extracts in pesticide multiresidue methods were octadecylsilyl (C18), aminopropyl (-NH_2) and primary-secondary amine (PSA), strong and weak anion-exchange columns, and graphitized carbon black (GCB). SPE columns in normal phase (-NH_2 and PSA) are the most effective for removing compounds from the matrices, especially fatty acids (palmitic, palmitooleic, oleic, linoleic and linolenic acids), while GCB removes pigments [5]. In the case of fruits and vegetables, the C18 and anion-exchange phases have proved to be inefficient, but C18 also removes lipids [15]. The latest advances within this technique can be grouped into:

- Size reduction through the 96 well plate format and the development of new polymeric sorbents.
- Synthesis of novel sorbents, such as monolithic polymers of butyl methacrylate and ethylene glycol dimethacrylate [103].

Magnetic solid-phase extraction (MSPE) is a recent contribution to SPE based on the use of magnetic materials (*e.g.* nanoparticles) as adsorbent dispersed in the sample (increasing the contact area between adsorbent and analytes). Then, these materials are isolated from the sample by applying a magnetic field. MSPE has been already applied to extract and/or clean-up organophosphorus pesticides in water, fruits and vegetables using superparamagnetic core-shells anchored onto graphene oxide grafted with phenylethylamine (Fe_3O_4@SiO_2@GO-PEA) [38]. The results indicate the suitability of this procedure to clean-up complex matrices.

Selectivity, during this clean-up step, can be obtained by using immunoaffinity sorbents or immunosorbents based on the use of specific antibodies tailored for the pesticide of interest. This type of clean-up is very specific in separation and originates high concentration factors thanks to the high specificity of antigen-antibody interactions as already demonstrated for numerous pesticides in complex samples. However, the development of these sorbents is long and they have a prohibitive cost being useful only for one (or few in some cases) pesticides. Hence, few of them are commercialized.

Other selective supports are the oligo-sorbents that consist of aptamers immobilized onto a solid support. Aptamers are small (usually from 20 to 60 nucleotides) single-stranded RNA or DNA oligonucleotides. The specific sequence of these oligonucleotides is able to bind the selected molecule with similar affinity to antibodies. The advantages of the use of oligonucleotides instead of antibodies are that aptamers are not immunogenic or toxic and their generation, once the sequence is available, is simple and inexpensive by an *in vitro* process called SELEX (Systematic evolution of ligands by exponential enrichment). Several DNA sequences have already been used in the sensor format for recognition of different pesticides from biological fluids and food samples but their application is still very restricted because a limited number of sequences is available and their stability is not well established. Unquestionably, they show high potential.

Selective purification of pesticides can also be carried out using molecularly im printed polymers (MIPs) based on synthetic polymers that have specific cavities in shape and size generated as a mold for a template molecule. The use of MIPs as selective sorbents was proposed more than 25 years ago by Sellergren *et al.* to extract pentamidine [104]. Since this first application, numerous MIPs were developed for the selective SPE [63, 93] and SPME [105, 106] of target pesticides in food and have already been successfully used in several other fields, mostly as recognition element in sensors [9]. They are thermally and chemically stable and their preparation is easy, cheap and fast. The most important problems are related to the molecule used as template, since its elimination is difficult and can interfere with the analyte.

Stir bar sorptive extraction (SBSE) and solid-phase micro-extraction (SPME) emerged as an alternative to SPE enabling to reduce even more the use of organic solvents. Both techniques adsorb apolar organic compounds in a stationary phase of polydimethylsiloxane (PDMS) from liquid samples. The stationary phase covers a fused silica fiber in the case of SPME and a magnetic stirring bar in the case of SBSE. Once adsorbed into the PDMS, the analytes can be thermally desorbed into the GC injector. The advantages are simplicity, efficacy, economy and elimination of organic solvents, and the disadvantages are the relatively long equilibrium time and the possible carry-over.

The most important innovation within SPME and SBSE is the research on novel coatings. The following new coatings have been recently described for pesticides analysis:

- Nanocomposites of montmorillonite (MMT) incorporated into a polyaniline-polyamide (PANI-PA) hybrid. This sorbent was prepared by *in situ* oxidative polymerization of polyaniline in the MMT-PA mixture, and the SBSE was coated by subsequent solvent exchange [13].
- Porous carbon nanotubes-silicon dioxide (CNTs-SiO_2) nanohybrids connected by glucose using a hydrothermal process. The SPME fiber was coated by electrospinning [50].

Dispersive Solid-Phase Extraction (dSPE)

This technique has been selected in approximately half of the methods reported in Tables **1** and **2**. dSPE is a type of SPE in which the solid phase is directly added to the unconditioned extract, and the clean-up is performed by agitation and centrifugation. This clean-up is used in the QuEChERS method. The most frequent adsorbents are PSA, C18, GCB, or their mixtures in different proportions.

As mentioned above, in conventional SPE, the most commonly used is PSA, which removes sugars and lipids. GCB is included in the case of samples with high pigment and sterol content while the C18 in the case of samples with high lipid content. Magnesium sulfate that removes water from the organic solvent is also standardized in dSPE, especially for GC applications. Furthermore, several new sorbents have been proposed, such as polyaniline (PANI)-modified zeolite NaY [59].

Recently, the implementation of magnetic dispersive solid-phase extraction (Md-SPE) has become a reality showing very promising results. Different types of pesticides, mostly organophosphorus, have been efficiently extracted using different magnetic sorbents from a wide range of foods [23].

Some alternative approaches, such as the automatic multiplug filtration cleanup (m-PFC) column were also proposed as alternative clean-up step to QuEChERS. This approach is based on the design of a filter that contains multiwalled carbon nanotubes (MWCNTs), PSA and $MgSO_4$. This process only requires tens of seconds with no additional concentration or solvent exchange steps [47].

Gel Permeation Chromatography (GPC)

GPC or size exclusion chromatography (SEC) separates analytes according to their size. This technique is often used for the clean-up of fatty extracts to determine pesticide residues [2]. However, this technique had little improvements in recent years.

Determination Techniques

Chromatographic Methods

The most important methods for pesticide residue determination in plants are based on two analytical strategies: (i) liquid chromatography (LC) and (ii) gas chromatography (GC). Both techniques are used to determine these residues since 1980 [5, 6].

GC combined with selective detectors was the preferred technique in the 80's and 90's for some types of pesticides, according to their chemical characteristics [107]. These detectors were electron capture detector (ECD) (for pesticides with halogenated atoms) [18, 19, 32], nitrogen-phosphorus detector (NPD) (organophosphorus and carbamates) [24, 25] and flame photometric detector (FPD) (organophosphorus) [20, 22, 28]. The specificity was top-priority since pesticide identity confirmation was needed. However, this technique fails with more polar and non-volatile pesticides. Conversely, LC was combined with universal detectors, such as UV or DAD, and it was able to detect the presence of a myriad of other compounds coming from the matrix, particularly, in the case of plant extracts. LC was also combined with specific detectors, such as fluorescence, but only the few pesticides that are fluorescent can be detected, the others require complex derivatization reactions that transform them into fluorescent compounds.

Nowadays these detectors have fallen into disuse because the preferred detection system is mass spectrometry (MS) combined either with GC and LC. The advent of MS solves the problem of LC selectivity and broadens the range of determinable pesticides by this technique. Some years ago, Alder *et al.* [108] compared LC-MS/MS and GC-MS for 500 pesticides and main metabolites of several chemical classes showing that LC-MS/MS presented a better performance for most families of pesticide residues, with the exception of organochlorine and pyrethroid pesticides. The first are almost obsolete but the last are still widely used (*e.g.* bifenthrin, cyfluthrin, cypermethrin, deltamethrin, esfenvalerate, ethofenprox, *etc.*).

More recently, Scherbaum *et al.* [109] compared the possibilities of GC-MS and LC-MS/MS by analyzing more than 6200 fruit and vegetable samples. The study initially included ~ 330 compounds that increased to ~ 540 at the end of the study. 276 pesticides were detected of which 255 (92%) were LC-MS/MS-amenable, 205 (74%) were GC-MS amenable and 189 (67%) could be determined by both techniques. This indicates that 71 pesticides (26%) would not have been detected if LC-MS/MS had not been used and 21 (8%) if GC-MS had not been used. This shows that LC-MS/MS can detect a wider range of pesticides than GC. In addition, for most pesticides, covered by both techniques, LC-MS/MS offers better sensitivity than GC-MS. Therefore, clearly LC-MS/MS is today considered a more versatile and universal technique.

In the last 10 years, there has been a spectacular advance in mass spectrometry, thus different interfaces and mass analyzers are available.

Gas Chromatography (GC)

Table **1** summarizes the GC methods reported in the last 4 years, including the conventional and the most innovative ones. Although ECD, NPD, FPD and even flame ionization detection (FID) are into disuse, a number of GC-based pesticide methods using these detectors has been reported in the last 4 years (2016-2019) [18, 19, 22, 25, 27, 28, 30, 32, 110]. Unlike the other selective detectors, FID is universal and able to determine almost all pesticides [13, 23, 26, 27]. The most commonly used mass analyzer for pesticide analysis in plants is still the single quadrupole (see Table **1**). The most common analysis by GC-MS involves splitless injection (with the Programmed Temperature Vaporizing – PTV – injector), fused silica capillary column separation (commonly 5% phenyl, 95% dimethylpolysiloxane; 30 m x 0.25 mm, 0.25 µm) followed by electronic ionization (EI, 70 eV) and MS analysis in selected ion monitoring (SIM) mode and full scan *m/z* 50-600. In SIM mode, at least 2-3 ions are selected to achieve proper identification of the analytes. In comparison with ECD, FPD, NPD and FID, MS has better accuracy, precision, sensitivity and specificity and its economic cost and maintenance (in its simplest version) makes it quite affordable.

Three types of injection systems have been applied in the analysis of pesticides: (i) conventional injection, (ii) large volume injection (LVI) and (iii) direct coupling of extraction to injection.

Within the first type of injection, the PTV injector is the most robust. This injector solves most of the problems presented by split/splitless and on-column injectors in the analysis of compounds at trace levels. It has a similar design to split/splitless but instead of a heating block, it has a vaporization system that heats or cools in a few seconds and can be programmed with different temperature plateau, times and rates. The PTV injector supports all split and splitless modes and, after the introduction of a special insert, can work as a cold on-column injector [111, 112].

LVI could be an interesting option, but only in those cases in which the expected concentrations are extremely low. However, pesticide residues are commonly at levels that most of the times do not require the use of these systems. The injection of large volumes (large volume in GC means hundreds of μL) is mostly carried out with the PTV injector. The solvent is evaporated at a temperature well below the boiling point of the analytes and eliminated through the open split valve. The analytes are retained in the liner and, once the solvent is evaporated, transferred from the liner to the GC column in closing the split valve. Availability of different liners, and systematic PTV parameter optimization broaden the scope of PTV LVI applications [44].

Other patented LVI system, exploited by Konik. is the through oven transfer adsorption (TOTAD). This consists of a PTV injector modified with an adsorbent that allows the introduction of milliliters (instead of μL) of water (instead of organic solvent) in GC. A high flow eliminates the solvent from the trap to a waste and trap the analytes in the PTV. One of the advantages of this interface is that aqueous extract can be injected in a GC-MS without deterioration of the column [113].

The TOTAD LVI interface is based on a quite simple principle and solves many problems associated with the other LVI systems. However, its use is not too extensive because many parameters need to be optimized to efficiently perform the LVI. The most important parameters requiring adjustment are:

- The adsorbent used and its inertness.
- Splitless transfer rate.
- Sample volume assumable by the adsorbent.
- Solvent-elimination time.

This laborious optimization of parameters complicates the determination of compounds, is time-consuming, and is the reason why this interface is not for general use.

Other interesting modification within GC-MS is the possibility of on-line coupling extraction and determination. One of these systems exploits the possibility to thermally desorb the analytes isolated by sorptive extraction methods (such as SPME, SBSE or in-tube SPME) into the GC system. This makes routine analysis of a large number of samples simpler, faster, and more ecological since it reduces the amount of organic solvents [12, 14, 43].

DI-SPME can be difficult in complex liquid food samples because the presence of matrix components (*e.g.* proteins, sugars and so on), as reported in the literature [13, 19]. Analysis of solid foods using DI-SPME presents additional problems, such as fiber fouling, and poor efficiency and reproducibility. However, if pesticides are relatively volatile, the use HS-SPME can eliminate many of these problems [13, 19].

The eluent from the sample clean-up can be introduced directly in the GC through LVI. However, this system is complicated and needs a careful optimization of many parameters [113].

In order to determine pesticide residues, the separation is carried out in the most conventional way, using traditional capillary columns. It is interesting to mention that there are some advances, such as fast GC (which uses shorter and narrower columns) or GC×GC, consisting of two different GC columns in series in a way that all analyte peaks emerging from the first column enter the second. These systems are applied to other fields but are not frequently applied to determine pesticide residues in food. Fast GC provides rapid separations but lower resolution than conventional methods. GC×GC is useful to separate complex mixtures but it is too time consuming and pesticide separation could be solved using less complicated systems [114].

Back flush is becoming increasingly used in pesticide residue analysis in food. Column back flush is a system that not only allows column replacement without the need to vent the mass spectrometer, but also extends the life of the column even if dirty samples are injected into the GC [29]. Back flushing allows less volatile matrix components of the GC column to be removed by reversing the column flow at a pressure junction point, such as, using a pressure controlled T (PCT). This system requires a short restriction capillary, and the PCT that is installed at the outlet of the column before the MS [35]. With this system, backwashing can only be done when the last analyte of interest is eluted from the column (retroactive taking after execution). Another more effective back flushing configuration is the so-called "concurrent back flushing", which installs the PCT between two columns (using 2 columns of 15 m) and the reverse column flow begins after the last analyte of interest is eluted from the first column [44].

Compared to the electron impact (EI), the other ionization methods are less used (see Table **1**). Chemical ionization (CI) is the second most used ionization method. EI can be applied to almost all pesticides (acid, basic and neutral). The positive CI also has a wide range of application (basic and neutral) while in negative CI, only those analytes that have an acid group or an electronegative group (such as halogen atoms) are able to stabilize the negative charge and give response providing better selectivity for pesticides possessing these electronegative groups [15].

The last addition to the ionization sources in GC has been atmospheric pressure chemical ionization (APCI), similar to the APCI sources coupled to LC-MS where corona discharges an electron current on the solvent spray producing primary ions. However, APCI coupling to GC is different to LC because the electron current instead of ionizing an electrosprayed solution, ionizes nitrogen that is able to provide ionization charge transfer and/or protonation. In the former, nitrogen radicals able to transfer the charge to the analytes form cation radicals. The latter requires traces of water to protonate molecules. Although there are some examples reported, APCI is not yet common in pesticide residue analysis because it has been recently introduced and the interface is extensive.

Nowadays, MS analyzers can be combined in many ways to perform tandem-MS (as in triple quadrupole – QqQ – analyzers) and high resolution-MS (HRMS) (as in quadrupole time-of-flight or quadrupole Orbitrap). The use of QqQ is currently routine for the determination of pesticide residues due to its appropriate sensitivity and robustness [16, 34, 35, 44]. The QqQ is often used in the Selected Reaction Monitoring (SRM) mode, which is very selective as it monitors a number of precursor ion → product ion transitions that are specific to the analytes to be determined and therefore the signal interferences of the matrix and background noise are reduced or eliminated completely, obtaining better detection limits. Proposed food methods use at least two SRMs for each pesticide to perform a proper identification. However, QqQ, when applied in combination with GC, presents several problems associated with the low *m/z* ratio of the pesticide and the use of EI as a source of ionization. EI produces a high fragmentation of the molecules, and the number and/or abundance of each ion may be low to obtain two adequate transitions making it difficult. The most suitable system is to combine QqQ with positive or negative CI instead of EI. The low *m/z* of the fragment can also decrease selectivity. For example, the presence of chlorine provides a *m/z* very specific but chlorine alone does not allow to discriminate among the different pesticides that contain it and have close retention times.

Another advance within the field is the coupling of GC to mass spectrometers with high or ultra-high mass resolution (HRMS). The first instrument introduced was the time-of-flight mass spectrometry (GC-HR-ToF-MS) that showed excellent possibilities due to its high sensitivity in full-spectrum-acquisition mode [29]. Commonly, ToF instruments when coupled to GC can work in two ways at high velocity sacrificing mass accuracy [40, 41] or preserving the HRMS at low velocity. These characteristics make this technique irreplaceable in qualitative analysis for wide-scope screening of pesticides in food. The possibility to obtain full MS avoids the need for managing analyte-specific information before the injection, identifying, at least in theory, any compound that could be present in samples. The progress in informatics ensure that data obtained could be stored and then, it is not necessary to re-inject the sample to perform retrospective analysis. GC can also be coupled to quadrupole time-of-flight (QToF), this detector offering the same advantages than ToF but additionally providing characteristic products ions of the precursors. As in LC, GC-QToF-MS commonly works in multiple selected reactions monitoring (MRM) mode [15, 17, 29].

Orbitrap is the last addition within the available mass analyzers based in trapping of ions in electrostatic fields. The trial version of this mass analyzer was introduced for the first time at a conference of the American Society for Mass Spectrometry in 1999 and the accurate and compact mass detector already commercially available and coupled to LC was presented in 2005 at the same conference. This system allows a higher mass resolution than QToF (>60000 vs 30000 FWHM, full width at half maximum) in addition to high accuracy and sensitivity, properties that have all become essential to meet analytical needs in numerous areas of research, as well as in routine analysis within this area of pesticide residues [115]. Although initially coupled to LC, in 2010, the coupling of GC to Orbitrap MS using EI was finally achieved providing resolving power higher than 100,000 FWHM. After this coupling, the GC-EI-QOrbitrap system was finally marketed in 2015 [57, 116]. GC-EI-full scan Orbitrap HRMS was evaluated for pesticide residue analysis by Mol *et al.* [117] who compared the spectra obtained with those of the NIST library. It was observed that both spectra are very similar but with a lower abundance of low *m/z* ions, especially those with *m/z* < 90.

Some characteristics of the working modes of this detector will be discussed later in the LC section.

Liquid Chromatography (LC)

Reversed phase (RP) is the most efficient LC separation mode for pesticides due to the high to medium range of polarity of most currently used pesticides. Octadecylsilane (C18) is the most common stationary phase and mixtures of water-methanol and water-acetonitrile are the most popular mobile phases for the analysis of pesticides with RP-LC. Several additives, which have to be volatile to use a mass spectrometry detector, are added to the mobile phase to select the optimum pH for the proper chromatography, improving the peak shape and increasing the ionization capacity. The most common additives are ammonium acetate or ammonium formate (at concentrations of 5-10 mM) and formic acid (at concentration of 0.1%). The mobile phases flow rates are low (200-300 μL min^{-1}) because the lower the flow of the mobile phase, the better the ionization, although sometimes it is at the cost of losing separation capacity. Injection volumes are also low (5 - 25 μL) to make them compatible with the mobile phase flow-rate. Differently to what happens in GC, in LC the use of other detectors, such as UV or DAD, is rare (more or less 10% of all the applications reported in the last 4 years) [58, 59, 63, 64]. In the other 90% of the applications, any type of LC-MS is commonly employed. The most usual ionization sources in LC-MS are the atmospheric pressure ionization (API) sources, which provide intact protonated molecules and few fragments, electrospray (ESI) and atmospheric pressure chemical ionization (APCI), in positive mode is the preferred one (in fact, all the studies compiled in Table **2** used ESI as ionization source) and tandem MS analyzers, such as the QqQ or Qtrap [4, 107]. Table **2** summarizes the most important references in which LC is applied.

The analytical column and the mobile phase in addition to separating the analytes from each other also separate the analytes (usually at low concentrations) from the co-extracted matrix present in the extract (at high concentrations), which is not always simple. Nowadays, LC analytical columns with smaller inner diameter (between 2 and 4.1 mm) and particle size (2.1-3 μm) are preferred because they improve peak resolution, sensitivity, and shorten analysis times. Ultra-high performance liquid chromatography (UHPLC) emerged from this tendency to reduce the internal diameter of the columns and the particle size of the stationary phase. UHPLC is an improvement of the instrument to reach higher pressures (>400 bar), which allows to work with columns with stationary phases that have a particle size between 1.7 and 1.8 μm. UHPLC is presently a routine instrument in the study of pesticides in food that in a very short time has almost replaced conventional LC. The advantages are better resolution, separation and lower solvent consumption. The main disadvantage is that instruments able to work at high pressures are more expensive. In addition, the high UHPLC separation capacity is not fully exploited in LC-MS systems, as mobile phase flow has to be low to maintain good ionization efficiency. However, speed and sensitivity are significantly improved [118].

The reduction of the inner column diameter is a trend within LC still important that will make miniaturization, the next step to UHPLC. Micro or micro-flow LC (μLC) based on the use of columns with inner diameters of 0.5-1.0 mm, capillary liquid chromatography (CLC) in which columns with inner diameters of 100-500 µm are used, and nano-liquid chromatography (nano-LC) with columns with inner diameters ≤100 µm are the future of LC. Some studies already show the prospects of µLC and nLC within the field [74]. Their main advantage is the greater sensitivity, particularly when combined with mass spectrometry due to the low mobile phase fluxes entering the source. The increase in sensitivity is such that, for example, it allows sample extracts to be diluted rather than concentrated, reducing or even eliminating matrix effects. Furthermore, the reduced flow of the mobile phase also reduces environmental waste and the costs of laboratory tests [74].

As mentioned before, LC is combined with MS through the API sources electrospray (ESI), atmospheric pressure chemical ionization (APCI) and atmospheric pressure photoionization (APPI). However, in pesticide residue analysis almost 100% of the samples used ESI [4, 101].

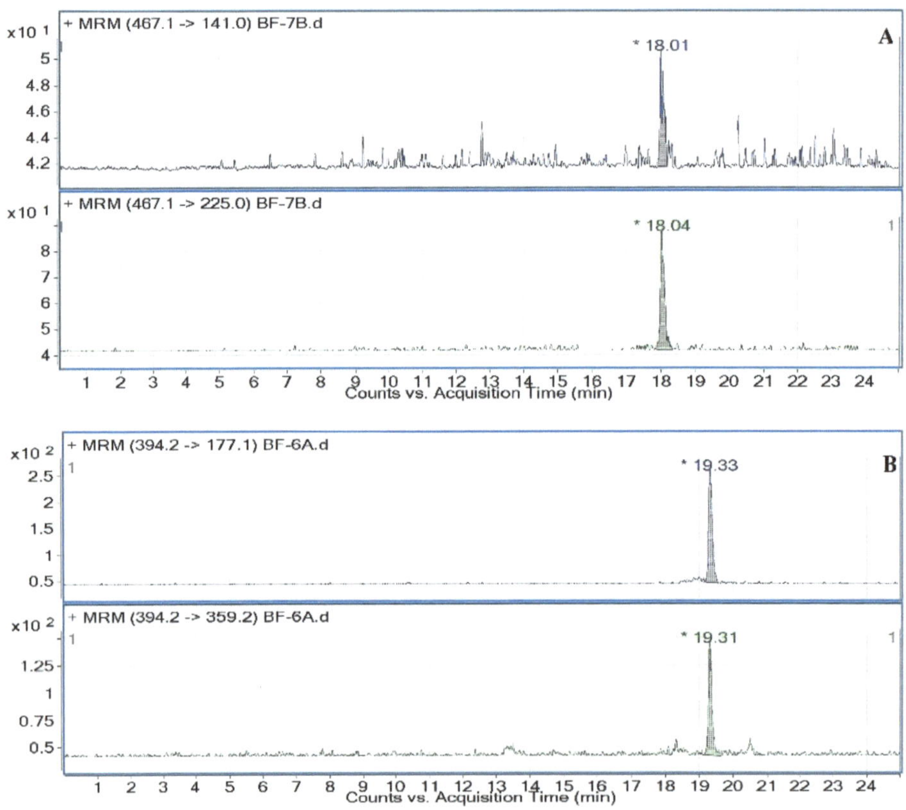

Fig. (2). LC-MS/MS chromatograms obtained in selective reaction monitoring (SRM) mode of baby food samples naturally contaminated with cyhalothrin at a level of 0.7 µg kg^{-1} (A) and etofenprox at a level of 0.6 µg kg^{-1} (B). Reprinted with permission from [119]. Copyright (2017) Elsevier.

Today, a large number of mass analyzers with different characteristics and properties have been developed. Those applicable for the determination of pesticide residues in food are triple quadrupole (QqQ) and quadrupole linear ion-trap (QLIT) which, in this order, are the most used (Table **2**). Both instruments are formed by a combination of several quadrupoles that provide the nominal mass of the molecular ion or fragments, and a targeted list of compounds can be determined using reference standards for their identification and quantification. Any other contaminant, present in the sample but not included in the list becomes undetectable because these instruments work by selecting transitions precursor ion → product ion of the monitored target analytes. These instruments also work in full scan, but SRM provides greater selectivity and sensitivity achieving a robust quantification that makes it suitable for determining these compounds at the low concentrations at which they are found. The EU criteria establish that two SRM transitions per analyte have to be recorded to ensure a correct identification and avoid false positives. These criteria also establish other parameters to be considered for a correct identification, such as the maximum deviation that can present retention time or the intensity ratio of product ions. As an example, Fig. **(2)** shows the detection of cyhalothrin and etofenprox in different samples at levels around the limit of quantification achieved for these compounds using a QqQ (0.7 and 0.6 μg kg^{-1}, respectively) [119]. LC-QqQ-MS/MS is just an alternative technique for pyrethrins that are more sensitive by GC-MS. However, these results pointed out that combining proper extraction techniques with LC, sensitivity much below the MRLs can be achieved.

High resolution mass spectrometry (HRMS) allows accurate mass (up to 6 decimals) and therefore, not only provides information about the molecular weight of the compound but also its empirical formula. Among all the high-resolution instruments available, QToF and Orbitrap platforms are those used in pesticide analysis. HRMS instruments enable fast detection and reliable identification of a wide range of compounds thanks to their operation in full-scan acquisition mode with accurate mass and high sensitivity combined with high resolving power (> 30,000 FWHM). Current instruments are in the range of 25,000-100,000 FWHM and moving up to higher values of resolution. In this sense, QToF has lower resolution values (ca. 30,000) but can complete a cycle more rapidly than Orbitrap. As a counterpart, Orbitrap-based mass analyzers reach easily 60,000-100,000 FWHM. Table **1** shows that both instruments are widely used in pesticide residue analysis.

This feature is very useful for pesticide residue determination in complex matrices. There are different workflows such as target analysis, suspected screening, non-target analysis, retrospective analysis that extend the compounds determined in a single injection and do not require the use of analytical standards for the identification of compounds. One of the main advantages of these instruments is that any matrix component gives a MS signal. Thus, they allow to ascertain the matrix compounds that are coextracted with the analytes. Fig. **(3)** illustrates the effectiveness of three sorbent mixtures employed to clean-up coffee leaf extracts overlapping full-scan total ion chromatograms (TICs) obtained by UHPLC-QToF-MS in positive-ion mode. As shown in Fig. **(3a)**, TICs obtained from coffee leaf extracts were different for the three sorbents tested. Fig. **(3b)** shows the three different extracts [56].

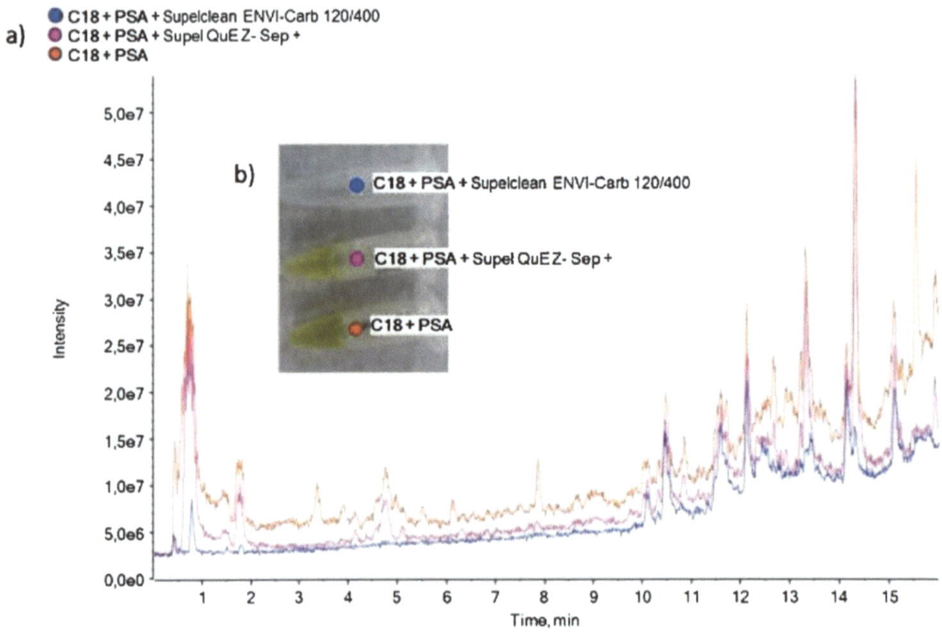

Fig. (3). (a) Overlapped total ion chromatograms obtained from UHPLC-QToF-MS analysis in full-scan mode (positive ion mode) of coffee leaves obtained after cleanup with different sorbents tested, (b) Extracts after cleanup with different sorbents. Reprinted with permission from [56]. Copyright (2017) Elsevier.

In target analysis, the targeted compounds are identified extracting the *m/z* of interest in a narrow window, commonly 20 mDa, which makes the identification very selective. In spite of the cliché that high resolution mass spectrometry does not provide accurate quantification, recent instruments have overcome this disadvantage and are able to quantify accurately.

Suspected screening is based on the identification with databases containing information on many compounds. The latest instruments or models have, as part of the software, databases that extract this information and identify non-target compounds. These databases contain experimental or theoretical information about the *m/z*, isotopic abundance and double bonds of a long list of compounds (extensible by the analyst) that is used to extract the ion chromatograms of each *m/z* with an error window. This information is combined with some other data such as the compound retention time and a MS/MS or fragment library.

Non-target screening is the most difficult workflow because there is no previous information on the possible identity of the compounds. In this case, the HRMS instrument provides a list with the potential empirical formulas that could match the accurate mass of the molecule and their mass error. A tentative structure could be assigned using the MS/MS and searching in databases such as Chemspider, Metlin, *etc*. Some HRMS instruments are aided by software providing theoretical fragmentation of a given structure and searching it in the on-line databases automatically. Fig. **(4)** shows the MS and MS/MS of cyprodinil that was identified without an analytical standard, just searching in the Metlin database and the match with the analytical standard [120].

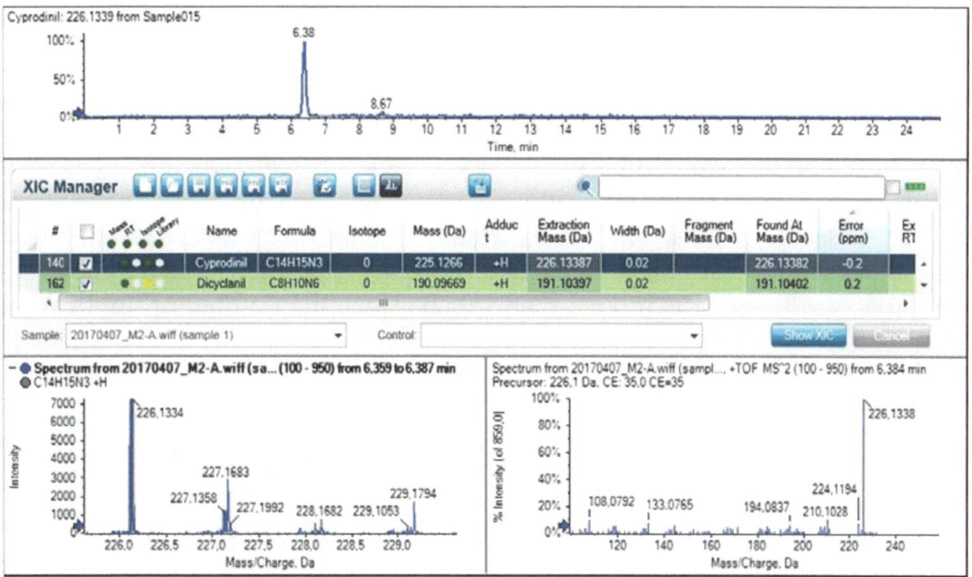

Fig. (4). Chromatogram, MS and MS/MS obtained for cyprodinil. Reprinted with permission from [120]. Copyright (2018) Elsevier.

Another important field of research is how MS and MS/MS information is recorded because both Orbitrap and QToF have limitations regarding cycle time (time spent to complete each scan/per time). The MS/MS spectrum can be recorded dependent or independent of the precursor ion data. In the data-dependent acquisition (DDA) mode, a precursor ion is selected for MS/MS, depending on its characteristics. Several criteria can be used to select a precursor ion for further fragmentation: (i) a specific nominal *m/z* or (ii) the intensity above a predefined threshold value (in this case, there is a maximum number of ions fragmented in each cycle). This strategy allows to exclude co-eluted matrix compounds or noisy peaks to improve identification and quantification making it easy. However, many pesticides that do not reach the selection criteria could remain unidentified.

Contrarily, data independent acquisition (DIA) involves that precursor ions are fragmented without a pre-selection. This produces alternating composite mass spectra of all intact molecular ions and their fragment ions, increasing the number of residues detected and the reproducibility. The most well-known DIA working flows are MS^E developed by Waters, all-ion fragmentation (AIF) employed in the Orbitrap instruments (Thermo) and sequential window acquisition of all theoretical spectra (SWATH). The most used is MS^E because it was the first developed and works by acquiring, in parallel, MS spectra at low and high collision energy. This is pseudo MS/MS because the ions are not isolated and fragmented but fragmented by induced collision dissociation (CID). AIF designed for the Orbitrap has the same principle that MS^E but analyte fragmentation takes place through a higher energy collisional dissociation (HCD) cell placed in the opposite side of the C-trap. The last addition to the working modes is SWATH, which is based on the acquisition in each cycle of a full scan and several MS/MS scans of a short *m/z* range (covering all precursor ions included in each *m/z* range). In this case, a real MS/MS is obtained. The three methods have been applied to determine pesticide residues in food [29, 74 - 80]. To this date, it is still difficult to say which one is more advantageous since the number of publications is rather small.

Other Chromatographic Techniques

Three additional chromatographic techniques, although less used, could play a role in pesticide residue determination: thin layer chromatography (TLC) [121], capillary electrophoresis (CE), and supercritical fluid chromatography (SFC) [122]. TLC presents commonly low sensitivity due to the lack of specific detection systems. Although it was recently combined with FID detection to determine chlorpyrifos from grapes [123], it does not provide advantages compared with LC or GC. CE (in any of its different working modes) presents the problem of sensitivity just as TLC, and therefore, it is not generally used. The problem with SFC is slightly different since this technique has always been considered promising but has never become of routine. These techniques are long enough in the market without reaching a maturity within pesticide residue determination in food to be able to establish that their drawbacks for this particular type of analysis are more than their advantages.

Ambient Mass Spectrometry (AMS)

This approach is most effective for fast screening and has shown promising potential in the determination of pesticide residues in food. This is mostly due to several advantages:

- Low sample preparation required without chromatography separation.
- Fast sample profiling allowing intra- and inter-sample differences.
- Ability to establish the identity of the compounds that make the difference in the profile.
- Available with a choice of HRMS systems.
- Possibility to work in image mode to locate analytes in the samples.

Some of the most reported AMS techniques are desorption electrospray ionization (DESI) based on ESI and direct analysis in real time (DART) based on APCI. DART —already commercialized by JEOL Inc. USA— has a gas stream (typically helium) that flows through a chamber where an electrical discharge produces ions, electrons and excited-state (meta-stable) atoms and molecules. The simplest ionization is Penning ionization involving transfer of energy from the excited gas M* to the analyte producing the protonated molecule plus an electron. There are several methods to expose the sample to the gas stream. The most well-known are: (i) direct exposure using a kind of tweezers; (ii) use of the named transmission module (a stain steel mesh where the sample is deposited) and (iii) the Dip-it® tips (a multi-sample rack that holds glass rods in the optimum position). Direct exposure lacks the required homogeneity for these analyses. Then, pesticide residues are previously extracted by QuEChERS to improve homogeneity and reducing matrix effects [124]. Additionally, with this interface, the sample can be directly exposed in interface using tweezers [125]. Fig. **(5)** shows an example of how the MS^n capabilities of the LTQ-Orbitrap could be exploited to confirm the identity of the compounds.

As a promising application, DART has been evaluated for the determination of a group of highly polar pesticides only extractable by QuPPe and difficult to determine using LC-MS. To establish its effectiveness, both Orbitrap and QToF, as well as Dip-it® tips and sample cards QuickStrip™ were used. QuPPe-DAR--HRMS was shown to be highly effective in determining 7 of the compounds studied -amitrole, cyromazine, propamocarb, melamine, diethanolamine, triethanolamine and 1,2,4-triazole. However, one of the drawbacks of this technique was the high fragmentation of some compounds [126].

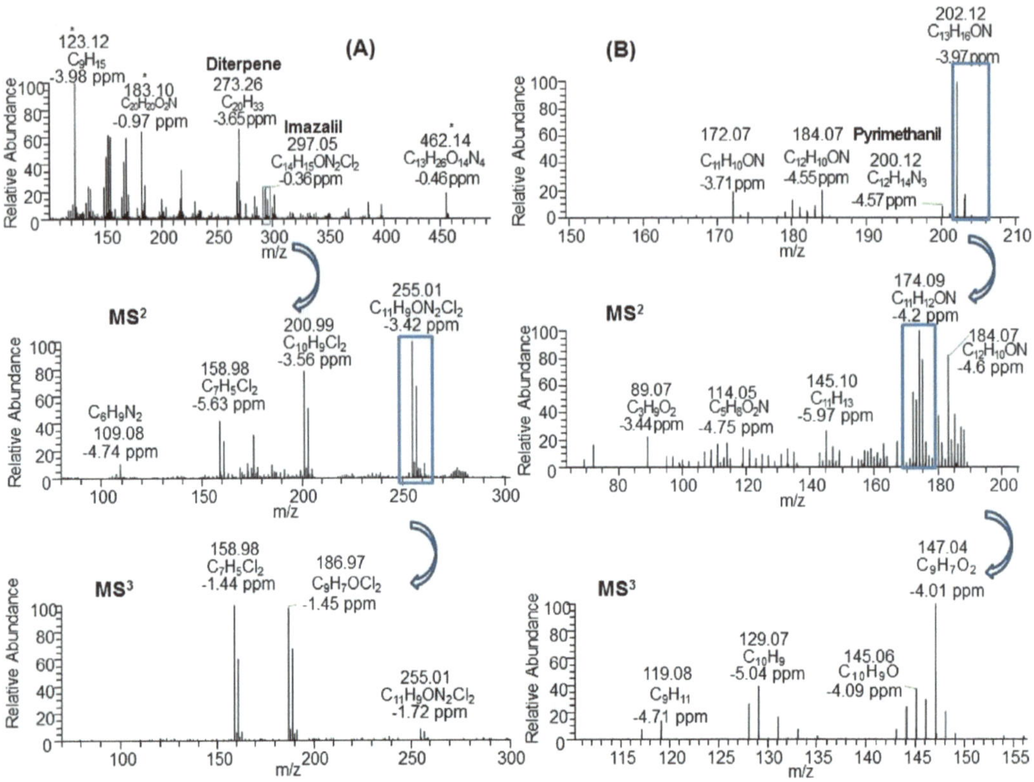

Fig. (5). DART-Orbitap-MS^n analysis of peel slide of (A) Valencia Late orange and (B) Royal Gala apple. Reprinted with permission from [125]. Copyright (2013) American Chemical Society.

An interesting application that offers advantages for analyzing complex matrices with high speed and efficiency is the combination of MSPE techniques with the DART probe. The system, already commercialized, is called solid-phase micro-extraction-transmission (SPME-TM) and allows an effective synergy between sample preparation and environmental ionization. Very briefly, the device consists of placing the solid phase (polymer particles) in a mesh that is exposed to the sample and allows the analytes to concentrate on the solid phase it contains and in turn acts as a support to expose the sample to the DART probe using transmission mode, achieving desorption of the analytes and their thermal ionization. The advantage of this system is its efficiency and improved detection limits [127].

DESI, commercially available by Prosolia, Inc., makes use of a fast-moving charged solvent stream, at an angle relative to the sample surface, to extract analytes from the surfaces and impel the secondary ions toward the mass analyzer. DESI approaches have already demonstrated qualitative and semi-quantitative capabilities to simultaneously determine multiple pesticides both directly in the sample surface or in an extract of the sample [128]. In addition, DESI-MS could be used in imaging mode. MS imaging (MSI) allows to study the distribution of pesticides throughout the sample. This technique has already been applied to investigate the distribution of contact and systemic pesticides in different parts of *Cotoneaster horizontalis* and *Kalanchoe blossfeldiana* with a spatial resolution of 50-100 μm in diameter [129]. Establishing the distribution of contact pesticides on plants is of great interest to the agrochemical industry, since in order to adequately protect crops from attack by fungi, insects and other pests, the pesticide application system must ensure that the distribution of pesticides on crops is homogeneous.

Other techniques based on the same theoretical principle as DESI are paper spray ionization (PS ionization), wooden-tip spray ionization (WTS ionization), probe electrospray ionization (PESI), coated blade spray ionization (CBS ionization), easy ambient sonic spray ionization, extractive electrospray ionization (ESI), and touch spray. PS ionization is the most widely applied to pesticide residues analysis among them. In this technique, a few μL of the sample are placed on the paper by whipping the paper into the sample or by dropping a few drops onto the paper. Then, an organic solvent is added and a high voltage applied which generates the ions [130]. The future development of disposable 3D microfluidic paper or the use of MIP membranes instead of paper is expected to be useful in these analyses [131, 132]. As for the other technique, it just reports small modifications of the ESI probe to adapt it to ambient mass spectrometry and, to our knowledge, it has not been applied to determine pesticide residues in plants yet. However, there are no technical limitations to do it.

There is also an alternative group of ambient techniques based on DART, such as desorption corona beam ionization (DCBI), dielectric barrier discharge ionization (DBDI), and atmospheric-pressure solids analysis probe (ASAP). DBDI consists of a dielectric layer between two electrodes applied with alternating voltages at atmospheric pressure. The dielectric layer reduces the average current density generated in the space occupied by the gas, forming a stable plasma that contains electrons at high energy but low temperature. Although this low-temperature microplasma cannot be used for direct determination in the sample, it was already used in combination with GC or LC for the analysis of pesticide residues in food [133, 134]

A recently developed technique of environmental ionization is Rapid Evaporative Ionization Mass Spectrometry (REIMS). This technique is based on the mechanical or thermal ablation of tissue material (which has a high percentage of water) and leads to the formation of an aerosol of charged particles. The ionization of the sample (*i.e.* biological tissues) takes place at the surgical site, in close conjunction with the electrosurgical dissection of the tissues. A vent line connected to the electrosurgical unit evacuated the formed aerosol with the gaseous ions to the mass spectrometer. The formation of charged droplets (of both polarities) during tissue evaporation provides the ions. This system is now available from Waters supplier that provides the so called iKnife hand-held sampling device. This system has been used in surgery, to identify tissues and cell cultures and is being modified to provide imaging mass spectrometry. Although it presents great prospects, no application to pesticide residues in food have been reported yet.

These techniques also have several disadvantages that have prevented their widespread use in routine analysis. The first downside is the cost since they must be connected with a HRMS (QToF or Orbitrap). In addition, direct analysis in the sample surface sometimes does not show homogeneous results and the required sensitivity is not always reached, needing some type of basic sample extraction in order to get homogeneous sample and increase sensitivity. Another problem is that there is no separation and the MS involves ions of all potential compounds present in the sample. Consequently, a trained operator in MS is required.

Other Spectroscopic Techniques

Other techniques that are less reported to determine pesticide residues in food are surface-enhanced Raman spectroscopy (SERS) thanks to the development of novel gold nanomaterial-based substrates, for rapid measurement and quantification of pesticides extracted from lemon, carrot, and mango juice [135] and apple [136].

Few spectrometric methods are still reported to determine pesticide residues in food [137]. They are developed for one pesticide, and commonly required highly complex extraction methods.

Immunoassays

An "immunoassay" is an analysis method for the identification or determination of analytes (pesticides in the case of this chapter) based on the use of an antibody (Ab) or a mixture of Abs as a reagent to recognize the pesticide in question by considering it as antigen.

The preceding paragraph defines the global methodology of immunoassay with a number of particular characteristics. However, within this methodology many types of techniques, procedures and approaches have been developed. The most common type of immunoassay for determining pesticides is the competitive one, in which the unlabeled pesticide present in the sample competes to bind to the antigen with an easy-to-detect labelled analyte.

Immunoassays can be differentiated based on the type of label used to detect pesticides. These labels could be radioisotopes [radioimmunoassays (RIA)], enzymes [enzyme-linked immunosorbent assays (ELISAs or EIAs)], fluorophores [fluoroimmnoassays (FIAs or PFIAs) and chemiluminescent compounds [chemiluminescent-linked immunoassays (CLIAs)]. There are some additional types of immunoassays, but they are not commonly used in pesticide analysis.

In the analysis of pesticides, the most used tests are ELISAs. The most employed enzymes are alkaline phosphatase, β-galactosidase and peroxidase, the latter derived from horseradish (HRP) being the most widely used. The wide use of chemical reactions catalyzed by an enzyme providing a colored substance as a product within the analysis of pesticides in foods is due to the fact that they present several advantages over other immunolabels. Compared to RIAs, ELISA avoids the use of radioactive material and provides better detection limits, while in comparison to FIAs or CLIAs, ELISA only requires inexpensive and easy-t--use spectrophotometers that measure quickly a stable colored product.

Although immunoassays have been used in many instances to determine pesticide residues and in a large number of cases they are commercially available, they are less used than the chemical techniques described above because they also have a number of drawbacks, *e.g.* the sensitivity and specificity of immunoassays designed for pesticides are always lower than those obtained for macromolecules due to weak immunogenicity and the number of pesticides that could be simultaneously determined is low.

The key feature of immunoassays is the very high specificity of the binding. Traditionally, an immunoassay attains detection of a single pesticide. This is currently a drawback reducing sample throughput and applicability in comparison to multiresidue methods. The current trend within immunoassay design is that each assay is applicable to at least several pesticides. To achieve this, there are two types of formats, one incorporates several antibodies in the same immunoassay so that each type of antibody recognizes a different compound and another (cheaper) is to obtain a single antibody capable of determining several pesticides. This type of antibody is called broad-specificity antibody.

Immunoassay has now evolved to immunochromatographic methods for detecting pesticide residues in food. This technique is a combination of chromatography (separation of sample components based on differences in their movement through a mobile and a stationary phases) and immunochemical reactions (biorecognition molecules, labels, detection systems). The most well-known devices are those performing the immunochromatography development in strips. Many strips and even labels have been reported.

Immunochromatographic tests are used to perform qualitative tests (presence/absence of a pesticide) and some even indicate whether the concentration of a pesticide exceeds a certain value. Research is currently underway to develop systems providing quantitative data. However, it is often at the cost of losing their simplicity.

Sensors

Biosensor technology consists of a biological sensing element connected to a transducer to convert the response of the biological system (when bound to a pesticide) into a measurable signal. The magnitude of the signal is proportional to the concentration of pesticide. A typical biosensor is represented in Fig. **(6)**; it consists of the following components:

- Analyte: the pesticide that should be determined.
- Bioreceptor or bioreceptor mimic is the biomolecule or biological element able to recognize or bind the pesticide. Some examples of biomolecules are enzymes, cells, aptamers, deoxyribonucleic acid (DNA) and antibodies. The biorecognition of the analyte and the bioreceptor initiate a process that generates a signal (in the form of light, heat, pH, charge or mass change, *etc.*).
- Transducer is the element that converts the signal of the biorecognition into a measurable signal. This involves just an energy conversion or signalization. The most typical type of transducers produces optical or electrical signals that are also proportional to the analyte-bioreceptor interactions.
- Electronics manage the transduced signal and display it in a readable format. Electronics are commonly formed by a complex electronic circuitry that amplifies and converts analogical signals to digital.
- Display is the user interpretation system (computer display or a direct printer) that shows the output signal in a format that can be numeric, graphic, tabular or an image using hardware and software that transform the results to make them understandable to the analyst.

Biosensors are portable and compact size devices that present a number of prospect in pesticide residue determination in food since they allow a continuous real time analysis that is specific, sensitive, rapid, and economic. Biosensors could be classified according to: (1) the type of bio-recognition elements (*i.e.* enzyme, antibody, nucleic acid (aptamer), whole cell, *etc.*) and (2) the signal transduction method (optical, electrochemical, piezoelectric, *etc.*).

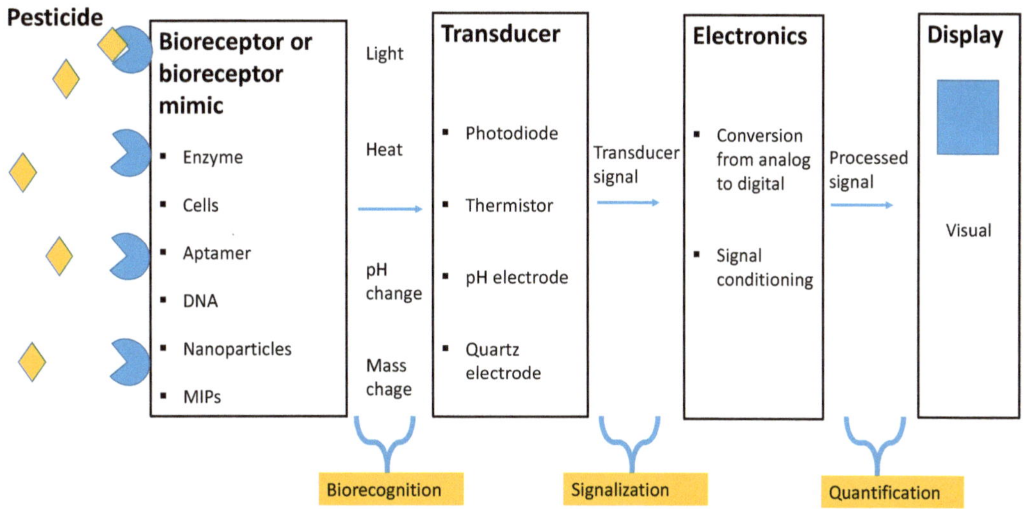

Fig. (6). Scheme of a biosensor.

With regard to transducers, the progress made in the design thanks to the developments in nanotechnology is worth mentioning. Nanoparticles as basic components and signaling elements have proved to have enormous potential leading to rapid progress in this field. Nanoparticles are characterized by their unique optical and electrochemical properties that attain functionality not achievable without them, particularly in terms of detection and imaging. In addition, nanoparticles have a large surface area, nanoscale size and unique morphology highly applicable in different detection methods (optics, electrochemistry, surface plasmon resonance (SPR), *etc.*). In recent years, different nanomaterials such as metallic nanoparticles (NPs), quantum dots (QDs), carbon nanotubes (CNTs) and silica NPs (SiO_2) have also been used as biorecognition elements with aptamers in biosensor design.

Biosensor technology is already used for the simultaneous determination of several classes of pesticides (organophosphates, organochlorides, carbamates, *etc.*) by incorporating various biocomponents with different transducers in the same device. Electrochemical acetyl cholinesterase biosensor based on MWCNTs/dicyclohexyl phthalate modified screen-printed electrode was used for the detection of chlorpyrifos and other organophosphorus pesticides [10]. The format of glassy carbon electrode (GCE) has also been used after being modified with polyzincon as a highly sensitive electrochemical sensor for the determination of organophosphorus pesticides [138].

Conversely, several site-heterologous haptens and competitive monoclonal antibody-based immunoassays have been used for pyrimethanil residue analysis in foodstuffs [139].

As an interesting biorecognition element called synthetic antibodies, MIPs have substituted antibodies to determine pesticides. Their synthesis, characterization and use as selective sorbents in biosensors for the selective recognition of organophosphorus pesticides have been described at length [9].

CONCLUDING REMARKS

This chapter shows that the analysis of pesticides is mainly carried out using multi-residue methods enabling the determination of a large number of pesticides. The most robust analysis schemes are based on QuEChERS extraction and LC-MS. These methods allow a fast, sensitive and reliable determination of pesticides. Sample extraction and purification remain as an important step in the determination of pesticides. However, with new HRMS techniques, data processing is gaining more weight within the analysis and taking a greater percentage of time. Despite the introduction of multiresidue methods, the large quantity of pesticides available does not allow the use of a single extraction process or a single determination technique.

Analytical methods to determine pesticide residues are well validated for all types of food matrices but results show that the analysis of high lipid content matrices (*e.g.* oils, avocados, *etc.*) is not yet well resolved. This is one of the gaps detected within this field. Hopefully, it will be resolved in the near future.

The determination of pesticide residues could be routinely performed using either GC or LC. Both techniques provide reliable results. However, most of the currently used pesticides can be determined by LC with higher sensitivity. So far, within the different LC-MS systems, QqQ is the most widely used because it can achieve maximum sensitivity and the most accurate quantification. However, HRMS instruments (QToF and Orbitrap) are gaining ground within the field. These HRMSs are not restricted to a list of predetermined compounds, but can determine any substance not contemplated a priori and also carry out retrospective analysis of samples. In addition, the new designs of these instruments have managed to diminish the disadvantages that were initially attributed to them as little sensitivity and difficulty in quantifying.

Biochemical tests have been relegated within pesticide residue determination. However, there is a growing field of application —development of biosensors. These systems are considered the future within the field. Nanotechnology has greatly contributed to the rapid advance of these systems and revolutionized the possibilities within the sector. Nanotechnology has about the development of new materials (*e.g.* nanoparticles, nanotubes, and dendrimers) with special properties, which confer them a high application in the field of biosensors. The advancement and progression of these systems in the relatively near future will surprise us all.

CONSENT FOR PUBLICATION

Not applicable.

CONFLICT OF INTEREST

The authors confirm that this chapter contents have no conflict of interest.

ACKNOWLEDGEMENT

This work has been supported by the Spanish Ministry of Economy and Competitiveness and the ERDF (European Regional Development Fund) through the project CLICIC (RTI2018-097158-B-C31) and by the Generalitat Valenciana through the project ANTROPOCEN@ (PROMETEO/2018/155).

REFERENCES

[1] EU Pesticides database. http://ec.europa.eu/food/plant/pesticides/eu-pesticides-database/public/?event=homepage&language=EN

[2] Barchanska H, Danek M, Sajdak M, Turek M. Review of Sample Preparation Techniques for the Analysis of Selected Classes of Pesticides in Plant Matrices. Crit Rev Anal Chem 2018; 48(6): 467-91. [http://dx.doi.org/10.1080/10408347.2018.1451297] [PMID: 29621408]

[3] Farina Y, Abdullah PB, Bibi N. Extraction procedures in gas chromatographic determination of pesticides. J Anal Chem 2016; 71(4): 339-50. [http://dx.doi.org/10.1134/S1061934816040092]

[4] Ghoniem IR, Attallah ER, Abo-Aly MM. Determination of acidic herbicides in fruits and vegetables using liquid chromatography tandem mass spectrometry (LC-MS/MS). Int J Environ Anal Chem 2017; 97(4): 301-12. [http://dx.doi.org/10.1080/03067319.2017.1306062]

[5] Grimalt S, Dehouck P. Review of analytical methods for the determination of pesticide residues in grapes. J Chromatogr A 2016; 1433: 1-23. [http://dx.doi.org/10.1016/j.chroma.2015.12.076] [PMID: 26803907]

[6] Gullick DR, Mott KB, Bartlett MG. Chromatographic methods for the bioanalysis of pyrethroid pesticides. Biomed Chromatogr 2016; 30(5): 772-89. [http://dx.doi.org/10.1002/bmc.3712] [PMID: 26916501]

[7] Hakme E, Lozano A, Ferrer C, Díaz-Galiano FJ, Fernández-Alba AR. Analysis of pesticide residues in olive oil and other vegetable oils. TrAC Trends Anal Chem 2018; 100: 167-79. [http://dx.doi.org/10.1016/j.trac.2017.12.016]

[8] Bendicho C, Costas-Mora I, Romero V, Lavilla I. Nanoparticle-enhanced liquid-phase microextraction. TrAC Trends Anal Chem 2015; 68: 78-87. [http://dx.doi.org/10.1016/j.trac.2015.02.007]

[9] Boulanouar S, Mezzache S, Combès A, Pichon V. Molecularly imprinted polymers for the determination of organophosphorus pesticides in complex samples. Talanta 2018; 176: 465-78. [http://dx.doi.org/10.1016/j.talanta.2017.08.067] [PMID: 28917777]

[10] Chen D, Liu Z, Fu J, *et al.* Electrochemical acetylcholinesterase biosensor based on multi-walled carbon nanotubes/dicyclohexyl phthalate modified screen-printed electrode for detection of chlorpyrifos. J Electroanal Chem (Lausanne Switz) 2017; 801: 185-91. [http://dx.doi.org/10.1016/j.jelechem.2017.06.032]

[11] Gajdar J, Horakova E, Barek J, Fischer J, Vyskocil V. Recent applications of mercury electrodes for monitoring of pesticides: A critical review. Electroanalysis 2016; 28(11): 2659-71. [http://dx.doi.org/10.1002/elan.201600239]

[12] Abdulra'uf LB, Tan GH. Chemometric approach to the optimization of HS-SPME/GC-MS for the determination of multiclass pesticide residues in fruits and vegetables. Food Chem 2015; 177: 267-73. [http://dx.doi.org/10.1016/j.foodchem.2015.01.031] [PMID: 25660885]

[13] Ayazi Z, Jaafarzadeh R, Matin AA. Montmorillonite/polyaniline/polyamide nanocomposite as a novel stir bar coating for sorptive extraction of organophosphorous pesticides in fruit juices and vegetables applying response surface methodology. Anal Met 2017; 9(31): 4547-57. [http://dx.doi.org/10.1039/C7AY00579B]

[14] Bagheri H, Es'haghi A, Es-Haghi A, Basiripour F. Microwave-assisted extraction and high-throughput monolithic-polymer-based micro-solid-phase extraction of organophosphorus, triazole, and organochlorine residues in apple. J Sep Sci 2016; 39(3): 576-83. [http://dx.doi.org/10.1002/jssc.201500978] [PMID: 26614571]

[15] Besil N, Uclés S, Mezcúa M, Heinzen H, Fernández-Alba AR. Negative chemical ionization gas chromatography coupled to hybrid quadrupole time-of-flight mass spectrometry and automated accurate mass data processing for determination of pesticides in fruit and vegetables. Anal Bioanal Chem 2015; 407(21): 6327-43. [http://dx.doi.org/10.1007/s00216-015-8514-8] [PMID: 25694145]

[16] Bilehal DC, Khetagoudar MC, Chetti MB. A multi residue analysis of 47 multiclass pesticide residues in mango (Mangiferae indica) samples by GC-MS/MS using a modified QuEChERS extraction method. Adv Chem Res 2017; 39: 161-86.

[17] Cheng Z, Dong F, Xu J, *et al.* Simultaneous determination of organophosphorus pesticides in fruits and vegetables using atmospheric pressure gas chromatography quadrupole-time-of-flight mass spectrometry. Food Chem 2017; 231: 365-73. [http://dx.doi.org/10.1016/j.foodchem.2017.03.157] [PMID: 28450019]

[18] Dar AA, Jan I, Wani AA, *et al.* Risk assessment, dissipation behavior and persistence of quinalphos in/on green pea by gas chromatography with electron capture detector. J Sep Sci 2018; 41(11): 2380-5. [http://dx.doi.org/10.1002/jssc.201701543] [PMID: 29505694]

[19] Dizdaş TN, Pelit FO, Korba K, *et al.* Preparation and characterization of a novel solid-phase microextraction material for application to the determination of pesticides. Anal Lett 2016; 49(7): 907-16. [http://dx.doi.org/10.1080/00032719.2015.1033720]

[20] Donkor A, Osei-Fosu P, Nyarko S, Kingsford-Adaboh R, Dubey B, Asante I. Validation of QuEChERS method for the determination of 36 pesticide residues in fruits and vegetables from Ghana, using gas chromatography with electron capture and pulsed flame photometric detectors. J Environ Sci Health B 2015; 50(8): 560-70. [http://dx.doi.org/10.1080/03601234.2015.1028833] [PMID: 26065516]

[21] Ersoy N, Tekinarslan O, Ozgür EA, Göktas U. Determination of pesticide residues in apricot (Prunus armeniaca L.) grown at good agricultural practices (GAPs) by LC-MS/MS and GC-MS. Erwerbs-Obstbau 2018; 60(4): 349-58. [http://dx.doi.org/10.1007/s10341-018-0383-9]

[22] Essumang DK, Asare EA, Dodoo DK. Determination of pesticides residue content in watermelon fruit from Ghana. Fruits 2017; 72(1): 55-63. [http://dx.doi.org/10.17660/th2017/72.1.6]

[23] Farajzadeh MA, Abbaspour M. Development of a new sample preparation method based on liquid-liquid-liquid extraction combined with dispersive liquid-liquid microextraction and its application on unfiltered samples containing high content of solids. Talanta 2017; 174: 111-21. [http://dx.doi.org/10.1016/j.talanta.2017.05.084] [PMID: 28738556]

[24] Farajzadeh MA, Afshar Mogaddam MR, Alizadeh Nabil AA. Polyol-enhanced dispersive liquid-liquid microextraction coupled with gas chromatography and nitrogen phosphorous detection for the determination of organophosphorus pesticides from aqueous samples, fruit juices, and vegetables. J Sep Sci 2015; 38(23): 4086-94. [http://dx.doi.org/10.1002/jssc.201500525] [PMID: 26420025]

[25] Farajzadeh MA, Khoshmaram L. A rapid and sensitive method for the analysis of pyrethroid pesticides using the combination of liquid-liquid extraction and dispersive liquid-liquid microextraction. *Clean - Soil, Air*. Water 2015; 43(1): 51-8.

[26] Farajzadeh MA, Mohebbi A, Feriduni B. Development of a simple and efficient pretreatment technique named pH-dependent continuous homogenous liquid-liquid extraction. Anal Met 2016; 8(28): 5676-83. [http://dx.doi.org/10.1039/C6AY00965D]

[27] Farajzadeh MA, Sattari Dabbagh M, Yadeghari A. Deep eutectic solvent based gas-assisted dispersive liquid-phase microextraction combined with gas chromatography and flame ionization detection for the determination of some pesticide residues in fruit and vegetable samples. J Sep Sci 2017; 40(10): 2253-60. [http://dx.doi.org/10.1002/jssc.201700052] [PMID: 28371329]

[28] Fernandes VC, Freitas M, Pacheco JPG, Oliveira JM, Domingues VF, Delerue-Matos C. Magnetic dispersive micro solid-phase extraction and gas chromatography determination of organophosphorus pesticides in strawberries. J Chromatogr A 2018; 1566: 1-12. [http://dx.doi.org/10.1016/j.chroma.2018.06.045] [PMID: 30017087]

[29] Fillâtre Y, Gray FX, Roy C. Pesticides in essential oils: Occurrence and concentration in organic and conventional orange essential oils from eleven geographical origins. Anal Chim Acta 2017; 992: 55-66. [http://dx.doi.org/10.1016/j.aca.2017.08.039] [PMID: 29054150]

[30] Freitas SS, Serafim FAT, Lanças FM. Determination of target pesticide residues in tropical fruits employing matrix solid-phase dispersion (MSPD) extraction followed by high resolution gas chromatography. J Braz Chem Soc 2018; 29(5): 1140-8. [http://dx.doi.org/10.21577/0103-5053.20180041]

[31] Jahanmard E, Ansari F, Feizi M. Evaluation of QuEChERS sample preparation and GC mass spectrometry method for the determination of 15 pesticide residues in tomatoes used in salad production plants. Iran J Public Health 2016; 45(2): 230-8. [PMID: 27114988]

[32] Jan I, Dar AA, Mubashir S, *et al.* Quantification, dissipation behavior and risk assessment of ethion in green pea by gas chromatography-electron capture detector. J Sep Sci 2018; 41(9): 1990-4. [http://dx.doi.org/10.1002/jssc.201701150] [PMID: 29377614]

[33] Kaveli IP, Sakkas VA, Albanis TA. Development of ultrasound-assisted emulsification microextraction for the determination of seven pesticides in apple juice by gas chromatography–mass spectrometry. Int J Environ Anal Chem 2015; 95(2): 145-57. [http://dx.doi.org/10.1080/03067319.2014.994618]

[34] Li P, Duan Y, Ge H, Zhang Y, Wu X. Multiresidue analysis of 113 pesticides in different maturity levels of mangoes using an optimized QuEChERS method with GC-MS/MS and UHPLC-MS/MS. Food Anal Met 2018; 1-16.

[35] Lin XY, Mou RX, Cao ZY, Cao ZZ, Chen MX. Analysis of pyrethroid pesticides in Chinese vegetables and fruits by GC–MS/MS. Chem Pap 2018; 72(8): 1953-62.

[http://dx.doi.org/10.1007/s11696-018-0447-1]

[36] Lozowicka B, Abzeitova E, Sagitov A, Kaczynski P, Toleubayev K, Li A. Studies of pesticide residues in tomatoes and cucumbers from Kazakhstan and the associated health risks. Environ Monit Assess 2015; 187(10): 609.
[http://dx.doi.org/10.1007/s10661-015-4818-6] [PMID: 26337756]

[37] Machado I, Gérez N, Pistón M, Heinzen H, Cesio MV. Determination of pesticide residues in globe artichoke leaves and fruits by GC-MS and LC-MS/MS using the same QuEChERS procedure. Food Chem 2017; 227: 227-36.
[http://dx.doi.org/10.1016/j.foodchem.2017.01.025] [PMID: 28274427]

[38] Mahpishanian S, Sereshti H, Baghdadi M. Superparamagnetic core-shells anchored onto graphene oxide grafted with phenylethylamine as a nano-adsorbent for extraction and enrichment of organophosphorus pesticides from fruit, vegetable and water samples. J Chromatogr A 2015; 1406: 48-58.
[http://dx.doi.org/10.1016/j.chroma.2015.06.025] [PMID: 26129984]

[39] Malhat F, Boulangé J, Abdelraheem E, Abd Allah O, Abd El-Hamid R, Abd El-Salam S. Validation of QuEChERS based method for determination of fenitrothion residues in tomatoes by gas chromatography-flame photometric detector: Decline pattern and risk assessment. Food Chem 2017; 229: 814-9.
[http://dx.doi.org/10.1016/j.foodchem.2017.03.017] [PMID: 28372248]

[40] Meghesan-Breja A, Cimpoiu C, Hosu A. Multiresidue analysis of pesticides and metabolites from fruits and vegetables by gas chromatography-time-of-flight mass spectrometry. Acta Chromatogr 2015; 27(4): 657-85.
[http://dx.doi.org/10.1556/AChrom.27.2015.4.6]

[41] Meghesan-Breja A, Cimpoiu C, Hosu A. Identification and quantification of some pesticide metabolites from vegetables by GC-TOF-MS and LC-MS-QQQ. Stud Univ Babes-Bolyai Chem 2017; 62(3): 19-34.
[http://dx.doi.org/10.24193/subbchem.2017.3.02]

[42] Mohapatra S. Residue levels and dissipation behaviors for trifloxystrobin and tebuconazole in mango fruit and soil. Environ Monit Assess 2015; 187(3): 95.
[http://dx.doi.org/10.1007/s10661-015-4324-x] [PMID: 25663402]

[43] Munitz MS, Medina MB, Montti MIT. Development and validation of an SPME-GC method for a degradation kinetics study of propiconazole I, propiconazole II and tebuconazole in blueberries in Concordia, the main production area of Argentina. Food Addit Contam Part A Chem Anal Control Expo Risk Assess 2017; 34(5): 793-9.
[http://dx.doi.org/10.1080/19440049.2017.1301682] [PMID: 28278123]

[44] Norli HR, Christiansen AL, Stuveseth K. Analysis of non-cleaned QuEChERS extracts for the determination of pesticide residues in fruit, vegetables and cereals by gas chromatography-tandem mass spectrometry. Food Addit Contam Part A Chem Anal Control Expo Risk Assess 2016; 33(2): 300-12.
[PMID: 26700312]

[45] Pano-Farias NS, Ceballos-Magaña SG, Muñiz-Valencia R, Jurado JM, Alcázar Á, Aguayo-Villarreal IA. Direct immersion single drop micro-extraction method for multi-class pesticides analysis in mango using GC-MS. Food Chem 2017; 237: 30-8.
[http://dx.doi.org/10.1016/j.foodchem.2017.05.030] [PMID: 28764000]

[46] Polyiem W, Naksen W, Prapamontol T. Gas chromatographic- flame photometric detection of organophosphate pesticide residues and its application in real vegetable and fruit samples from Chiang Mai city, Thailand. Warasan Khana Witthayasat Maha Witthayalai Chiang Mai 2018; 45(4): 1933-43.

[47] Qin Y, Zhang J, He Y, *et al.* Automated multiplug filtration cleanup for pesticide residue analyses in kiwi fruit (Actinidia chinensis) and kiwi juice by gas chromatography-mass spectrometry. J Agric

Food Chem 2016; 64(31): 6082-90.
[http://dx.doi.org/10.1021/acs.jafc.5b06027] [PMID: 26809284]

[48] Rai S, Singh AK, Srivastava A, Yadav S, Siddiqui MH, Mudiam MKR. Comparative evaluation of QuEChERS method coupled to DLLME extraction for the analysis of multiresidue pesticides in vegetables and fruits by gas chromatography-mass spectrometry. Food Anal Methods 2016; 9(9): 2656-69.
[http://dx.doi.org/10.1007/s12161-016-0445-2]

[49] Rashidi Nodeh H, Sereshti H, Gaikani H, Kamboh MA, Afsharsaveh Z. Magnetic graphene coated inorganic-organic hybrid nanocomposite for enhanced preconcentration of selected pesticides in tomato and grape. J Chromatogr A 2017; 1509: 26-34.
[http://dx.doi.org/10.1016/j.chroma.2017.06.032] [PMID: 28634067]

[50] Saraji M, Jafari MT, Mossaddegh M. Carbon nanotubes@silicon dioxide nanohybrids coating for solid-phase microextraction of organophosphorus pesticides followed by gas chromatography-corona discharge ion mobility spectrometric detection. J Chromatogr A 2016; 1429: 30-9.
[http://dx.doi.org/10.1016/j.chroma.2015.12.008] [PMID: 26709024]

[51] Słowik-Borowiec M, Szpyrka E. Multiresidue analysis of pesticides in wine and grape using gas chromatography with microelectron capture and nitrogen–phosphorus detection. Food Anal Met 2018; 11(12): 3516-0.
[http://dx.doi.org/10.1007/s12161-018-1329-4]

[52] Stocka J, Biziuk M, Namieśnik J. Analysis of pesticide residue in fruits and vegetables using analytical protocol based on application of the QuEChERS technique and GC-ECD system. Int J Glob Environ Issues 2016; 15(1-2): 136-50.
[http://dx.doi.org/10.1504/IJGENVI.2016.074361]

[53] Su WH, Sun DW, He JG, Zhang LB. Variation analysis in spectral indices of volatile chlorpyrifos and non-volatile imidacloprid in jujube (Ziziphus jujuba Mill.) using near-infrared hyperspectral imaging (NIR-HSI) and gas chromatograph-mass spectrometry (GC–MS). Comput Electron Agric 2017; 139: 41-55.
[http://dx.doi.org/10.1016/j.compag.2017.04.017]

[54] Uclés S, Lozano A, Sosa A, Parrilla Vázquez P, Valverde A, Fernández-Alba AR. Matrix interference evaluation employing GC and LC coupled to triple quadrupole tandem mass spectrometry. Talanta 2017; 174: 72-81.
[http://dx.doi.org/10.1016/j.talanta.2017.05.068] [PMID: 28738648]

[55] Vijaya Bhaskar Reddy A, Yusop Z, Jaafar J, *et al.* Simultaneous determination of three organophosphorus pesticides in different food commodities by gas chromatography with mass spectrometry. J Sep Sci 2016; 39(12): 2276-83.
[http://dx.doi.org/10.1002/jssc.201600155] [PMID: 27095506]

[56] Trevisan MTSO. R.W.; Calatayud-Vernich, P.; Breuer, A.; Picó, Y., Pesticide analysis in coffee leaves using a quick, easy, cheap, effective, rugged and safe approach and liquid chromatography tandem mass spectrometry: Optimization of the clean-up step. J Chromatogr A 2017; 1512: 9.
[http://dx.doi.org/10.1016/j.chroma.2017.07.033]

[57] Tienstra M, Mol HGJ. Application of gas chromatography coupled to quadrupole-orbitrap mass spectrometry for pesticide residue analysis in cereals and feed ingredients. J AOAC Int 2018; 101(2): 342-51.
[http://dx.doi.org/10.5740/jaoacint.17-0408] [PMID: 29141711]

[58] Adak T, Mukherjee I. Dissipation kinetics of spinosad from tomato under sub-tropical agro-climatic conditions. Environ Monit Assess 2016; 188(5): 299.
[http://dx.doi.org/10.1007/s10661-016-5291-6] [PMID: 27094056]

[59] Arnnok P, Patdhanagul N, Burakham R. Dispersive solid-phase extraction using polyaniline-modified zeolite NaY as a new sorbent for multiresidue analysis of pesticides in food and environmental

samples. Talanta 2017; 164: 651-61.
[http://dx.doi.org/10.1016/j.talanta.2016.11.003] [PMID: 28107986]

[60] Abad-Fuentes A, Ceballos-Alcantarilla E, Mercader JV, Agulló C, Abad-Somovilla A, Esteve-Turrillas FA. Determination of succinate-dehydrogenase-inhibitor fungicide residues in fruits and vegetables by liquid chromatography-tandem mass spectrometry. Anal Bioanal Chem 2015; 407(14): 4207-11.
[http://dx.doi.org/10.1007/s00216-015-8608-3] [PMID: 25796526]

[61] Abdallah OI, Alamer SS, Alrasheed AM. Monitoring pesticide residues in dates marketed in Al-Qassim, Saudi Arabia using a QuEChERS methodology and liquid chromatography-tandem mass spectrometry. Biomed Chromatogr 2018; 32(6): e4199.
[http://dx.doi.org/10.1002/bmc.4199] [PMID: 29377192]

[62] Andrade GCRM, Monteiro SH, Francisco JG, Figueiredo LA, Botelho RG, Tornisielo VL. Liquid chromatography-electrospray ionization tandem mass spectrometry and dynamic multiple reaction monitoring method for determining multiple pesticide residues in tomato. Food Chem 2015; 175: 57-65.
[http://dx.doi.org/10.1016/j.foodchem.2014.11.105] [PMID: 25577051]

[63] Bayat M, Hassanzadeh-Khayyat M, Mohajeri SA. Determination of diazinon pesticide residue in tomato fruit and tomato paste by molecularly imprinted solid-phase extraction coupled with liquid chromatography analysis. Food Anal Methods 2015; 8(4): 1034-41.
[http://dx.doi.org/10.1007/s12161-014-9984-6]

[64] Bedassa T, Gure A, Megersa N. Modified QuEChERS method for the determination of multiclass pesticide residues in fruit samples utilizing high-performance liquid chromatography. Food Anal Methods 2015; 8(8): 2020-7.
[http://dx.doi.org/10.1007/s12161-014-0081-7]

[65] Besil N, Cesio V, Heinzen H, Fernandez-Alba AR. Matrix effects and interferences of different citrus fruit coextractives in pesticide residue analysis using ultrahigh-performance liquid chromatography-high-resolution mass spectrometry. J Agric Food Chem 2017; 65(23): 4819-29.
[http://dx.doi.org/10.1021/acs.jafc.7b00243] [PMID: 28541668]

[66] Botero-Coy AM, Marín JM, Serrano R, Sancho JV, Hernández F. Exploring matrix effects in liquid chromatography-tandem mass spectrometry determination of pesticide residues in tropical fruits. Anal Bioanal Chem 2015; 407(13): 3667-81.
[http://dx.doi.org/10.1007/s00216-015-8586-5] [PMID: 25772561]

[67] Cao YQ, Ai LF, Wang Q, *et al.* Reusable 2-(Dimethylamino)ethyl methacrylate polymers on-line solid-phase extraction coupled to liquid chromatography tandem mass spectrometry for rapid determination of acidic pesticides in foods. Food Anal Met 2018; 1-10.

[68] Chamkasem N. Rapid determination of polar pesticides and plant growth regulators in fruits and vegetables by liquid chromatography/tandem mass spectrometry. J Environ Sci Health B 2018; 53(9): 622-31.
[http://dx.doi.org/10.1080/03601234.2018.1473977] [PMID: 29787687]

[69] Christia C, Bizani E, Christophoridis C, Fytianos K. Pesticide residues in fruit samples: comparison of different QuEChERS methods using liquid chromatography-tandem mass spectrometry. Environ Sci Pollut Res Int 2015; 22(17): 13167-78.
[http://dx.doi.org/10.1007/s11356-015-4456-0] [PMID: 25929454]

[70] Danezis GP, Anagnostopoulos CJ, Liapis K, Koupparis MA. Multi-residue analysis of pesticides, plant hormones, veterinary drugs and mycotoxins using HILIC chromatography - MS/MS in various food matrices. Anal Chim Acta 2016; 942: 121-38.
[http://dx.doi.org/10.1016/j.aca.2016.09.011] [PMID: 27720116]

[71] Fillatre Y, Rondeau D, Daguin A, Communal PY. A workflow for multiclass determination of 256 pesticides in essential oils by liquid chromatography tandem mass spectrometry using evaporation and

dilution approaches: Application to lavandin, lemon and cypress essential oils. Talanta 2016; 149: 178-86.
[http://dx.doi.org/10.1016/j.talanta.2015.11.052] [PMID: 26717829]

[72] Fu Y, Yang T, Zhao J, Zhang L, Chen R, Wu Y. Determination of eight pesticides in Lycium barbarum by LC-MS/MS and dietary risk assessment. Food Chem 2017; 218: 192-8.
[http://dx.doi.org/10.1016/j.foodchem.2016.09.014] [PMID: 27719897]

[73] Gao J, Wang J, Zuo M, *et al.* A highly sensitive method for simultaneous determination of the quaternary ammonium pesticides chlormequat and mepiquat in pears and potatoes by modified QuEChERS-high performance liquid chromatography-tandem mass spectrometry. RSC Advances 2015; 5(8): 5895-903.
[http://dx.doi.org/10.1039/C4RA10698A]

[74] Moreno-González D, Pérez-Ortega P, Gilbert-López B, Molina-Díaz A, García-Reyes JF, Fernández-Alba AR. Evaluation of nanoflow liquid chromatography high resolution mass spectrometry for pesticide residue analysis in food. J Chromatogr A 2017; 1512: 78-87.
[http://dx.doi.org/10.1016/j.chroma.2017.07.019] [PMID: 28720221]

[75] Sivaperumal P, Salauddin A, Ramesh Kumar A, Santhosh K, Rupal T. Determination of Pesticide Residues in Mango Matrices by Ultra High-Performance Liquid Chromatography Coupled with Quadrupole Time-of-Flight Mass Spectrometry. Food Anal Methods 2017; 10(7): 2346-57.
[http://dx.doi.org/10.1007/s12161-016-0779-9]

[76] Pérez-Ortega P, Lara-Ortega FJ, Gilbert-López B, Moreno-González D, García-Reyes JF, Molina-Díaz A. Screening of over 600 pesticides, veterinary drugs, food-packaging contaminants, mycotoxins, and other chemicals in food by ultra-high performance liquid chromatography quadrupole time-of-flight mass spectrometry (UHPLC-QTOFMS). Food Anal Methods 2017; 10(5): 1216-44.
[http://dx.doi.org/10.1007/s12161-016-0678-0]

[77] Qi P, Yuan Y, Wang Z, *et al.* Use of liquid chromatography- quadrupole time-of-flight mass spectrometry for enantioselective separation and determination of pyrisoxazole in vegetables, strawberry and soil. J Chromatogr A 2016; 1449: 62-70.
[http://dx.doi.org/10.1016/j.chroma.2016.04.051] [PMID: 27133864]

[78] Sivaperumal P, Anand P, Riddhi L. Rapid determination of pesticide residues in fruits and vegetables, using ultra-high-performance liquid chromatography/time-of-flight mass spectrometry. Food Chem 2015; 168: 356-65.
[http://dx.doi.org/10.1016/j.foodchem.2014.07.072] [PMID: 25172721]

[79] Rajski Ł, Díaz Galiano FJ, Cutillas V, Fernández-Alba AR. Coupling ion chromatography to Q-orbitrap for the fast and robust analysis of anionic pesticides in fruits and vegetables. J AOAC Int 2018; 101(2): 352-9.
[http://dx.doi.org/10.5740/jaoacint.17-0410] [PMID: 29141721]

[80] Zomer P, Mol HGJ. Simultaneous quantitative determination, identification and qualitative screening of pesticides in fruits and vegetables using LC-Q-Orbitrap™-MS. Food Addit Contam Part A Chem Anal Control Expo Risk Assess 2015; 32(10): 1628-36.
[http://dx.doi.org/10.1080/19440049.2015.1085652] [PMID: 26301946]

[81] Parrilla Vázquez P, Lozano A, Ferrer C, Martínez Bueno MJ, Fernández-Alba AR. Improvements in identification and quantitation of pesticide residues in food by LC-QTOF using sequential mass window acquisition (SWATH®). Anal Met 2018; 10(24): 2821-33.
[http://dx.doi.org/10.1039/C8AY00678D]

[82] Golge O, Kabak B. Evaluation of QuEChERS sample preparation and liquid chromatography-tripl--quadrupole mass spectrometry method for the determination of 109 pesticide residues in tomatoes. Food Chem 2015; 176: 319-32.
[http://dx.doi.org/10.1016/j.foodchem.2014.12.083] [PMID: 25624240]

[83] Hanot V, Goscinny S, Deridder M. A simple multi-residue method for the determination of pesticides

in fruits and vegetables using a methanolic extraction and ultra-high-performance liquid chromatography-tandem mass spectrometry: optimization and extension of scope. J Chromatogr A 2015; 1384: 53-66.
[http://dx.doi.org/10.1016/j.chroma.2015.01.040] [PMID: 25660522]

[84] Hazer O, Akkbik M, Demir D, Turhan Y. Determination of carbendazim and chlorpyrifos in selected fruits and vegetables samples using QuEChERS-HPLC-FD. Eurasian J Anal Chem 2017; 12(2): 17-30.
[http://dx.doi.org/10.12973/ejac.2017.00151a]

[85] He Z, Peng Y, Wang L, Luo M, Liu X. Unequivocal enantiomeric identification and analysis of 10 chiral pesticides in fruit and vegetables by QuEChERS method combined with liquid chromatography-quadruple/linear ion trap mass spectrometry determination. Chirality 2015; 27(12): 958-64.
[http://dx.doi.org/10.1002/chir.22531] [PMID: 26392120]

[86] Hu M, Liu X, Dong F, *et al.* Determination of ametoctradin residue in fruits and vegetables by modified quick, easy, cheap, effective, rugged, and safe method using ultra-performance liquid chromatography/tandem mass spectrometry. Food Chem 2015; 175: 395-400.
[http://dx.doi.org/10.1016/j.foodchem.2014.11.158] [PMID: 25577097]

[87] Kakitani A, Yoshioka T, Nagatomi Y, Harayama K. A rapid and sensitive analysis of dithiocarbamate fungicides using modified QuEChERS method and liquid chromatography-tandem mass spectrometry. J Pestic Sci 2017; 42(4): 145-50.
[http://dx.doi.org/10.1584/jpestics.D17-025] [PMID: 30363097]

[88] Kemmerich M, Bernardi G, Prestes OD, Adaime MB, Zanella R. Comprehensive method validation for the determination of 170 pesticide residues in pear employing modified QuEChERS without clean-up and ultra-high performance liquid chromatography coupled to tandem mass spectrometry. Food Anal Methods 2018; 11(2): 556-77.
[http://dx.doi.org/10.1007/s12161-017-1026-8]

[89] Kim SW, Abd El-Aty AM, Rahman MM, *et al.* Detection of pyridaben residue levels in hot pepper fruit and leaves by liquid chromatography-tandem mass spectrometry: effect of household processes. Biomed Chromatogr 2015; 29(7): 990-7.
[http://dx.doi.org/10.1002/bmc.3383] [PMID: 25402259]

[90] Kim YA, Abd El-Aty AM, Rahman MM, *et al.* Method development, matrix effect, and risk assessment of 49 multiclass pesticides in kiwifruit using liquid chromatography coupled to tandem mass spectrometry. J Chromatogr B Analyt Technol Biomed Life Sci 2018; 1076: 130-8.
[http://dx.doi.org/10.1016/j.jchromb.2018.01.015] [PMID: 29406026]

[91] Lawal A, Wong RCS, Tan GH, Abdulra'uf LB. Determination of pesticide residues in fruit and vegetables by high-performance liquid chromatography–tandem mass spectrometry with multivariate response surface methodology. Anal Lett 2018; 52(2): 231-48.

[92] Lawal A, Wong RCS, Tan GH, Abdulra'uf LB, Alsharif AMA. Multi-pesticide residues determination in samples of fruits and vegetables using chemometrics approach to QuEChERS-dSPE coupled with ionic liquid-based DLLME and LC–MS/MS. Chromatographia 2018; 81(5): 759-68.
[http://dx.doi.org/10.1007/s10337-018-3511-7]

[93] Li D, Zhang X, Kong F, Qiao X, Xu Z. Molecularly imprinted solid-phase extraction coupled with high-performance liquid chromatography for the determination of trace trichlorfon and monocrotophos residues in fruits. Food Anal Methods 2017; 10(5): 1284-92.
[http://dx.doi.org/10.1007/s12161-016-0687-z]

[94] Li H, Wang C, Zhu Q, *et al.* Reduction of matrix effects through a simplified QuEChERS method and using small injection volumes in a LC-MS/MS system for the determination of 28 pesticides in fruits and vegetables. Anal Met 2016; 8(25): 5061-9.
[http://dx.doi.org/10.1039/C6AY00080K]

[95] Li M, Dai C, Wang F, *et al.* Chemometric-assisted QuEChERS extraction method for post-harvest pesticide determination in fruits and vegetables. Sci Rep 2017; 7: 42489.

[http://dx.doi.org/10.1038/srep42489] [PMID: 28225030]

[96] Li N, Chen J, Shi YP. Magnetic graphene solid-phase extraction for the determination of carbamate pesticides in tomatoes coupled with high performance liquid chromatography. Talanta 2015; 141: 212-9.
[http://dx.doi.org/10.1016/j.talanta.2015.04.018] [PMID: 25966405]

[97] Anastassiades MK, D.I. Eichhorn,, E. Benkenstein,, *et al.* Quick method for the analysis of numerous highly polar pesticides in foods of plant origin *via* LC-MS/MS involving simultaneous extraction with methanol (QuPPe-Method); Fellbach, Germany 3/2, DE-7073: CVUA Stuttgart, Schaflandstr 2015.

[98] Clifford PA, Bassen JL, Mills PA. Chlorinated organic pesticide residues in fluid milk. Public Health Rep 1959; 74: 1109-14.
[PMID: 13810699]

[99] Madej K, Kalenik TK, Piekoszewski W. Sample preparation and determination of pesticides in fat-containing foods. Food Chem 2018; 269: 527-41.
[http://dx.doi.org/10.1016/j.foodchem.2018.07.007] [PMID: 30100469]

[100] Picó Y, Alfarham A, Barceló D. Analysis of emerging contaminants and nanomaterials in plant materials following uptake from soils. TrAC Trends Anal Chem 2017; 94: 173-89.
[http://dx.doi.org/10.1016/j.trac.2017.07.016]

[101] Lawal A, Wong RCS, Tan GH, Abdulra'uf LB, Alsharif AMA. Recent modifications and validation of QuEChERS-dSPE coupled to LC-MS and GC-MS instruments for determination of pesticide/agrochemical residues in fruits and vegetables: review. J Chromatogr Sci 2018; 56(7): 656-69. [Review].
[http://dx.doi.org/10.1093/chromsci/bmy032] [PMID: 29688338]

[102] Andraščíková M, Matisová E, Hrouzková S. Liquid phase microextraction techniques as a sample preparation step for analysis of pesticide residues in food. Sep Pur Rev 2015; 44(1): 1-18.
[http://dx.doi.org/10.1080/15422119.2013.872125]

[103] Vichapong J, Burakham R, Srijaranai S. Alternative liquid–liquid microextraction as cleanup for determination of neonicotinoid pesticides prior HPLC analysis. Chromatographia 2016; 79(5-6): 285-91.
[http://dx.doi.org/10.1007/s10337-016-3022-3]

[104] Sellergren B. Direct drug determination by selective sample enrichment on an imprinted polymer. Anal Chem 1994; 66(9): 1578-82.
[http://dx.doi.org/10.1021/ac00081a036]

[105] Li JW, Wang YL, Yan S, Li XJ, Pan SY. Molecularly imprinted calixarene fiber for solid-phase microextraction of four organophosphorous pesticides in fruits. Food Chem 2016; 192: 260-7.
[http://dx.doi.org/10.1016/j.foodchem.2015.07.018] [PMID: 26304345]

[106] Hashemi-Moghaddam H, Jedi DJ. Solid-phase microextraction of chlorpyrifos in fruit samples by synthesised monolithic molecularly imprinted polymer fibres. Int J Environ Anal Chem 2015; 95(1): 33-44.
[http://dx.doi.org/10.1080/03067319.2014.983495]

[107] Płonka M, Walorczyk S, Miszczyk M. Chromatographic methods for the determination of active substances and characterization of their impurities in pesticide formulations. TrAC Trends Anal Chem 2016; 85: 67-80.
[http://dx.doi.org/10.1016/j.trac.2016.03.011]

[108] Alder L, Greulich K, Kempe G, Vieth B. Residue analysis of 500 high priority pesticides: better by GC-MS or LC-MS/MS? Mass Spectrom Rev 2006; 25(6): 838-65.
[http://dx.doi.org/10.1002/mas.20091] [PMID: 16755599]

[109] Scherbaum, E. A. M, Schüle E, Bauer N, Ellendt K, Wieland M. GC-MS or LC-MS(/MS) - Which technique is more essential? EPRW. 2008.

[110] Shida SS, Nemoto S, Matsuda R. Simultaneous determination of acidic pesticides in vegetables and fruits by liquid chromatography--tandem mass spectrometry. J Environ Sci Health B 2015; 50(3): 151-62.
[http://dx.doi.org/10.1080/03601234.2015.982381] [PMID: 25602148]

[111] Romero Hernández J A, Amaya Chávez A, Miranda Rivera M G, García Fabila M M. Chromatographic methods for the determination of endosulfan in food Rev Int Cotam Amb 2018; 34(Special Issue 1): 81-94.

[112] Samsidar A, Siddiquee S, Shaarani SM. A review of extraction, analytical and advanced methods for determination of pesticides in environment and foodstuffs. Trends Food Sci Technol 2018; 71: 188-201.
[http://dx.doi.org/10.1016/j.tifs.2017.11.011]

[113] Špánik I, Machyňáková A. Recent applications of gas chromatography with high-resolution mass spectrometry. J Sep Sci 2018; 41(1): 163-79.
[http://dx.doi.org/10.1002/jssc.201701016] [PMID: 29111584]

[114] Wei H, Wang Y, Jin HY, Ma SC. Research progress in plant growth regulators and their application, content determination in planting traditional chinese medicine. Chung Kuo Yao Hsueh Tsa Chih 2016; 51(2): 81-5.

[115] Yaqub G, Iqbal K, Sadiq Z, Hamid A. Rapid determination of residual pesticides and polyaromatic hydrocarbons in different environmental samples by HPLC. Pak J Agric Sci 2017; 54(2): 355-61.
[http://dx.doi.org/10.21162/PAKJAS/17.5057]

[116] Lozano A, Uclés S, Uclés A, Ferrer C, Fernández-Alba AR. Pesticide residue analysis in fruit-And vegetable-based baby foods using GC-Orbitrap MS. J AOAC Int 2018; 101(2): 374-82.
[http://dx.doi.org/10.5740/jaoacint.17-0413] [PMID: 29141710]

[117] Mol HGJ, Tienstra M, Zomer P. Evaluation of gas chromatography - electron ionization - full scan high resolution Orbitrap mass spectrometry for pesticide residue analysis. Anal Chim Acta 2016; 935: 161-72.
[http://dx.doi.org/10.1016/j.aca.2016.06.017] [PMID: 27543025]

[118] Zhang YV, Wei B, Zhu Y, Zhang Y, Bluth MH. Liquid chromatography-tandem mass spectrometry: An emerging technology in the toxicology laboratory. Clin Lab Med 2016; 36(4): 635-61.
[http://dx.doi.org/10.1016/j.cll.2016.07.001] [PMID: 27842783]

[119] Petrarca MH, Ccanccapa-Cartagena A, Masiá A, Godoy HT, Picó Y. Comparison of green sample preparation techniques in the analysis of pyrethrins and pyrethroids in baby food by liquid chromatography-tandem mass spectrometry. J Chromatogr A 2017; 1497: 28-37.
[http://dx.doi.org/10.1016/j.chroma.2017.03.065] [PMID: 28372842]

[120] Picó Y, El-Sheikh MA, Alfarhan AH, Barceló D. Target *vs* non-target analysis to determine pesticide residues in fruits from Saudi Arabia and influence in potential risk associated with exposure. Food Chem Toxicol 2018; 111: 53-63.
[http://dx.doi.org/10.1016/j.fct.2017.10.060] [PMID: 29109044]

[121] Sherma J. Review of thin-layer chromatography in pesticide analysis: 2014–2016. J Liq Chromatogr Relat Technol 2017; 40(5-6): 226-38.
[http://dx.doi.org/10.1080/10826076.2017.1298024]

[122] Cutillas V, Galera MM, Rajski Ł, Fernández-Alba AR. Evaluation of supercritical fluid chromatography coupled to tandem mass spectrometry for pesticide residues in food. J Chromatogr A 2018; 1545: 67-74.
[http://dx.doi.org/10.1016/j.chroma.2018.02.048] [PMID: 29496188]

[123] Chauhan V, Tomar S, Saini Y, Tripathi R M. Method development for determination of residual Chlorpyrifos in the grapes by Tlc-fid Egyptian J Foren Sci 2017; 7(1)

[124] Guo T, Fang P, Jiang J, *et al.* Rapid screening and quantification of residual pesticides and illegal

adulterants in red wine by direct analysis in real time mass spectrometry. J Chromatogr A 2016; 1471: 27-33.
[http://dx.doi.org/10.1016/j.chroma.2016.09.073] [PMID: 27769534]

[125] Farré M, Picó Y, Barceló D. Direct peel monitoring of xenobiotics in fruit by direct analysis in real time coupled to a linear quadrupole ion trap-orbitrap mass spectrometer. Anal Chem 2013; 85(5): 2638-44.
[http://dx.doi.org/10.1021/ac3026702] [PMID: 23356415]

[126] Lara FJ, Chan D, Dickinson M, Lloyd AS, Adams SJ. Evaluation of direct analysis in real time for the determination of highly polar pesticides in lettuce and celery using modified Quick Polar Pesticides Extraction method. J Chromatogr A 2017; 1496: 37-44.
[http://dx.doi.org/10.1016/j.chroma.2017.03.020] [PMID: 28366571]

[127] Gómez-Ríos GA, Gionfriddo E, Poole J, Pawliszyn J. Ultrafast screening and quantitation of pesticides in food and environmental matrices by solid-phase microextraction-transmission mode (SPME-TM) and direct analysis in real time (DART). Anal Chem 2017; 89(13): 7240-8.
[http://dx.doi.org/10.1021/acs.analchem.7b01553] [PMID: 28540722]

[128] Gerbig S, Stern G, Brunn HE, Düring RA, Spengler B, Schulz S. Method development towards qualitative and semi-quantitative analysis of multiple pesticides from food surfaces and extracts by desorption electrospray ionization mass spectrometry as a preselective tool for food control. Anal Bioanal Chem 2017; 409(8): 2107-17.
[http://dx.doi.org/10.1007/s00216-016-0157-x] [PMID: 28035435]

[129] Gerbig S, Brunn HE, Spengler B, Schulz S. Spatially resolved investigation of systemic and contact pesticides in plant material by desorption electrospray ionization mass spectrometry imaging (DESI-MSI). Anal Bioanal Chem 2015; 407(24): 7379-89.
[http://dx.doi.org/10.1007/s00216-015-8900-2] [PMID: 26229027]

[130] Evard H, Kruve A, Lõhmus R, Leito I. Paper spray ionization mass spectrometry: Study of a method for fast-screening analysis of pesticides in fruits and vegetables. J Food Compos Anal 2015; 41: 221-5.
[http://dx.doi.org/10.1016/j.jfca.2015.01.010]

[131] Damon DE, Maher YS, Yin M, *et al.* 2D wax-printed paper substrates with extended solvent supply capabilities allow enhanced ion signal in paper spray ionization. Analyst (Lond) 2016; 141(12): 3866-73.
[http://dx.doi.org/10.1039/C6AN00168H] [PMID: 27121269]

[132] Pereira I, Rodrigues MF, Chaves AR, Vaz BG. Molecularly imprinted polymer (MIP) membrane assisted direct spray ionization mass spectrometry for agrochemicals screening in foodstuffs. Talanta 2018; 178: 507-14.
[http://dx.doi.org/10.1016/j.talanta.2017.09.080] [PMID: 29136855]

[133] Mirabelli MF, Wolf JC, Zenobi R. Atmospheric pressure soft ionization for gas chromatography with dielectric barrier discharge ionization-mass spectrometry (GC-DBDI-MS). Analyst (Lond) 2017; 142(11): 1909-15.
[http://dx.doi.org/10.1039/C7AN00245A] [PMID: 28443843]

[134] Mirabelli MF, Wolf JC, Zenobi R. Pesticide analysis at ppt concentration levels: coupling nano-liquid chromatography with dielectric barrier discharge ionization-mass spectrometry. Anal Bioanal Chem 2016; 408(13): 3425-34.
[http://dx.doi.org/10.1007/s00216-016-9419-x] [PMID: 26898206]

[135] Alsammarraie FK, Lin M, Mustapha A, *et al.* Rapid determination of thiabendazole in juice by SERS coupled with novel gold nanosubstrates. Food Chem 2018; 259: 219-25.
[http://dx.doi.org/10.1016/j.foodchem.2018.03.105] [PMID: 29680047]

[136] Fan Y, Lai K, Rasco BA, Huang Y. Determination of carbaryl pesticide in Fuji apples using surface-enhanced Raman spectroscopy coupled with multivariate analysis. Lebensm Wiss Technol 2015; 60(1): 352-7.

[http://dx.doi.org/10.1016/j.lwt.2014.08.011]

[137] Altunay N, Ülüzger D, Gürkan R. Simple and fast spectrophotometric determination of low levels of thiabendazole residues in fruit and vegetables after pre-concentration with ionic liquid phase microextraction. Food Addit Contam Part A Chem Anal Control Expo Risk Assess 2018; 35(6): 1139-54.
[http://dx.doi.org/10.1080/19440049.2018.1444284] [PMID: 29470140]

[138] Ensafi AA, Rezaloo F, Rezaei B. Electrochemical determination of fenitrothion organophosphorus pesticide using polyzincon modified-glassy carbon electrode. Electroanalysis 2017; 29(12): 2839-46.
[http://dx.doi.org/10.1002/elan.201700406]

[139] Esteve-Turrillas FA, Mercader JV, Agulló C, Abad-Somovilla A, Abad-Fuentes A. Site-heterologous haptens and competitive monoclonal antibody-based immunoassays for pyrimethanil residue analysis in foodstuffs. Lebensm Wiss Technol 2015; 63(1): 604-11.
[http://dx.doi.org/10.1016/j.lwt.2015.03.074]

CHAPTER 7

Perfluoroalkyl Substances (PFASs) in Foodstuffs and Human Dietary Exposure

Qian Wu and **Kurunthachalam Kannan***

Wadsworth Center, New York State Department of Health, and Department of Environmental Health Sciences, University at Albany, State University of New York, Albany, NY 12201, USA

Abstract: Perfluoroalkyl substances (PFASs) have been used as surfactants and surface protectors in many industrial materials and consumer products. PFASs have been reported to be associated with numerous adverse health outcomes in humans. Americans have the highest levels of PFASs in their bodies in comparison with populations from other countries. To our knowledge, data on the sources and pathways of human exposure to PFASs are limited. In this study, we determined PFASs in a wide variety of samples (water, food, indoor dust), and calculated exposure dose from various environmental sources including diet. A mass balance analysis was performed by comparison of calculated exposure doses (environmental sources) with modeled doses (biomonitoring results). PFASs occurred widely in drinking water, food, and indoor dust. Breast milk is the major source of exposure to PFASs in breast-fed infants. For PFOS and PFOA, indoor dust and diet are the major sources of exposure in adults. The results of mass balance analysis showed a good agreement between exposure doses calculated based on external sources and those modeled from biomonitoring studies.

Keywords: Drinking water, Exposure assessment, biomonitoring, Foodstuffs, Perfluoroalkyl substances, PFASs.

INTRODUCTION

Sources of Human Exposure to PFASs

Perfluoroalkyl substances (PFASs) are a class of man-made chemicals, with a fully fluorinated hydrocarbon chain (tail) and a hydrophilic functional group (head). The fluorinated hydrocarbon moiety is both lipophobic and hydrophobic. Due to this unique property, PFASs have been used as surface protectors and surfactants in many industrial applications and consumer products such as textiles,

* **Corresponding author Kurunthachalam Kannan:** Wadsworth Center, New York State Department of Health, and Department of Environmental Health Sciences, University at Albany, State University of New York, Albany, NY 12201, USA; Tel: +1-518-474-0015; E-mail: kurunthachalam.kannan@health.ny.gov

Belén Gómara & María Luisa Marina (Eds.)
Current and Future Developments in Food Science (Vol. 1)

leathers, waterproof clothing, carpets, specialty papers including food packing, cleaning agents, floor polishes, fire-fighting foams, and insecticides [1]. Two processes have been used for the production of PFASs: electrochemical fluorination (ECF) that produces a mixture of linear and branched isomers; and telomerization process that produces only linear products. Perfluorooctan-esulfonyl fluoride (POSF) is a product of ECF and is a precursor to produce POSF-based PFASs such as perflurooctanesulfonate (PFOS). PFOS can also be metabolized from many other POSF-based compounds including perfluorooctane sulfonamide (PFOSA) and perfluorooctane sulfonamidoalcohols, which are often referred to as "precursors". PFOA and its salts are also produced by telomerization process. PFOA is an emulsifier in the production of fluoropolymers and fluoroelastomers, and is also a degradation product of fluorotelomer alcohols (FTOHs; also referred to as "precursors of PFOA"). Fluoropolymers are used in various applications including construction, automobile, electronics, telecommunication, non-stick coating, thread sealant tape, and breathable clothing, for their stability, strength, and durability. FTOHs are used in surfactants and surface protectors, in carpets, textiles, painting, papers, non-stick cookware coating, and fire-fighting foams. It is estimated that over 4000 per- and polyfluoroalkyl substances are, or have been, on the global market [2], but only a limited number of PFASs including PFOS and PFOA have been studied for over the past two decades due their predominance in environmental and biological samples [3]. The chemical formulae for the most commonly studied perfluoroalkyl sulfonates (PFSAs) and perfluoroalkyl carboxylates (PFCAs) and their major precursor compounds are listed in Table **1**.

Table 1. Chemical formula of commonly studied PFASs and their precursors.

Name	Abbreviation	Formula
Perfluorobutanesulfonate	PFBS	$C_4F_9SO_3^-$
Perfluorohexanesulfonate	PFHxS	$C_6F_{13}SO_3^-$
Perfluorooctanesulfonate	PFOS	$C_8F_{17}SO_3^-$
Perflorodecanesulfonate	PFDS	$C_{10}F_{21}SO_3^-$
Perfluorooctane sulfonamide	PFOSA	$C_8F_{17}SO_2NH_2^-$
Perfluorooctanesulfonyl fluoride	POSF	$C_8F_{17}SO_2F$
N-methyl perfluorooctane sulfonamidoethanol	N-MeFOSE	$C_{11}H_8F_{17}NO_3S$
N-ethyl perfluorooctane sulfonamidoethanol	N-EtFOSE	$C_{12}H_{10}F_{17}NO_3S$
N-methyl perfluorooctane sulfonamido ethylacrylate	N-MeFOSA	$C_{14}H_{10}F_{17}NO_4S$
N-ethyl perfluorooctane sulfonamido ethylacrylate	N-EtFOSA	$C_{15}H_{12}F_{17}NO_4S$
Perfluorohexanoic acid	PFHxA	$C_5F_{11}COOH$
Perfluoroheptanoic acid	PFHpA	$C_6F_{13}COOH$

(Table 1) cont.....

Name	Abbreviation	Formula
Perfluorooctanoic acid	PFOA	$C_7F_{15}COOH$
Perfluorononanoic acid	PFNA	$C_8F_{17}COOH$
Perfluorodecanoic acid	PFDA	$C_9F_{19}COOH$
Perfluoroundecanoic acid	PFUnDA	$C_{10}F_{21}COOH$
Perfluorododecanoic acid	PFDoDA	$C_{11}F_{23}COOH$
4:2 Fluorotelomer alcohol	4:2 FTOH	$C_4H_5F_9O$
6:2 Fluorotelomer alcohol	6:2 FTOH	$C_8H_5F_{13}O$
8:2 Fluorotelomer alcohol	8:2 FTOH	$C_{10}H_5F_{17}O$
10:2 Fluorotelomer alcohol	10:2 FTOH	$C_{12}H_5F_{21}O$

In 2001, following the discovery of global distribution of PFASs [3], the 3M Company, the major manufacturer of PFOS-based chemistries, announced phase-out of ECF production of all POSF-based compounds in the U.S. However, the production of PFOS continued in Europe, Japan, and China [4]. The production of PFOA by telomerization process, by other manufactures continued [5]. Following the phase-out of POSF-based chemistry by 3M Company in 2001, the global production of PFASs was thought to be dropped. In May 2009, PFOS and its salts were listed under the Stockholm Convention as Persistent Organic Pollutants (POPs) for their persistent, bioaccumulative, toxic (PBT), and long-range transportation properties [6]. Currently, China is reported to continue the production POSF based compounds.

PFOS and PFOA are bioaccumulative and, PFOS particularly can biomagnify in the food chain. PFOS has been detected at higher concentrations in top predators (bald eagles, dolphins and polar bears) than in animals at the lower trophic levels in the food chain [3, 7]. PFOA is more frequently detected in aquatic medium (*e.g.*, water) than that of PFOS, and PFOA is relatively more water soluble than PFOS [8, 9]. Several precursors of PFASs such as perfluoroalkyl sulfonamides and FTOHs have been reported to be transformed into perfluorinated acids in biological and environmental media [10, 11]. Several PFASs, especially PFOS and PFOA, have been detected globally in air, water, soil, fish, birds, marine mammals, and humans [3, 11 - 16].

Diet has been suggested as an important source of PFAS exposures in humans. Following the release into the environment, PFASs can concentrate and accumulate in plants and animals at the bottom of the food chain, which are then consumed by animals at the higher trophic levels [7, 17, 18]. One of the main sources of PFASs to humans is food producing animals and plants. PFASs were reported to occur in drinking water [19 - 23]. Discharge of wastewater has been identified as the main source of PFASs in the water cycle [24, 25]; recently, communities living around certain chemical plants, airports and military bases in New York State were shown to be exposed to PFOS and/or PFOA through contaminated drinking water [26]. In addition, the use of sewage sludge or biosolids as fertilizer in agriculture contributes to the contamination of soil and food products [27]. Bioaccumulation of PFASs in the food chain contributes to their occurrence in animal-origin foodstuffs. For example, the bioconcentration factors (BCFs) of PFOS in scabbardfish and croaker were 390 – 537 and 12 – 18, respectively, whereas the BCFs of PFOA in the two fish species were 2.2 – 11 and 18 – 96, respectively [28]. Studies have suggested that fish consumption is an important source of PFAS exposures in humans [29].

Apart from seafood, PFASs have been measured in other food categories. PFOS, PFOA and their precursors have been determined in fish, meat, popcorn, dairy products, vegetables, and pizza [20, 29 - 33]. Several PFASs and their precursors were detected in non-stick cookware and food packaging [30, 32, 34]. In samples collected from a Spanish market in 2006, PFASs were detected in fish, seafood, pork, chicken, eggs, vegetables, and dairy products [20]. Concentrations of PFOS ranged from below the limit of detection (LOD) to 0.82 ng g^{-1} wet wt, whereas PFOA and PFHpA were only found in whole milk at mean concentrations of 0.056 and 0.015 ng g^{-1}, respectively. Dietary intake of PFOS was estimated to be 1.07 ng kg^{-1} body weight (bw) day^{-1} for Spanish adults and 2 – 2.5 ng kg^{-1} bw day^{-1} for Spanish children. Following this study, 20 additional samples including veal, pork, chicken, lamb, salmon, and lettuce were analyzed from Spanish markets in 2008. PFOS was found in 8 of the 20 samples analyzed, at concentrations of up to 0.33 ng g^{-1} in lamb liver; PFHxS, PFHxA, and PFOA were also detected in samples at concentrations ranging from 0.01 to 0.18 ng g^{-1} [19]. In fish, mollusks, crabs, and shrimp collected from China, the wet weight concentrations of PFOS ranged from 0.3 to 13.9 ng g^{-1}, PFOA from <0.25 to 1.67 ng g^{-1}, and PFUnDA from <0.25 to 0.77 ng g^{-1} [31]. In a later study on food samples collected from China, frequently consumed animal-origin foods including pork, beef, chicken, and goat, and their livers and blood cakes were analyzed. The highest mean concentration of PFOA was found in chicken meat (12.5 ng g^{-1}), and the highest mean concentration of PFOS was found in pork liver (1.99 ng g^{-1}). The highest total PFAS concentration was found in chicken meat (12.7 ng g^{-1}), followed by pork meat (4.43 ng g^{-1}) [35]. In a Canadian dietary survey, PFOS was detected at concentrations ranging from <1 to 2.7 ng g^{-1}, and PFOA was detected at concentrations ranging from <0.5 to 3.6 ng g^{-1} in fish, beef, pizza, and microwave popcorn. The total daily dietary intake of PFOS and PFOA for adults in Canada was estimated at 110 ng day^{-1} and 70 ng day^{-1}, respectively [36]. In addition to PFOS and PFOA, PFOSA and several other precursors including N-methyl perfluorooctane sulfonamido ethylacrylate (N-MeFOSA), and N-ethyl perfluorooctane sulfonamido ethylacrylate (N-EtFOSA) were detected in bakery products, dairy products, eggs, fish, seafood, meat, fast-food, and packaged food, contributing approximately to 73 ng day^{-1} of total perfluorooctanesulfonamide intake by adults [33]. Based on a duplicate diet study in Germany, the average total daily dietary intake of PFOS and PFOA was estimated at 4.3 ng kg^{-1} bw day^{-1} for adults [37]. There are very few studies that report PFASs in foodstuffs from the U.S. Schecter *et al.* [38] reported the occurrence of PFBS, PFHxS, and PFOA in food items collected from Texas, U.S.; PFOS and other seven PFASs were not detected at concentrations above the LOD in these samples. PFOA had the highest detection frequency (17 of 31 samples), at concentrations ranging from 0.07 ng g^{-1} in potatoes to 1.8 ng g^{-1} in olive oil. PFBS and PFHxS were only detected in cod sample at 0.12 and 0.07 ng g^{-1}, respectively. No comprehensive study on PFASs in foodstuffs from the U.S. is available to date. Our research group analyzed a large number of food samples collected from the U.S. Table **2** summarizes the reported concentrations of PFASs in food items from around the world.

Table 2. Summary of reported mean (range) of concentrations of PFASs in food items collected from around the world (unit of concentration: ng g^{-1} for solid food or ng mL^{-1} for liquid food).

Country	Year	Sample Types	PFOS (ng g^{-1})	PFOA (ng g^{-1})	Other PFASs	Ref.
Canada	1992-2004	Beef stick	<0.6	2.6	<LOQ	[33, 36]
		Ground beef	2.1	<0.4	<LOQ	
		Luncheon meats, cold cuts	0.5	<0.4	<LOQ	
		Marine fish	2.6	<0.5	<LOQ	
		Freshwater fish	1.5 – 2.0	<0.5 – 2	<LOQ – 5 (∑PFOSA)	
		Pizza	<1	0.74	<LOQ – 27.3 (∑PFOSA)	
		Microwave popcorn	0.98	3.6	<LOQ – 18.9 (∑PFOSA)	
		Egg breakfast sandwich	<LOD	<LOD	<LOQ – 11.9 (∑PFOSA)	
		French fries	<LOD	<LOD	<LOQ – 9.72 (∑PFOSA)	
		Chicken nuggets	<LOD	<LOD	<LOQ – 5.87 (∑PFOSA)	
		Fish burger	<LOD	<LOD	<LOQ – 3.82 (∑PFOSA)	
Spain	2006	Vegetables	0.022	<0.027	<LOQ	[20]

(Table 2) cont.....

Country	Year	Sample Types	PFOS (ng g^{-1})	PFOA (ng g^{-1})	Other PFASs	Ref.
		Cereals	<0.027	<0.045	<LOQ	
		Pulse	<0.069	<0.080	<LOQ	
		White fish	0.407	<0.065	<LOQ	
		Seafood	0.148	<0.029	<LOQ	
		Tinned fish	0.271	<0.126	<LOQ	
		Blue fish	0.654	<0.132	<LOQ	
		Pork	0.045	<0.053	<LOQ	
		Chicken	0.021	<0.004	<LOQ	
		Veal	0.028	<0.003	<LOQ	
		Lamb	0.040	<0.012	<LOQ	
		Eggs	0.082	<0.005	<LOQ	
		Dairy products	0.121	<0.007	<LOQ	
		Whole milk	<0.014	<0.015	<LOQ	
		Semiskimmed milk	<0.019	<0.004	<LOQ	
		Fruits	<0.017	<0.004	<LOQ	
		Margarine	<0.034	<0.014	<LOQ	
		Oil	<0.099	<0.035	<LOQ	
		Raw veal	<0.015	<0.415	0.118 (PFHxA)	[19]
		Grilled veal	<0.008	<0.068	<LOQ	
		Fried veal	<0.018	<0.073	<LOQ	
		Raw pork	<0.008	<0.085	<LOQ	
		Grilled pork	0.011	<0.075	<LOQ	
		Fried pork	<0.008	<0.088	<LOQ	
		Raw chicken	<0.008	<0.390	<LOQ	
		Grilled chicken	0.012	<0.420	<LOQ	
		Fried chicken	0.010	<0.355	<LOQ	
		Packaged fried chicken nuggets	<0.003	<0.200	0.080 (PFHxA)	
		Black pudding	<0.015	<0.063	<LOQ	
		Lamb liver	0.330	<0.60	<LOQ	
		Pork liver	<0.083	<0.225	<LOQ	
		Foie gras of duck	<0.008	<0.440	<LOQ	
		Frankfurt sausages	<0.015	<0.183	0.079 (PFHxA)	

(Table 2) cont.....

Country	Year	Sample Types	PFOS (ng g^{-1})	PFOA (ng g^{-1})	Other PFASs	Ref.
		Homemade marinated salmon	0.026	<0.163	0.024 (PFHxA); 0.014 (PFHxS)	
		Packaged marinated salmon	0.054	0.179	<LOQ	
		Lettuce	0.010	<0.185	<LOQ	
		Packaged lettuce	0.034	<0.200	0.012 (PFHxA)	
		Packaged common salt	<0.001	<0.675	<LOQ	
China	2009	Pork meat	0.06	6.17	<LOQ	[35]
		Beef meat	0.05	4.26	<LOQ – 0.15 (PFHpA)	
		Chicken meat	0.08	12.5	<LOQ	
		Goat meat	0.08	1.56	<LOQ – 0.02 (PFNA)	
		Pork liver	1.99	2.02	<LOQ – 0.20 (PFUnDA)	
		Beef liver	0.75	0.48	<LOQ – 0.66 (PFNA)	
		Chicken liver	0.32	0.52	<LOQ – 0.17 (PFUnDA)	
		Duck liver	1.07	0.45	<LOQ – 0.31 (PFNA)	
		Goat liver	1.00	0.43	<LOQ – 0.18 (PFDS)	
		Pork blood cake	0.08	0.27	<LOQ – 0.12 (PFNA)	
		Beef blood cake	0.06	0.06	<LOQ	
		Duck blood cake	<0.12	0.58	<LOQ – 0.04 (PFDA)	
		Goat blood cake	<0.12	0.21	<LOQ – 0.04 (PFDA)	
		Chicken eggs	0.08	0.26	<LOQ – 0.02 (PFDA)	
		Duck eggs	0.34	0.12	<LOQ – 0.30 (PFUnDA)	
Japan	2004	1-day duplicate diet	0.008 – 0.087	0.008 – 0.040	<LOQ	[39]
Germany	2007	7-day duplicate diet	<0.1 – 1.0	<0.1 – 118	<LOQ – 3.2 (PFHxA)	[37]
Norway	2008-2009	Lettuce	0.00017	0.0018	<LOQ – 0.0013 (PFDoDA)	[40]
		Carrot	0.00067	0.002	<LOQ	
		Potato	0.001	0.0053	<LOQ – 0.0031 (PFHxA)	
		Cheese	0.012	0.013	<LOQ – 0.016 (PFNA)	

(Table 2) cont.....

Country	Year	Sample Types	PFOS (ng g-1)	PFOA (ng g-1)	Other PFASs	Ref.
		Margarine	0.0023	0.012	<LOQ – 0.0025 (PFHxA)	
		Milk	0.007	0.0047	<LOQ – 0.004	
		Bread	0.070	0.051	<LOQ – 0.017 (PFDA)	
		Strawberry jam	0.003	0.014	<LOQ – 0.0087 (PFDA)	
		Pork meat	0.017	0.015	<LOQ – 0.016 (PFDA)	
		Beef	0.060	0.012	<LOQ – 0.023 (PFDA)	
		Chicken meat	0.021	0.052	<LOQ – 0.020 (PFHpA)	
		Egg	0.039	0.030	<LOQ – 0.013 (PFHxA)	
		Fish sticks	0.013	0.049	<LOQ – 0.021 (PFHpA)	
		Canned mackerel	0.043	0.024	<LOQ – 0.019 (PFUnDA)	
		Salmon	0.055	0.046	<LOQ – 0.026 (PFDA)	
		Cod	0.100	0.030	<LOQ – 0.021 (PFUnDA)	
		Cod liver	0.31	0.051	<LOQ – 0.230 (PFUnDA)	
Netherlands	2009	Fatty fish	0.061	0.008	<LOQ – 0.041 (PFTrDA)	[41]
		Lean fish	0.308	0.023	<LOQ – 0.229 (PFTrDA)	
		Crustaceans	0.582	0.046	<LOQ – 0.268 (PFTrDA)	
		Butter	0.033	0.016	<LOQ – 0.020 (PFHxA)	
		Cheese	<0.085	<0.019	<LOQ – 0.008 (PFDA)	
		Milk	0.010	0.001	<LOQ – 0.043 (PFBA)	
		Eggs	0.029	<0.032	<LOQ – 0.011 (PFDA)	

(Table 2) cont.....

Country	Year	Sample Types	PFOS (ng g^{-1})	PFOA (ng g^{-1})	Other PFASs	Ref.
		Pork	0.082	0.015	<LOQ – 0.006 (PFHpA)	
		Beef	0.082	<0.005	<LOQ – 0.006 (PFDA)	
		Chicken/poultry	<0.005	<0.005	<LOQ – 0.003 (PFHxS)	
		Bakery products	0.004	0.005	<LOQ – 0.006 (PFHxs)	
		Vegetables/fruit	<0.047	0.005	<LOQ – 0.130 (PFBA)	
		Flour	<0.009	0.017	<LOQ – 0.018 (PFHxS)	
		Vegetable oil	<0.003	<0.003	<LOQ – 0.001 (PFHpA)	
		Industrial oil	<0.012	0.006	<LOQ – 0.007 (PFHxS)	
Texas, USA	2009	Hamburger		0.15		[38]
		Bacon		0.24		
		Sausages		0.09		
		Ham		0.02		
		Chicken breast		0.02		
		Canned chili		0.02		

Human breast milk is an important source of exposure to PFASs for infants from the U.S. and other countries [42 - 44]. A Swedish study found that the concentrations of PFOS in breast milk were approximately 1% of the corresponding concentrations in the serum of mothers [45]. Similarly, PFAS concentrations in breast milk samples collected from China were 1 – 2% of the concentrations found in the serum of women [46]. PFOS was found in breast milk collected from Hungary at median concentrations of 330 pg mL^{-1} [47]. In Catalonia, Spain, PFOS and PFHxS were found in breast milk at concentrations of 120 and 40 pg mL^{-1}, respectively [48]. In the U.S., 45 human breast milk samples, collected from Massachusetts in 2004 contained PFOS and PFOA at mean concentrations of 131 and 43.8 pg mL^{-1}, respectively [43]. The estimated average and highest daily intakes of total PFASs by breast-fed infants in the U.S. were 23.5 and 87.1 ng kg^{-1} bw day^{-1}, respectively [43]. In 184 human breast milk samples collected from seven Asian countries, PFOS was the predominant PFAS, followed by PFHxS and PFOA. Median PFOS concentrations ranged from 39.4

pg mL^{-1} for India to 196 pg mL^{-1} for Japan. PFOS and PFHxS were detected in infant formula and dairy milk purchased from the U.S., at concentrations of up to 11.3 and 3.82 pg mL^{-1}, respectively [44]. Table **3** summarizes the reported PFAS concentrations in human breast milk from around the world.

Table 3. Summary of reported mean (range) PFAS concentrations in human breast milk and infant formula (unit of concentration: ng mL^{-1}).

Country	Year	N	PFOS	PFOA	PFHxS	Ref.
China	2004	19	0.10 (0.05 – 0.36)	0.11 (0.05 – 0.21)	0.01 (0.004 – 0.10)	[46]
China	2007	24 pools	0.049	0.0345		[49]
Japan	1999	24	0.232 (0.14 – 0.523)	0.078 (0.043 – 0.17)	0.008 (0.002 – 0.018)	[44]
Cambodia	2000	24	0.067 (0.017 – 0.327)	0.043 – 0.089	0.002 – 0.019	
Vietnam	2000-2001	40	0.076 (0.017 – 0.393)	0.043 – 0.089	0.007 (0.002 – 0.011)	
Philippines	2000, 2004	24	0.098 (0.027 – 0.208)	<0.043	0.016 (0.002 – 0.059)	
Indonesia	2001	20	0.084 (0.025 – 0.256	<0.043	0.025 – 0.256	
India	2002, 2004, 2005	39	0.046 (0.011 – 0.12)	<0.043 – 0.355	<0.002 – 0.013	
Malaysia	2003	13	0.121 (0.049 – 0.35)	<0.043 – 0.17	<0.002 – 0.013	
Korea	2007	17	0.011	0.042		[50]
Korea	2013	264	0.049 (0.031 – 0.077)	0.071 (0.052 – 0.110)		[51]
Belgium	2006	22 pools	<0.4 – 28.2	<0.3 – 3.5		[52]
Czech Republic	2010	50	0.033 (0.007 – 0.114)	0.05 (0.012 – 0.128)	<0.006 – 0.022	[53]
Norway	2007-2008	19	0.087	0.025		[54]
Norway	2001-2009	66	0.11	0.05		[55]
Sweden	2004	12	0.201 (0.06 – 0.47)	<0.209 – 0.492	0.085 (0.03 – 0.17)	[45]
Sweden	1972-2008	20 pools	0.023 – 0.237	0.019 – 0.148	<0.005 – 0.028	[56]
Bavaria, Germany	2006	57	0.12 (0.03 – 0.31)	(<0.20 – 0.29)		[47]

(Table 3) cont.....

Country	Year	N	PFOS	PFOA	PFHxS	Ref.
Hungary	1996-1997	13	0.33 (0.096 – 0.639)			
Spain	2007	10	0.12 (0.070 – 0.22)	<0.5	0.04 (0.020 – 0.10)	[39]
Spain	2008	20	0.122 (0.028 – 0.865)	0.332 (0.015 – 0.907)		[57]
USA	2004	45	0.131 (<0.032 – 0.617)	0.438 (<0.030 – 0.161)	0.145 (0.030 – 0.161)	[43]
NC, USA	2004-2005	34	<0.06	<0.03	<0.03	[58]
Infant formula						
Washington DC, US	2007	21	<0.011 – 0.0113	<0.0483	<0.0014 – 0.0036	[44]
Dairy milk						
Albany, NY	2008	12	<0.011	<0.00483	<0.014 – 0.038	

Drinking water is an important source of PFAS exposure to the general population. In locations where there is a point source (near PFAS manufacturing plants, military bases and airports), drinking water has been the major source of PFAS exposure [59]. PFOA and PFOS were found in surface waters on the order of several ng L^{-1} levels [8, 14, 60]. In a U.S. study, PFOA was found at concentrations of 25 – 26 ng L^{-1} and PFOS at concentrations of 55 – 63 ng L^{-1} in tap water from Columbus, MO; nevertheless, in other five cities studied, both PFOS and PFOA were below the detection limits (2.5 ng L^{-1} for PFOS and 7.5 ng L^{-1} for PFOA) [61]. In drinking water samples collected from various locations in Japan, concentrations of PFOA as high as 47.3 ng L^{-1} were found whereas majority of the sampling sites had concentrations below 4 ng L^{-1} [62]. In a study of municipal drinking water from Catalonia, Spain, PFOS and PFHxS were the most frequently detected compounds, with the maximum concentrations of 58.1 and 5.30 ng L^{-1}, respectively [19]. PFBS, PFHxA, and PFOA were also detected in the samples, with maximum concentrations of 69.4, 8.55, and 57.4 ng L^{-1}, respectively. Mak *et al.* [63] determined PFASs in tap water collected from China, Japan, India, Canada, and the U.S. The highest total PFAS concentration was found in drinking water samples from Shanghai, China, at 130 ng L^{-1}, whereas the lowest was from Toyama, Japan, at 0.62 ng L^{-1}. The mean concentration of PFOS and PFOA in tap water samples collected from Albany, NY, U.S., was 1.4 and 1.2 ng L^{-1}, respectively. A survey on mineral and tap water from Germany showed that PFAS contamination in drinking water system is ubiquitous, with 52% of 177 samples contained PFAS at concentrations above 1 ng L^{-1}, and the maximum total PFASs concentration in tap water was 42.7 ng L^{-1}. Median PFOS and PFOA concentrations in bottled water, spring water, and tap water collected from Germany were 1.5 and 1.6 ng L^{-1}, <LOQ and 1.4 ng L^{-1}, and 1.3 and 2.6 ng L^{-1}, respectively [21]. In drinking water treatment plants in Japan, ranges of concentrations of PFOS and PFOA were 0.26 – 22 ng L^{-1} and 5.2 – 92 ng L^{-1}, respectively, in raw water, and were 0.16 – 22 ng L^{-1} and 2.3 – 84 ng L^{-1}, respectively, in treated water, suggesting that the removal efficiency in water treatment was low [64].

Drinking water from Little Hocking, OH, U.S., was contaminated by PFOA from a known source of industrial release near the well fields [59]. Elevated concentrations of PFOA were found in drinking water from Little Hocking, at concentrations ranging from 1700 to 7200 ng L^{-1}, which were the highest levels for PFOA ever reported in drinking water up to date. Median PFOA concentration in serum from the affected population was 354 ng mL^{-1}, which was 100-fold higher than that reported for the U.S. general population in the mid-2000s. Similarly, PFOA was detected at concentrations ranging from 22 to 519 ng L^{-1} in drinking water from North Rhie-Westphalia, Germany, which was contaminated by a known source of PFOA release [27]. Table **4** summarizes the reported concentrations of PFOS and PFOA in drinking water.

Table 4. Summary of reported mean (range) concentrations of PFOS and PFOA in drinking water (unit of concentration: ng L^{-1}).

Location	PFOS (ng L^{-1})	PFOA (ng L^{-1})	Ref.
West Virginia, USA		3550	[65]
Albany, NY, USA	1.4	1.2	[63]
Alberta, Canada		0.2	[66]
British Columbia, Canada		0.2	
Niagara on the lake, Canada	3.3	2.1	[63]
United Kingdom	<11 – 45	<24 – 240	[67]
Orebro, Sweden	0.55	0.65	[66]
Poland	0.098 – 0.11	0.005 – 0.013	[68]
Rhine-Ruhr, Germany	0 – 22	<1 – 519	[27]
Catalonia, Spain	0.61 (0.4 – 0.9)	2.2 (0.3 – 6.3)	[19]
Lake Maggiore, Italy	8.1	2.4	[69]
Osaka and Tohoku, Japan	<0.1 – 12	0.7 – 50	[22]
Tokyo and Kyoto, Japan	0.1 – 51		[62]
Iwate, Japan	0.26	0.70	[66]
Ibaraki, Japan	0.44	4.2	

(Table 4) cont.....

Location	PFOS (ng L^{-1})	PFOA (ng L^{-1})	Ref.
Osaka, Japan	1.6	18	[63]
Tokyo, Japan	1.6	4.0	
Toyama, Japan	0.066	0.18	
Shanghai, China	7.6	78	
Wuhan, China	0.35	5.3	
Nanjing, China	0.94	5.9	
Shenzhen, China	11	1.0	
Macau, China	6.2	1.3	
Hong Kong, China	7.0	1.1	
Taipei, China	5.4	3.7	
Xiamen, China	0.76	2.6	
Shenyang, China	0.39	0.79	
Beijing, China	0.042	0.44	
Chennai, India	8.4	2.0	
Hanoi, Vietnam	<0.05 – 0.1	<0.1 – 0.2	
Bangkok, Thailand	1.9	4.6	
Khon Kean, Thailand	0.13	1.2	
Kota Kinablu, Malaysia	0.025	0.025	

Indoor dust has been a major source of human exposure to several contaminants. Because PFASs are widely used in many consumer products, these chemicals are present in the indoor environments. Similar to polybrominated diphenyl ethers (PBDEs), PFASs were found at notable concentrations, on the order of several hundreds of nanograms to micrograms per gram levels in house dust [35]. Carpet protection products (*e.g.*, Scotchguard®) and textiles have been reported as possible sources of PFASs in the indoor environment [70]. Non-stick cookware contained residual PFASs and FTOHs and heating of these products at normal cooking temperatures released PFASs into the indoor air [32]. In addition, PFASs are also used in floor wax, shampoos, and conditioners [71]. Some cosmetic products including skin milk and foundations contain PFASs [72]. Studies suggested that inhalation and ingestion of house dust can be a source of human exposure to PFASs [35, 73, 74].

Studies have reported the concentrations of PFASs in indoor dust samples collected from around the world. PFOS and PFOA were the main PFASs found in indoor dust and the concentrations typically ranged from low ng g^{-1} to μg g^{-1} [75 - 77]. Median concentrations of PFOS and PFOA in indoor dust from Ottawa,

Canada, were 37.8 ng g^{-1} and 19.7 ng g^{-1}, respectively [75]. Median concentration of PFOS in 16 house dust samples from Japan was 24.5 ng g^{-1}, whereas median concentration of PFOA was 165 ng g^{-1} [78]. Similarly, indoor dust samples from China contained a median concentration of 4.86 ng g^{-1} PFOS, and 205 ng g^{-1} PFOA [35]. In the U.S., high median concentrations of PFOS at 201 ng g^{-1}, and PFOA at 142 ng g^{-1} were found in house dust samples collected from several cities in Ohio and North Carolina [77]; this study also reported the occurrence of other PFASs including PFBS, PFHxS, and long-chain PFCAs including PFNA, PFDA, PFUnDA, and PFDoDA at median concentrations that ranged from 6.7 to 50.2 ng g^{-1} in indoor dust. The highest reported concentrations in house dust from the U.S. were 12,100 ng g^{-1} for PFOS, 35,700 ng g^{-1} for PFHxS, and 1,960 ng g^{-1} for PFOA. FTOHs and perfluorooctanesulfonamides were also found in house dust. Median concentrations were 87 ng g^{-1} for the sum of 6:2, 8:2, and 10:2 FTOHs, and 251 ng g^{-1} for the sum of MeFOSE, EtFOSE, EtFOSA, and MeFOSEA [76, 77]. Table **5** summarizes the reported PFAS concentrations in indoor dust from around the world.

Table 5. Summary of reported mean (median or geometric mean) (range) concentrations of PFASs in indoor dust from around the world (unit of concentration: ng g^{-1}).

Country	Year	PFOS	PFOA	PFHxS	Ref.
Norway	2008	11 (3) (1.2 – 95)	20 (18) (6.2 – 56)	8.4 (0.6) (0.21 – 142)	[74]
Canada	2007-2008	280 (1.5 – 4700)	97 (2.0 – 1400)		[79]
Norway	2007-2008	11 (9.1) (4.4 – 24)	39 (39) (10 – 80)	(1.4)	[80]
Belgium	2008	9.4 (0.5) (<0.1 – 211)	6.4 (0.7) (<0.05 – 109)		[81]
UK	2007-2009	450 (140) (3.5 – 7400)	310 (190) (<1 – 4100)	450 (210) (7.6 – 6100)	[82]
Australia	2007-2009	1000 (170) (6.5 – 8100)	530 (120) (15 – 2900)	240 (180) (19 – 110)	
Canada	2007-2009	290 (140) (42 – 1300)	270 (69) (<0.98 – 4000)	150 (120) (<0.09 – 850)	
France	2007-2009	330 (160) (54 – 1700)	52 (31) (15 – 220)	130 (77) (54 – 320)	
Germany	2007-2009	310 (170) (47 – 1000)	290 (300) (19 – 730)	290 (150) (16 – 790)	
Kazakhstan	2007-2009	56 (59) (<0.03 – 130)	49 (<0.98) (<0.98 – 220)	94 (120) (1 – 180)	
Thailand	2007-2009	32 (16) (3 – 130)	45 (18) (<0.98 – 290)	25 (16) (3 – 84)	
USA	2007-2009	420 (310) (110 – 930)	370 (240) (27 – 1800)	270 (240) (86 – 560)	
Japan	NA	200 (25) (11 – 2500)	380 (165) (70 – 3700)		[78]

(Table 5) cont.....

Country	Year	PFOS	PFOA	PFHxS	Ref.
Canada	2002-2003	440 (38) (2.3 – 5100)	110 (20) (1.2 – 1200)	390 (23) (2.3 – 4300)	[75]
USA, Germany, UK, Australia	2004	761 (201) (<8.9 – 12100)	296 (142) (<10 – 1960)	874 (46) (<13 – 35700)	[83]
China	2009	4.86	205	0.17	[35]
USA	2001-2002	761 (201) (<8.9 – 12000)	296 (142) (<10 – 2000)	874 (46) (<13 – 35700)	[77]
Japan		(7 – 41)	(18 – 89)		[84]
Germany		(16) (3 – 342)	(11) (2 – 141)		[85]
USA	2008	(47) (8.7 – 1100)	(44) (6.5 – 420)	(16) (2.1 – 1000)	[86]
Canada	2008/2012	(184 – 1090)	(88 – 550)	(253 – 2900)	[87]
Spain	2009	(1.1 – 12)	(1.5 – 36)	(0.17 – 5.3)	[88]

Concentrations of PFASs in indoor air usually exceeded those of outdoor air, and a person typically spends 90% of the time in the indoor environment; thus, indoor air is a source of human exposure to PFASs [73]. Very few studies have reported indoor air concentrations of PFASs. In Norway, FTOHs were found in the air of homes and offices at concentrations that ranged from 21 pg m^{-3} for 4:2 FTOH to 10,005 pg m^{-3} for 8:2 FTOH; FOSAs and FOSEs concentrations ranged from 17 to 1211 pg m^{-3} [80]. In Canada, concentrations of 8:2 FTOH ranged from 261 to 47,000 pg m^{-3}, and of FOSAs and FOSEs ranged from 67 to 11,000 pg m^{-3} [79, 89]. In the same study, PFOA was found to be the dominant PFAS in indoor air at a geometric mean concentration of 28 pg m^{-3}.

Fromme *et al.* [90] summarized monitoring data for PFASs in food, drinking water, house dust, indoor and outdoor air, as well as biomonitoring data on PFAS levels in human blood, breast milk, and other tissues. The meta-analysis of data showed that humans are directly exposed to PFASs through food, drinking water, ingestion of house dust, and inhalation of indoor air. The average and upper bound daily exposure from all potential sources were also estimated. For PFOS and PFOA, the average (and upper bound) daily intake was estimated to be 1.6 ng kg^{-1} bw day^{-1} (8.8 ng kg^{-1} bw day^{-1}) and 2.9 ng kg^{-1} bw day^{-1} (12.6 ng kg^{-1} bw day^{-1}), respectively.

Select Exposure Related Epidemiological Studies of PFASs

Serum PFOS and PFOA concentrations in 3512 workers employed by 3M Company in Decatur, Alabama, were as high as 10 μg mL^{-1}. The high PFOS concentrations in workers were associated with high rates of bladder cancer but not liver cancer [91]. A cross-sectional analysis of PFOS and PFOA concentrations in serum of 3M employees in Decatur and Antwerp was performed [92]. There were no significant changes in hematological, hepatic, thyroid, or urinary clinical parameters measured in employees from Antwerp and Decatur plants when the data were examined separately. However, when the data were combined for employees from both plants, a significant increase of triiodothyronine (T3) level and a decrease of thyroid hormone binding ratio were associated with PFOS exposure; PFOA was negatively associated with high-density lipoprotein (HDL); triglycerides and T3 were positively associated with PFOS, PFOA, and total organic fluorine. In obese men, high liver enzymes were associated with occupationally exposed PFOA [93]. A study examined PFASs in serum samples and reproductive hormones and semen quality in 105 Danish men [94]. The study showed that high PFOS and PFOA levels were significantly associated with fewer normal sperm, and contributed to observed low sperm quality in young men. The study did not find association between PFOS or PFOA concentrations and reproductive hormones to be significant.

The effects of placental exposures to PFOS and PFOA on human fetal development have been reviewed [95]. Inoue *et al.* [96] first reported maternal and cord serum concentrations of PFOS and PFOA in 15 Japanese mothers. No associations between PFOS concentration and birth weight, gender, TSH, or free T4 in newborn blood were found. However, the study was limited by small sample size. A larger cross-sectional study on the analysis of 293 cord blood from Baltimore, MD, compared PFOS and PFOA levels with birth outcomes including gestational age, birth weight, birth length, head circumference, and ponderal index [97, 98]. Negative associations between concentrations of PFOS and PFOA and birth weight, ponderal index, and head circumference were found. Though the associations were not statistically significant for most of the parameters and statistical significance was found for ponderal index. Another study involving maternal blood from Danish National Birth Cohort showed a relationship of birth weight and gestational age to maternal PFOS and PFOA concentrations [99]. Maternal concentrations of PFOA were negatively associated with birth weight in Danish newborns [100]. Further study showed a statistically significant negative association between abdominal circumference and maternal PFOA concentration, a statistically significant positive association between ponderal index to maternal PFOA in obese women before pregnancy, and a non-significant negative correlation between head circumference and maternal PFOA serum concentrations [99]. These findings suggested that PFOA might affect the growth of organs and skeleton of infants. In a study from Japan, PFOS concentration was found significantly correlated with birth weight, but not other measurements [101]. A study involving occupationally exposed women from a POSF-manufacturing plant did not show associations between gestational exposure to both PFOS and PFOA and birth weight [102]. A highly exposed population to PFOA in Little

Hocking Water district in Ohio, U.S., did not show an association between maternal serum PFOA levels and birth weight or gestational age [103]. Similarly, neither birth weight nor gestational age was found to be associated with PFOS and PFOA in maternal or cord serum in 101 mothers from Hamilton, Ontario, Canada [104]. These studies suggested that PFOS and PFOA can have adverse effects on fetal growth and neonatal development. However, the literature results are inconsistent and suggested the influence of several confounding factors. Studies have shown that PFASs were associated with attention deficit/hyperactivity disorder (ADHD) in children. Hoffman *et al.* [105] conducted a cross-sectional study using samples from 1999-2000 and 2003-2004 NHANES ($n = 571$; 12 -15 years of age) to investigate the association between serum PFOS concentrations and ADHD. The study showed a positive ($p < 0.5$) dose response relationship between parent-reported ADHD and maternal serum PFOS concentrations modeled as continuous predictors. The adjusted odds ratio (OR) for reported ADHD in association with a 1 ng mL^{-1} increment in blood serum PFOS was 1.03. Gump *et al.* [106] showed that high levels of blood PFOS, PFNA, PFDA, PFHxS, and PFOSA in children were associated significantly with shorter inter-response times during the differential reinforcement of low rates of responding task. The result suggested an association between PFAS exposure and children's impulsivity.

Among several PFASs that are in commerce, PFOS and PFOA have received considerable public health concern because of their persistence, toxicity, and accumulation in humans. The major manufacturer of PFOS, The 3M Company, announced a phase out of PFOS production in North America in May 2000 [107]. Other fluorochemical manufacturers including DuPont, committed to reduce environmental release of PFASs during manufacturing processes [5].

The European Food Safety Authority (EFSA) proposed a tolerable daily intake value (TDI) of 1500 ng kg^{-1} bw day^{-1} for PFOA and 150 ng $^{-1}$kg bw day^{-1} for PFOS in 2008 [108]. PFOS and its salts have been listed as POPs under the Stockholm Convention for global elimination since May 2009 [6]. Nevertheless, PFOS is still being produced in some countries such as China. In January 2009, the USEPA set a Provisional Health Advisory level for PFOS and PFOA at 0.2 and 0.4 μg L^{-1}, in drinking water, respectively [109]. In May 2016, following the public health concerns over PFOA in communities in New York and some other states, the USEPA revised the health advisory level for combined PFOS and PFOA concentration of 70 ng L^{-1} in drinking water [110]. A tolerable exposure dose of 20 ng kg^{-1} bw day^{-1} has been suggested for PFOS in 2018.

Biomonitoring studies in the United States (*e.g.*, NHANES) have shown widespread human exposure to PFASs, particularly PFOS and PFOA. The

Americans have some of the highest concentrations of PFASs in the world [111, 112]. The CDC's NHANES program determines PFASs in serum collected from a representative U.S. population. Geometric mean concentrations of PFOS, PFOA, PFHxS, and PFNA in 1999 – 2000 were 30.4, 5.21, 2.13, and 0.557 ng mL^{-1}, respectively, whereas the concentrations in 2009 – 2010 were 9.32, 3.07, 1.66, and 1.54 ng mL^{-1}, respectively. Serum concentrations of PFOS determined in the NHANES were applied in a pharmacokinetic model to estimate the daily exposure dose. PFOS intake dose was estimated as 160 ng day^{-1} for adults, and 50 ng day^{-1} for children in the U.S [113]. The estimation of human exposure doses on the basis of biomonitoring data could provide a total dose (*i.e.*, body burden), but not contribution from various sources, and pathways. The assessment and validation of chemical exposures can be made based on the direct analysis of contaminants in sources such as food, drinking water, indoor air and dust; a combination of environmental and biomonitoring approaches to assess exposures would further refine and validate the use of one of these approaches in exposure assessments. We used environmental monitoring to describe exposure to PFASs from food, drinking water, air and dust. A major challenge in environmental monitoring studies is that they involve collection and analysis of contaminants in a wide range of matrices (air, water, foods of various kinds). The exposure doses calculated from this approach may not reflect actual body burden because of factors such as bioavailability, intestinal absorption, and metabolism, loss and transformation of contaminants during cooking and culinary practices. Furthermore, the wide variety of food matrices requires multiple sample preparation and analytical methods for the measurement of contaminants. We assessed sources, pathways, and relative contribution of each pathway to PFASs exposures in Albany, New York, U.S. We also compared measured exposure doses (through environmental monitoring) with modeled doses estimated from biomonitoring studies.

METHODS

Food Sample Collection

Food samples (n=139) were collected from local grocery stores in Albany, New York, in 2008 and 2011 (Table **6**). According to the United States Environmental Protection Agency (USEPA) published Exposure Factors Handbook (2011 edition), food items were categorized into beverages, condiments, dairy products, fish & seafood, meat & poultry, grains, fruits, vegetables, and infant food. A portion of the food samples was homogenized in a DuPont Sorvall® Omni-mixer (Newtown, CT) and transferred into 50-mL polypropylene (PP) tubes for further analysis.

Table 6. Sample information of foodstuffs collected from Albany, NY, USA, for PFAS analysis.

Category	n	Food Items (numbers in the brackets are the number of samples for each subcategory)
Beverages	20	Bottled water (1), tea (3), carbonated soft drink (7), juice (4), beer (3), wine (1), sports drink (1)
Condiments	7	Tomato sauce (2), salad dressing (5)
Dairy & milk	13	Yogurt (3), cheese (2), butter (2), egg (2), cream (2), pudding (1), milk (1)
Fish & seafood	13	Marine fish (7), freshwater fish (3), shellfish (3)
Grains	13	Rice (1), noodle (3), cereal (1), bread (1), pizza (3), cookies (4)
Meat	27	Beef (7), pork (7), chicken (9), turkey (4)
Fruits	11	Pear (2), peach (2), apple (2), grapefruit (1), palm (1), cantaloupe (1), mixed fruit (2)
Vegetables	25	Onion (2), mushroom (3), lettuce (2), cabbage (3), radish (1), tofu (2), green bells (1), squash (1), potato (1), corn (1), celery (1), rhubarb (1), beets root (1), cucumber (1), tomato (1), spinach (1), mixed vegetables (2)
Infant food	7	Infant formula (2), rice cereal (1), fruit purees (4)

Drinking (tap) water samples (n=23) were collected in 15-mL PP-tubes from homes, schools, restaurants, hotels, airports, and playgrounds in the following cities across New York state from 2011 to 2012: Albany, NY (n=10); New York, NY (n=4); Buffalo, NY (n=9). Indoor dust samples (n=23) were collected from homes in Albany, NY, in 2006 and 2010. Dust samples were obtained from vacuum cleaner bags, transferred to pre-cleaned glass jars, and transported to the laboratory. Samples were sieved through a ≤2 mm sieve, homogenized, and packed in a pre-cleaned aluminum foil for storage.

PFAS Analysis in Food

Food samples including beverages, condiments, dairy products, fish & seafood, meat & poultry, grains and grain products, fruits, vegetables, and infant food were analyzed by ion-pair extraction method. The complex nature of food matrices imposed considerable challenge in the analysis of PFASs. Ion-pair extraction method was originally developed for the analysis of PFASs in soft tissues or blood. The application of this method for food samples required some modifications. Food matrices including beverages, condiments, dairy products, fish & seafood, meat & poultry, grains and grain products, fruits, vegetables, and infant food were extracted using an ion-pair extraction method with some modifications [13, 60]. One milliliter (1 mL) of beverage or approximately one gram (1 g) of solid food, known amount of stable isotopic labeled internal standards ($^{13}C_4$-PFOA, $^{13}C_4$-PFOA, $^{18}O_2$-PFHxS, $^{13}C_2$-PFHxA, $^{13}C_2$-PFNA, $^{13}C_2$-PFDA, $^{13}C_2$-PFUnDA, and $^{13}C_2$-PFDoDA), 2 mL of 0.25 M sodium carbonate/bicarbonate buffer, and 1 mL of 0.5 M tetrabutylammonium hydrogen sulfate (TBAHS) solution (pH adjusted to 10) were mixed in a 15-mL PP tube. Sample was then extracted with 5 mL of methyl-*tert*-butyl ether (MTBE) by shaking vigorously in an orbital shaker for 45 min. The MTBE layer was separated by centrifugation at 3500 x g for 5 min and then transferred into another PP tube. The extraction was repeated with another 3 mL of MTBE, and the two MTBE extracts were combined. The MTBE extract was evaporated to near-dryness under a gentle stream of nitrogen and then reconstituted with 1 mL of methanol. The samples were frozen at -20 °C overnight. The samples were then filtered through a 0.2 µm nylon filter (Phenomenex, Torrance, CA) into amber autosampler vials immediately. The samples were then analyzed by HPLC-MS/MS.

PFAS Extraction in House Dust

House dust samples were extracted using methanol after spiking of internal standards. Briefly, approximately 0.1 g of sieved indoor dust sample was weighed in a PP-tube. Known amounts of stable isotope labeled internal standards ($^{13}C_4$-PFOA, $^{13}C_4$-PFOA, $^{18}O_2$-PFHxS, $^{13}C_2$-PFHxA, $^{13}C_2$-PFNA, $^{13}C_2$-PFDA, $^{13}C_2$-PFUnDA, and $^{13}C_2$-PFDoDA) were spiked into samples, followed by the addition of 5 mL of methanol. Samples were sonicated in a sonicator (Branson, Danbury, CT) for 45 min. Sample was then extracted with 5 mL of methanol by shaking vigorously in an orbital shaker for 45 min. The methanol layer was separated by centrifugation at 3500 x g for 5 min and then transferred into another PP tube. The extraction was repeated with another 3 mL of methanol, and the two methanol extracts were combined. The methanol extract was evaporated to near-dryness under a gentle stream of nitrogen and then the volume adjusted 1 mL. The samples were filtered through a 0.2 µm nylon filter (Phenomenex; Torrance, CA) into an amber autosampler vial for instrumental analysis. Similar to food matrices, the method was validated for dust samples by performing recovery tests with internal standards and matrix spikes. Isotopic dilution method was used for quantification of PFAS concentrations.

PFAS Analysis in Water

Water samples were analyzed using direct injection after spiking of internal standards. Briefly, 0.9 mL of water sample was transferred into a PP-tube. Known amounts of stable isotope labeled internal standards ($^{13}C_4$-PFOA, $^{13}C_4$-PFOA, $^{18}O_2$-PFHxS, $^{13}C_2$-PFHxA, $^{13}C_2$-PFNA, $^{13}C_2$-PFDA, $^{13}C_2$-PFUnDA, and $^{13}C_2$-PFDoDA) were spiked into samples in 0.1 mL methanol. The samples were vortexed for 1 min then filtered through a 0.2 µm nylon filter (Phenomenex; Torrance, CA) into an amber autosampler vial for instrumental analysis.

Method Validation

Eight internal standards, $^{18}O_2$-PFHxS,$^{13}C_4$-PFOS, $^{13}C_2$-PFHxA, $^{13}C_4$-PFOA, $^{13}C_2$-PFNA, $^{13}C_2$-PFDA,$^{13}C_2$-PFUnDA, and $^{13}C_2$-PFDoDA, were available at the time of analysis to validate the analytical method. For the quantification of eight PFASs in food matrices, isotopic dilution method was applied. The recoveries of internal standards spiked into food, water and dust matrices as well as standards spiked into matrix are shown in Tables **7** and **8**, respectively.

Table 7. Internal standard recoveries (%) of spiked isotopic labeled PFASs into different sample matrices.

		$^{18}O_2$-PFHxS	$^{13}C_4$-PFOS	$^{13}C_2$-PFHxA	$^{13}C_4$-PFOA	$^{13}C_2$-PFNA	$^{13}C_2$-PFDA	$^{13}C_2$-PFUnDA	$^{13}C_2$-PFDoDA
Food N=139	Mean	108	88	121	103	86	85	111	123
	SD	27	22	35	34	24	28	36	34
Water N=23	Mean	117	93	123	111	107	87	69	59
	SD	10	20	11	11	16	22	23	13
Dust N=23	Mean	74	75	59	70	62	59	52	59
	SD	10	11	14	13	15	13	8	8

Table 8. Recoveries (%) of PFASs spiked into matrices (at 10 ng) (recoveries of PFHxS, PFOS, PFHxA, PFOA, PFNA, PFDA, PFUnDA, and PFDoDA were corrected by internal standard recoveries).

		PFBS	PFHxS	PFOS	PFDS	PFOSA	PFHxA	PFHpA	PFOA	PFNA	PFDA	PFUnDA	PFDoDA
Food N=19	**Mean**	74	99	115	63	56	114	57	104	98	104	100	105
	SD	27	7	25	16	17	6	7	5	4	6	5	12
Water N=4	**Mean**	82	95	102	75	79	112	69	102	105	98	97	95
	SD	5	7	5	10	8	8	10	7	7	6	6	11
Dust N=5	**Mean**	72	92	98	73	60	108	58	97	93	105	98	96
	SD	13	5	7	12	16	10	13	8	9	10	13	9

Instrumental Analysis

PFASs in house dust samples were analyzed using an Agilent 1100 series high performance liquid chromatography (HPLC) coupled with an Applied Biosystems API 2000 electrospray triple quadrupole tandem mass spectrometer (ESI-MS/MS). PFASs in food matrices were analyzed with an Agilent 1100 series HPLC coupled with an Applied Biosystems API 5500 ESI-MS/MS. The chromatographic separation was similar for the two matrices, except for the injection volume, which was ten microliters for dust and five microliters for foods. Sample extract was injected onto a 100 × 2.1 mm Betasil® C18 column (5µm; Thermo Electron Corporation, Bellefonte, PA) with a Javelin™ guard column (20 × 2.1 mm) connected serially. The mobile phase was 2 mM ammonium acetate/methanol at a flow rate of 300 µL min^{-1}. The mobile phase gradient started at 10% methanol and increased to 99% methanol in 10 min, and was held for 2 min before reverting to 10% methanol. The API 2000 and API 5500 mass spectrometers were operated in electrospray negative ionization mode, and the target compounds were determined by multiple reaction monitoring (MRM). Compound specific MS/MS parameters and mass transitions for each analyte are shown in Table **9**.

Table 9. Optimized mass spectrometric parameters for PFAS analysis in environmental and biological samples.

	PFBS	PFHxS	PFOS	PFDS	PFOSA	PFHxA	PFHpA	PFOA	PFNA	PFDA	PFUnDA	PFDoDA
MS/MS	299>80	399>80	499>99	599>99	498>78	312>269	363>319	413>369	463>419	513>469	563>519	613>569
IS MS/MS		403>84	503>99			315>270		417>372	465>420	515>470	564>520	615>570
IS Voltage	-4500	-4500	-4500	-4500	-4500	-4500	-4500	-4500	-4500	-4500	-4500	-4500
Collision Gas	5	5	5	5	5	5	5	5	5	5	5	5
Declustering Potential (DP)	-35	-55	-60	-65	-70	-10	-10	-14	-10	-10	-10	-10
Focusing Potential (FP)	-350	-350	-350	-350	-350	-350	-350	-350	-350	-350	-350	-350
Entrance Potential (EP)	-10	-9	-9	-9	-10	-6	-6	-3	-6	-6	-6	-6
Collision Energy (CE)	-55	-75	-75	-80	-60	-15	-15	-12	-15	-15	-11	-11
Collision Cell Exit Potential (CXP)	-5	-5	-5	-5	-15	-12	-12	-14	-12	-12	-17	-17

Background Contamination

PFOA was found in procedural blanks at trace levels. The solvents and reagents used for ion-pair extraction were analyzed to identify the sources of PFOA contamination. TBAHS solution was found to contain PFOA. In general, 0.5 M TBAHS solution contained ~0.2 ng mL^{-1} of PFOA, which was at the same level as that found in procedural blanks. This trace background contamination did not affect the analysis of indoor dust, as the limit of quantification (LOQ) was set at 3 times the concentrations found in blanks, and the reporting levels were subtracted from the blank levels. However, this trace level of PFOA background contamination was much higher than the levels found in food samples; therefore, it was necessary to develop a protocol to remove the background level of PFOA for the analysis of food samples. The extraction/purification of TBAHS solution with MTBE considerably reduced the contamination level, to below the detection limit of 0.01 ng mL^{-1}. In this study, all TBAHS solution used in the extraction of samples was pre-extracted/purified with MTBE.

Several parts of the HPLC-MS/MS instrumentation can introduce PFOA contamination as they are made up of polytetrafluoroethylene (PTFE) (*e.g.*, degasser and tubing) [114]. High levels of PFOA were found in the first few injections of pure methanol. It was found necessary to inject methanol at least 3 times prior to the injection of samples in order to reduce the background levels of PFOA present in the HPLC parts. Flow of methanol through the instrument for at least 30 minutes prior to the analysis of samples can reduce the background levels of PFASs. We also removed degasser from the HPLC system, as this was not always required for the operation of the instrument.

Analytical standards also contained impurities of PFASs at trace levels. For example, a few PFAS standards purchased from certain vendors, including PFOS standard from the 3M Company, contained up to 2% PFOA as an impurity. Furthermore, autosampler vial caps made of Teflon contained PFOA, and therefore pre-cleaned aluminum foil was used as caps for autosampler vials to prevent the contamination of samples.

Matrix Effects and Internal Standards

Matrix effects, including ion enhancement and ion suppression, are an issue in the analysis of complex matrices by HPLC-MS/MS. Low (below 100%) or high (above 100%) recoveries of native and internal PFAS standards passed through the analytical methods have been reported [15, 35, 106]. Matrix effects of samples prepared by ion-pair extraction method with minimal cleanup procedures are usually greater than those methods that involve rigorous cleanup steps, including activated carbon cleanup. The ion-pair extraction method can extract many ionic components and matrix components, in addition to PFASs, present in samples. These matrix components co-elute with PFASs during chromatographic separation, and interfere with the ionization of PFASs due to the competition for charge and space on the aerosols formed from sample extracts in the electrospray interface of HPLC-MS/MS. Thus, enhancement or suppression of PFAS signal occurs depending on the magnitude and type of ionic interferences that are present in sample matrices. Matrix effect can vary from sample to sample or even within a sample processed by different methods of extraction and purification. Matrix effect can result in high or low recoveries of internal or native standards spiked into samples. To minimize errors introduced by the matrix effect, use of isotopic labeled internal standard or matrix matched calibration standards is recommended as a choice of quantification method. A matrix matched calibration standard can be prepared by spiking of various levels of target compounds in a sample matrix that is free of the target compounds. However, due to the ubiquitous contamination by PFOS and PFOA in environmental and biological samples, it is unlikely to find a sample that is free of PFASs. Alternatively, matrix that contains very low concentrations of target compounds and spiking at levels, at least 2 times the concentration found in the matrix is frequently used. When isotopic labeled internal standards are available, matrix matched calibration curves may not be required, since the matrix effects can be corrected by the recoveries of internal standard. Nevertheless, the use of both, matrix matched calibration curves and isotope labeled standards for each of the target chemicals is recommended for HPLC-MS/MS analysis of environmental chemicals.

Matrix effects were monitored in this study by preparing matrix spikes for each batch of samples analyzed. Recoveries of PFASs in matrix spikes were reported for each of the matrices analyzed. Data reported in this study were not corrected for the matrix spike recoveries, but corrected for isotope labeled internal standard recoveries. For those compounds for which internal standards were not available, the data were reported as semi-quantitative. In general, matrix effects are related to impure sample matrix, although, instrumental drift in retention time of the analytes, dirty HPLC column and MS ion source can also reduce the response of the target compounds, thus increasing the detection limits.

The use of isotopic internal standards (IS) allows for the correction of loss of target compounds during sample extraction, cleanup, and concentration/dilution, in addition to matrix effects. In the present study, ISs were used in the analysis of all samples and they were spiked in blanks, matrix spikes, matrix spike duplicates, and matrix matched calibrations. Recoveries of IS were acceptable (100±30%) in most cases and the reported concentrations were corrected by individual IS recoveries.

Matrix spikes were prepared for each matrix type analyzed. Matrix spike duplicates were also prepared when adequate sample volume was available. On average, for every 15 to 20 samples analyzed, a set of matrix spike and matrix spike duplicate was prepared.

Instrumental Performance and Limit of Quantification (LOQ)

Calibration standards were injected daily before and after the injection of samples. A mid-point calibration standard was injected after every ten samples as a check for the instrumental drift in response. Samples were re-injected when the coefficient of variation (CV) was more than 30% between the two calibration standard injections. Methanol was injected at least three times before the injection of a sequence of standards and samples, to minimize carryover and to reduce the background levels of contamination. Methanol was also injected after every ten samples as a check for the carryover of target analytes between samples. Methanol was injected at the end of the injection of a batch of samples.

The relative response of native PFAS standard to the stable isotopic labeled internal standard was used for quantification (*i.e.*, isotope dilution). The use of a relative response factor minimizes variations in signal strength due to the instrumental drift or matrix effects. The limit of quantification (LOQ) was determined based on the linear range of the calibration curve prepared in methanol. LOQ was determined as the lowest acceptable standard if it was back calculated to be within 30% of the theoretical value when evaluated *versus* the 1/x weighted curve, and the peak area of the standard was at least 3 times greater than that present in the blank. If the blank level was higher than 1/3 of the lowest acceptable concentration in the calibration standards, the LOQ was calculated as 3 times the average concentration found in blanks. Reported concentrations in samples were subtracted from the blank values, when applicable.

HUMAN EXPOSURE ASSESSMENT OF PFASS

Studies on human exposure to PFASs in the United States are meager [113]. In this study, two approaches were used to assess human exposure to PFASs. First, PFAS concentrations in external sources of human exposure were determined and the dose was calculated based on ingestion rates of various sources and the second approach was based on the measured concentrations in human blood using a simple pharmacokinetic approach.

External/Environmental Sources Method

PFAS concentrations were measured in each source of human exposure (water, food, and indoor dust), and the rate of intake/ingestion of each source was multiplied by the measured PFAS concentrations, which were then summed to obtain the total daily intake. The daily intake of PFASs by the general population was estimated by the following equation:

$$PFAS\ Dose = \frac{\sum_{i=1}^{n} C_i Q_i}{bw} \times f_{uptake} \quad \textbf{(1)}$$

where EDI (ng kg^{-1} bw day^{-1}) is the total estimated daily intake from diet, drinking water, dust ingestion, and air inhalation; Ci (ng g^{-1} or ng mL^{-1}) is the PFAS concentration in each environmental matrix analyzed; Qi (g day^{-1} or mL day^{-1}) is the average daily ingestion rate of food, drinking water, dust, and air; bw is the average body weight (kg); f is the fraction absorbed into the body (percentage); in the present study, a value of 100% was used. The daily ingestion rates for each food category, drinking water, dust, air, and body weight for each age group were obtained from the USEPA *Exposure Factors Handbook: 2011 edition*, and *Child-Specific Exposure Factors Handbook* [115, 116].

Internal Dose/Biomonitoring Method

Egeghy and Lorber [113] proposed a simple, one-compartment, first-order pharmacokinetic model, which predicts concentration of a chemical in the blood serum as a function of dose, elimination rate, and volume of distribution. The model is presented in the form of eq. (2):

$$\frac{d(CP)}{dt} = \frac{DP(t)}{Vd} - kP \times CP(t) \quad \textbf{(2)}$$

where CP is the serum concentration (ng mL^{-1}), DP is the daily absorbed dose (ng kg^{-1} bw day^{-1}), Vd is the volume of distribution (mL kg^{-1}), and kP is the first-order elimination rate in the body (day^{-1}). This model has been validated for adults only, for the reason that the volume distribution is validated only for adults. Assuming a steady state condition, the daily intake dose is calculated as in eq. (3):

$$DP = CP \times kP \times Vd \quad \textbf{(3)}$$

The volume of distribution is defined as the total amount of a substance in the body divided by the concentration of that substance in the blood or serum (Vd [mL kg^{-1} bw] = mass in body [ng kg^{-1} bw] / concentration in blood or serum [ng mL^{-1}]) [117]. The Vd likely has a physiological underpinning specific to the contaminant of concern and the animal species from which the value was derived. This parameter can influence the results considerably and needs to be carefully assigned. Thompson *et al.* [55] calibrated values of Vd for PFOA using data obtained from two communities where drinking water supplies had been contaminated with PFOA from a fluoropolymer manufacturing facility. The higher than background serum PFOA concentrations of residents in these two communities have been related to exposure to PFOA-contaminated drinking water, and a well characterized association between serum and drinking water concentrations. A value of 170 mL kg^{-1} bw was used for Vd, as has been reported in Thompson *et al.* [55]. Assuming a difference of 35% between PFOS and PFOA, the Vd for PFOS was estimated to be 230 mL kg^{-1} bw. It was assumed that PFAS levels in human body are in steady state, which implies that background exposures have been occurring in an individual for a long period [113]. Therefore, the model may not be applicable for children as the rapid growth rate may affect PFAS levels in blood and the Vd values. The first-order elimination rate constants for PFOS and PFOA were assumed to be 0.00039 day^{-1} and 0.0008 day^{-1}, respectively, based on the reported median half-life of PFOS and PFOA of approximately 4.8 years and 2.3 years [118, 119]. Concentrations of PFASs in serum of the U.S. general population were derived from the NHANES report published in 2013 [120].

RESULTS

PFAS Concentrations in Food and Beverage Samples

Twelve PFASs including PFBS, PFHxS, PFOS, PFDS, PFOSA, PFHxA, PFHpA, PFOA, PFNA, PFDA, PFUnDA, and PFDoDA were analyzed in 139 food and beverage samples collected from Albany, NY. Among 139 samples analyzed, 121 (87%) samples contained at least one of the 12 PFAS species. Among the 12 PFASs analyzed, PFOS was detected in 72% of the samples followed by PFOA, PFHxA, and PFHxS, at 55%, 29%, and 27%, respectively. PFUnDA and PFDoDA were not detected in any of the 139 samples at concentrations above the LOQ. PFDA, PFHpA, PFOSA, and PFBS were detected only in <10% of the samples analyzed. For the compounds with > 50% detection frequency (*i.e.*, PFOS at 87% and PFOA at 55%), descriptive statistics were applied in two ways to describe concentrations below the LOQ. First, LOQ was substituted with a value of zero and second, the LOQ was substituted at ½ LOQ. Because of the less frequent detection of PFASs other than PFOS and PFOA in food samples, no further discussions were made for these compounds.

Food samples were categorized as beverages, condiments, dairy products, fish & seafood, fruits, grains, infant food, meat & poultry, oil, and vegetables. A summary of mean, SD, median, and GM concentrations of PFOS and PFOA for each food categories are given in Table **10** and Figs. (**1** and **2**).

Table 10. Concentrations of PFOS and PFOA (pg g^{-1}, ppt wet wt) in food, water (pg mL^{-1}), and indoor dust (ng g^{-1} dry wt) samples from New York State, U.S. Concentrations below the LOQ were assigned a value of zero and ½ LOQ.

Category	N	PFOS						PFOA					
		<LOQ = 0			<LOQ = ½ LOQ			<LOQ = 0			<LOQ = ½ LOQ		
		mean	SD	median	mean	SD	GM	mean	SD	median	mean	SD	GM
Beverages	20	7.07	11.5	ND	10.3	9.52	7.88	9.20	11.0	5.60	12.3	8.49	10.3
Condiments	7	11.5	11.9	10.1	13.0	10.4	10.5	15.0	11.1	16.5	16.8	8.44	14.6
Dairy	13	21.8	12.5	21.3	22.2	11.8	19.0	16.2	18.8	13.6	19.1	16.2	13.8

(Table 10) cont.....

Category	N	PFOS						PFOA					
		<LOQ = 0			<LOQ = ½ LOQ			<LOQ = 0			<LOQ = ½ LOQ		
Fish & seafood	13	33.4	8.18	11.8	35.4	81.0	13.3	12.5	23.1	ND	16.8	20.6	10.4
Fruits	11	13.0	9.80	14.3	14.4	7.88	12.3	14.9	7.78	16.8	16.0	5.47	14.8
Grains	13	41.3	52.2	22.2	41.7	51.9	27.0	24.0	25.1	20.6	26.4	22.7	17.9
Infant food	7	18.4	6.70	17.1	18.4	6.70	17.4	12.0	20.5	ND	16.4	17.5	10.7
Meat	27	102	357	24.5	103	357	29.5	34.0	25.6	35.4	35.4	23.7	25.9
Oil	3	9.78	10.2	8.98	11.5	7.95	9.74	8.83	8.44	9.66	10.9	5.42	10.0
Vegetables	25	9.87	11.8	6.22	11.9	10.2	9.03	11.2	17.7	ND	15.2	15.2	10.7
Overall foods	139	34.4	162	14.3	35.8	162	14.7	17.6	20.9	13.4	20.4	18.7	14.1
Tap water	23	ND		ND	ND		ND	6.96	3.85	7.24	7.23	3.54	6.44
Indoor dust	23	391	745	185	391	745	116	353	925	94.5	353	925	96.0

* ND= Not detected.

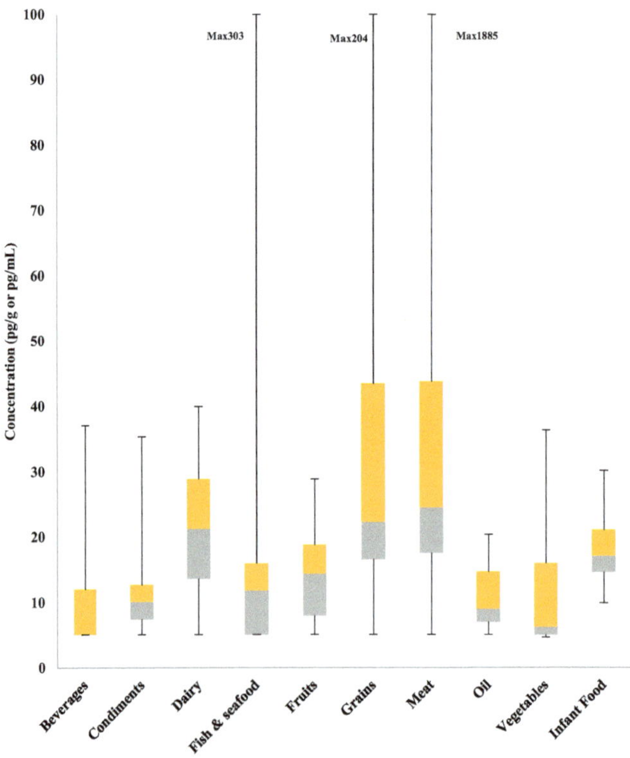

Fig. (1). Boxplot of PFOS concentrations (ppt) in foods from the U.S., by categories; error bars and columns show minimum, 25th percentile, median, 75th percentile, and maximum.

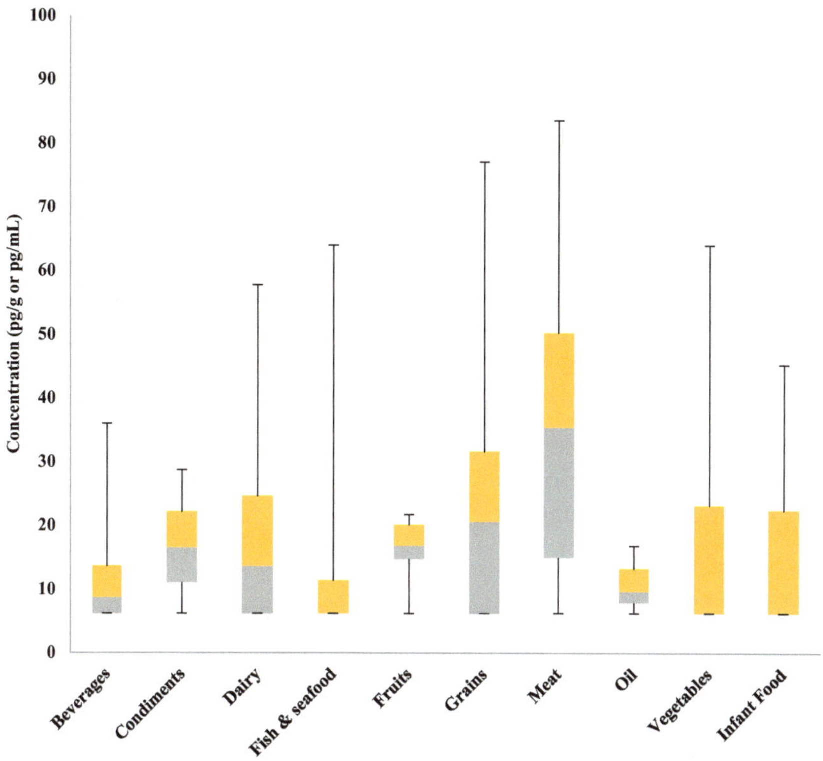

Fig. (2). Boxplot of PFOA concentrations (ppt) in foods from the U.S. by categories; error bars and columns show minimum, 25th percentile, median, 75th percentile, and maximum.

PFOS concentrations in beverage samples ranged from <LOQ to 0.037 ng mL^{-1}, and PFOA from <LOQ to 0.036 ng mL^{-1}. The highest PFOS and PFOA concentration was found in an apple juice sample and a green tea sample, respectively. PFOS concentration in condiments ranged from <LOQ to 0.035 ng g^{-1}, and PFOA from <LOQ to 0.029 ng g^{-1}. The highest PFOS and PFOA concentration was found in a pancake syrup sample and a marinade sample, respectively. PFOS concentrations in dairy products ranged from <LOQ to 0.040 ng g^{-1}, and PFOA from <LOQ to 0.058 ng g^{-1}. The highest PFOS and PFOA concentration was found in a pudding sample and an egg sample, respectively. PFOS concentrations in fish and seafood ranged from <LOQ to 0.303 ng g^{-1}, and PFOA from <LOQ to 0.064 ng g^{-1}. The highest PFOS and PFOA concentration was found in canned anchovies and raw shrimp, respectively. PFOS concentrations in fruit samples ranged from <LOQ to 0.029 ng g^{-1}, and PFOA from <LOQ to 0.022 ng g^{-1}. The highest PFOS and PFOA concentration was found in cantaloupe and canned mixed fruit sample, respectively. PFOS concentrations in grains ranged from <LOQ to 0.204 ng g^{-1}, and PFOA from

<LOQ to 0.077 ng g^{-1}. the highest PFOS and PFOA concentration was found in a cereal and a baking mix sample, respectively. PFOS concentrations in meat products ranged from <LOQ to 1.89 ng g^{-1}, and PFOA from <LOQ to 0.083 ng g^{-1}. When concentrations below the LOQ were assigned a value of zero, the mean and median PFOS concentrations were 0.102 and 0.025 ng g^{-1}, respectively and the mean and median PFOA concentrations were 0.034 and 0.035 ng g^{-1}, respectively. When concentrations below the LOQ were assigned ½ LOQ, the mean and GM PFOS concentrations were 0.103 and 0.030 ng g^{-1}, respectively, and the mean and GM PFOA concentrations were 0.035 and 0.026 ng g^{-1}, respectively. The highest PFOS and PFOA concentration was found in a beef liver sample and a hotdog sample, respectively. PFOS concentrations in vegetables ranged from <LOQ to 0.036 ng g^{-1}, and PFOA from <LOQ to 0.064 ng g^{-1}. The highest PFOS and PFOA concentration was found in a readymade vegetable soup sample and a tofu sample, respectively. PFOS concentrations in infant formula ranged from <LOQ to 0.030 ng g^{-1}, and PFOA from <LOQ to 0.045 ng g^{-1}. The highest PFOS and PFOA concentration was found in a rice cereal sample and a homogenate food sample, respectively.

Overall, meat products contained the highest mean concentrations of PFOS and PFOA, and the results were similar to those reported in previous studies on PFASs in food samples [19, 20, 35, 41].

PFASs in Tap Water

PFASs were also detected in drinking water samples collected from New York State. Some PFSAs (C4, C6, C8, and C10) were not detected in any samples at concentrations above the LOQ. For PFCAs, PFHxA and PFHpA were not detected in any of the tap water samples at concentrations above the LOQ; however, PFOA, PFNA, PFDA, PFUnDA, and PFDoDA were detected frequently in tap water samples. The detection frequency was 22/23 for PFOA, 19/23 for PFNA, 15/23 for PFDA, 22/23 for PFUnDA, and 11/23 for PFDoDA. PFOA concentrations in tap water ranged from <LOQ to 0.0174 ng mL^{-1} (17.4 pg mL^{-1}); PFNA concentrations ranged from <LOQ to 38.4 pg mL^{-1}; PFDA concentrations ranged from <LOQ to 202 pg mL^{-1}; PFUnDA ranged from <LOQ to 2100 pg mL^{-1}; PFDoDA ranged from <LOQ to 418 pg mL^{-1}. When concentrations below the LOQ were assigned a value of zero, the mean ± SD concentration was 7.0 ± 3.8 pg mL^{-1} for PFOA, 7.7 ± 8.1 pg mL^{-1} for PFNA, 15.0 ± 42.0 pg mL^{-1} for PFDA, 215 ± 420 pg mL^{-1} for PFUnDA, and 40.4 ± 90.1 pg mL^{-1} for PFDoDA. When concentrations below the LOQ were assigned a value of ½ LOQ, the mean and geometric mean concentration was 7.2 and 6.4 pg mL^{-1} for PFOA, 8.3 and 6.4 pg mL^{-1} for PFNA, 18.3 and 8.9 pg mL^{-1} for PFDA, 215 and 104 pg mL^{-1} for PFUnDA, and 43.0 and 14.4 pg mL^{-1} for PFDoDA. Table **11** shows the

concentrations of PFASs in drinking water analyzed in this study.

Table 11. Concentrations of PFOA, PFNA, PFDA, PFUnDA, and PFDoDA (pg mL^{-1}) in drinking water samples from New York.

	PFOA	PFNA	PFDA	PFUnDA	PFDoDA
ALB1	8.40	38.4	202	2100	418
ALB2	5.12	5.42	ND	75.2	ND
ALB3	7.57	4.96	ND	156	ND
ALB4	3.24	7.06	ND	103	12.1
ALB5	7.64	2.67	ND	116	ND
ALB6	5.07	6.16	12.8	161	39.0
ALB7	2.56	2.02	ND	82.1	21.4
ALB8	ND	ND	ND	100	ND
ALB9	4.91	7.61	20.3	258	94.1
NYC1	11.0	8.76	ND	45.2	ND
NYC2	17.4	10.8	ND	102	ND
NYC3	5.01	8.35	ND	31.1	ND
NYC4	7.24	5.54	ND	45.1	ND
BUF1	8.89	7.22	14.0	188	13.0
BUF2	3.32	ND	ND	ND	ND
BUF3	4.85	2.27	ND	10.0	ND
BUF4	2.39	7.56	ND	158	59.5
BUF5	10.9	ND	ND	98.4	90.9
BUF6	8.55	ND	ND	61.8	14.9
BUF7	11.8	14.5	32.8	330	130
BUF8	5.48	15.5	20.2	263	ND
BUF9	10.6	13.6	25.8	257	ND
BUF10	8.20	8.54	18.2	205	35.8

PFASs in Indoor Dust

Twenty-three indoor dust samples were analyzed for 12 PFASs. PFOS, PFOA, PFDS, PFHxA, PFNA, PFDA, and PFDoDA were detected in all 23 samples at concentrations above the LOQ, whereas PFHxS was detected in 22 of the 23 samples. PFHpA was not detected in any of the 23 samples at concentrations above the LOQ, and PFOSA was detected in 3 of the 23 samples. Fig. (**3**) shows total concentrations of PFASs in each indoor dust collected from Albany, NY.

Concentrations ranged from <LOQ to 14.4 ng g^{-1} for PFBS, from <LOQ to 3417 ng g^{-1} for PFHxS, from 11 to 2920 ng g^{-1} for PFOS, from 0.81 to 274 ng g^{-1} for PFDS, from <LOQ to 15.5 ng g^{-1} for PFOSA, from 7.7 to 4460 ng g^{-1} for PFOA, from 1.0 to 603 ng g^{-1} for PFNA, from 1.8 to 1667 ng g^{-1} for PFDA, from <LOQ to 282 for PFUnDA, and from 0.86 to 1120 ng g^{-1} for PFDoDA. Overall, the total PFAS concentrations ranged from 59.3 to 9070 ng g^{-1} in all samples. The overall mean ± SD for concentration of total PFASs in indoor dust samples was 1441 ± 2300 ng g^{-1}. Fig. (**4**) shows the composition profiles of PFASs in indoor dust samples collected from Albany, NY. PFOS and PFOA were the predominant PFASs in indoor dust. However, PFHxS also contributed significantly to total PFAS concentrations in some samples. A recent study showed that carpet and carpet treatments contain high levels of PFHxS [87].

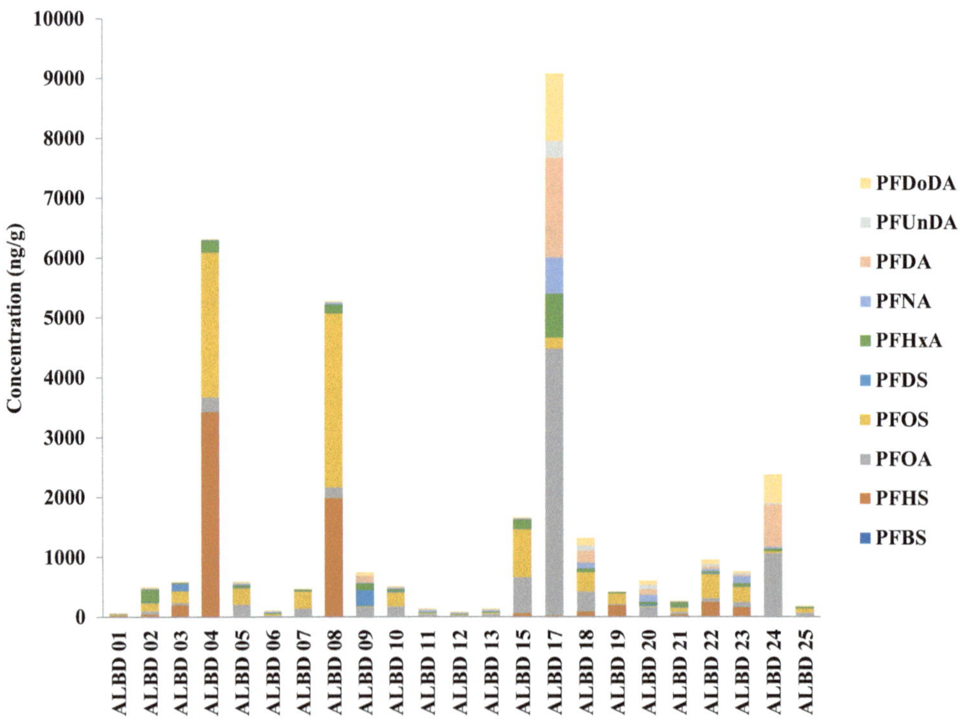

Fig. (3). Total concentrations (ng g^{-1} dw) of PFASs in indoor dust samples collected from Albany, NY.

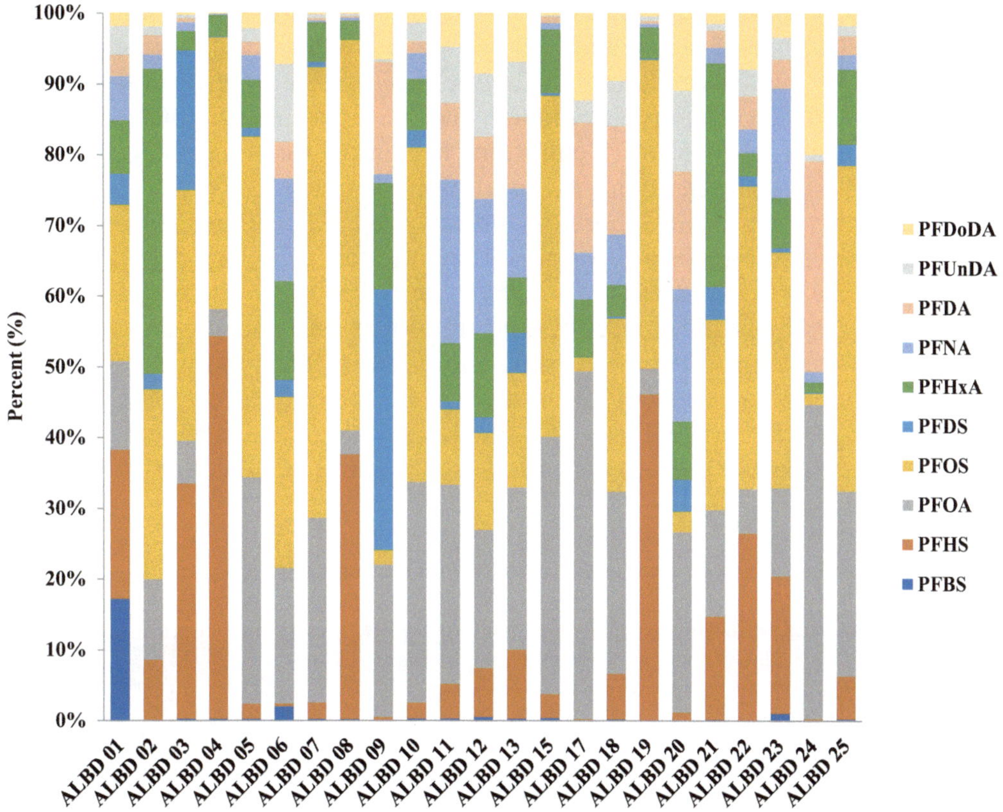

Fig. (4). Composition profiles of PFASs in indoor dust samples collected from Albany, NY.

Human Exposure to PFASs through Environmental Exposures

Human exposure doses to PFASs through food, drinking water, and indoor dust were calculated for nine age groups as follows; 0 - 3 months, 3 - 6 months, 6 - 12 months, 1 - 3 years, 3 - 6 years, 6 - 11 years, 11 - 16 years, 16 - 21 years, and >21 years. According to the *Child-Specific Exposure Factors Handbook* [115], the average duration of breastfeeding in the U.S. was 11.4 months (SD = 5.8), and the percentage of "infants still breast-fed" at 3, 6, and 12 months of age were 100%, 80%, and 38%, respectively. In this assessment, the diet sources for the age groups of 0 - 3, 3 - 6, and 6 - 12 months were considered to be breast milk and infant food. The total PFAS exposure doses for these three age groups were calculated as the sum of exposures from tap water, indoor dust, air inhalation, infant food, and breast milk. For the assessment, concentrations of PFASs in breast milk analyzed previously in our laboratory were used [43]. Tao *et al.* [43] determined nine PFASs in 45 human breast milk samples collected in 2004 from Massachusetts, U.S. PFOS and PFOA were the predominant PFASs found at respective mean and median concentrations of 131 and 106 pg mL^{-1} for PFOS, and 43.8 and 36.1 pg mL^{-1} for PFOA. The mean and median PFOS and PFOA concentrations determined in human breast milk from Massachusetts were used in the calculation of exposure of infants below 1 year. Very few studies have reported PFAS concentrations in indoor and outdoor air samples. In this assessment, the data for PFAS concentrations in outdoor air were obtained from the reported PFAS concentrations in air samples from Albany, NY, and the samples were analyzed in our laboratory previously [121]; the respective mean and median concentrations were 1.70 and 1.42 pg m^{-3} for PFOS, and 3.16 and 2.86 pg m^{-3} for PFOA. Concentrations of PFOS and PFOA in indoor air were approximately 20 times the concentrations found in outdoor air, as has been reported by Shoeib *et al.* [76]. According to Lorber (2007), the time spent by children in indoor and outdoor activity was 19 and 5 h day^{-1}, respectively, and for adults it was 21 h day^{-1} and 3 h day^{-1}. Table **12** lists the daily ingestion rates of each category of exposure sources for various age groups (calculated or cited from [115, 116]).

Table 12. Daily ingestion rates of different exposure sources for various age groups.

Overall Intake (Per Capita)	**Age**	**0 to 3 m**	**3 to 6 m**	**6 to 12 m**	**0 to 1 Year**	**1 to 3 Years**	**3 to 6 Years**	**6 to 11 Years**	**11 to 16 Years**	**16 to 21 Years**	**>21 Years**
Tap water (mL kg^{-1} day^{-1})	Mean	50	52	41	48.25	23	18	14	10	9	13
	95th	218.5	159	126	180.5	65.5	61	43	32	31.5	40
Beverages (mL kg^{-1} day^{-1})	Mean				18.32	13.35	10.05	5.88	4.04	5.49	4.55
	95th				54.96	40.05	30.15	17.64	12.12	16.47	13.65
Soil and Dust (g kg^{-1} day^{-1})	Mean	0.01121	0.00811	0.00652	0.00879	0.00794	0.00538	0.00314	0.00176	0.0014	0.00063
	95th	0.01869	0.01351	0.01087	0.01465	0.01587	0.01075	0.00629	0.00352	0.00279	0.00126
Inhalation (m^3 kg^{-1} day^{-1})	Mean	0.664	0.554	0.587	0.608	0.567	0.543	0.377	0.268	0.228	0.185
	95th	1.201	0.824	1	0.989	0.908	0.751	0.522	0.386	0.344	0.245
Fruits (g kg^{-1} day^{-1})	Mean				6.2	7.8	4.6	2.3	0.9	0.9	1.15
	95th				23	21.3	14.9	8.7	3.5	3.5	4.05
Vegetables (g kg^{-1} day^{-1})	Mean				5	6.7	5.4	3.7	2.3	2.3	2.55
	95th				16.2	15.6	13.4	10.4	5.5	5.5	6
Fish and Seafood (g kg^{-1} day^{-1})	Mean				0.03	0.26	0.24	0.21	0.13	0.13	0.225
	95th				0	1.2	1.4	1.3	0.7	0.7	1.4
Meat (g kg^{-1} day^{-1})	Mean				1.2	4	3.9	2.8	2	2	1.6
	95th				5.4	10	8.5	6.4	4.7	4.7	3.6

(Table 12) cont.....

Overall Intake (Per Capita)	Age	0 to 3 m	3 to 6 m	6 to 12 m	0 to 1 Year	1 to 3 Years	3 to 6 Years	6 to 11 Years	11 to 16 Years	16 to 21 Years	>21 Years
Dairy products (g kg^{-1} day^{-1})	Mean				10.1	43.2	24	12.9	5.5	5.5	3.4
	95th				43.2	94.7	51.1	31.8	16.4	16.4	9.95
Grains (g kg^{-1} day^{-1})	Mean				3.1	6.4	6.2	4.4	2.4	2.4	1.95
	95th				9.5	12.4	11.1	8.2	5	5	4.05
Condiments (g kg^{-1} day^{-1})	Mean				2.2	1.19	0.81	0.47	0.26	0.21	0.19
	95th				6.6	3.57	2.43	1.41	0.78	0.63	0.57
Infant Food (g kg^{-1} day^{-1})	Mean	18	28	56							
	95th	50.5	65	134							
Breastmilk (mL kg^{-1} day^{-1})	Mean	145	110	83							
	95th	200	150	130							

Assessment of Exposure to PFOS

Table **13** shows calculated human exposure dose to PFOS from different sources. The sources of exposure include tap water, indoor dust, air inhalation, infant food, and human milk for the three age groups that are less than 1-year old (0 - 3 months, 3 - 6 months, and 6 - 12 months). For the six age groups above 1 year (1 to 3 years, 3 to 6 years, 6 to 11 years, 11 to 16 years, 16 to 21 years, and older than 21 years), tap water, beverages, indoor dust, air inhalation, fruits, vegetables, fish & seafood, meat & poultry, dairy, grains, and condiments were included in the exposure calculation.

Table 13. PFOS exposure dose (ng kg^{-1} bw day^{-1}) and total exposure (ng day^{-1}) calculated for various age groups in the U.S.

Dose (ng kg^{-1} bw day^{-1})	PFOS <LOQ = 0				PFOS <LOQ = ½ LOQ			
Exposure (ng day^{-1})	**Mean**	**Median**	**95th mean**	**95th median**	**Mean**	**GM**	**95th mean**	**95th GM**
Total dose (0-3 m)	23.7	17.8	34.5	25.6	23.7	17.0	34.5	24.3
Total exposure (0-3 m)	128	95.9	186	138	128	91.8	186	131
Total dose (3-6 m)	18.1	13.7	26.2	19.5	18.1	13.1	26.2	18.6
Total exposure (3-6 m)	134	101	193	145	134	96.9	193	138
Total dose (6-12 m)	14.5	11.0	23.8	18.1	14.5	10.5	23.8	17.4
Total exposure (6-12 m)	133	101	219	167	133	97.0	219	160
Total dose (1-3 y)	5.02	2.81	10.62	5.95	5.11	2.32	10.86	5.06

(Table 13) cont.....

Dose (ng kg^{-1} bw day^{-1})	PFOS <LOQ = 0				PFOS <LOQ = ½ LOQ			
Exposure (ng day^{-1})	Mean	Median	95th mean	95th median	Mean	GM	95th mean	95th GM
Total exposure (1-3 y)	63.2	35.4	134	75.0	64.4	29.3	137	63.8
Total dose (3-6 y)	3.50	1.86	7.27	3.89	3.56	1.57	7.45	3.37
Total exposure (3-6 y)	65.0	34.7	135	72.3	66.2	29.2	138.7	62.7
Total dose (6-11 y)	2.11	1.09	4.56	2.41	2.15	0.93	4.67	2.13
Total exposure (6-11 y)	67.0	34.8	145	76.7	68.2	29.7	149	67.0
Total dose (11-16 y)	1.19	0.58	2.65	1.34	1.21	0.51	2.72	1.21
Total exposure (11-16 y)	67.5	33.1	150	76.0	68.9	28.8	154	68.7
Total dose (16-21 y)	1.06	0.52	2.39	1.20	1.09	0.48	2.47	1.16
Total exposure (16-21 y)	75.7	36.9	171	85.9	77.7	34.0	177	82.8
Total dose (>21 y)	0.65	0.31	1.51	0.75	0.68	0.32	1.59	0.79
Total exposure (>21 y)	51.6	24.8	120	59.1	53.5	25.3	125	62.7

PFOS Exposure in Infants below 1-Year Old

For infants between 0 and 3 months, the average total dose and average total exposure were 23.7 ng kg^{-1} bw day^{-1} and 128 ng day^{-1}, respectively. Tap water and air inhalation contributed below 1% of the total exposure, whereas infant food, indoor dust, and breast milk contributed 1%, 19%, and 80%, respectively, of the total exposure. For infants of ages between 3 and 6 months, the average total dose and average total exposure were 18.1 ng kg^{-1} bw day^{-1} and 134 ng day^{-1}, respectively. Tap water and air inhalation contributed less than 1% of the total exposure, whereas infant food, indoor dust, and breast milk contributed 3%, 17%, and 80% of the total exposures. For infants of ages between 6 and 12 months, the average total dose and average total exposure were 14.5 ng kg^{-1} bw day^{-1} and 133 ng day^{-1}, respectively. Tap water and air inhalation contributed less than 1% of the total exposure, whereas infant food, indoor dust, and breast milk contributed 7%, 18%, and 75% of the total exposure. The total exposure (ng day^{-1}) was similar across all three age groups below 1 year. Overall, breast milk is the major contributor to PFOS exposure in infants, followed by indoor dust.

PFOS Exposure in Children, Adolescents, and Adults

The sources of PFAS exposure in children are tap water, beverages, indoor dust, air inhalation, fruits, vegetables, fish & seafood, meat & poultry, dairy products, grains, and condiments. The total average intake dose of PFOS was 5.02 ng kg^{-1} bw day^{-1}, for children between 1 and 3 years. Indoor dust was the predominant source of PFOS exposure in this age group, accounting for 61% of the total exposure, followed by dairy (19%), meat & poultry (8%), and grains (5%).

The total average intake of PFOS was 3.5 ng kg^{-1} bw day^{-1}, for children of 3 to 6 years. Indoor dust was the predominant source of PFOS exposure in children of 3 to 6 years, accounting for 59% of the total exposure, followed by dairy (15%), meat & poultry (11%), and grains (7%).

The total average intake dose for PFOS was 2.11 ng kg^{-1} bw day^{-1}, for children of 6 to 11 years. Indoor dust was the predominant source of PFOS exposure, accounting for 57% of the total exposure, followed by dairy (13%), meat & poultry (13%), and grains (9%).

The total average PFOS intake was 1.19 ng kg^{-1} bw day^{-1}, for adolescents from 11 to 16 years. Indoor dust was the predominant source of PFOS exposure for this age group, accounting for 57% of the total exposure, followed by meat & poultry (17%), dairy (10%), and grains (8%).

The total average intake for PFOS was 1.06 ng kg^{-1} bw day^{-1}, for adolescents of 16 to 21 years. Indoor dust was the predominant source of PFOS exposure in adolescents of this age group, accounting for 50% of the total exposure, followed by meat & poultry (19%), dairy (11%), and grains (9%).

The total average intake dose for PFOS was 0.65 ng kg^{-1} bw day^{-1}, for adults > 21 years. Indoor dust was the predominant source of PFOS exposure in adults > 21 years old, accounting for 36% of the total exposure, followed by meat & poultry (24%), grains (12%), and dairy (10%).

In general, the total dose of PFAS exposure in children of 1 to 3 years was 5.02 ng kg^{-1} bw day^{-1}, which was 1/3 of that calculated for infants between 6 to 12 months (14.5 ng kg^{-1} bw day^{-1}). The total PFOS exposure in children of 1 to 3 years was half of the value calculated for infants of 6 to 12 months (63.2 ng day^{-1}*versus* 133 ng day^{-1}). These results suggest that PFOS exposure drops significantly with the cessation of breast-feeding. Breast milk is a significant source of PFOS exposure in infants and children. With the increase in age, the total PFAS exposure remained uniform at ~60 ng day^{-1}, until 16 years of age. The total PFOS exposure increased to 75.7 ng day^{-1} in adolescents (16 to 21 years), and then dropped to 51.6 ng day^{-1} in adults (> 21 years). The exposure dose of PFOS gradually decreased from 5.02 ng kg^{-1} bw day^{-1} in children (1 to 3 years) to 0.65 ng kg^{-1} bw day^{-1} for adults (> 21 years). The results indicate that at a constant exposure rate, body growth contributes to "dilution" of internal PFOS doses. The overall effect of high PFOS "loading" *via* breast milk and the "dilution" by body mass would explain higher PFOS concentration in the blood of children than in adults. Indoor dust ingestion was the major source of PFOS exposure in children and adults. However, the contribution of indoor dust to PFOS exposure decreases with age. The reason for that is, first, the ingestion rate of indoor dust decreases with age (*i.e.*, children ingest more dust than adults); second, the age-dependent increment in body weight dilutes the internal PFOS exposure dose. Dairy products are the second largest contributors of PFOS exposure and meat including poultry is the third largest source of PFOS exposure in children. In adults, meat and poultry are the second largest sources of PFOS exposure. This reflects the consumption pattern, in which children consume more dairy products than adults, and adults consume more meat and poultry than children.

The results of this investigation are similar to the findings of Egeghy and Lorber [113]. Egeghy and Lorber [113] used median PFOS intakes calculated from studies published over the past decade from several countries, and summed the doses. The estimated doses of PFOS exposure were 160 ng day^{-1} and 2200 ng day^{-1} for adults and 50 and 640 ng day^{-1} for children under typical and high exposure scenarios, respectively. Fromme *et al.* [90] summarized studies reporting the concentrations of PFASs in various media, and calculated the average PFOS exposure to be 1.6 ng kg^{-1} bw day^{-1}. In our calculations, the PFOS exposures in adults were about half the doses reported in other two studies. The samples analyzed in our study were collected within the recent 3 to 6 years. Due to the phase out of PFOS production by the 3M Company in the U.S., PFOS concentrations in human blood have decreased in the past ten years [120]. Fig. (**5**) shows the source profile in percentage for human exposure to PFOS among nine age groups in the U.S.

PFOS Exposure Dose Calculation from Biomonitoring Data

To calculate PFAS exposure for the U.S. general population, data were obtained from the NHANES Fourth National Exposure Report (for samples collected in 2009-2010), Updated Tables, March 2013 [120]. In this report, age groups were simplified as 12-19 years and >20 years. The 50th and 95th percentile of serum PFOS concentrations for both age groups were used in the calculation (Table **14**).

Table 14. Serum PFOS concentrations (ng mL^{-1}) reported in the CDC's NHANES Fourth Report (serum samples collected in 1999-2010).

Age Group	GM (95% CI)	Selected Percentiles (95% CI)				Sample Size
		50th	75th	90th	95th	
12-19 years						
99-00	29.1 (26.2-32.4)	29.5 (26.9-34.2)	39.0 (35.9-45.0)	52.7 (46.0-56.2)	57.4 (52.7-66.5)	543
03-04	19.3 (17.5-21.4)	19.9 (17.8-22.0)	27.1 (23.7-30.2)	36.5 (28.6-45.6)	42.6 (35.1-52.1)	640

(Table 14) cont.....

Age Group	GM (95% CI)	Selected Percentiles (95% CI)				Sample Size
		50th	75th	90th	95th	
05-06	15.0 (14.3-15.7)	14.9 (13.6-16.6)	22.7 (19.7-24.9)	30.6 (27.8-34.1)	38.5 (33.0-44.6)	640
07-08	11.3 (10.3-12.3)	11.3 (10.3-13.0)	15.9 (15.1-17.7)	21.7 (17.7-28.2)	28.0 (22.0-32.2)	357
09-10	6.84 (5.81-8.06)	6.90 (6.00-8.40)	10.7 (8.90-12.5)	14.4 (12.4-18.1)	18.1 (13.5-26.0)	364
20 years and older						
99-00	30.6 (27.1-34.4)	30.3 (27.9-33.9)	44.4 (37.9-48.0)	58.0 (50.1-75.7)	78.0 (59.9-107)	1019
03-04	20.9 (19.3-22.5)	21.4 (19.8-22.8)	30.4 (28.1-33.0)	42.7 (35.7-53.3)	57.8 (45.7-69.4)	1454
05-06	17.4 (16.2-18.7)	18.0 (17.1-19.4)	27.8 (25.3-30.8)	40.2 (35.6-44.5)	49.6 (42.8-60.7)	1480
07-08	13.5 (12.4-14.6)	14.0 (13.0-15.5)	21.7 (19.5-24.6)	33.9 (29.9-39.0)	42.8 (37.3-50.3)	1743
09-10	9.72 (8.45-11.2)	10.1 (8.90-11.2)	15.7 (13.4-18.4)	25.3 (19.3-32.9)	34.1 (23.4-52.7)	1869

Table 15. Model parameters used in the assessment of human exposure to PFOS from the 2009-2010 NHANES data and calculated exposure doses. DP values are in ng kg^{-1} bw day^{-1} and CP values are in ng mL^{-1}.

Age Group						
12 - 19 years	kP (day^{-1})	Vd (mL kg^{-1} bw)	CP (50th)	CP (95th)	DP (50th)	DP (95th)
05-06	0.00039	200	14.9	38.5	1.2	3.0
07-08	0.00039	200	11.3	28.0	0.9	2.2
09-10	0.00039	200	6.9	18.1	0.5	1.4
20 years and older						
05-06	0.00039	200	18.0	49.6	1.4	3.9
07-08	0.00039	200	14.0	42.8	1.1	3.3
09-10	0.00039	200	10.1	34.1	0.8	2.7

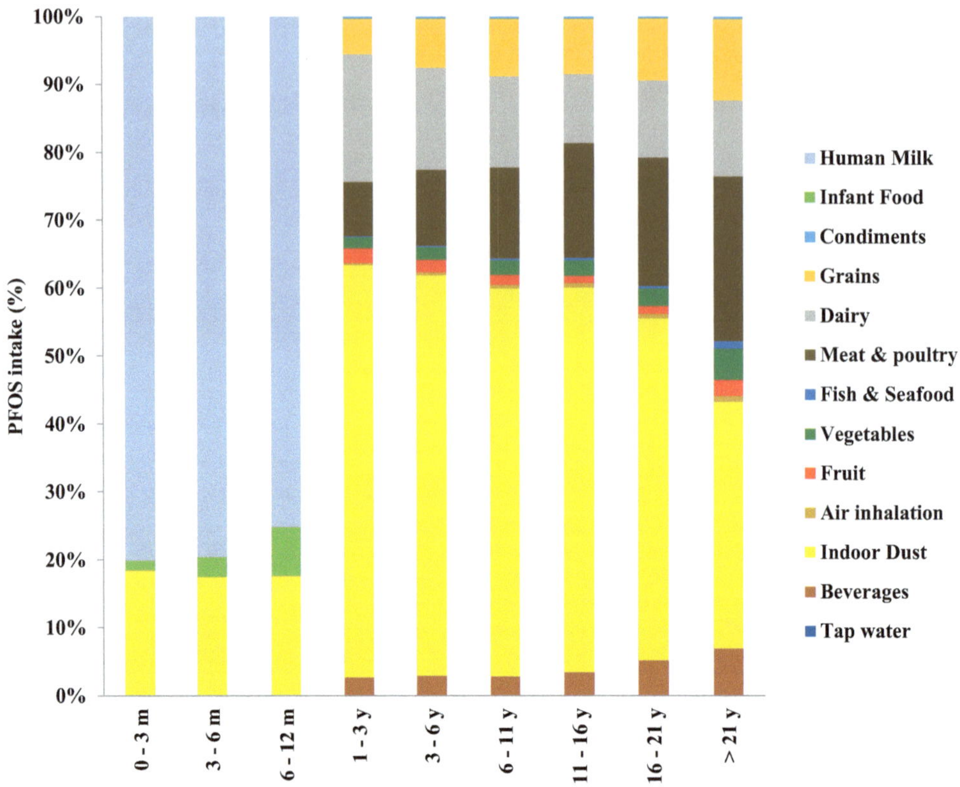

Fig. (5). Source exposure profile (in percentage) for human exposure to PFOS among nine age groups in the U.S.

The calculated human exposure dose to PFOS from the NHANES biomonitoring data (Table **15**) showed that 50th and 95th percentile values were 0.5 and 1.4 ng kg^{-1} bw day^{-1} for the age group 12-19 years, and 0.8 and 2.7 ng kg^{-1} bw day^{-1} for the age group >20 years. Since the model is only validated for adults, the exposure doses calculated from environmental sources for adult population can be compared. The calculated mean and 95th percentile dose for PFOS intakes were 0.65 and 1.51 ng kg^{-1} bw day^{-1}, respectively, which are similar to the modeled exposure dose for the U.S. general population.

Assessment of Exposure to PFOA

Table **16** shows calculated exposure doses to PFOA from various sources of exposure. The method used in the calculation was similar to that described for PFOS. The sources of exposure considered for this assessment include tap water, indoor dust, air inhalation, infant food, and human milk for the three age groups below 1-year old (0-3, 3-6, and 6-12 months). For the six age groups above 1 year (1-3, 3-6, 6-11, 11-16, 16-21, and > 21 years), tap water, beverages, indoor dust, air inhalation, fruits, vegetables, fish and seafood, meat including poultry, dairy, grains, and condiments were included in the assessment of sources of exposure.

Table 16. PFOA exposure dose (ng kg^{-1} bw day^{-1}) and total exposure (ng day^{-1}) calculated for various age groups in the U.S.

Dose (ng kg^{-1} bw day^{-1})	PFOA <LOQ = 0				PFOA <LOQ = ½ LOQ			
Exposure (ng day^{-1})	**Mean**	**Median**	**95th mean**	**95th median**	**Mean**	**GM**	**95th mean**	**95th GM**
Total dose (0-3 m)	10.9	6.68	17.6	10.6	11.0	6.85	17.8	11.0
Total exposure (0-3 m)	58.9	36.1	94.8	57.3	59.4	37.0	96.2	59.4
Total dose (3-6 m)	8.41	5.14	13.3	7.87	8.54	5.41	13.6	8.46
Total exposure (3-6 m)	62.2	38.0	98.2	58.3	63.2	40.0	101	62.6
Total dose (6-12 m)	6.93	3.93	12.1	6.67	7.18	4.51	12.7	8.02
Total exposure (6-12 m)	63.7	36.2	111	61.4	66.1	41.5	117	73.8
Total dose (1-3 y)	4.32	2.03	9.21	4.55	4.55	2.09	9.76	4.73
Total exposure (1-3 y)	54.4	25.6	116	57.4	57.3	26.4	123	59.6
Total dose (3-6 y)	2.96	1.40	6.35	3.17	3.11	1.44	6.72	3.31
Total exposure (3-6 y)	55.1	26.1	118	59.0	57.9	26.9	125	61.5
Total dose (6-11 y)	1.78	0.86	3.92	2.02	1.87	0.88	4.17	2.11
Total exposure (6-11 y)	56.5	27.3	125	64.4	59.4	28.0	133	67.1
Total dose (11-16 y)	1.00	0.49	2.28	1.21	1.05	0.50	2.42	1.25
Total exposure (11-16 y)	57.0	27.8	129	69.0	59.8	28.5	137	71.2
Total dose (16-21 y)	0.88	0.45	2.05	1.16	0.93	0.47	2.21	1.22
Total exposure (16-21 y)	63.0	32.4	147	83.2	66.9	33.9	158	87.5
Total dose (>21 y)	0.57	0.35	1.40	0.92	0.62	0.37	1.53	0.98
Total exposure (>21 y)	45.4	28.0	111	72.7	49.0	29.5	121	77.5

PFOA Exposure to Infants Below 1 Year Old

For infants of 0 - 3 months, the average total dose and average total exposure dose to PFOA were 10.9 ng kg^{-1} bw day^{-1} and 58.9 ng day^{-1}, respectively. Air inhalation contributed less than 1% of the total PFOA exposure, whereas tap water, infant food, indoor dust, and human breast milk contributed 3%, 3%, 36%, and 58% of the total PFOA exposures. For infants of 3 - 6 months, the average total dose and average total exposure dose to PFOA were 8.41 ng kg^{-1} bw day^{-1} and 62.2 ng day^{-1}, respectively. Air inhalation contributed less than 1% of the total exposure, whereas tap water, infant food, indoor dust, and human breast milk contributed

4%, 5%, 34%, and 57% of the total exposures. For infants of 6 - 12 months, the average total dose and average total exposure dose were 6.93 ng kg^{-1} bw day^{-1} and 63.7 ng day^{-1}, respectively. Air inhalation contributed less than 1% of the total exposure, whereas tap water, infant food, indoor dust, and human breast milk contribute 4%, 13%, 32%, and 51% of the total exposure. The total exposure (ng day^{-1}) was similar across all three age groups below 1 year. Overall, breast milk is the major contributor to PFOA exposure in infants, followed by indoor dust.

PFOA Exposure in Children, Adolescents, and Adults

Similar to that described for PFOS, the sources of exposure considered were tap water, beverages, indoor dust, air inhalation, fruits, vegetables, fish and seafood, meat and poultry, dairy, grains, and condiments. The method used in the calculation was similar to that described above, for infants. For children of 1 to 3 years, the total average PFOA dose was 4.32 ng kg^{-1} bw day^{-1}, and the total exposure dose was 54.4 ng day^{-1}. Indoor dust was the predominant source of PFOA exposure in children of 1 to 3 years, accounting for 62% of the total exposure, followed by dairy products (18%). For children of 3 to 6 years, the total average PFOA dose was 2.96 ng kg^{-1} bw day^{-1}, and the total exposure was 55.1 ng day^{-1}. Indoor dust was the predominant source of PFOA exposure in children of 3 to 6 years, accounting for 61% of the total exposure, followed by dairy products (15%). For children of 6 to 11 years, the total average PFOA dose was 1.78 ng kg^{-1} bw day^{-1}, and the total exposure was 56. 5 ng day^{-1}. Indoor dust was the predominant source of PFOA exposure in children of 6 to 11 years, accounting for 59% of the total exposure, followed by dairy products (13%). For children aged 11 to 16 years, the total average PFOA dose was 1.00 ng kg^{-1} bw day^{-1}, and the total exposure was 56.9 ng day^{-1}. Indoor dust was the predominant source of PFOA exposure in children of 11 to 16 years, accounting for 59% of the total exposure, followed by dairy products (10%), tap water (7%), and meat & poultry (7%).

For adolescents, the total average PFOA dose was 0.88 ng kg^{-1} bw day^{-1}, and the total exposure was 62.9 ng day^{-1}. Indoor dust was the predominant source of PFOA exposure in population of 16 to 21 years, accounting for 53% of the total exposure, followed by dairy products (11%), meat & poultry (8%), tap water (7%), grains (7%), and beverages (7%).

For adults > 21 years old, the total average PFOA dose was 0.57 ng kg^{-1} bw day^{-1}, and the total exposure was 45.4 ng day^{-1}. Indoor dust was the predominant source of PFOA exposure in adults, accounting for 36% of the total exposure, followed by tap water (15%), dairy products (10%), meat & poultry (9%), beverages (9%), grains (8%), and vegetables (6%).

For age groups of 1 year and older, the daily intake of PFOA (~50 ng day^{-1}) did not change significantly in comparison with what was found for infants below 1-year old (~60 ng day^{-1}). Unlike PFOS, which showed a reduction in daily intake rates with age, PFOA daily intake was constant at ~50 ng day^{-1}, for all age groups. For children, indoor dust was the major source for PFOA exposure, followed by diet. Among foodstuffs, dairy products were the major source of PFOA exposure. The contribution of indoor dust to PFOA exposure decreased with age. For adults above 21 years, diet was the major source of PFOA exposure. Among foodstuffs, dairy products, grains, meat and poultry, and beverages contributed ~10% each of the exposure to PFOA. Fig. (**6**) shows the source profile in percentage for human exposure to PFOA among the nine age groups in the U.S.

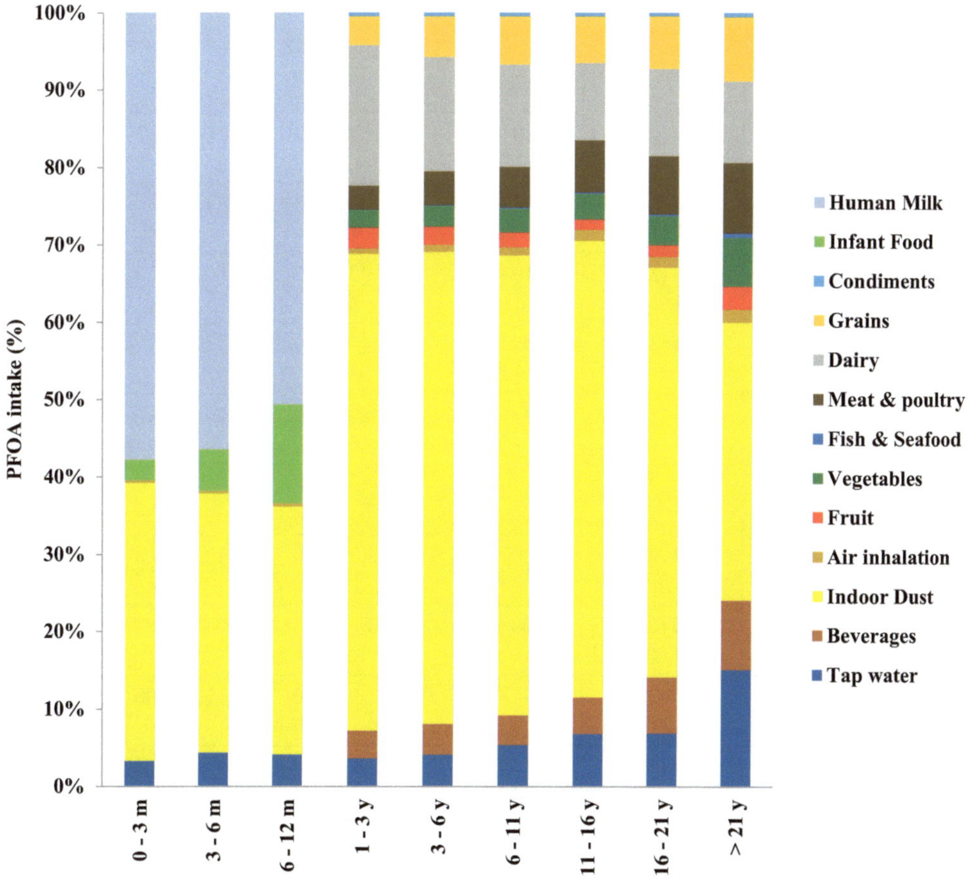

Fig. (6). Source exposure profile (in percentage) for human exposure to PFOA among the nine age groups in the U.S.

PFOA exposure dose calculation from biomonitoring data

To calculate the human exposure dose for PFOA, data were obtained from the NHANES Fourth National Exposure Report, Updated Tables, March 2013 (serum data for the samples collected in 2009-2010) [120]. In this report, age groups were simplified as 12-19 years and >20 years. The 50th and 95th percentile of serum PFOA concentrations for both age groups were used in the calculation (Table **17**).

Table 17. Serum PFOA concentrations (ng mL^{-1}) reported in the NHANES Fourth Report for the U.S. general population.

Age Group	GM (95% CI)	Selected Percentiles (95% CI)				Sample Size
		50th	75th	90th	95th	
12-19 years						
99-00	5.46 (4.98-5.99)	5.60 (4.90-6.10)	6.90 (6.20-7.70)	9.40 (7.70-11.0)	11.6 (10.2-12.5)	543
03-04	3.89 (3.47-4.35)	4.00 (3.50-4.50)	5.40 (4.60-6.10)	7.00 (5.60-9.20)	8.60 (5.90-12.6)	640
05-06	3.59 (3.26-3.96)	3.80 (3.30-4.20)	5.40 (4.60-5.90)	6.90 (6.30-7.90)	8.40 (7.30-10.1)	640
07-08	3.91 (3.71-4.12)	4.00 (3.70-4.40)	5.00 (4.70-5.50)	6.10 (5.70-6.70)	7.30 (6.20-8.00)	357
09-10	2.74 (2.46-3.05)	2.90 (2.60-3.10)	3.80 (3.30-4.20)	4.80 (4.10-5.30)	5.00 (4.40-7.20)	364
20 years and older						
99-00	5.16 (4.66-5.72)	5.20 (4.70-5.70)	7.00 (6.30-7.80)	9.40 (8.20-11.1)	12.0 (10.8-14.0)	1019
03-04	3.96 (3.67-4.27)	4.10 (3.90-4.40)	5.90 (5.40-6.50)	7.80 (6.80-9.60)	9.90 (7.60-14.2)	1454
05-06	3.97 (3.51-4.49)	4.20 (3.80-4.60)	6.40 (5.40-7.50)	9.30 (7.40-11.9)	11.6 (9.00-14.8)	1480
07-08	4.15 (4.02-4.30)	4.30 (4.20-4.60)	6.10 (5.80-6.50)	8.10 (7.70-8.70)	9.80 (9.00-10.4)	1743
09-10	3.12 (2.84-3.43)	3.30 (3.00-3.60)	4.70 (4.20-5.30)	6.40 (5.40-7.50)	7.70 (6.30-10.2)	1869

Table 18. Model parameters used in the assessment of exposure to PFOA from the 2009-2010 NHANES data and calculated exposure doses. DP values are in ng kg^{-1} bw day^{-1} and CP values are in ng mL^{-1}.

Age Group						
12 - 19 years	**kP (day^{-1})**	**Vd (mL kg^{-1} bw)**	**CP (50th)**	**CP (95th)**	**DP (50th)**	**DP (95th)**
05-06	0.00080	170	3.8	8.4	0.52	1.14
07-08	0.00080	170	4.0	7.3	0.54	0.99
09-10	0.00080	170	2.9	5.0	0.39	0.68
20 years and older						
05-06	0.00080	170	4.2	11.6	0.57	1.58
07-08	0.00080	170	4.3	9.8	0.58	1.33
09-10	0.00080	170	3.0	7.7	0.41	1.05

The calculated exposure doses to PFOA based on the NHANES biomonitoring data (Table **18**) at 50th and 95th percentile values were 0.39 and 0.68 ng kg^{-1} bw day^{-1} for the age group 12-19 years, and 0.41 ng kg^{-1} bw day^{-1} and 1.05 ng kg^{-1} bw day^{-1} for the age group >20 years. The model was only validated for adults, the therefore exposure was calculated only for the adult population. On the basis of environmental monitoring, the calculated mean and 95th percentile human exposure doses for PFOA were 0.57 and 1.40 ng kg^{-1} bw day^{-1}, which were similar to those estimated from the pharmacokinetic model.

CONCLUDING REMARKS

Biomonitoring studies conducted in the U.S., by the NHANES, have shown widespread exposure of humans to PFASs. For infants below 1 year, PFOS exposure dose ranged from 14.5 to 23.7 ng kg^{-1} bw day^{-1}. Human breast milk was the major contributor to PFOS exposure in infants, followed by indoor dust. For age groups above 1 year, the total dose of PFOS exposure ranged from 0.65 ng kg^{-1} bw day^{-1} in adults to 5.02 ng kg^{-1} bw day^{-1} in children. The results indicated that PFOS exposure significantly drops after the cessation of breastfeeding. Breastfeeding results in significant "loading" of PFOS in infants and children. The exposure dose remained uniform at ~60 ng day^{-1}, until 16 years. The total PFOS exposure increased to 75.7 ng day^{-1} in adolescents (16 to 21 years), and then dropped to 51.6 ng day^{-1} in adults (>21 years). The dose of PFOS exposure gradually dropped from 5.02 ng kg^{-1} bw day^{-1} for children (1 to 3 years) to 0.65 ng kg^{-1} bw day^{-1} for adults (> 21 years). Indoor dust was the predominant source of PFOS exposure in children. Dairy product is the second largest source of PFOS exposure and meat including poultry is the third largest source of PFOS exposure in children. In adults, meat including poultry is the second largest source of PFOS exposure. The median human exposure doses to PFOS estimated from the

NHANES biomonitoring data were 0.5 ng kg^{-1} bw day^{-1} for the age group 12-19 years, and 0.8 ng kg^{-1} bw day^{-1} for the age group >20 years. The pharmacokinetic parameters are validated only for adults, and therefore the exposures were calculated only for adults. The calculated mean and 95th percentile human exposure doses for PFOS were 0.65 and 1.51 ng kg^{-1} bw day^{-1}, respectively, which were similar to the values derived from urinary biomonitoring data for the U.S. general population.

For infants below 1 year, PFOA exposure dose from ranged from 6.93 ng kg^{-1} bw day^{-1} to 10.9 ng kg^{-1} bw day^{-1}. Human breast milk was the major contributor to PFOA exposure in infants, followed by indoor dust. For age groups above 1 year, the total dose of PFOA exposure ranged from 0.57 ng kg^{-1} bw day^{-1} in adults to 4.32 ng kg^{-1} bw day^{-1} in children. The results indicated that PFOA exposure significantly drops after the discontinuation of breast-feeding. Breast-feeding results in significant "loading" of PFOA in infants and children. The total exposure remained uniform at ~55 ng day^{-1}, until 16 years. The total PFOA exposure increased to 63.0 ng day^{-1} in adolescents (16 to 21 years), and then dropped to 45.4 ng day^{-1} for adults (>21 years). The dose of PFOA exposure gradually dropped from 4.32 ng kg^{-1} bw day^{-1} for children (1 to 3 years) to 0.57 ng kg^{-1} bw day^{-1} for adults (> 21 years). Indoor dust was the predominant source of PFOA exposure in children. Dairy product was the second largest source of PFOA exposure and tap water was the third largest source for PFOA exposure in children. In adults, tap water was the second largest source of PFOA exposure. The median human exposure doses to PFOA estimated from the NHANES biomonitoring data were 0.39 ng kg^{-1} bw day^{-1} for the age group 12-19 years, and 0.41 ng kg^{-1} bw day^{-1} for the age group >20 years. The calculated mean and 95th percentile human exposure doses for PFOA in the present study were 0.57 and 1.40 ng kg^{-1} bw day^{-1}, respectively, which were similar to the values derived from the urinary biomonitoring data for the U.S. general population.

CONSENT FOR PUBLICATION

Not applicable.

CONFLICT OF INTEREST

The authors confirm that this chapter contents have no conflict of interest.

ACKNOWLEDGEMENTS

Declared none.

REFERENCES

[1] OECD. Results of survey on production and use of PFOS, PFAS and PFOA, related substances and products/mixtures containing these substances. 2005.

[2] Wang Z, DeWitt JC, Higgins CP, Cousins IT. A never-ending story of per- and polyfluoroalkyl substances (PFASs)? Environ Sci Technol 2017; 51(5): 2508-18. [http://dx.doi.org/10.1021/acs.est.6b04806] [PMID: 28224793]

[3] Giesy JP, Kannan K. Global distribution of perfluorooctane sulfonate in wildlife. Environ Sci Technol 2001; 35(7): 1339-42. [http://dx.doi.org/10.1021/es001834k] [PMID: 11348064]

[4] Prevedouros K, Cousins IT, Buck RC, Korzeniowski SH. Sources, fate and transport of perfluorocarboxylates. Environ Sci Technol 2006; 40(1): 32-44. [http://dx.doi.org/10.1021/es0512475] [PMID: 16433330]

[5] USEPA. Fact Sheet: 2010/2015 PFOA Stewardship Program 2018.

[6] Wang T, Wang Y, Liao C, Cai Y, Jiang G. Perspectives on the inclusion of perfluorooctane sulfonate into the Stockholm Convention on Persistent Organic Pollutants. Environ Sci Technol 2009; 43(14): 5171-5. [http://dx.doi.org/10.1021/es900464a] [PMID: 19708337]

[7] Kannan K, Tao L, Sinclair E, Pastva SD, Jude DJ, Giesy JP. Perfluorinated compounds in aquatic organisms at various trophic levels in a Great Lakes food chain. Arch Environ Contam Toxicol 2005; 48(4): 559-66. [http://dx.doi.org/10.1007/s00244-004-0133-x] [PMID: 15883668]

[8] Taniyasu S, Kannan K, Horii Y, Hanari N, Yamashita N. A survey of perfluorooctane sulfonate and related perfluorinated organic compounds in water, fish, birds, and humans from Japan. Environ Sci Technol 2003; 37(12): 2634-9. [http://dx.doi.org/10.1021/es0303440] [PMID: 12854699]

[9] Conder JM, Hoke RA, De Wolf W, Russell MH, Buck RC. Are PFCAs bioaccumulative? A critical review and comparison with regulatory lipophilic compounds. Environ Sci Technol 2008; 42(4): 995-1003. [http://dx.doi.org/10.1021/es070895g] [PMID: 18351063]

[10] Hagen DF, Belisle J, Johnson JD, Venkateswarlu P. Characterization of fluorinated metabolites by a gas chromatographic-helium microwave plasma detector the biotransformation of 1H, 1H, 2H, 2H-perfluorodecanol to perfluorooctanoate. Anal Biochem 1981; 118(2): 336-43. [http://dx.doi.org/10.1016/0003-2697(81)90591-1] [PMID: 7337230]

[11] Tomy GT, Budakowski W, Halldorson T, *et al.* Fluorinated organic compounds in an eastern Arctic marine food web. Environ Sci Technol 2004; 38(24): 6475-81. [http://dx.doi.org/10.1021/es049620g] [PMID: 15669302]

[12] Barber JL, Berger U, Chaemfa C, *et al.* Analysis of per- and polyfluorinated alkyl substances in air samples from Northwest Europe. J Environ Monit 2007; 9(6): 530-41. [http://dx.doi.org/10.1039/b701417a] [PMID: 17554424]

[13] Kannan K, Corsolini S, Falandysz J, *et al.* Perfluorooctanesulfonate and related fluorochemicals in human blood from several countries. Environ Sci Technol 2004; 38(17): 4489-95. [http://dx.doi.org/10.1021/es0493446] [PMID: 15461154]

[14] Sinclair E, Mayack DT, Roblee K, Yamashita N, Kannan K. Occurrence of perfluoroalkyl surfactants in water, fish, and birds from New York State. Arch Environ Contam Toxicol 2006; 50(3): 398-410. [http://dx.doi.org/10.1007/s00244-005-1188-z] [PMID: 16435086]

[15] Tao L, Kannan K, Kajiwara N, *et al.* Perfluorooctanesulfonate and related fluorochemicals in albatrosses, elephant seals, penguins, and polar skuas from the Southern Ocean. Environ Sci Technol 2006; 40(24): 7642-8.

[http://dx.doi.org/10.1021/es061513u] [PMID: 17256507]

[16] Yamashita N, Kannan K, Taniyasu S, Horii Y, Petrick G, Gamo T. A global survey of perfluorinated acids in oceans. Mar Pollut Bull 2005; 51(8-12): 658-68. [http://dx.doi.org/10.1016/j.marpolbul.2005.04.026] [PMID: 15913661]

[17] Ghisi R, Vamerali T, Manzetti S. Accumulation of perfluorinated alkyl substances (PFAS) in agricultural plants: a review. Environ Res 2018. [PMID: 30502744]

[18] Rylander C, Sandanger TM, Brustad M. Associations between marine food consumption and plasma concentrations of POPs in a Norwegian coastal population. J Environ Monit 2009; 11(2): 370-6. [http://dx.doi.org/10.1039/B811868J] [PMID: 19212595]

[19] Ericson I, Domingo JL, Nadal M, *et al.* Levels of perfluorinated chemicals in municipal drinking water from Catalonia, Spain: public health implications. Arch Environ Contam Toxicol 2009; 57(4): 631-8. [http://dx.doi.org/10.1007/s00244-009-9375-y] [PMID: 19685096]

[20] Ericson I, Martí-Cid R, Nadal M, Van Bavel B, Lindström G, Domingo JL. Human exposure to perfluorinated chemicals through the diet: intake of perfluorinated compounds in foods from the Catalan (Spain) market. J Agric Food Chem 2008; 56(5): 1787-94. [http://dx.doi.org/10.1021/jf0732408] [PMID: 18251500]

[21] Gellrich V, Brunn H, Stahl T. Perfluoroalkyl and polyfluoroalkyl substances (PFASs) in mineral water and tap water. J Environ Sci Health A Tox Hazard Subst Environ Eng 2013; 48(2): 129-35. [http://dx.doi.org/10.1080/10934529.2013.719431] [PMID: 23043333]

[22] Saito N, Harada K, Inoue K, Sasaki K, Yoshinaga T, Koizumi A. Perfluorooctanoate and perfluorooctane sulfonate concentrations in surface water in Japan. J Occup Health 2004; 46(1): 49-59. [http://dx.doi.org/10.1539/joh.46.49] [PMID: 14960830]

[23] Takagi S, Adachi F, Miyano K, *et al.* Fate of perfluorooctanesulfonate and perfluorooctanoate in drinking water treatment processes. Water Res 2011; 45(13): 3925-32. [http://dx.doi.org/10.1016/j.watres.2011.04.052] [PMID: 21628066]

[24] Schultz MM, Higgins CP, Huset CA, Luthy RG, Barofsky DF, Field JA. Fluorochemical mass flows in a municipal wastewater treatment facility. Environ Sci Technol 2006; 40(23): 7350-7. [http://dx.doi.org/10.1021/es061025m] [PMID: 17180988]

[25] Sinclair E, Kannan K. Mass loading and fate of perfluoroalkyl surfactants in wastewater treatment plants. Environ Sci Technol 2006; 40(5): 1408-14. [http://dx.doi.org/10.1021/es051798v] [PMID: 16568749]

[26] NYSDOH. Drinking Water Response Activities to Address Local Water Supply Concerns 2018.

[27] Skutlarek D, Exner M, Färber H. Perfluorinated surfactants in surface and drinking waters. Environ Sci Pollut Res Int 2006; 13(5): 299-307. [http://dx.doi.org/10.1065/espr2006.07.326] [PMID: 17067024]

[28] Quinete N, Wu Q, Zhang T, Yun SH, Moreira I, Kannan K. Specific profiles of perfluorinated compounds in surface and drinking waters and accumulation in mussels, fish, and dolphins from southeastern Brazil. Chemosphere 2009; 77(6): 863-9. [http://dx.doi.org/10.1016/j.chemosphere.2009.07.079] [PMID: 19744696]

[29] Falandysz J, Taniyasu S, Gulkowska A, Yamashita N, Schulte-Oehlmann U. Is fish a major source of fluorinated surfactants and repellents in humans living on the Baltic Coast? Environ Sci Technol 2006; 40(3): 748-51. [http://dx.doi.org/10.1021/es051799n] [PMID: 16509313]

[30] Begley TH, White K, Honigfort P, Twaroski ML, Neches R, Walker RA. Perfluorochemicals: potential sources of and migration from food packaging. Food Addit Contam 2005; 22(10): 1023-31. [http://dx.doi.org/10.1080/02652030500183474] [PMID: 16227186]

[31] Gulkowska A, Jiang Q, So MK, Taniyasu S, Lam PKS, Yamashita N. Persistent perfluorinated acids in seafood collected from two cities of China. Environ Sci Technol 2006; 40(12): 3736-41. [http://dx.doi.org/10.1021/es060286t] [PMID: 16830535]

[32] Sinclair E, Kim SK, Akinleye HB, Kannan K. Quantitation of gas-phase perfluoroalkyl surfactants and fluorotelomer alcohols released from nonstick cookware and microwave popcorn bags. Environ Sci Technol 2007; 41(4): 1180-5. [http://dx.doi.org/10.1021/es062377w] [PMID: 17593716]

[33] Tittlemier SA, Pepper K, Edwards L, Cao XL. Concentrations of perfluorooctanesulfonamides in Canadian total diet study composite food samples collected between 1992 and 2004 (vol 54, pg 8385, 2006). J Agric Food Chem 2006; 54(24): 9277-7. [http://dx.doi.org/10.1021/jf068015p]

[34] Schaider LA, Balan SA, Blum A, *et al.* Fluorinated Compounds in U.S. Fast Food Packaging. Environ Sci Technol Lett 2017; 4(3): 105-11. [http://dx.doi.org/10.1021/acs.estlett.6b00435] [PMID: 30148183]

[35] Zhang T, Sun HW, Wu Q, Zhang XZ, Yun SH, Kannan K. Perfluorochemicals in meat, eggs and indoor dust in China: assessment of sources and pathways of human exposure to perfluorochemicals. Environ Sci Technol 2010; 44(9): 3572-9. [http://dx.doi.org/10.1021/es1000159] [PMID: 20377175]

[36] Tittlemier SA, Pepper K, Seymour C, *et al.* Dietary exposure of Canadians to perfluorinated carboxylates and perfluorooctane sulfonate via consumption of meat, fish, fast foods, and food items prepared in their packaging. J Agric Food Chem 2007; 55(8): 3203-10. [http://dx.doi.org/10.1021/jf0634045] [PMID: 17381114]

[37] Fromme H, Midasch O, Twardella D, Angerer J, Boehmer S, Liebl B. Occurrence of perfluorinated substances in an adult German population in southern Bavaria. Int Arch Occup Environ Health 2007; 80(4): 313-9. [http://dx.doi.org/10.1007/s00420-006-0136-1] [PMID: 16915390]

[38] Schecter A, Colacino J, Haffner D, *et al.* Perfluorinated compounds, polychlorinated biphenyls, and organochlorine pesticide contamination in composite food samples from Dallas, Texas, USA. Environ Health Perspect 2010; 118(6): 796-802. [http://dx.doi.org/10.1289/ehp.0901347] [PMID: 20146964]

[39] Kärrman A, Harada KH, Inoue K, Takasuga T, Ohi E, Koizumi A. Relationship between dietary exposure and serum perfluorochemical (PFC) levels a case study. Environ Int 2009; 35(4): 712-7. [http://dx.doi.org/10.1016/j.envint.2009.01.010] [PMID: 19250678]

[40] Haug LS, Salihovic S, Jogsten IE, *et al.* Levels in food and beverages and daily intake of perfluorinated compounds in Norway. Chemosphere 2010; 80(10): 1137-43. [http://dx.doi.org/10.1016/j.chemosphere.2010.06.023] [PMID: 20599247]

[41] Noorlander CW, van Leeuwen SPJ, Te Biesebeek JD, Mengelers MJB, Zeilmaker MJ. Levels of perfluorinated compounds in food and dietary intake of PFOS and PFOA in the Netherlands. J Agric Food Chem 2011; 59(13): 7496-505. [http://dx.doi.org/10.1021/jf104943p] [PMID: 21591675]

[42] Macheka-Tendenguwo LR, Olowoyo JO, Mugivhisa LL, Abafe OA. Per- and polyfluoroalkyl substances in human breast milk and current analytical methods. Environ Sci Pollut R 2018. [http://dx.doi.org/10.1007/s11356-018-3483-z]

[43] Tao L, Kannan K, Wong CM, Arcaro KF, Butenhoff JL. Perfluorinated compounds in human milk from Massachusetts, U.S.A. Environ Sci Technol 2008; 42(8): 3096-101. [http://dx.doi.org/10.1021/es702789k] [PMID: 18497172]

[44] Tao L, Ma J, Kunisue T, Libelo EL, Tanabe S, Kannan K. Perfluorinated compounds in human breast milk from several Asian countries, and in infant formula and dairy milk from the United States.

Environ Sci Technol 2008; 42(22): 8597-602.
[http://dx.doi.org/10.1021/es801875v] [PMID: 19068854]

[45] Kärrman A, Ericson I, van Bavel B, *et al.* Exposure of perfluorinated chemicals through lactation: levels of matched human milk and serum and a temporal trend, 1996-2004, in Sweden. Environ Health Perspect 2007; 115(2): 226-30.
[http://dx.doi.org/10.1289/ehp.9491] [PMID: 17384769]

[46] So MK, Yamashita N, Taniyasu S, *et al.* Health risks in infants associated with exposure to perfluorinated compounds in human breast milk from Zhoushan, China. Environ Sci Technol 2006; 40(9): 2924-9.
[http://dx.doi.org/10.1021/es060031f] [PMID: 16719092]

[47] Völkel W, Genzel-Boroviczény O, Demmelmair H, *et al.* Perfluorooctane sulphonate (PFOS) and perfluorooctanoic acid (PFOA) in human breast milk: results of a pilot study. Int J Hyg Environ Health 2008; 211(3-4): 440-6.
[http://dx.doi.org/10.1016/j.ijheh.2007.07.024] [PMID: 17870667]

[48] Kärrman A, Domingo JL, Llebaria X, *et al.* Biomonitoring perfluorinated compounds in Catalonia, Spain: concentrations and trends in human liver and milk samples. Environ Sci Pollut Res Int 2010; 17(3): 750-8.
[http://dx.doi.org/10.1007/s11356-009-0178-5] [PMID: 19458971]

[49] Liu J, Li J, Zhao Y, Wang Y, Zhang L, Wu Y. The occurrence of perfluorinated alkyl compounds in human milk from different regions of China. Environ Int 2010; 36(5): 433-8.
[http://dx.doi.org/10.1016/j.envint.2010.03.004] [PMID: 20400180]

[50] Kim SK, Lee KT, Kang CS, *et al.* Distribution of perfluorochemicals between sera and milk from the same mothers and implications for prenatal and postnatal exposures. Environ Pollut 2011; 159(1): 169-74.
[http://dx.doi.org/10.1016/j.envpol.2010.09.008] [PMID: 20932617]

[51] Kang H, Choi K, Lee H-S, *et al.* Elevated levels of short carbon-chain PFCAs in breast milk among Korean women: Current status and potential challenges. Environ Res 2016; 148: 351-9.
[http://dx.doi.org/10.1016/j.envres.2016.04.017] [PMID: 27111244]

[52] Roosens L, D'Hollander W, Bervoets L, *et al.* Brominated flame retardants and perfluorinated chemicals, two groups of persistent contaminants in Belgian human blood and milk. Environ Pollut 2010; 158(8): 2546-52.
[http://dx.doi.org/10.1016/j.envpol.2010.05.022] [PMID: 20573431]

[53] Lankova D, Lacina O, Pulkrabova J, Hajslova J. The determination of perfluoroalkyl substances, brominated flame retardants and their metabolites in human breast milk and infant formula. Talanta 2013; 117: 318-25.
[http://dx.doi.org/10.1016/j.talanta.2013.08.040] [PMID: 24209347]

[54] Haug LS, Huber S, Becher G, Thomsen C. Characterisation of human exposure pathways to perfluorinated compounds--comparing exposure estimates with biomarkers of exposure. Environ Int 2011; 37(4): 687-93.
[http://dx.doi.org/10.1016/j.envint.2011.01.011] [PMID: 21334069]

[55] Thomsen C, Haug LS, Stigum H, Froshaug M, Broadwell SL, Becher G. Changes in concentrations of perfluorinated compounds, polybrominated diphenyl ethers and polychlorinated biphenyls in norwegian breast-milk during twelve months of lactation (vol 44, pg 9550-6, 2010). Environ Sci Technol 2011; 45(7): 3192.
[http://dx.doi.org/10.1021/es200682w]

[56] Sundström M, Ehresman DJ, Bignert A, *et al.* A temporal trend study (1972-2008) of perfluorooctanesulfonate, perfluorohexanesulfonate, and perfluorooctanoate in pooled human milk samples from Stockholm, Sweden. Environ Int 2011; 37(1): 178-83.
[http://dx.doi.org/10.1016/j.envint.2010.08.014] [PMID: 20880590]

[57] Llorca M, Farré M, Picó Y, Teijón ML, Alvarez JG, Barceló D. Infant exposure of perfluorinated compounds: levels in breast milk and commercial baby food. Environ Int 2010; 36(6): 584-92. [http://dx.doi.org/10.1016/j.envint.2010.04.016] [PMID: 20494442]

[58] von Ehrenstein OS, Fenton SE, Kato K, Kuklenyik Z, Calafat AM, Hines EP. Polyfluoroalkyl chemicals in the serum and milk of breastfeeding women. Reprod Toxicol 2009; 27(3-4): 239-45. [http://dx.doi.org/10.1016/j.reprotox.2009.03.001] [PMID: 19429402]

[59] Emmett EA, Shofer FS, Zhang H, Freeman D, Desai C, Shaw LM. Community exposure to perfluorooctanoate: relationships between serum concentrations and exposure sources. J Occup Environ Med 2006; 48(8): 759-70. [http://dx.doi.org/10.1097/01.jom.0000232486.07658.74] [PMID: 16902368]

[60] Hansen KJ, Johnson HO, Eldridge JS, Butenhoff JL, Dick LA. Quantitative characterization of trace levels of PFOS and PFOA in the Tennessee River. Environ Sci Technol 2002; 36(8): 1681-5. [http://dx.doi.org/10.1021/es010780r] [PMID: 11993863]

[61] USEPA. Environmental Mornitoring-Multi-City Study: water, sludge, sediment, POTW effluent and landfill leachate samples 2001.

[62] Harada K, Saito N, Sasaki K, Inoue K, Koizumi A. Perfluorooctane sulfonate contamination of drinking water in the Tama River, Japan: estimated effects on resident serum levels. Bull Environ Contam Toxicol 2003; 71(1): 31-6. [http://dx.doi.org/10.1007/s00128-003-0126-x] [PMID: 12945838]

[63] Mak YL, Taniyasu S, Yeung LWY, *et al.* Perfluorinated compounds in tap water from China and several other countries. Environ Sci Technol 2009; 43(13): 4824-9. [http://dx.doi.org/10.1021/es900637a] [PMID: 19673271]

[64] Takagi S, Adachi F, Miyano K, *et al.* Perfluorooctanesulfonate and perfluorooctanoate in raw and treated tap water from Osaka, Japan. Chemosphere 2008; 72(10): 1409-12. [http://dx.doi.org/10.1016/j.chemosphere.2008.05.034] [PMID: 18602659]

[65] Paustenbach DJ, Panko JM, Scott PK, Unice KM. A methodology for estimating human exposure to perfluorooctanoic acid (PFOA): a retrospective exposure assessment of a community (1951-2003). J Toxicol Environ Health A 2007; 70(1): 28-57. [http://dx.doi.org/10.1080/15287390600748815] [PMID: 17162497]

[66] Fujii S, Polprasert C, Tanaka S, Hong L. N. P.; Qiu, Y., New POPs in the water environment: distribution, bioaccumulation and treatment of perfluorinated compounds – a review paper. J Water Supply 2007; 56(5): 313-26. [http://dx.doi.org/10.2166/aqua.2007.005]

[67] Clare Atkinson S B. Paul Rumsby Survey of the prevalence of perfluorooctane sulphonate (PFOS), perfluorooctanoic acid (PFOA) and related compounds in drinking water and their sources: Foundation for Water Research 2008.

[68] Rostkowski P, Taniyasu S, Yamashita N, *et al.* Survey of perfluorinated compounds (PFCs) in surface waters of Poland. J Environ Sci Health A Tox Hazard Subst Environ Eng 2009; 44(14): 1518-27. [http://dx.doi.org/10.1080/10934520903263330] [PMID: 20183509]

[69] Loos R, Wollgast J, Huber T, Hanke G. Polar herbicides, pharmaceutical products, perfluorooctanesulfonate (PFOS), perfluorooctanoate (PFOA), and nonylphenol and its carboxylates and ethoxylates in surface and tap waters around Lake Maggiore in Northern Italy. Anal Bioanal Chem 2007; 387(4): 1469-78. [http://dx.doi.org/10.1007/s00216-006-1036-7] [PMID: 17200857]

[70] Mawn MP, McKay RG, Ryan TW, Szostek B, Powley CR, Buck RC. Determination of extractable perfluorooctanoic acid (PFOA) in water, sweat simulant, saliva simulant, and methanol from textile and carpet samples by LC/MS/MS. Analyst (Lond) 2005; 130(5): 670-8. [http://dx.doi.org/10.1039/b414403a] [PMID: 15852136]

[71] Hori H, Yamamoto A, Hayakawa E, *et al.* Efficient decomposition of environmentally persistent perfluorocarboxylic acids by use of persulfate as a photochemical oxidant. Environ Sci Technol 2005; 39(7): 2383-8.
[http://dx.doi.org/10.1021/es0484754] [PMID: 15871280]

[72] Fujii Y, Harada KH, Koizumi A. Occurrence of perfluorinated carboxylic acids (PFCAs) in personal care products and compounding agents. Chemosphere 2013; 93(3): 538-44.
[http://dx.doi.org/10.1016/j.chemosphere.2013.06.049] [PMID: 23932147]

[73] Harrad S, de Wit CA, Abdallah MAE, *et al.* Indoor contamination with hexabromocyclododecanes, polybrominated diphenyl ethers, and perfluoroalkyl compounds: an important exposure pathway for people? Environ Sci Technol 2010; 44(9): 3221-31.
[http://dx.doi.org/10.1021/es903476t] [PMID: 20387882]

[74] Haug LS, Huber S, Schlabach M, Becher G, Thomsen C. Investigation on per- and polyfluorinated compounds in paired samples of house dust and indoor air from Norwegian homes. Environ Sci Technol 2011; 45(19): 7991-8.
[http://dx.doi.org/10.1021/es103456h] [PMID: 21417377]

[75] Kubwabo C, Stewart B, Zhu J, Marro L. Occurrence of perfluorosulfonates and other perfluorochemicals in dust from selected homes in the city of Ottawa, Canada. J Environ Monit 2005; 7(11): 1074-8.
[http://dx.doi.org/10.1039/b507731c] [PMID: 16252056]

[76] Shoeib M, Harner T, Wilford BH, Jones KC, Zhu J. Perfluorinated sulfonamides in indoor and outdoor air and indoor dust: occurrence, partitioning, and human exposure. Environ Sci Technol 2005; 39(17): 6599-606.
[http://dx.doi.org/10.1021/es048340y] [PMID: 16190217]

[77] Strynar MJ, Lindstrom AB. Perfluorinated compounds in house dust from Ohio and North Carolina, USA. Environ Sci Technol 2008; 42(10): 3751-6.
[http://dx.doi.org/10.1021/es7032058] [PMID: 18546718]

[78] Moriwaki H, Takatah Y, Arakawa R. Concentrations of perfluorooctane sulfonate (PFOS) and perfluorooctanoic acid (PFOA) in vacuum cleaner dust collected in Japanese homes. J Environ Monit 2003; 5(5): 753-7.
[http://dx.doi.org/10.1039/b307147m] [PMID: 14587845]

[79] Shoeib M, Harner T, M Webster G, Lee SC. Indoor sources of poly- and perfluorinated compounds (PFCS) in Vancouver, Canada: implications for human exposure. Environ Sci Technol 2011; 45(19): 7999-8005.
[http://dx.doi.org/10.1021/es103562v] [PMID: 21332198]

[80] Huber S, Haug LS, Schlabach M. Per- and polyfluorinated compounds in house dust and indoor air from northern Norway - a pilot study. Chemosphere 2011; 84(11): 1686-93.
[http://dx.doi.org/10.1016/j.chemosphere.2011.04.075] [PMID: 21632089]

[81] D'Hollander W, de Voogt P, De Coen W, Bervoets L. Perfluorinated substances in human food and other sources of human exposure. Rev Environ Contam Toxicol 2010; 208: 179-215.
[http://dx.doi.org/10.1007/978-1-4419-6880-7_4] [PMID: 20811865]

[82] Goosey E, Harrad S. Perfluoroalkyl compounds in dust from Asian, Australian, European, and North American homes and UK cars, classrooms, and offices. Environ Int 2011; 37(1): 86-92.
[http://dx.doi.org/10.1016/j.envint.2010.08.001] [PMID: 20810169]

[83] Kato K, Calafat AM, Needham LL. Polyfluoroalkyl chemicals in house dust. Environ Res 2009; 109(5): 518-23.
[http://dx.doi.org/10.1016/j.envres.2009.01.005] [PMID: 19261270]

[84] Nakata AKT, Iwasaki Y, Ito R, Saito K, Izumi S. Measurement of perfluorinated compounds in human milk and house dust. Organohalogen Compd 2007; 69: 2844-6.

[85] Fromme HNL, Kiranoglu M, Albrecht M. Perfluorinated substances in house dust in Bavaria, Germany. Organohalogen Compd 2008; 70: 001048-50.

[86] Knobeloch L, Imm P, Anderson H. Perfluoroalkyl chemicals in vacuum cleaner dust from 39 Wisconsin homes. Chemosphere 2012; 88(7): 779-83. [http://dx.doi.org/10.1016/j.chemosphere.2012.03.082] [PMID: 22542201]

[87] Beesoon S, Genuis SJ, Benskin JP, Martin JW. Exceptionally high serum concentrations of perfluorohexanesulfonate in a Canadian family are linked to home carpet treatment applications. Environ Sci Technol 2012; 46(23): 12960-7. [http://dx.doi.org/10.1021/es3034654] [PMID: 23102093]

[88] Ericson Jogsten I, Nadal M, van Bavel B, Lindström G, Domingo JL. Per- and polyfluorinated compounds (PFCs) in house dust and indoor air in Catalonia, Spain: implications for human exposure. Environ Int 2012; 39(1): 172-80. [http://dx.doi.org/10.1016/j.envint.2011.09.004] [PMID: 22208757]

[89] Shoeib M, Harner T, Lee SC, Lane D, Zhu J. Sorbent-impregnated polyurethane foam disk for passive air sampling of volatile fluorinated chemicals. Anal Chem 2008; 80(3): 675-82. [http://dx.doi.org/10.1021/ac701830s] [PMID: 18179247]

[90] Fromme H, Tittlemier SA, Völkel W, Wilhelm M, Twardella D. Perfluorinated compounds exposure assessment for the general population in Western countries. Int J Hyg Environ Health 2009; 212(3): 239-70. [http://dx.doi.org/10.1016/j.ijheh.2008.04.007] [PMID: 18565792]

[91] Alexander BH, Olsen GW. Bladder cancer in perfluorooctanesulfonyl fluoride manufacturing workers. Ann Epidemiol 2007; 17(6): 471-8. [http://dx.doi.org/10.1016/j.annepidem.2007.01.036] [PMID: 17448680]

[92] Olsen GW. B M, Burris J, Mandel J, A cross-sectional analysis of serum perfluorooctanesulfonate (PFOS) and perfluorooctanoate (PFOA) in relation to clinical chemistry, thyroid hormone, hematology and urinalysis results from male and female employee participants of the 2000 Antwerp and Decatur fluorochemical medical surveillance program Final Report Administrative Record AR-226-1087 Washington. DC: US Environmental Protection Agency 2001.

[93] Gilliland FD, Mandel JS. Serum perfluorooctanoic acid and hepatic enzymes, lipoproteins, and cholesterol: a study of occupationally exposed men. Am J Ind Med 1996; 29(5): 560-8. [http://dx.doi.org/10.1002/(SICI)1097-0274(199605)29:5<560::AID-AJIM17>3.0.CO;2-Z] [PMID: 8732932]

[94] Joensen UN, Bossi R, Leffers H, Jensen AA, Skakkebaek NE, Jørgensen N. Do perfluoroalkyl compounds impair human semen quality? Environ Health Perspect 2009; 117(6): 923-7. [http://dx.doi.org/10.1289/ehp.0800517] [PMID: 19590684]

[95] Olsen GW, Butenhoff JL, Zobel LR. Perfluoroalkyl chemicals and human fetal development: an epidemiologic review with clinical and toxicological perspectives. Reprod Toxicol 2009; 27(3-4): 212-30. [http://dx.doi.org/10.1016/j.reprotox.2009.02.001] [PMID: 19429401]

[96] Inoue K, Okada F, Ito R, *et al.* Perfluorooctane sulfonate (PFOS) and related perfluorinated compounds in human maternal and cord blood samples: assessment of PFOS exposure in a susceptible population during pregnancy. Environ Health Perspect 2004; 112(11): 1204-7. [http://dx.doi.org/10.1289/ehp.6864] [PMID: 15289168]

[97] Apelberg BJ, Goldman LR, Calafat AM, *et al.* Determinants of fetal exposure to polyfluoroalkyl compounds in Baltimore, Maryland. Environ Sci Technol 2007; 41(11): 3891-7. [http://dx.doi.org/10.1021/es0700911] [PMID: 17612165]

[98] Apelberg BJ, Witter FR, Herbstman JB, *et al.* Cord serum concentrations of perfluorooctane sulfonate (PFOS) and perfluorooctanoate (PFOA) in relation to weight and size at birth. Environ Health Perspect

2007; 115(11): 1670-6.
[http://dx.doi.org/10.1289/ehp.10334] [PMID: 18008002]

[99] Fei C, McLaughlin JK, Lipworth L, Olsen J. Prenatal exposure to perfluorooctanoate (PFOA) and perfluorooctanesulfonate (PFOS) and maternally reported developmental milestones in infancy. Environ Health Perspect 2008; 116(10): 1391-5.
[http://dx.doi.org/10.1289/ehp.11277] [PMID: 18941583]

[100] Fei C, McLaughlin JK, Tarone RE, Olsen J. Perfluorinated chemicals and fetal growth: a study within the Danish National Birth Cohort. Environ Health Perspect 2007; 115(11): 1677-82.
[http://dx.doi.org/10.1289/ehp.10506] [PMID: 18008003]

[101] Washino N, Saijo Y, Sasaki S, *et al.* Correlations between prenatal exposure to perfluorinated chemicals and reduced fetal growth. Environ Health Perspect 2009; 117(4): 660-7.
[http://dx.doi.org/10.1289/ehp.11681] [PMID: 19440508]

[102] Grice MM, Alexander BH, Hoffbeck R, Kampa DM. Self-reported medical conditions in perfluorooctanesulfonyl fluoride manufacturing workers. J Occup Environ Med 2007; 49(7): 722-9.
[http://dx.doi.org/10.1097/JOM.0b013e3180582043] [PMID: 17622844]

[103] Nolan LA, Nolan JM, Shofer FS, Rodway NV, Emmett EA. The relationship between birth weight, gestational age and perfluorooctanoic acid (PFOA)-contaminated public drinking water. Reprod Toxicol 2009; 27(3-4): 231-8.
[http://dx.doi.org/10.1016/j.reprotox.2008.11.001] [PMID: 19049861]

[104] Monroy R, Morrison K, Teo K, *et al.* Serum levels of perfluoroalkyl compounds in human maternal and umbilical cord blood samples. Environ Res 2008; 108(1): 56-62.
[http://dx.doi.org/10.1016/j.envres.2008.06.001] [PMID: 18649879]

[105] Hoffman K, Vieira V, Webster T, White R. Exposure to polyfluoroalkyl chemicals and attention deficit hyperactivity disorder in US children aged 12-15 years. Epidemiology 2009; 20(6): S70-0.
[http://dx.doi.org/10.1097/01.ede.0000362918.40325.e9]

[106] Gump BB, Wu Q, Dumas AK, Kannan K. Perfluorochemical (PFC) exposure in children: associations with impaired response inhibition. Environ Sci Technol 2011; 45(19): 8151-9.
[http://dx.doi.org/10.1021/es103712g] [PMID: 21682250]

[107] USEPA. EPA and 3M Announce Phase Out of PFOS 2000.

[108] EFSA. Opinion of the scientific panel on fontaminants in the food chain on perfluorooctane sulfonate (PFOS), perfluorooctanoic acid (PFOA) and their salts. EFSA J 2008; (653): 1-131.

[109] USEPA. Provisional Health Advisories for Perfluorooctanoic Acid (PFOA) and Perfluorooctane Sulfonate (PFOS). Washington, D. C.: Office of Water, United States Environmental Protection Agency 2009.

[110] USEPA. Fact Sheet: PFOA and PFOS Drinking Water Health Advisories 2016.

[111] Calafat AM, Kuklenyik Z, Caudill SP, Reidy JA, Needham LL. Perfluorochemicals in pooled serum samples from United States residents in 2001 and 2002. Environ Sci Technol 2006; 40(7): 2128-34.
[http://dx.doi.org/10.1021/es0517973] [PMID: 16646443]

[112] Calafat AM, Kuklenyik Z, Reidy JA, Caudill SP, Tully JS, Needham LL. Serum concentrations of 11 polyfluoroalkyl compounds in the u.s. population: data from the national health and nutrition examination survey (NHANES). Environ Sci Technol 2007; 41(7): 2237-42.
[http://dx.doi.org/10.1021/es062686m] [PMID: 17438769]

[113] Egeghy PP, Lorber M. An assessment of the exposure of Americans to perfluorooctane sulfonate: a comparison of estimated intake with values inferred from NHANES data. J Expo Sci Environ Epidemiol 2011; 21(2): 150-68.
[http://dx.doi.org/10.1038/jes.2009.73] [PMID: 20145679]

[114] Yamashita N, Kannan K, Taniyasu S, *et al.* Analysis of perfluorinated acids at parts-per-quadrillion

levels in seawater using liquid chromatography-tandem mass spectrometry. Environ Sci Technol 2004; 38(21): 5522-8.
[http://dx.doi.org/10.1021/es0492541] [PMID: 15575267]

[115] USEPA. Child-Specific Exposure Factors Handbook. Washington, DC: National Center for Environmental Assessment 2008.

[116] USEPA. Exposure Factors Handbook. 2011 Edition., Washington, DC: National Center for Environmental Assessment 2011.

[117] DJ. Pharmacokinetics made easy: McGraw-Hill Australia 1998.

[118] Bartell SM, Calafat AM, Lyu C, Kato K, Ryan PB, Steenland K. Rate of decline in serum PFOA concentrations after granular activated carbon filtration at two public water systems in Ohio and West Virginia. Environ Health Perspect 2010; 118(2): 222-8.
[http://dx.doi.org/10.1289/ehp.0901252] [PMID: 20123620]

[119] Olsen GW, Burris JM, Ehresman DJ, *et al.* Half-life of serum elimination of perfluorooctanesulfonate, perfluorohexanesulfonate, and perfluorooctanoate in retired fluorochemical production workers. Environ Health Perspect 2007; 115(9): 1298-305.
[http://dx.doi.org/10.1289/ehp.10009] [PMID: 17805419]

[120] CDC, Fourth National Report on Human Exposure to Environmental Chemicals, Updated Tables, March 2013. Atlanta, GA: Centers for Disease Control and Prevention 2013.

[121] Kim SK, Kannan K. Perfluorinated acids in air, rain, snow, surface runoff, and lakes: relative importance of pathways to contamination of urban lakes. Environ Sci Technol 2007; 41(24): 8328-34.
[http://dx.doi.org/10.1021/es072107t] [PMID: 18200859]

CHAPTER 8

Mercury

Zoyne Pedrero Zayas*

CNRS/UNIV PAU & PAYS ADOUR, Institut des Sciences Analytiques et de Physicochimie pour l'Environnement et les Matériaux, UMR 5254, 64000, Pau, France

Abstract: Mercury (Hg) pollution is an acknowledged major environmental problem. Considering its extreme toxicity, Hg has recently been included in the top ten list of chemicals of major public health concern according to the World Health Organization. Once released into the environment, it is transformed in aquatic ecosystems by microorganisms into the neurotoxic methylmercury. The hazardous effect is then biomagnified through the trophic/food chain. Diet is considered the main exposure pathway of Hg in humans. Therefore, safety values have been established by food safety authorities in order to protect consumers. Seafood, followed by rice, is the primary source of Hg in the human diet. A variety of analytical methodologies are available for the analysis of Hg and its species in food. This chapter presents recent advances in the determination of Hg in foodstuffs. Special attention is given to innovative Hg (species) extraction and preconcentration systems assisted by nanoparticles. Non-chromatographic approaches, as an alternative to classical chromatographic approaches used for speciation are detailed. The potential and limitations of Hg isotopic analysis in food are also discussed.

Keywords: Certified reference materials, Diet, Fish, Food, GC, HPLC, ICP-MS, Isotopic dilution analysis, Isotopic fractionation, MC-ICP-MS, Methylmercury, Mercury, Mercury species, Non-chromatographic methods, Rice, Speciation.

INTRODUCTION

Mercury (Hg) pollution is considered a major environmental and public health concern. Because of its toxicity, Hg has been recently included in the top ten hazardous chemicals by the World Health Organization (WHO). Pregnant women and children in early life are considered the most vulnerable population to Hg harmfulness. Toxic effects can be lethal and include infections of the nervous, digestive and immune systems, and lungs and kidneys.

* **Corresponding author Zoyne Pedrero Zayas:** CNRS/UNIV PAU & PAYS ADOUR, Institut des Sciences Analytiques et de Physicochimie pour l'Environnement et les Matériaux, UMR 5254, 64000, Pau, France; Tel: +33 (0)5 40 17 5027; E-mail: zoyne.pedrerozayas@univ-pau.fr

Belén Gómara & María Luisa Marina (Eds.)
Current and Future Developments in Food Science (Vol. 1)

Taking into account the capability of Hg to bind thiols [1], its interaction with essential proteins and enzymes leading to their dysfunction seems to be the origin of such toxicity.

Despite occupational exposure (*i.e.*, miners, dentists) - dental amalgams being undisputable sources of Hg - diet appears as the main exposure pathway of Hg in humans. In general, seafood consumption is recognized as the most common pathway of Hg human exposure. It is especially troubling considering the recent and significant increase of Hg in oceanic waters [2]. Anthropogenic activities, such as mining and coal burning are responsible for the increased Hg levels in the atmosphere and in oceanic surfaces [3]. Microorganisms in aquatic ecosystems play a crucial role since they biotransform inorganic Hg (iHg) into methylmercury (MeHg which is present in its free form as CH_3Hg^+). The latter exhibits high levels of toxicity and it is easily bioaccumulated through the food chain resulting in serious social and health effects.

Considering Hg toxicity, food safety authorities set the maximal acceptable levels for Hg in food. The established Hg values in foodstuffs depend on their nature. For food supplements, the maximum level is as high as 0.10 mg kg^{-1} in the final product. In the case of fishery products comprising crustaceans and muscle meat of fish (except predatory ones), it is fixed at 0.5 mg kg^{-1} wet weight and for predatory fish species as bonito, eel, marlin, sharks and tuna, among others, it is 1 mg kg^{-1} wet weight (COMMISSION REGULATION (EC) No. 1881/2006 of 19 December 2006 setting maximum levels for certain contaminants in foodstuffs). These regulations are effective for fresh and processed fish.

A tolerable weekly intake (TWI) has been set as a safe consumption threshold in order to avoid Hg exposure risks and is regularly reevaluated. The European Food Safety Authority set in 2012 a new MeHg TWI at 1.3 μg kg^{-1} bodyweight, lower than that established by the Joint Food and Agriculture Organization (FAO)/WHO Expert Committee on Food Additives of 1.6 μg kg^{-1} bodyweight. However, due to seafood consumption, a significant part of the global population, mainly from developing countries, is exposed to higher Hg levels than the thresholds established by food safety authorities [3].

Taking into account that diet is the principal source of exposure to Hg, its toxicity is strongly related to its speciation, and Hg species are toxic at low concentrations, the analytical chemistry community is continuously seeking for advances in foodstuffs analysis methods. Currently, the main goals of the new and trendy analytical approaches are the development of sensitive, cost-effective and green methods for the determination of Hg and its species. Isotopic fractionation analysis also appears as a fresh strategy for the identification and discrimination of Hg sources in food products, adding another dimension to total Hg quantification and speciation. In this chapter, current and promising approaches for Hg analysis in food are described.

FOOD MATRICES WHERE HG IS OFTEN DETERMINED

According to a recent report of FAO (2016), fisheries and aquaculture are very important sources of food, nutrition, income and livelihood for hundreds of millions of people around the world. Driven by rising domestic income, consumers in emerging economies (where consumption was previously based on locally available products) are experiencing a diversification of the types of available fish through an increase in fishery imports. The significant growth in fish consumption has enhanced people's diets around the world through diversified and nutritious food. Fish consumption represents in many countries the dominant source of proteins. Therefore, the high seafood consumption could lead to significant risks due to MeHg ingestion.

Since seafood is considered a major contributor of Hg through diet, the quantification of Hg and its species in such products has grabbed the attention of the analytical chemistry community. As a consequence, most of the Hg speciation studies in foodstuffs correspond to the analysis of fish and other seafood. The new methodologies developed for Hg speciation in fish and seafood are presented all along the text.

Rice is a dominant global crop, recognized to be one of the most important sources of Hg in human diet. Microbial Hg methylation is considered the main source of this organomercurial species in paddy soils. The traditional rice culture practice involves several flooding processes, which lead to anaerobic conditions facilitating iHg methylation by sulfate reducing bacteria [4]. In addition, the use of iodomethane as fumigant, enhances Hg methylation in soil under sunlight, increasing MeHg exposure from rice [5]. After soil uptake by the plant, Hg is transported to the edible part [6, 7]. It constitutes a potential risk in Hg polluted areas like Hg-contaminated mining regions, where Hg values reach up to 500 ng g^{-1} [7]. In such regions, the most important MeHg exposure source is not fish, but rice consumption [8]. MeHg intake through rice ingestion has been reflected on the levels of MeHg in hair of inhabitants of such areas [7, 9].

Rice seeds consist of a hull and a nutritious bran coat layer surrounding the endosperm and inner embryo. Brown rice is the result of removing the hull (inedible) and can be consumed in this state. Further processing yields to "white" or "polished" rice. The distribution of Hg species varies according to the fraction of the rain grains. Mostly, iHg is located in hull and bran, while MeHg is found in edible white rice. Rice processing leads to a release of up to 78% of iHg, which is contrasting with MeHg that remains almost untouched [10]. The analysis of commercial rice and baby-food rice from European markets (2 samples of each type) reveals a higher proportion of MeHg in babyfood. It is attributed to the processing practice that includes rice washing and cooking. It is suggested that iHg could be lost during these steps, leading to higher MeHg percentage [11].

Hg speciation in rice is a challenge principally taking into consideration the difficult extraction from the starch matrix of Hg species without provoking their degradation/interconversion and the lack of rice CRMs (Certified Reference Materials) for MeHg. In the absence of certified MeHg values in rice matrices, other biological CRMs are used for method validation despite the remarkable matrix mismatching [12, 13]. Species specific isotope dilution is definitely an elegant solution for the validation of analytical methods in rice. The addition of the isotopically enriched species allows tracking potential species degradations/transformations and losses during the analytical procedure [11, 14].

Alkaline and acid digestion including several subsequent extractions and back-extractions steps are commonly used [13]. Two fast and sensitive methodologies based on both, acid and basic microwave (MW) extraction of Hg species followed by their derivatization (ethylation) and analysis by gas chromatography (GC) coupled to Atomic Fluorescence Spectrometry (AFS) via pyrolysis, were developed. Basic digestion uses 25% tetramethylammonium hydroxide while 6N HNO_3 is chosen for acid treatment. Compared to traditional methods for Hg speciation in this matrix, it implies a noticeable time reduction (it is less time consuming) (from sample preparation to analysis). The sum of Hg(II) (being present in its free form as Hg^{2+}) and MeHg in the CRMs was similar for acid and basic digestions. However, it should be taken into account that adequate blanks are only obtained for acid MW digestion, being more suitable for samples with low Hg contents. Limits of detection (LODs) were in the range of ng g^{-1} (0.1-0.7 ng g^{-1}) [13].

Considering the economic rises on developing countries, there is an urgent need for developing inexpensive and simple but accurate and sensitive Hg speciation analysis methods [15]. Inspired by this urgency, a fresh ultrasensitive Hg speciation method by headspace solid-phase microextraction (HS-SPME) using porous carbons (PCs) and GC-Dielectric Barrier Discharge (DBD) Optical Emission Spectrometry (OES) was developed.

The applicability of the method was demonstrated by the analysis of 12 rice samples from both, unpolluted and abandoned Hg mining districts in China. The protocol for Hg species extraction from rice comprised the addition of 10.0 mL of 25% (w/v) potassium hydroxide-methanol to 1 g of grounded samples and the mixture was maintained in a water bath at 65 °C during 4 h. The resulting digest was diluted to 25 mL with methanol. Hg(II), ethylmercury (EtHg, $CH_3CH_2Hg^+$) and MeHg were derivatized with $NaBPH_4$ to their volatile species and preconcentrated by HS-SPME assisted by PCs. In-house prepared PCs SPME presented superior qualities than commercial SPME considering the simultaneous and outstanding extraction efficiency for the three Hg species that was maintained even after 200 extraction cycles. The comparison of Hg species quantification by the new GC-DBD-OES method and conventional high-performance liquid chromatography (HPLC)- inductively coupled plasma mass spectrometry (ICP-MS) showed evidence of a complete matching of the results. The sum of the Hg(II) and MeHg coincided with the total Hg concentration determined by ICP-MS after MW assisted acid digestion. LODs were 0.5, 0.75, and 1.0 µg kg^{-1} for Hg(II), MeHg, and EtHg, respectively, with a recovery between 90 and 105% [15].

DBD-OES benefits from low power consumption, simple set up, high electron temperature and weak background radiation. Considering the small size and compactness of DBD-OES, it can be directly connected to the GC capillary column by a T tube conveniently introducing Ar discharge gas. Direct coupling without using transport tubes prevents loss of Hg and memory effect. This represents a pioneer application of GC-DBD-OES on elemental speciation, used before for volatile carbon containing compounds and halohydrocarbons [15]. Despite the PCs HS-SPME-GC-DBD-OES method was developed and applied on rice samples, its accuracy validation using fish protein (DORM-4) and lobster hepatopancreas (TORT-3) CRMs reveals the potential of this approach on the analyses of complex matrices as seafood and other foodstuffs. The potential coupling of DBD-OES with miniaturized GC should be promising in the field of Hg speciation due to its compactness, and low power and gas consumption [15].

Rice is one of the scarce food matrices where the analysis of Hg is not restricted to the quantification of its most common species, Hg(II) and MeHg. In view of the public health concern associated to the consumption of this staple product, several studies focus on the understanding of Hg accumulation pathways by different approaches as molecular speciation [10, 14], imaging [10], and isotopic analysis [9, 14].

Hg imaging analyses in food are not common and one of the rare examples corresponds to Hg characterization in rice. Synchrotron radiation microscopic X-ray fluorescence (SR-µXRF) provides the Hg map in rice grains, as well as for other elements as Ca, Cd, Se, Cu, Fe, Mn, Zn and K. The principal Hg location is the surface of brown rice grains, corresponding to the pericarp and aleurone layer. The elements K, Ca, Fe, Mn and Cd are co-localized at the surface of the rice grain. It is attributed to the absorption of atmospheric iHg [10]. Speciation by X-ray absorption near-edge spectroscopy (XANES) is limited to bran samples due to their higher Hg content. Results indicate that Hg(II) and MeHg are both binding cysteine (Cys). Although the structure of the Cys-rich bioligands binding the species are not identified, it has been suggested that Hg(II) could be associated to phytochelatins and MeHg to proteins. These differences on speciation could justify the higher mobility of MeHg-Cys, actively transported to the endosperm during seed ripening [10].

Breast milk is a food matrix particularly interesting that allows the Hg exposure estimation to lactating women and newborns, who constitute an extremely sensitive population. MeHg is able to cross the mammary gland, being present in the colostrum and breast milk [16]. Hg content in breast milk is strongly linked to maternal exposure. Mothers' diet and their dental amalgams are the main sources of Hg in breast milk [17 - 19]. The percentage of MeHg varies in a wide range according to the diet, where highest values correspond to fish-consuming populations [19, 20].

Most of the analytical methods for Hg quantification in breast milk involve acid digestions and relatively long sample treatments [21, 22]. The characterization of Hg content and speciation in large sets of samples (*i.e.*, breast milk banks, epidemiological studies) claims for sensitive, cost-effective and validated analytical approaches. Hg quantification by advanced mercury analyzer demonstrated to be a good alternative to expensive ICP-MS analysis, with a simple sample treatment based on freeze-drying [23].

GC coupled to different detection systems like cold vapor (CV)-AFS [19, 20, 22] and electron capture detection [21] has been proposed for the separation of Hg species in breast milk after their derivatization into volatile forms. ICP-MS has recently been hyphenated to chromatographic techniques for Hg speciation in breast milk in order to exploit the advantages of isotopic dilution (ID) analysis, which enhance accuracy by the correction of potential losses and/or species transformation during the analytical procedure [23].

MeHg speciation methods in breastmilk based on both GC-isotopic dilution (ID)-ICP-MS and HPLC-ID-ICP-MS, exhibit similar quantification limits (~1 $\mu g\ kg^{-1}$) and accuracy (94-97%). Although GC is more efficient considering the time required for species separation, this method involves a tedious derivatization step. In contrast, sample treatment for HPLC is simpler, limited to a preconcentration by freeze-drying and dissolution of the dried breastmilk on a minimum volume of water [23]. Taking into account the similarity on the analytical characteristics, both approaches can be used for the analysis of this complex matrix. HPLC-ID-

ICP-MS was chosen in a recent French epidemiological study of 180 breast milk samples. Hg and MeHg values reach up to 16.9 μg Hg kg^{-1} and 0.43 μg Hg kg^{-1} wet weight, respectively [23].

In addition to quantification of Hg and its species, a metalloproteomic study has also been carried out on breast milk. Quantification of Hg in the breast milk proteome, obtained using two-dimensional electrophoresis after protein precipitation with acetone, indicates that Hg is principally associated to proteins with molecular masses in the range from 14 to 26 kDa [24].

Nuts in general are an important source of essential minerals, proteins and unsaturated fat. They are consumed worldwide and often recommended to be included in diet in order to reduce risks of coronary diseases. They are included in the list of foodstuffs analyzed in national dietary studies since they became habitually consumed [25]. Hg content in such products can be higher when grown in contaminated regions. The content of Hg is attributed to the soil-plant interaction since nuts are isolated by the shell, therefore excluding the potential Hg atmospheric contribution. The concentration in nut peel is up to 10 times higher than in the enclosed nuts, endorsed to the shield role of peel, preventing the transfer of toxic elements as Hg and Cd to nuts [26]. Acid digestion by using HNO_3 or HNO_3/HF mixture of nuts samples in a MW system is largely used for the quantification of Hg, as part of a multi-elemental analysis, by ICP-MS [25, 26]. The fat fraction in edible nuts varies between 25 to 60% and does not contain Hg [27]. Defatting strategies are often included in the sample treatment protocol. For ICP-MS analysis, a defatting step consisting of vigorously shaking 20 g of grounded sample with 100 mL of a (2:1, v/v) chloroform-methanol solution, repeated three times, is included before acid digestion on a MW system [28].

Direct mercury analyzer (DMA) appears as an alternative for Hg quantification by a cost-effective method. Despite one of the main advantages of this technique is the non-requirement of sample pretreatment, in the case of nuts, the presence of fat causes interferences during the pyrolysis processes due to the production of high volumes of fumes, conducing to erroneous Hg measurements. Defatting is guaranteed by vigorous shaking during 10 min of 3 g of samples with 15 mL of (2:1, v/v) chloroform-methanol solution. The LOD for the analysis of 100 mg sample by DMA is 0.08 μg kg^{-1} (LOQ: 0.3 μg kg^{-1} and RSD 2-14%) [27].

Hg concentration in diverse types of nuts from different geographical origin is commonly lower than 5 μg kg^{-1} [25 - 27]. In Brazil nuts, Hg reaches up to 8000 μg kg^{-1}. It is principally associated (72%) to the NaOH extractable fraction, mainly binding biomolecules between 0.36 and 14.4 kDa [28].

Honey's benefits as antioxidant and anti-inflammatory raise the interest in its consumption. Honey is worldwide largely consumed due to its exceptional nutritional value. The increase on popularity of such natural product is also accompanied by authenticity and geographical origin concerns. The safety quality of honey depends, among others, on the level of some toxic elements as Hg. Honey, considered a well-known environmental indicator, is able to accumulate metals and metalloids. Hg is often included in a wide list comprising other mineral and trace elements analyzed in honey in order to determine the quality and safety of honey, but also for the assessment of biological and geographical origin by chemometric methods. It is therefore usually analyzed by multi-elemental analytical techniques like ICP-MS after nitric acid MW digestion [25, 29 - 31]. The DMA alternative avoids any sample treatment and a complete analysis of Hg in five min [32]. According to the Hg values reported in honey, an ordinary consumption of this product does not seem to represent a risk for human health.

ADVANCES ON THE DETERMINATION OF TOTAL HG IN FOOD

Hg usually belongs to a long list of toxic elements to be analyzed in food products in order to evaluate the dietary exposure of a population to food contaminants. ICP-MS is therefore the most popular technique chosen by food control laboratories/institutions due to its multi-elemental character and its sensitivity. Hg is commonly quantified, after MW assisted digestion, in a wide type of matrices including cereals, eggs (and eggs products), baby food, drinks, snacks, ice creams, spices, meat, sweeteners, honey, and confectionery [25, 29, 33]. In spite of the undeniable capacity of this technique for Hg determination in food, the analytical community actively works on the development of innovative approaches dealing with specific situations like the treatment of great numbers of samples, and environmentally friendly and miniaturized solutions, among others. Representative advances on the analysis of Hg in food are presented hereafter.

Advanced or direct mercury analyzers (AMA or DMA) are techniques for Hg quantification in food that can be used for solid and liquid samples. One of the main advantages of direct mercury analyzer compared with the primary techniques used for Hg analysis in food -atomic absorption spectrometry (AAS), AFS, inductively coupled plasma atomic emission spectrometry (ICP-AES) and mass spectrometry (ICP-MS)- is the possibility of analyzing Hg at ultra-trace levels without any sample pre-treatment. Short analysis time, easy use, and cost effectiveness associated are some of the main advantages of the method. DMA is based on the thermal decomposition of the sample followed by Hg amalgamation by gold system subsequent atomic absorption detection. It has successfully been applied to diverse food matrices like honey, nuts, infant food, seafood and animal edible tissues as equine muscle, bovine kidney and swine kidney, and poultry muscle [34].

DMA is the technique chosen by European countries authorities to carry out national diet studies on infant and toddlers, resulting on the analysis of more than 370 infant food samples [35, 36]. Accurate dietary exposure assessment claims for the detection and quantification limits as low as possible. The most recent DMA method, developed and validated for (infant) food analysis, exhibits detection and quantification limits of 0.3 μg kg^{-1} fresh weight and 0.6 μg kg^{-1}, respectively [35].

Although DMA is principally associated to solid samples, it has been validated for the quantification of Hg in breast milk. The comparison of ICP-MS and DMA in this food matrix shows that quantification limits for both analytical approaches are in the range between 2 and 5 μg Hg kg^{-1}, tending to be lower by DMA. Considering the similar sensitivity, Hg quantification by direct analysis represents a better option due to the superior sample throughput. ICP-MS analysis implies a previous digestion of the sample, in this case based on mineralization with aqua regia on a heating block during 3 h. In contrast, DMA does not require any sample treatment. Breast milk samples are exclusively freeze-dried before analysis and Hg is quantified by the standard addition method [23].

Unquestionably, DMA has an enormous potential for Hg analysis in food. Nevertheless, the main limitation of this technique is its exclusive success on total Hg concentration, not extended to speciation. A few recent examples suggest the possibility of using thermal release for speciation in solid samples [37], not yet widely extended. The recent combination of a thiourea derivative polymer with DMA leads to the screening of Hg(II) and MeHg in food, specifically in fish [38].

Electro thermal vaporization (ETV) appears as a direct sampling technique with a versatile formation and high sample introduction efficiency. It has been coupled to diverse detection systems as AAS, AFS, ICP-OES/MS. The gas phase enrichment method guarantees an optimum interference separation by trapping the analyte. A gold amalgamator is frequently used for selective Hg trapping, separating the element from other pyrolysis gases and concentrating it before thermal induced release into the detector. In a similar way, a tungsten coil is used for the separation of Cd from the matrix interferences [39].

Gold and tungsten coil traps were integrated for the first time into one ETV-AAS system for the quantification of two toxic elements usually simultaneously tracked in foodstuffs. Considering the noticeable differences on vaporization temperature between Hg and Cd, a sequential vaporization of these elements in food matrices was obtained by using an improved on-line ashing furnace. The whole analytical procedure is carried out within 10 min, including sample pre-treatment, with a LOD of 0.7 pg for Hg (and 0.5 pg for Cd). Mushrooms, green tea, corn, milk and spinach are some of the food samples already analyzed using this method. The coupling to miniaturized detection AAS or AFS systems will promote the application of this easy, green and digestion free method for *in situ* monitoring of Hg (and Cd) in food [39].

Electrochemistry represents a fast, simple, and sensitive alternative, suitable for Hg routine analysis. In addition to the low cost, compared with other analytical techniques (ICP-MS, CVAAS, CVAFS), it is portable and appropriate for *in situ* determination. In general terms, Hg is preconcentrated on the electrode followed by stripping, mainly by anodic stripping voltammetry (ASV). Electrochemical techniques for Hg detection are a hot topic. They are characterized by a high pre-concentration capacity, achieving low LODs. Most of the recent advances on this subject correspond to the development of electrode surfaces allowing the preconcentration of lowest amounts of Hg. DNA-based assays and nanostructured electrodes are the most used methodologies in the last years [40].

DNA-based electrodes exploit the strong bound between Hg(II) and thymine (T) DNA bases, allowing the preconcentration of Hg. DNA-based electrodes exhibit extremely low LODs, but they are less stable and more time-consuming than nanostructured electrodes. In general, the longer analysis time and the stability difficulties of DNA-based assays hamper their application in complex samples [40].

Electrodes modified with lead nanoparticles with thiol-functionalized polysiloxilane [41], nickel nanoparticles with carbon composites [42], among a long list, have been used for Hg determination. The high affinity of gold towards Hg justifies the larger application of gold nanoparticles (AuNPs) on electrode surface for Hg determination [40]. In general, nanostructured electrode surfaces with high electroactive areas and excellent surface properties guarantee a sensitive detection in a short time. Nanoporous materials appear as a prominent approach, characterized by shorter analysis times and relatively simpler procedures. Screen-printed electrodes are expected to be used for the development of future commercial Hg sensors, principally due to their low-cost, disposable and miniaturized properties [40].

Most of the applications of electrochemical detection correspond to Hg analysis in water so far [40, 43]. An ultrasensitive and reusable electrochemical biosensor for the determination of mercury ions (Hg(II)) in water in the range of 10 ng L^{-1} to 1 μg L^{-1} has been developed based on thymine modified gold nanoparticles/reduced graphene oxide nanocomposites [44]. The development of new assays and the analytical improvement of electrochemical methods should open new perspectives for the analysis of more matrices, like biological and food samples. Some examples are already present in the literature, describing successful Hg characterization in urine and blood [40, 43]. Indium-tin oxide surfaces have been used to develop nanoporous gold electrodes for the detection of Hg. They have been primarily used in water but also in milk analysis achieving a detection limit of 0.03 μg L^{-1} [45].

The application of electrochemical Hg detection in food is incipient. A variety of food samples comprising cockles [46, 47], seaweed [47], milk [45], breast milk [48] and vegetable origin samples like red chili and tomato [46] has been successfully analyzed. However, the existing studies on foodstuffs focus mainly in fish samples [46, 49 - 51]. Under optimal conditions, anodic stripping voltammetry gold nanoparticle-modified glassy carbon electrode achieves a detection limit of 0.001 μg L^{-1} with a short deposition time of 2 min in fish. The analytical performance of the electrochemical approach is comparable to conventional DMA analysis [50] Compared to DMA, the main drawback of electrochemical techniques is the need of longer analysis time and a wet sample digestion which implies the generation of waste solution with possible loss of analyte by vaporization [50].

Sample Treatment for Total Hg Determination Assisted by Solid-Phase Extraction

Effective isolation of the analyte and preconcentration are common sample treatment requirements for trace element analysis. In general terms, solid-phase extraction (SPE) is the technique that conquered the preconcentration of Hg from real samples. Low reagent consumption, simplicity and minimum cost are their main advantages. Assisted by the tremendous progress in nanomaterials, there are noticeable advances on SPE for Hg determination in food.

Dispersed solid-phase extraction represents an alternative to classical SPE. It consists in the direct addition of the solid phase to the agitated sample solution during an optimized time. Therefore, no special equipment is required for sample treatment. The enriched solid phase is mechanically recovered (*i.e.*, centrifugation, filtration) and the analyte is therefore eluted with an adequate solvent for further analysis. This SPE alternative is robust, fast, low cost and simply operated. Matrix solid-phase dispersion (MSPD) is particularly interesting in sample preparation of solid, semisolid and highly viscous samples. The sample is blended with a solid support in a mortar with a pestle. The solid support plays an abrasive role leading to the disruption of sample architecture. Sample components are consequently dispersed on the surface of the support particles. The solid homogeneous mixture is commonly packaged into a cartridge for analyte elution with a suitable solvent [52].

The association of matrix solid phase dispersion assisted by multiwall carbon nanotubes (MWCNTs-MSPD) with single drop solute on electrode glow discharge induced (SD-SEGD) chemical vapor generation constitutes an important advancement on Hg quantification in micro amount samples. The MWCNTs-MSPD consists in blending 1 mg of the solid sample with 0.5 mg of MWCNTs for 5 min using an agate pestle. The homogeneous solid mixture is placed in the MSPD column and Hg species retained are extracted by 100 µL of eluent containing 0.5% L-Cys and 4% HCOOH. Hg species are efficiently converted to Hg^0 by SD-SEGD-CVG and further transported to AFS for detection. The accuracy of the method was validated by the analysis of several biological CRMs and it was applied to fish samples [53]. LOD improved 100 fold with this robust SD-SEGD-CVG-AFS method, in comparison to conventional CVG-AFS where the micro amount of sample is significantly diluted by the carrier solution. In addition, memory effect, typically associated to CVG or pneumatic nebulization ICP-MS is not observed. The absence of memory effect is attributed to several factors as: the low sample volume (20µL), the inhibition of adsorption in connecting tubes by the presence of L-Cys and the minimized transport tube length [53]. The combination of MWCNTs-MSPD and SD-SEGD-CVG-AFS is an absolutely promising approach for Hg quantification in foodstuffs. It could be particularly interesting for Hg analysis in rare and/or luxurious food products such as caviar, saffron and truffles, where minimal amounts of samples are available for analysis.

Magnetic SPE (MSPE) is a new kind of solid-phase extraction based on the combination of a non-magnetic adsorbent with a magnetic material [54]. Magnetic nanoparticles (MNPs) have a high surface area and sorption capacity. Iron, nickel, cobalt and their oxides constitute the core of most of the MNPs. The surface of the MNPs is frequently covered by molecular imprinted polymers, silica or graphene, in order to avoid nanoparticles aggregation [55]. Once the sorption of the analyte is complete on the magnetic sorbent, they are separated by an external magnetic field. Therefore, any additional step of physical separation (*i.e.*, centrifugation, filtration) is required. The analyte is eluted by magnetic separation of the regenerated solvent, resulting in a simple and fast methodology [55]. Considering the multiple advantages of MSPE, there is an unquestionable trend on its application to Hg (and other heavy metal ions). So far, it has been particularly used for water analysis, but several examples of Hg analysis in food show the analytical potential of the method (Table **1**) [56].

Fe_3O_4 nanoparticles are excellent supports for SPE due to their magnetic properties, making easy the separation from the solution by using a magnet. They exhibit high thermal and chemical stability and can be modified to develop selective sorbents [57]. However, they tend to form aggregates and are susceptible to oxidation. Surface decoration with specific coating avoids these drawbacks and potentially increases their selectivity even in complex matrices. Several examples show its potential for the analysis of Hg in food [57 - 60].

Magnetized polymers, including both ion imprinted and non-imprinted polymers have been used to determine Hg(II) in food [57, 61]. Uptake time, sample pH, and nanosorbent amount influence the sorption procedure. Type and concentration of the eluent, volume and time of elution are the main parameters to be considered for elution optimization.

Silica is one of the most ideal shells, since it enhances mechanical and chemical stability in extreme acid/basic media. Polymeric materials have been used to modify the surface of such MNPs used on the extraction and preconcentration of Hg(II). They have been applied to the determination of Hg in complex food matrices [61]. The surface of Fe_3O_4@SiO_2 NPs was modified by polythiophene using a self-assembly of thiophene onto Fe_3O_4 NPs. The sorbent, resulting on the modification of Fe_3O_4@SiO_2 presents a sorption capacity of 59 mg g^{-1}. Quantification by CV-AAS of the eluted Hg(II) provides a detection limit of 0.02 ng mL^{-1}. The method was validated by the good correlation between the Hg certified and experimentally obtained values in fish CRMs, DOLT 4 and DORM 2. The magnetic nanosorbent was therefore used for the quantification of Hg in shrimp, fish and canned tuna [61].

Ion imprinted polymers (IIPs) appear as an advanced generation of polymeric materials. The presence of specific binding sites from the metal ion/ligand guarantees an exceptional selectivity. IIPs particles are grafted on various substrates in order to increase active surface area and favor the extraction efficiency. IIPs are obtained by several polymerization methods as precipitation, suspension and bulk polymerization [56]. Surface imprinting of MNPs is a modern and promising alternative for Hg analysis that combines a magnetic core with a highly selective surface (IIP) [57].

A novel IIP based on N-(pyridin-2-ylmethyl)ethenamine (V-Pic) was grafted on Fe_3O_4 NPs. It was used with success for the quantification of Hg in fish. Hg was extracted from the sample solution (pH adjusted at 8) by 2 min contact with 0.001 g of the sorbent. After isolation of the MNPs containing Hg, by the action of a magnet, Hg ions were eluted with 5 mL solution of 0.1 mol L^{-1} of HCl and 0.1 mol L^{-1} of EDTA. Finally, the eluent was analyzed by ICP-OES. The procedure exhibited excellent analytical performance, with a relative standard deviation (RSD) of 1.5% and a 0.03 ng mL^{-1} detection limit [57].

Metal organic frameworks (MOF) have also been used to modify Fe_3O_4 NPs, for the extraction of several toxic metal ions [56]. MOFs have highly porous-crystalline structures based on various coordinative connections among metal ions and organic linkers, resulting on materials of outstanding features such as diverse geometrical configurations, tunable pore size, large surface area, and mechanical/thermal stability. The shapes and sizes of the porous determine their selectivity, which justify the application on SPE of such nanocomposites, relatively quickly prepared [56].

The combination of magnetic SPE and MOF lead to superior high-yield extraction in complex matrices. Magnetic MOFs are easily dispersed and magnetically isolated from the extraction medium. Fe_3O_4 NPs have been modified with 4-(5--imidazoledithiocarboxylic acid and then reacted with trimesic acid and Cu(II) acetate to form the metal-organic framework capable of extracting Hg(II). This material has been employed for the separation and preconcentration of Hg(II) in fish and canned tuna samples. It has a great sorption capacity (254 mg g^{-1}) and is reusable during up to 12 sorption/elution cycles, without decreasing extraction efficiency. Hg(II) is eluted with a 1.1 mol L^{-1} thiourea solution and subsequently analyzed by CV AAS. The detection limit and the relative standard deviation of the method are 10 ng L^{-1} and <8.3%, respectively [59].

Magnetic carbon nanomaterials exhibit exclusive physicochemical properties, exceptional thermal/chemical stability and large sorption capacity [56]. The versatile structure of graphene and graphene-based materials explains the exceptional adsorption of metal ions (*i.e.*, Hg(II)), among others [54, 58]. The most recent example of magnetized carbonaceous materials corresponds to the incorporation of $ZnFe_2O_4$ nanoparticles on graphene sheets. The integration of this mixed oxide, $ZnFe_2O_4$, prevents agglomeration of graphene sheets and improves their stability and adsorption capacity. This material, with a preconcentration factor of 30 and a high reusability (50 cycles) was used for the fast extraction of Hg(II) from water and fish samples. Hg was quantified by CV-AAS exhibiting a detection limit of 10 ng L^{-1} and a precision (RSD) of 2.7% [54].

The application of MSPE *on column* minimizes reagent volume consumption and sample handling in comparison to MSPE in batch. However, there is a limited number of applications so far, using MSPE *on column*. Hg isolation from food of animal and vegetable origin (*i.e.*, fish, cabbage, potatoes, green pepper, etc.) was achieved by an *on column* method with a new biosorbent based on the use of fungi *Coprinus micaceus* and γ-Fe_2O_3. Biomagnetic-sorbents, that match with low cost and green procedures, have been used in MSPE. Bacteria, yeast, fungi, and algae are some of the biomaterials employed. In general terms, structure stability is one of the main limitations of this approach. These materials are not widely used for Hg(II) extraction up to now. The fungal biomass immobilized with γ-Fe_2O_3 NPs with a sorption capacity of 26.2 mg $g^{-1}$was incorporated to a polyethylene column (1x10 cm size). The sample solution, resulting on the digestion of the mentioned foodstuffs, was passed through the MSPE column by using a peristaltic pump. Hg(II) ions eluted with 5 mL of 1 mol L^{-1} HCl and were analyzed by ICP-OES. The method, validated with water and fish muscle CRMs presents excellent LODs [62].

The experience related with the use of on-line SPE for Hg detection is scarce, but based on the examples corresponding to other elements some challenges have been identified. The main limitations are the high backpressure on the SPE column and the feasibility of reconditioning/exchange after degradation [56]. In general, backpressure difficulties are due to decreased column permeability with NPs. Off-line Hg detection, performed by a large choice of analytical techniques (Table **1**), is commonly used after MSPE. However, the handling could lead to sample contamination and there is also a risk of non-quantitative recovery of the analyte due to losing aliquots of the sorbent. The on-line hyphenation of SPE, is expected to allow automatization and consequently accuracy and precision. The on-line coupling minimizes the use of solvents and the efficient factor is favored.

The immobilization of the magnetic SPE material onto the inner walls of a knotted reactor placed in the injection valve of the flow injection manifold allows the development of an *on-line* detection system [63]. The flow injection magnetic solid-phase microextraction coupled to CV-ETAAS was used for Hg determination in seafood (fish, mussels). The proposed configuration does not present backpressure problems and represents a promising solution in routine analysis. MNPs were coated with silica, which is recognized by its unique features of ordered pore network, high surface area and versatility regarding further functionalization. The resulting silica coated MNPs were modified with 1,5-bis(di-2-pyridyl)methylene thiocarbohydrazide. The Hg(II) adsorption capacity of the material was 5.22 mg g^{-1}. This green analytical approach is rapid, selective, easy to use and the analytical performances (LOD: 7.8 ng L^{-1}; RSD: 1.7%) are comparable to similar methods for Hg determination [63].

Despite the irrefutable popularity of MSPE for the analysis of Hg(II) in food and environmental samples, its use in Hg speciation is limited to a few applications in food analysis [11, 60, 64, 65].

RECENT ADVANCES IN SPECIATION ANALYSIS OF MERCURY IN FOOD

Hg speciation in food is principally restricted to the quantification of two species: Hg(II) and MeHg, since food safety authorities exclusively regulate the content of such species. In general terms, the most common analytical methods for Hg speciation consist in the hyphenation of a highly effective separation technique and a sensitive elemental detection (*i.e.*, ICP-MS, ICP-OES, AAS, AFS). The most popular separation techniques are unequivocally GC and HPLC. ICP-MS detection is favored by the possibility of multi-elemental analysis, high sensitivity, wide linear range, and isotopic information.

Table 1. Application of magnetized nanomaterials for MSPE of Hg in food.

Magnetic Solid-phase	Analyte	Isolation	Detection	Elution Solvent	Contact Time (min)	Elution Time (min)	pH	Nano-sorbent Amount (mg)	Sorption Capacity (mg g^{-1})	LOD (ng L^{-1})	RSD (%)	Cycles	Matrix	Ref.
Fe_3O_4@SiO_2@polythiophene	Hg(II)	batch	CV-AFS *(off-line)*	3.0 mL, 0.3 mol L^{-1} thiourea in 0.34 mol L^{-1} HCl solution	6.8	8.4	5.2	26	59	20	< 9.2	-	Shrimp, fish, canned tuna	[61]
Fe_3O_4@IIP based on N-(pyridin-2-ylmethyl)ethenamine (V-Pic)	Hg(II)	batch	ICP-OES *(off-line)*	5 mL solution 0.1 mol L^{-1} HCl and 0.1 mol L^{-1} EDTA	2	10	8	10	147	30	1.47		Fish	[57]
GO/Fe_3O_4@ polythiophene	Hg(II)	batch	FI-CV-AAS *(off-line)*	2.5 mL HCl 1.7 mol L^{-1}	21	2	6.5	20	1	25	<10	-	Seafood	[58]
MNPs functionalized with 1,5-bis(d- -2- pyridyl)methylene thiocarbohydrazide	Hg(II)	on column	CV-ETAAS *(on-line)*	0.6 mL min^{-1}, thiourea 6%, $NaBH_4$ 0.002%, NaOH 0.5%	2	1.7	5	50	5.22	7.8	1.7	At least 550	Mussel, fish	[63]
graphene/$ZnFe_2O_4$	Hg(II)	batch	CV-AAS *(off-line)*	5 mL 2.0 mol L^{-1} HCl	3	3	6	100	4.1	10	2.7	50	Fish	[54]
Fe_3O_4@DTIM-HKUST-1(Cu)	Hg(II)	batch	CV-AAS *(off-line)*	3.5 mL 1.1 mol L^{-1} thiourea	8	11	6	24	254	10	<8.3	12	Fish, canned tuna	[59]
bio-MSPE (*C. micaceus* immobilized with γ-Fe_2O_3)	Hg(II)	on column	ICP-OES *(off-line)*	5 mL 1.0 mol L^{-1} HCl	-	1.7	5	100	26.2	40	-	35	Potato, cabbage, ketchup, green pepper, meat, fish, chicken, milk	[62]
Fe_3O_4@SiO_2@γ-mercaptopropyltrimethoxysilane	Hg(II), MeHg, PhHg	batch	HPLC-ICP-MS *(off-line)*	0.5 mL 0.1 mol L^{-1} HNO_3 and 4% (m/v) thiourea	15	2	4	5	29.8 (for Hg(II)), 98.1 (for MeHg)	0.74 (for Hg(II)), 0.67 (for MeHg), 0.49 (for PhHg)	<10		Fish	[60]
poly-*L*-methionine-MWCNTs	Hg(II), MeHg	on column	HPLC-ICP-MS *(on-line)*	10% HCl + 20% methanol	-	10	9	3	-	15 (for Hg(II)), 17 (for MeHg)	<5	575	Fish oil based dietary supplements	[64]

There is a clear upward trend in analytical chemistry towards the development of green and environmentally friendly methods. Hg analysis in food is also concerned by this purpose. The progress on nanomaterials undeniably assists with the development of miniaturized, sensitive and green approaches for the speciation of Hg in foodstuffs. Several non-chromatographic methods have recently been developed for Hg speciation in food (Fig. **1**).

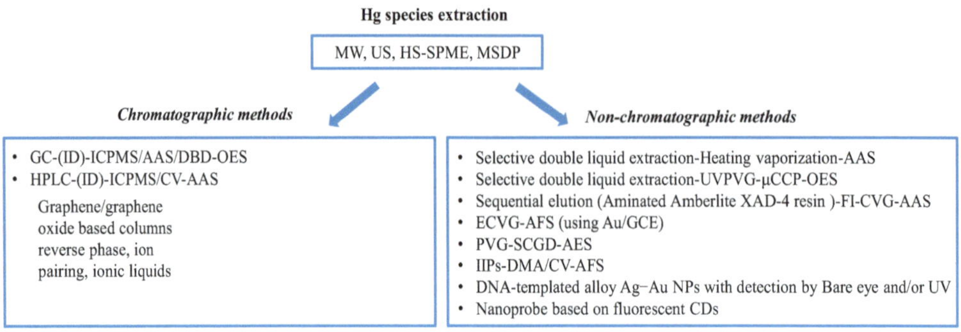

Fig. (1). Chromatographic and non-chromatographic approaches for Hg speciation in food.

GC is one of the traditional separative techniques used for Hg speciation. The hyphenation of GC-ICP-MS is recognized as a powerful tool for Hg speciation. The detection by ICP-MS allows isotopic dilution analysis enhancing accuracy and precision on Hg species determination [23, 66]. A recent scheme for ultrasensitive Hg speciation (Hg(II), MeHg, and EtHg) analysis in rice proposes the hyphenation of GC to DBD-OES. Compactness and cost-effectiveness are some of the main advantages of this approach thanks to the small size, low power, and low gas consumption of DBD-OES [15].

In general terms, GC is characterized for shorter separation times and higher resolution in comparison to HPLC. Nevertheless, contrary to HPLC separation, GC requires the derivatization of the Hg species present in food to volatilize them. This crucial step of sample preparation processes is long, tedious and strongly influenced by the sample matrix. For these reasons, HPLC is often chosen for Hg speciation in food [23].

Hg Speciation in Food by Using High Pressure Liquid Chromatography

HPLC is widely used for Hg speciation in food. The main advances in this regard are largely assisted by nanomaterials, from species extraction [53, 64, 67] to interfaces on the detection system [68 - 70].

Chromatographic Separation Assisted by Solid-Phase Extraction

The key role of SPE on recent methods for Hg analysis in food is not limited to total Hg quantification approaches, but extended to Hg speciation studies. Dispersed solid-phase extraction has a recognized potential on the extraction of elemental species despite being principally applied in the isolation of organic compounds. In comparison to conventional species extraction methods like sonication, heating treatments or MW radiation, MSPD is carried out at room temperature and atmospheric pressure reducing the risks of species instability/degradation. Additionally, this quantitative and fast extraction approach can be considered environmentally friendly due to the low waste generated and the use of reagents with low toxicity. A few but promising analytical methods exploit the advantages of this sample treatment approach for Hg speciation [52, 67].

The first Hg speciation study using MSPD, achieved in 2013, consisted in blending for 3 min 0.2 g of fish sample with 0.5 g of SiO_2, acting as solid support. Hg species were quantitatively extracted with 5 mL of 0.5 mol L^{-1} NaCl and 4.2 mol L^{-1} HCl solution with 1 min of stirring time. After derivatization with sodium tetraphenylborate, separation and determination of Hg(II) and MeHg were performed by GC coupled to mass spectrometry (MS). Considering the matrix effect, quantification was carried out by matrix-matched calibration. After method validation with CRMs, Hg speciation was carried out in tuna, shark and guitar fish. The relative standard deviation was lower than 9.5% and LODs for MeHg and Hg(II) were 0.06 and 0.12 μg g^{-1}, respectively [52].

The hyphenation of MSPD with HPLC represents an evolution with respect to GC speciation considering the non-requirement of derivatization steps. The coupling of these techniques constitutes the basis for a simple on-line solid sampling platform for Hg speciation in fish. MSPD step was improved by the use of multiwall carbon nanotubes (MWCNTs) considered an excellent solid support for Hg species extraction [53, 64, 67]. This carbonaceous nanomaterial is characterized by a large surface area and outstanding mechanical properties. 4 mg of MWCNTs and 1 mg of fish were blended for MSPD, using a solution containing 2% (v/v) HCl and 1.5% (m/v) L-Cys as eluent. Extracted Hg species were directed to the HPLC-ICP-MS system and separated by reversed-phase liquid chromatography. The LODs were 9.9 ng g^{-1} for Hg(II) and 8.4 ng g^{-1} for MeHg, whereas the LOQs were 21.5 ng g^{-1} for Hg(II) and 18.3 ng g^{-1} for MeHg (RSD <10% for Hg(II) and <12.5% for MeHg) [67]. The *on-line* speciation method assisted by MWCNTs is superior to previous MSPD approaches from several points of view. It involves high sample throughput, a low sample (1 mg) and solid support (4 mg) consumption without any derivatization of the extracted species minimizing the contamination risk [67].

Functionalization of MWCNTs is a simple process that increases their selectivity. MWCNTs are decorated with poly-L-methionine, a sulfur-based substrate, on a novel *on-line* SPE-HPLC-ICP-MS system for Hg(II) and MeHg preconcentration from fish muscle and fish oil-based dietary supplements [64]. The preconcentration microcolumn contained a mixture of the functionalized MWCNTs with inert particles of a low density polyethylene wax (Epolene) (1:10) to avoid aggregation and back pressure. The stable sorbent has a life time of at least 575 cycles with an enrichment factor up to 190. Hg(II) and MeHg eluted from the preconcentration microcolumn were separated by HPLC on a C18 column in approximately 10 min using a ternary mixture of 0.5% formic acid, 0.2% 2-mercaptoethanol and 20% methanol as mobile phase [64]. This approach is very attractive considering the LODs (15-17 ng L^{-1}) and the high sample throughput (5 samples h^{-1}), using minimum amount of substrate without derivatization of the mercurial species [64].

MNPs are suitable adsorbents for micro MSPE of different Hg species depending on the reagent used for functionalization [65]. Magnetic core-modified silver nanoparticles (Fe_3O_4@Ag) functionalized with the sodium salt of 2-mercaptoethane-sulphonate adsorb Hg(II) selectively. However, a functionalization of the NPs with L-Cys (Fe_3O_4@Ag@Cys) leads to the retention of the inorganic species, but also monomethyl mercury, dimethyl mercury, ethyl mercury, phenyl mercury and diphenyl mercury. Once eluted with KI, total Hg was measured by ETAAS using silver nitrate and potassium permanganate solutions for chemical modification. Organomercurial species are desorbed in two fractions (MeHg + dimethylmercury (Me_2Hg, $(CH_3)_2Hg$) + EtHg and phenylmercury (PhHg, $C_6H_5Hg^+$) + diphenylmercury (Ph_2Hg, $(C_6H_5)_2Hg$)). Therefore, this method cannot be considered a complete Hg speciation tool, but an easy screening of the organic and inorganic fractions. It was applied to the analysis of edible oils and the low amount of sorbent required (3 mg) facilitated the miniaturization of the procedure. Inorganic and organic Hg levels in vegetal edible oils are lower than the LOD (0.01 µg L^{-1}), contrasting with fish oils where the concentration ranged between 0.08 and 0.45 µg kg^{-1} [65].

An advanced method, by using the same magnetic-core NPs, enabled precise quantification of organomercurial species. Fe_3O_4@SiO_2@γ-mercaptopropyltrimethoxysilane MNPs were used for Hg speciation in water and fish samples. After ultrasound-based extraction of Hg species from fish muscle, the MNPs efficiently retained Hg(II), MeHg and PhHg. A solution of 0.1 mol L^{-1} HNO_3 and 4% (m/v) thiourea was used for analytes desorption. After magnet isolation of the sorbent, Hg species were quantified by HPLC-ICP-MS. The proposed method is characterized by outstanding preconcentration factors (around 175 enhancement factor), fast separation (8 min) and high sensitivity (LOD: 0.49-0.74 ng L^{-1}) [60].

A robust, cost-effective, and sensitive method was developed for MeHg speciation in rice and rice products without derivatization. It consists of a simple sample preparation followed by on-line SPE preconcentration coupled to HPLC-CV-AFS [11]. Rice and baby-food rice samples (0.3 g) were digested with 3 mL THAMH (25% (*m/v*)) for 20 min at 55 °C and 20 min at 60 °C and 1600 W in open vessels. After addition of 2 mL HCl, the mixture was heated again in the MW for 20 min at 55 °C and 20 min at 60 °C. Approximately 60% of the filtered supernatant was diluted up to 50 mL, of which 35 mL were loaded on the SPE column (2.1 x 3 mm) at a speed of 5 mL min^{-1}. A commercial thiol/thiourea silica material was used as sorbent for SPE. Hg species were eluted with 1.5 mmol L^{-1} ammonium pyrrolidine dithiocarbamate in 75% (v/v) methanol and separated *on-line* by a C8 HPLC column. Post column addition of bromine as oxidant and UV irradiation assured oxidation of MeHg to Hg(II). Conventional reduction with acidic tin (II) chloride was used to convert Hg(II) to Hg^0. The sample throughput of the SPE preconcentration coupled to HPLC-CV-AFS method is four per hour [11].

Contrary to fish Hg (speciation) analysis that benefits from a wide range of CRMs for method validation, there are no CRMs for MeHg in rice. Therefore, an elegant solution was cross-validation against standard addition and species specific isotope dilution (SSID)-GC-ICP-MS, which showed no significant differences versus the external calibration with SPE-HPLC-CV-AFS [11, 12]. Isotopically enriched $^{199}Hg(II)$ spike was added to the sample in order to track abiotic MeHg formation in this organic sample. The isotopic standard was enriched in isotope 199 at 98% and was spiked in excess (20-folds) in order to clearly see any alteration on the natural Hg isotopic ratio, attributed to abiotic MeHg formation. The natural isotopic ratio of MeHg remained unalterable, which guaranteed the absence of artificial MeHg formation during the whole analytical procedure- from digestion until determination. Therefore, the MeHg found in rice corresponded to the occurrence of this organomercurial species in food products [11].

The LOD for MeHg by this simple and reliable approach is 0.4 ng L^{-1}. In the analyzed rice grains and baby-food rice samples from local European shops, the concentration varied between 1.56 and 2.69 µg kg^{-1} with a RSD ranging between 5.2 and 16.7%. It is a cost effective method for Hg speciation based on the detection by CV-AFS in comparison with GC/HPLC-ICP-MS approaches. Therefore, it is attractive for routine food control laboratories [11].

Trends in the Choice of Stationary and Mobile Phases in HPLC for Hg Analysis

HPLC is largely used for Hg speciation analysis, mainly using reversed-phase (RP) chromatographic mode (Table **2**). Mobile phases in RP may contain high salt levels and considerable amount of organic solvents. It has negative consequences on the operations of some detectors, such as nebulizer obstruction and interface deposition of carbon and salt in ICP-MS. Species separation by HPLC implies long separation times reducing analytical efficiency and throughput [71]. The large volumes of mobile phase waste, rich on toxic organic compounds, place this technique at a great distance of green chemistry. Hg speciation by ion-pairing--P-ICP-MS using aqueous mobile phases containing minimal quantities of salts, promises to solve the previously mentioned problems.

Two advanced ion-pairing HPLC methods were developed based on positively and negatively charged ion-pairing reagents (tetrabutylammonium hydroxide –TBAH and sodium dodecylbenzene sulfonate –SDBS), where sodium 3-mercapto-1-propylsulfonate (MPS) and L-Cys were included respectively in the mobile phases to transform mercury species into negative and positive Hg-complexes for good resolution. The addition of phenylalanine (Phe) enhanced the analytical efficiency and reduced waste production through reducing the baseline separation time in combination with short C18 guard columns (5 μm, 12.5 mm × 2.1 mm i.d.) [71]. Sample preparation consisted of ultrasonication (15 min at 40 °C) of 0.5 g of fish with two successive additions of 10 mL of a solution containing 5 mol L^{-1} HCl and 2.0 mmol L^{-1} Cys. After centrifugation and 0.45 μm filtration, 5 μL of the extract were injected. The analysis of five freshwater fish samples evidenced that this green and efficient method could be exploited for (food) routine analysis laboratories [71]. Hg(II), MeHg, EtHg and PhHg were baseline separated by two consecutive C18 guard columns with optimized mobile phases (2.0 mmol L^{-1} SDBS + 2.0 mmol L^{-1} Cys + 1.0 mmol L^{-1} Phe (pH 3.0) and 4.0 mmol L^{-1} TBAH + 2.0 mmol L^{-1} MPS + 2.0 mmol L^{-1} Phe (pH 6.0)). The LODs were 0.015, 0.014, 0.028 and 0.042 μg L^{-1} for Hg(II), MeHg, EtHg and PhHg, respectively. The analysis of fish tissues CRMs demonstrated a good recovery of Hg species (91-106%) and the accuracy of the method [71].

Rice consumption is one of the most important sources of exposure to arsenic (As). Therefore, speciation methods enabling the simultaneous speciation of As and Hg in this food product are particularly interesting. Ion-pairing reversed-phase liquid chromatography coupled to ICP-MS was used for the simultaneous speciation of four As species as well as Hg(II), EtHg and MeHg. Taking into account the compromise of best extraction for Hg (60% of total Hg) and As species, 1% HNO_3 MW assisted extraction was chosen. The species were separated by ion pairing-RP liquid chromatography within 20 min (including 10 min of re-equilibration) by using tetrabutylammonium hydroxide (TBAH) as ion-pairing reagent [72]. Hg species were quantified by external calibration. The LODs of the simultaneous As/Hg speciation method were 0.04, 0.01 and 0.03 μg L^{-1} for Hg(II), EtHg, and MeHg, respectively [72]. The main advantage of this approach is the simultaneous speciation of three Hg and four As species in a single extraction and chromatographic run without any requirement of organic solvents in the mobile phase.

Ionic liquids (ILs) have demonstrated to be a fresh alternative to conventional ion-pairing agents, such as tetraalkylammonium salts, for Hg speciation by RP-HPLC-CV-AFS. MeHg, EtHg, and Hg(II) were separated using a C18 column and a gradient developed by mixing methanol and a solution composed of 0.4% (v/v) 1-octyl-3-methylimidazolium chloride [C8mim]Cl, 100 mmol L^{-1} NaCl and 20 mmol L^{-1} buffer citric acid/citrate (pH 2.0) within 12 min. LODs were 0.05, 0.06 and 0.11 μg L^{-1} for MeHg, EtHg and Hg(II), respectively, which are comparable to other methods [73]. The method was applied to a variety of samples, including fresh seafood, canned fish, yeast and garlic. Hg species were extracted from food samples (300 mg) by sonication during 10 min with 5 mL of a 5 mol L^{-1} solution. 5-fold dilution for seafood samples and 10-fold for yeast and garlic was proposed to avoid possible matrix effect. The extraction yield varied according to the matrix, from 51% for yeast up to 87% for seafood. Hg species were separated by HPLC and organic species were photo-oxidized *on-line* in a UV digestion unit by using a 0.75% (w/v) $K_2S_2O_8$ solution [73]. In the analyzed foodstuffs, Hg levels were in a range between 0.3 and 0.7 μg g^{-1} and Hg species distribution varied according to the food sample. MeHg was the major species in seafood while Hg(II) dominated in yeast and garlic [73].

For HPLC analysis, the development of new stationary phases based on nanomaterials opens new perspectives on Hg speciation. A recent example is the successful use of graphene as stationary phase in HPLC. Graphene is a new class of carbon-based nanomaterial, commonly obtained by graphite oxidation and subsequent reduction processes; it does not contain metal impurities. This nanomaterial has outstanding sorptive properties and therefore a great trace metal preconcentration capacity [74]. Both surfaces of the planar sheet are accessible for sorption, resulting in a huge specific surface area. Graphene oxide is oxidized graphene, keeping the singular structures and properties of graphene.

In general terms, the application of graphene to trace metal speciation is rare so far. The unique adsorption properties of graphene and graphene oxide have been exploited for the first time on the development of HPLC stationary phase for Hg speciation. Hg species (inorganic and organic ones) complexes with 2-thiosalicylic acid were attracted by graphene oxide due to π electron stacking. The coupling of the graphene oxide based column to ICP-MS provided LODs of 0.016, 0.027, 0.032 and 0.068 μg L^{-1}, for Hg(II), MeHg, EtHg and PhHg, respectively [75]. Graphene and graphene oxide based columns (4.6 mm i.d. and 5 cm long) did not show different results regarding retention time and peak resolution. The separation of inorganic and organic mercury species was achieved in 12 min, by using 2 mmol L^{-1} TSA and 5 mmol L^{-1} phenylalanine solution at pH 8 with a flow rate of 1.0 mL min^{-1} [75].

The comparison with a commercial C8 column (5 μm, 4.6 mm i.d. and 5 cm long) demonstrated the superior analytical performance of the graphene based developed column [75]. The GO@SiO_2 based column, successfully used for Hg speciation in fish, showed its feasibility for routine Hg speciation analysis by HPLC-ICP-MS. The success of the graphene oxide based column on the retention of Hg species evidenced the potential of this carbon-based nanomaterial as a powerful adsorbent in preconcentration procedures [75].

Vapor Generation of Hg Species Assisted by Nanomaterials

A considerable number of applications of nanomaterials on Hg speciation corresponds to their use in efficient interfaces between liquid chromatography and the detection by CV-AFS (Table **3**). Photochemical vapor generation is an advanced approach that can also be successfully hyphenated to chromatographic and flow injection on one side and atomic detection methods, as AFS, on the other side. It consists of the generation of reducing radicals from low molecular weight organic compounds under UV irradiation. Laser and microwave irradiation are also effective on the generation of free radicals with similar analytical purposes [76, 77].

AFS is a convenient method for Hg determination characterized by its sensitivity, selectivity, and relatively low cost instrumentation. The sensitivity, and consequently the LODs, are strongly influenced by the sample introduction efficiency. Vapor/hydride generation is largely used, guaranteeing matrix separation and an efficient analyte transport. Vapor/hydride generation by using tetrahydroborate is the most frequent approach for Hg determination [68]. The chemical vapor generation method based on the use of tetrahydroborate suffers of interferences that degrade the sensitivity and reproducibility of the method, and the large amount of hydrogen generated can dramatically affect the plasma/flame source.

The photocatalytic properties of TiO_2 are exploited on the development of an efficient interface for Hg speciation by HPLC-CV-AFS [70]. The method allows direct vapor generation of Hg species on nano TiO_2 under UV irradiation in the presence of a mixture of formic acid and sodium formate, acting as a hole scavenger. Compared to the most frequently used system KBH_4/NaOH-HCl, the direct photocatalysis vapor generation on nano TiO_2 clearly exhibited better analytical performance [70]. The method was applied on the investigation of Hg species (Hg(II), MeHg, EtHg and PhHg) in fish, oysters and others seafood, with a LOD between 10 and 70 ng L^{-1}, according to the mercurial species. The principle of *on-line* photocatalyst-assisted vapor generation using nano TiO_2, has also successfully been used in the hyphenation of HPLC with ICP-MS [78].

Table 2. Some recent examples of Hg speciation in food by methods based on HPLC separation.

Chromatographic Method	Column	Mobile Phase	Hg Species	Detection System	Time (min)	LOD (ng L^{-1})	Samples	Ref.
RP	C8 (4.6 x 150 mm, 5 µm)	1.5 mmol L^{-1} ammonium pyrrolidine dithiocarbamate in 75% (v/v) methanol	MeHg	CV-AFS	14	0.4 (MeHg)	Rice, baby-food rice	[11]
Ion-pairing-RP	C18 (4.6 x 150 mm, 5 µm)	Gradient: (A) 5 mmol L^{-1} TBAH, 10 mmol L^{-1} $NH_4H_2PO_4$; (B) 5% (v/v) methanol, 0.1% (m/v) L-Cys, 0.06 mmol L^{-1} CH_3COONH_4 (pH 7.1)	Hg(II), EtHg, MeHg	ICP-MS	20	10 (EtHg), 40 (Hg(II)), 30 (MeHg)	Rice, rice flour standard	[72]
Ion-pairing-RP	C18 guard columns (5 µm, 12.5 mm x 2.1 mm i.d.)	2.0 mmol L^{-1} SDBS + 2.0 mmol L^{-1} Cys + 1.0 mmol L^{-1} Phe (pH 3.0); 4.0 mmol L^{-1} TBAH + 2.0 mmol L^{-1} MPS + 2.0 mmol L^{-1} Phe (pH 6.0)	Hg(II), MeHg, EtHg, PhHg	ICP-MS	3	15 (Hg(II)), 14 (MeHg), 28 (EtHg), 42 (PhHg)	Fish	[71]
Ion-pairing-RP	C18 (150 mm x 4.6 mm)	Gradient: (A) 100% (v/v) methanol (B) 0.4% (v/v) [C8mim]Cl; pH 2.0; 0.02 mol L^{-1} citric acid/citrate buffer; 0.1 mol L^{-1} NaCl	Hg(II), MeHg, EtHg	CV-AFS	12	11 (Hg(II)), 50 (MeHg), 60 (EtHg)	Fresh seafood, canned fish, yeast, garlic	[73]

Chromatographic Method	Column	Mobile Phase	Hg Species	Detection System	Time (min)	LOD (ng L^{-1})	Samples	Ref.
Graphene oxide column	GO@SiO_2 column (500 mm x 4.6 mm)	2 mmol L^{-1} TSA and 5 mmol L^{-1} phenylalanine (pH 8)	Hg(II), MeHg, EtHg, PhHg	ICP-MS	12	16 (Hg(II)), 27 (MeHg), 32 (EtHg), 68 (PhHg)	Fish	[75]

Nanosized TiO_2 has also been included in the integrated MW/UV interface for Hg species determination by HPLC-AFS. Hg(II), EtHg, MeHg were converted to Hg^0 by the combined action of UV and MW in presence of formic acid and Cys in the mobile phase. This method allows the replacement of the commonly used tetrahydroborate system by an environmentally friendly alternative for Hg speciation with LODs between 0.15 and 0.35 μg L^{-1} in fish [68].

The use of ZrO_2-NPs improves the efficiency of the photocatalytic system, compared to nanosized TiO_2 [55]. A UV/nano-ZrO_2/HCOOH photocatalytic reduction unit was developed as an on-line interface between HPLC and AFS for Hg speciation. Hg species were transformed into atomic Hg on nanosized ZrO_2 upon UV irradiation. The sensitive novel system reached LODs of 13, 16 and 24 ng L^{-1} for MeHg, EtHg and Hg(II), respectively, and was successfully applied to the analysis of fish and other seafood samples [69].

Advanced oxidation using Fe_3O_4 MNPs constitutes another example of how nanoparticles assist Hg speciation. Hydride generation/cold vapor inactive species were transformed in active ones by efficient *on-line* post-column advanced oxidation using Fe_3O_4 MNPs without MW/UV irradiation. The LODs of this simple, efficient and green approach were 0.7, 0.8, 0.9 and 1.1 μg L^{-1}, for Hg(II), EtHg, PhHg and MeHg, respectively. Peak broadening was reduced due to the very short length of Fe_3O_4 MNPs column (2.5 cm). This approach, which was applied to Hg speciation in fish, does not require the complex experimental set up of conventional oxidation techniques and eliminates the use of toxic oxidation reagents [79].

Hg Speciation in Food by Non-Chromatographic Methods

Chromatographic methods are absolutely well established on Hg speciation in food. However, the high cost of the instrumentation and analysis as well as the complex sample pre-treatment used in these conventional methods motivated the development of non-chromatographic approaches. Non-chromatographic methods are based on the individual chemical and physical properties of the different species. Low solvent consumption, promptness, and preconcentration are some of the advantages of such methods. The functionalization of nanomaterials opens new perspectives on the development and application of non-chromatographic approaches. Some of these recent and promising methods are presented hereafter.

Table 3. Some examples of nanomaterials used as efficient interfaces between liquid chromatography and CV-AFS detection for Hg speciation in foodstuffs.

Hg Species	Matrix	NPs (action)	Analytical Technique	Mobile Phase	LOD ($\mu g\ L^{-1}$)	Ref.
Hg(II), EtHg, MeHg	Fish	TiO_2 (photocatalysis)	HPLC-AFS	1% methanol, 1% formic acid, 5 mmol L^{-1} Cys, 0.15 mol L^{-1} $NaClO_4$.	0.15-0.35	[68]
Hg(II), EtHg, MeHg, PhHg	Fish, oysters and others seafood	TiO_2 (UV/TiO_2 photocatalysis reaction device)	HPLC-AFS	acetonitrile and water (65:35) containing 1.5 mmol L^{-1} ammonium pyrrolidine dithiocarbamate (pH 5.5)	0.01-0.07	[70]
Hg(II), EtHg, MeHg	Fish, oysters and others seafood	ZrO_2 (UV/ZrO_2/HCOOH photocatalytic reduction unit)	HPLC-AFS	10% HCOOH, pH 3	0.013-0.024	[69]
Hg(II), MeHg, EtHg, PhHg	Fish	Fe_3O_4 (*on-line* advanced oxidation)	HPLC-AFS	acetonitrile and water (10:90) at pH 6.8 containing 0.12% (m/v) L-Cys	0.7-1.1	[79]

Electrochemical vapor generation (ECVG) is an important sample introduction technique based on the reaction between H· derived from the cathode of the electrolytic cell and the analyte. In comparison to photochemical vapor generation (PVG), ECVG is more efficient and environmentally friendly considering that it only requires a simple electrolytic cell and a power supply [80]. ECVG has been coupled to spectroscopic techniques as AAS and AFS for Hg quantification in a wide type of samples. However, its application to speciation is relatively new [81]. The first example corresponds to the development of a L-Cys modified graphite cathode, which efficiently converts Hg(II) and MeHg to Hg vapor [80]. Taking into account that at low current mode (0.2 A) exclusively Hg(II) is efficiently converted to Hg vapor and at high current mode (2.2 A), both MeHg and Hg(II) can be reduced efficiently, a sensitive speciation method was established relying on the current control. The novel non-chromatographic speciation method was validated by the analysis of several CRMs and applied on

Hg speciation in seafood. The superiority of ECVG, compared to conventional CVG was demonstrated since it avoids sodium tetraethylborate and oxidizing agents consumption, minimizing costs and potential environmental pollution risks [80].

An innovator compact method based on flow injection photochemical vapor generation (PVG) coupled with miniaturized solution cathode glow discharge atomic emission spectroscopy (SCGD-AES) was developed for the determination of Hg and its species [82]. This novel method combines the small size, independence of compressed gases and low power consumption of SCGD-AES with the simplicity, miniaturization feasibility and green analytical characteristic of PVG. The SCGD was generated between a miniature hollow titanium tube and a solution emerging from a glass capillary. An automated stopped-flow operation allowed increasing the UV irradiation time on-line. Hg vapor, generated by PVG, was transported to the SCGD for excitation, and finally the emission signals were recorded by a miniaturized spectrograph. Inorganic and organic Hg species (MeHg and EtHg) exhibited dissimilar responses at different wavelengths and powers from the UV lamp. Therefore, Hg speciation was possible by the variation of wavelengths and powers from the UV lamp and irradiation times. Inorganic and organic species were converted to Hg^0 for the determination of total Hg with 8 W/254 nm UV lamp and 60 s irradiation time; while only Hg(II) can be reduced to Hg^0 and determined selectively with 4 W/365 nm UV lamp and 20 s irradiation time. Consequently, organic mercury concentration was calculated by the difference on total Hg and inorganic Hg concentrations. Hg(II), less toxic than organic species, was successfully used as a primary standard for the quantification of Hg considering the similar sensitivity at 8 W/254 nm. This pioneer PVG-SCGD-AES approach results on a 365-fold improvement in Hg(II) LOD (0.2 $\mu g\ L^{-1}$) compared to SCGD-AES. The method was validated for water and biological samples including fish and could be extrapolated to other elements [82]. Nickel demonstrated to be an optimum enhancement reagent for the PVG efficiency of Hg(II) and MeHg. Under optimized conditions, the addition of nickel (10 mg L^{-1}) increased the signal intensity of Hg(II) and MeHg around 1.5 times compared to that without nickel ion. Separative techniques like GC and HPLC could be potentially combined to PVG-SCGD-AES, opening new possibilities for Hg speciation analysis [82].

Double liquid-liquid extractions and/or selective derivatization are commonly used for Hg speciation by non-chromatographic methods [83]. A recent example is the introduction of methyl isobutyl ketone on the sample treatment of seafood as degassing agent leaving Hg into the water phase. MeHg was extracted from an aliquot of the aqueous phase with HBr, $CuCl_2$ and toluene, and back-extracted into a L-Cys–sodium acetate solution. This method was proposed for quantification of both total Hg and MeHg in two consecutive steps using a single Hg commercial standard solution by heating vaporization AAS. Despite most analytical studies focused on fish muscle, liver and gonads are also consumed. The analysis revealed remarkable differences between the MeHg fraction in sea urchins and squids glands [84]. This protocol, validated by the analysis of several CRMs, is much simpler and cost-effective in comparison with chromatographic techniques [84].

A non-chromatographic Hg speciation method based on double liquid-liquid extraction and measurement by flow injection UV photochemical vapor generation capacitively coupled plasma microtorch optical emission spectrometry (UVPVG-μCCP-OES) was developed for Hg determination in fish and others seafood [83]. In brief, Hg species extracted from 0.2 g of lyophilized fish and seafood samples were confined in 10 mL hydrobromic acid (47%), followed by MeHg extraction in toluene and back-extraction in 1% L-Cys solution. MeHg was quantified by UV-PVG-μCCP-OES and external calibration against Hg(II) standards in formic acid. LOD and LOQ (2 μg kg^{-1} and 6 μg kg^{-1}, respectively) were improved in comparison to thermal decomposition atomic absorption spectrometry (TDAAS) reference method and other methods used for Hg speciation in seafood [83]. Hg(II) was determined by the difference between total Hg and MeHg concentrations [85]. This new miniaturized approach involves a low power/Ar consumption plasma microtorch (100 mL min^{-1}) and a low resolution microspectrometer. The developed method is simple and safe concerning sample preparation through elimination of conventional, harmful reductants and attractive for using unsophisticated instrumentation [85].

A modification of this method, based on ultrasound assisted extraction of Hg(II) species with formic acid go further on its green concept. In that case the use of toluene is no longer required, formic acid being the unique reagent. Concentration of MeHg was determined by subtraction of Hg(II) amount from total Hg. The eco-scale non-chromatographic method, with LODs and LOQs of 4.8 and 14.4 μg kg^{-1} respectively, was applied on Hg speciation in fish [86].

Non-Chromatographic Speciation Assisted by SPE

Amberlite XAD-series resins, easy derivatized and highly stable in a wide pH range, are on the top of sorbents used for speciation. Aminated Amberlite XAD-4 resin was proposed as a solid-phase extractant for Hg speciation in seafood (fish and mussel). Hg(II) and MeHg were sequentially eluted from the aminated XAD-4 resin column coupled to flow injection cold vapor generation AAS. 400 mg of AAXAD-4 were placed into a SPE cartridge column (3 mL volume, 8 mm i.d., 65 mm length) and sample solution passed through the column at a flow rate of 1.5

mL min^{-1}. Hg(II) and MeHg eluted consecutively with 10 mL of 0.1% (w/v) thiourea in 3% (v/v) HCl and 10 mL of 6.0 mol L^{-1} HCl solution at 1.5 mL min^{-1} flow rate, respectively. Hg(II) was directly quantified by FI-CVG-AAS, while MeHg was previously oxidized before analysis with 20 µL of 1% (m/v) $KMNO_4$. The enhancement factor for Hg(II) and MeHg was 11.7 and 17.2, respectively, and the column did not show significant decay in the Hg species extraction efficiency up to 60 cycles. This method (LOD: 0.148 and 0.157 µg L^{-1} for Hg(II) and MeHg, respectively) allowed semiautomatic quantification of Hg species in seafood [87].

IIPs also assist Hg speciation in food [38, 88, 89]. For example, one of the few Hg speciation studies in wine consists of the high selectivity and preconcentration towards Hg(II) of a non-chromatographic method by using a core-shell ion-imprinted sorbent. Airborne particulate matter deposition and the grapevine intake from the soil/groundwater system explained the presence of Hg, and other toxic elements in wine [88]. Considering the low Hg content, usually at sub µg L^{-1}, the speciation required preconcentration steps.

The Hg(II) ion-imprinted sorbents were synthesized via copolymerization using MAA as a functional monomer, as a crosslinking agent in the presence of Hg-complexes with 1-pyrrolidinedithiocarboxylic acid, as chelating agent. Silica gel acted as supporting material, chemically modified with 3-(trimethoxysilyl)propyl methacrylate. The sorbent exhibited a high specificity for the retention of Hg(II) without interferences of the organic matrix and benefits from a fast kinetics of sorption and desorption processes. Hg speciation in white and red wines (20-50 mL) was carried out in a column solid-phase extraction scheme using 100 mg of sorbent. Hg(II) was quantified in column eluates by CV-AAS and MeHg concentration was determined by the difference between total Hg and Hg(II) concentrations [88]. The LOD for Hg(II) (0.02 µg L^{-1}), was comparable to those obtained by GC-ICP-MS in wine without requirement of derivatization steps. This sorbent exhibited a great chemical and mechanical stability, unalterable after more than 50 cycles of sorption/desorption [88].

The combination of a thiourea derivative polymer with DMA was proposed for screening and pre-concentration of Hg species. The method was based on the use of a new selective polymer, synthesized by using 2-(methacryloylamino) ethyl 2-methyl acrylate as cross-linker, 1-phenyl-3-(-3-vinyl phenyl) thiourea as monomer and azobisisobutyronitrile as initiator. Hg species were extracted from fish by ultrasonication and loaded on the highly selective polymer. The speciation strategy consisted of the specific release of MeHg at 120 °C, followed by the standard DMA temperature ramp up to 600 °C for the quantification of Hg(II). On the first heating step (at 60 °C), MeHg was recovered up to 50%, being the rest of the organomercurial species, measured in conjunction with Hg(II) after application of higher temperatures [38]. The method was therefore considered for rapid screening, and not for precise quantification of Hg species. This innovative approach provides a fast MeHg screening in food with simple handling and sample pretreatment.

A more advanced approach was based on the use of a new polymeric material that allowed, for the first time, the retention of both species, Hg(II) and MeHg. After sequential elution, Hg(II) and MeHg were quantified by CV-AFS. The polymer was synthesized by using vinyl derivative of 8-hydroxiquinoline as monomer, and 2-(methacryloylamino) ethyl 2-methyl acrylate as co-monomer in presence of azobisisobutyronitrile, as initiator. At least 20 cycles of Hg species sorption/desorption can be carried out, with a reproducibility of 5% [89]. The analytical procedure, with a quantification limit of 0.015 mg kg^{-1}, was validated with a tuna CRM (BCR 464) and applied to commercial fish. Hg species extraction from lyophilized fish samples was achieved with 5 mol L^{-1} HCl under ultrasonication. Highly acidic fish extracts were diluted before loading on the polymer cartridge in order not to disturb the optimum sorption loading, at neutral pH. 1 mL of the fish extract was loaded on a conventional glass solid-phase extraction cartridge. Hg(II) was eluted by the addition of 3 mL of 2 mol L^{-1} HCl in methanol for Hg (II), followed by the MeHg elution with 3 mL of NaClO 20% in 1 mol L^{-1} HCl. The proposed method involves minimum handling and could be exploited for Hg speciation in routine laboratories [89]. This rapid, low cost, and simple Hg speciation approach also reduces considerably the reagents waste compared with conventional speciation approaches.

Non-Chromatographic Methods Assisted by Nanomaterials

An innovative recent technology proposes the use of gold particles deposited glassy carbon electrode (Au/GCE) as cathode for ECVG combined with AFS for Hg speciation. The Au/GCE exhibited outstanding catalytic property for electrochemical conversion of MeHg to gaseous mercury. Highly consistent or distinct difference between the AFS signals of MeHg and Hg(II) could be achieved by control of ECVG electrolytic parameters. The method, based on the divergent electrochemical reactivity of Hg(II) and MeHg showed high sensitivity (LOD 0.44 µg kg^{-1}) and good repeatability (<6%, RSD) in fish analysis [81]. This sensitive, green and simple technique has an enormous potential in food routine laboratories.

A visual detection method of MeHg and EtHg in fish samples has been developed based on DNA-templated alloy Ag–Au NPs. The speciation approach relies on the ability of DNA sequences containing T to combine with organomercurial species. The different and tunable binding force of DNA sequence to Hg species by modifying the number and locations of T bases explains the application of T-rich DNA on Hg speciation [90].

Two T-rich aptamers were designed: H_{T5} for exclusive recognition of MeHg and H_{T7} that recognizes MeHg and EtHg. Therefore, quantification of EtHg was carried out by subtraction of MeHg obtained with H_{T5} from the total MeHg and EtHg concentration obtained with H_{T7}. In presence of Ag^+ and Au^+, the aptamers preferentially bind the organomercurial species and consequently the formation of alloy Ag-Au NPs after reduction producing the color change of the solution (from colorless to purple). The presence of Hg(II) and other metallic ions, even at higher concentration than MeHg and EtHg did not interfere the detection of these organic species. It demonstrates the excellent recognition specificity of the designed aptamers [90].

This method allows bare eye detection of the organomercurial species at 1.0 mg kg^{-1}. LODs are around 10 times lower by using UV-Vis spectrometer as detector. The method was applied to the detection and quantification of organomercurial species in fish, extracted by MW acid digestion with a recovery of 101-109% (RSD <8%). The instrument-free visual discrimination and detection of MeHg and EtHg constitutes a simple, rapid, and cost effective alternative for Hg speciation in food [90].

Carbon dots (CDs), a new alternative to quantum dots, are fluorescent nanoparticles with excellent optical probes (highly photo stables, strong fluorescence and tunable color emission). They are attractive also considering that in contrast to quantum dots, their simple synthesis, using eco-friendly carbon sources (*i.e.*, glucose, sucrose) does not involve precursors containing toxic elements, becoming a green alternative [74, 91].

The applications of CDs in elemental speciation are still scarce but promising. A fast and simple fluorescent assay has been recently developed for MeHg detection using D-fructose as carbon source. Ultrasonication was used for simultaneous CDs fabrication and selective recognition of the target analyte. The assay was accomplished within 1 min, exhibiting a LOD of 5.9 nmol L^{-1} of MeHg (RSD 2.2%). Hg species were extracted from marine animal tissues CRMs (0.4 g) by ultrasonication during 4 min at 12% of amplitude in 5 mol L^{-1} HCl solution. MeHg was quantified by the fluorescent assay after filtration (0.45 μm) and clean up (C18 cartridges) of the fish extracts. MeHg recognition was attributed to the hydrophobicity of this species and its capability to facilitate a non-radiative electron/hole recombination [91]. The validation of this environmentally friendly method with water and marine animal tissues CRMs guarantees its successful application in (sea)food and beverages.

ADVANCES AND TRENDS IN THE USE OF HG STABLE ISOTOPES

The use of stable isotope tracers is a great support for developing analytical methods for Hg speciation. The addition of isotopically enriched species, usually ^{199}Hg(II) and ^{201}MeHg, allows monitoring and correction of losses and/or species transformation from the sample preparation steps. Isotopic dilution analysis is a powerful tool in analytical quality assurance and quality control. Its application in Hg analysis in food is extended to a wide variety of samples as seafood [66, 92], rice [11, 14] and breast milk [23].

Species specific isotopic dilution is a smart solution to the quantification of Hg species in absence of CRMs. In complex rice and breast milk samples, it proves the validation of speciation methods based on both GC and HPLC hyphenated to ICP-MS [11, 14, 23].

The potential of Hg stable isotopes in food is not restricted to isotopic dilution analysis. Hg isotopic fractionation-the natural isotopic abundance variation-can be also exploited in dietary studies.

Hg is recognized as a "model element" for isotopic fractionation studies. It has several stable isotopes that exhibit mass-dependent and mass-independent fractionation [93], which, combined, provide precious information on sources discrimination, reactivity and fate of Hg in a biogeochemical context. The improvement of analytical methods and the development of multiple collector inductively coupled mass spectrometry (MC-ICP-MS) has drastically expanded the research on non-traditional stable isotope deviated from traditional geological applications.

The great relevance of Hg isotopic signature on a large variety of subjects such as the elucidation of metabolic pathways [14, 94 - 97], tracking of pollution sources [98, 99], discrimination of human exposure pathways [100 - 103], and biogeochemical processes [93, 104 - 107] has been demonstrated. The potential of the Hg isotopic pattern can be extended to food products.

Hg isotopic composition has been determined in rice and fish. Hg isotopic pattern in fish is an excellent tool for tracking Hg (pollution) sources in aquatic ecosystems. Moreover, the isotopic pattern of Hg in this foodstuff is extremely useful for the determination of Hg human exposure. In general, an enrichment of heavier isotopes of approximately 2‰ (δ^{202} Hg) is observed in hair as compared to the levels of these isotopes in consumed fish. Additionally, the preservation of mass independent fractionation signatures in human's hair and urine allows the discernment between occupational exposure (gold mining, dental professionals) and fish ingestion [100, 102]. The clear differences between Oceanic and coastal fish Hg isotopic patterns provide the hints to successfully identify Oceanic and mixed seafood consumers by the analysis of their hair [103]. More recently, the characterization of Hg stable isotopes enabled the discernment of different populations of European seabass, *Dicentrarchus labrax*, what can be extrapolated to high resolution dietary discrimination in a local scale [108].

In the case of rice, it has provided, among others, crucial information about sources of Hg contamination. The fraction of atmospheric Hg in the plants follows the trend: leaf > stem > seeds > roots [97]. The combination of Hg speciation and isotopic composition revealed the dissimilar uptake and translocation pathways of Hg species from paddy soil to the edible part. The fact that seeds, containing higher MeHg proportion, exhibit significantly heavier Hg isotopes is understood as a facilitated MeHg uptake and transport to the seed in comparison to inorganic Hg species [14].

In general, most of Hg isotopic studies are so far restricted to elemental isotopic pattern determination. The major drawbacks of Hg isotopic analysis are: 1) the high cost of the multicollector instrument, being exclusive of few laboratories; 2) the reduced number of CRM for the validation of Hg elemental isotopic characterization and the absolute lack of such materials regarding the isotopic composition of Hg species and 3) the elevated concentrations usually required for a precise analysis.

The analysis is based on a complete digestion of the sample, usually by acid digestion, followed by isotopic determination through continuous CV-MC-I-P-MS. In order to diminish the problems associated with Hg isotopic characterization at ultra-trace levels, an on-line pre-concentration method was developed based on the hyphenation of a commercially available cold vapor generation dual gold amalgamation system to a multi-collector ICP-MS. The automated method supports a rapid throughput and consumes less than 100 mL of sample, representing a reduction of the sample volume of more than two orders of magnitude in comparison to existing off-line pre-concentration strategies. This method proved to be accurate and precise for the determination of total Hg isotopic composition of fish [109].

The relatively recent coupling of GC to MC-ICP-MS [110] constitutes a remarkable evolution on isotopic analysis since it allows the characterization of both Hg species, named compound-specific stable isotope analysis (CSIA) method. An automated on-line preconcentration system, which allows injections volumes up to 100 folds higher than in a conventional GC-MC-ICP-MS scheme has been introduced in order to be able to analyze samples with low Hg content. Preconcentration and transfer of Hg species (Hg(II) and MeHg) to the column was achieved with a programmed temperature vaporization injector. Hg species were efficiently preconcentrated and fish reference materials with concentrations down to 150 ng g^{-1} were successfully analyzed.

Another alternative to the isotopic characterization of MeHg is via a selective extraction method for MeHg, followed by continuous CV-MC-ICP-MS [111]. It has been applied to the analysis of rice samples. Briefly, 0.25-0.5 g were extracted with 5 mL of 30% w/w NaBr in 4 mol L^{-1} H_2SO_4, 10 mL of 2.5% w/w $CuSO_4$ and 10 mL of toluene. The mixture was shaked for 1 h at 420 rpm, in order to transfer MeHgBr from the matrix to the toluene phase, followed by a back-extraction with aqueous sodium thiosulfate solution (0.005 mol L^{-1}) to convert the MeHgBr into a stable MeHg–thiosulfate complex. Hg isotopic characterization of MeHg was then carried out by CV-MC-ICP-MS after dilution of the MeHg-thiosulfate solution with a 3:1 v/v mixture of HNO_3 and HCl. Concerning the stable isotopic composition of Hg(II), it was determined by isotopic mass balance difference between total Hg and MeHg isotopic patterns. The method was validated with a fish CRM and applied to the analysis of rice and human hair samples in order to understand human exposure pathways [9].

A promising chromatographic methodology has recently been validated for Hg isotope ratio measurement of MeHg in fish tissues at environmentally relevant levels. It is based on the isocratic separation of Hg(II) and MeHg in a reversed-phase HPLC column with a mobile phase containing L-Cys as complexing agent. One of the main advantages of this approach is the non-requirement of derivatization steps. The MeHg fraction isolated from the chromatographic columns was completely oxidized, prior to CV-MC-ICP-MS, due to the challenges driven by the non-reactivity of MeHg with $SnCl_2$ and, interferences caused by the presence of reduced sulfur in the mobile phase [112]. Considering that for the moment there are not commercially available CRMs for isotopic composition of Hg species, it is extremely valuable to have the possibility of contrasting the measurements obtained by different methods, such as GC and HPLC coupled to MC-ICP-MS.

CONCLUDING REMARKS

The urgency to develop sensitive and environmentally friendly methodologies promotes the advances in the analysis of Hg in food. Such development is greatly supported by the progress in nanomaterials that assist in the whole procedure comprising Hg (species) extraction, preconcentration, separation and detection. Hg isotopic fractionation emerges as a powerful tool on dietary studies, allowing the discernment among several Hg sources of exposure. In addition, the identification of pollution sources in food is also possible through the Hg isotopic signature. It would be extremely valuable on the development of strategies that alleviate the impact of Hg contamination sources in foodstuffs.

Seafood monopolizes the attention of the analytical community that validated a wide variety of analytical approaches for such matrix. However, the global consumption of rice, considered an important source of Hg, claims for adequate methodologies to this complex samples.

Hg analysis in food is motivated by the human risk associated to dietary exposure. Consumer's guidelines are based on the assumption that MeHg from food is absolutely absorbed after human ingestion. However, it seems to overestimate the real assimilation of MeHg, since the nature of the matrix, fish species, cooking processes and the co-ingestion of certain food components seems to play a relevant role on the potential human assimilation of ingested Hg [113 - 118]. For example, cooking practices like, steaming, boiling, grill and frying have been evaluated [113, 114, 118, 119], resulting in a reduction of the potential absorption at the intestinal epithelium level up to 90% [114], The co-ingestion of polyphenol-rich beverages like tea and coffee or pure polyphenols are leading to a decrease on Hg bioaccessibility [116, 118, 119].

The existing methods for Hg analysis in food focus on the quantification of Hg and its species based on the consumer guidelines, which set safe consumption thresholds. Nevertheless, some new risk assessment criteria are under development. A new seafood safety index named Selenium health benefit value (HBVSe) is a good case in point, which is based on the observation that Se is an antagonist of Hg toxicity. It considers both, MeHg and Se concentrations in fish, and shows more reliable accuracy than predictions based exclusively on MeHg [120]. In summary, the future and perspectives of Hg analysis in food will be absolutely influenced by the trends and evolution of consumption guidelines.

LIST OF ABBREVIATIONS

AAS	Atomic Absorption Spectrometry
AFS	Atomic Fluorescence Spectrometry
Au/GCE	Gold Particle Deposited Glassy Carbon Electrode
ASV	Anodic Stripping Voltammetry
CDs	Carbon Dots
CRMs	Certified Reference Materials
CSIA	Compound-Specific Stable Isotope Analysis
CV	Cold Vapor

Cys	Cysteine
DMA	Direct Mercury Analyzer
DBD	Dielectric Barrier Discharge
ECVG	Electrochemical Vapor Generation
ETAAS	Electro Thermal Atomic Absorption Spectrometer
ETV	Electro Thermal Vaporization
GC	Gas Chromatography
GFAAS	Graphite Furnace AAS
HPLC	High Performance Liquid Chromatography
HS-SPME	Headspace Solid-phase Microextraction
ICP-AES	Inductively Coupled Plasma Atomic Emission Spectrometry
ICP-MS	Inductively Coupled Plasma Mass Spectrometry
ID	Isotopic Dilution
IIP	Ion Imprinted Polymers
ILs	Ionic Liquids
LOD	Limit of Detection
LOQ	Limit of Quantification
MS	Mass Spectrometry
MPS	3-mercapto-1-propylsulfonate
MSPE	Magnetic Solid-Phase Extraction
MSPD	Matrix Solid-Phase Dispersion
MC-ICP-MS	Multiple Collector Inductively Coupled Mass Spectrometry
MW	Microwave
MWCNTs	Multiwall Carbon Nanotubes
NPs	Nanoparticles
OES	Optical Emission Spectrometry
PCs	Porous Carbons
RSD	Relative Standard Deviation
TWI	Tolerable Weekly Intakes
PVG	Photochemical Vapor Generation
RP	Reversed Phase
SD-SEGD	Single Drop Solution Electrode Glow Discharge Induced
SDBS	Sodium Dodecylbenzene Sulfonate
HBVSe	Selenium Health Benefit Value
SPE	Solid-Phase Extraction

SSID	Species Specific Isotope Dilution
SR-μXRF	Synchrotron Radiation Microscopic X-ray Fluorescence
T	Thymine
TBAH	Tetrabutylammonium Hydroxide
T-Hg	Total Mercury
UV	Ultraviolet
UVPVG-μCCP-OES	UV Photochemical Vapor Generation Capacitively Coupled Plasma Microtorch Optical Emission Spectrometry
WHO	World Health Organization
XANES	X-ray Absorption Near-Edge Spectroscopy

CONSENT FOR PUBLICATION

Not applicable.

CONFLICT OF INTEREST

The author confirms that this chapter contents have no conflict of interest.

ACKNOWLEDGEMENTS

The author thanks The Agence Nationale de la Recherche (MERSEL Project No. ANR-18-CE34-0004-01) for funding and R. Saz for English editing.

REFERENCES

[1] Harris HH, Pickering IJ, George GN. The chemical form of mercury in fish. Science 2003; 301(5637): 1203.
[http://dx.doi.org/10.1126/science.1085941] [PMID: 12947190]

[2] Lamborg CH, Hammerschmidt CR, Bowman KL, *et al.* A global ocean inventory of anthropogenic mercury based on water column measurements. Nature 2014; 512(7512): 65-8.
[http://dx.doi.org/10.1038/nature13563] [PMID: 25100482]

[3] Lavoie RA, Bouffard A, Maranger R, Amyot M. Mercury transport and human exposure from global marine fisheries. Sci Rep 2018; 8(1): 6705.
[http://dx.doi.org/10.1038/s41598-018-24938-3] [PMID: 29712952]

[4] Meng B, Feng X, Qiu G, *et al.* Distribution patterns of inorganic mercury and methylmercury in tissues of rice (*Oryza sativa L.*) plants and possible bioaccumulation pathways. J Agric Food Chem 2010; 58(8): 4951-8.
[http://dx.doi.org/10.1021/jf904557x] [PMID: 20369851]

[5] Yin Y, Li Y, Tai C, Cai Y, Jiang G. Fumigant methyl iodide can methylate inorganic mercury species in natural waters. Nat Commun 2014; 5: 4633.
[http://dx.doi.org/10.1038/ncomms5633] [PMID: 25137238]

[6] Feng X, Qiu G. Mercury pollution in Guizhou, southwestern China - an overview. Sci Total Environ 2008; 400(1-3): 227-37.
[http://dx.doi.org/10.1016/j.scitotenv.2008.05.040] [PMID: 18617222]

[7] Horvat M, Nolde N, Fajon V, *et al.* Total mercury, methylmercury and selenium in mercury polluted areas in the province Guizhou, China. Sci Total Environ 2003; 304(1-3): 231-56. [http://dx.doi.org/10.1016/S0048-9697(02)00572-7] [PMID: 12663187]

[8] Feng X, Li P, Qiu G, *et al.* Human exposure to methylmercury through rice intake in mercury mining areas, Guizhou province, China. Environ Sci Technol 2008; 42(1): 326-32. [http://dx.doi.org/10.1021/es071948x] [PMID: 18350916]

[9] Li P, Du B, Maurice L, *et al.* Mercury isotope signatures of methylmercury in rice samples from the wanshan mercury mining area, china: Environmental implications. Environ Sci Technol 2017; 51(21): 12321-8. [http://dx.doi.org/10.1021/acs.est.7b03510] [PMID: 28958148]

[10] Meng B, Feng X, Qiu G, Anderson CWN, Wang J, Zhao L. Localization and speciation of mercury in brown rice with implications for pan-Asian public health. Environ Sci Technol 2014; 48(14): 7974-81. [http://dx.doi.org/10.1021/es502000d] [PMID: 24925231]

[11] Brombach CC, Ezzeldin MF, Chen B, Corns WT, Feldmann J, Krupp EM. Quick and robust method for trace determination of MeHg in rice and rice products without derivatisation. Anal Methods 2015; 7: 8584-9. [http://dx.doi.org/10.1039/C5AY01640A]

[12] Brombach C-C, Manorut P, Kolambage-Dona PPP, *et al.* Methylmercury varies more than one order of magnitude in commercial European rice. Food Chem 2017; 214: 360-5. [http://dx.doi.org/10.1016/j.foodchem.2016.07.064] [PMID: 27507486]

[13] Jiménez-Moreno M, Lominchar MÁ, Sierra MJ, Millán R, Martín-Doimeadios RCR. Fast method for the simultaneous determination of monomethylmercury and inorganic mercury in rice and aquatic plants. Talanta 2018; 176: 102-7. [http://dx.doi.org/10.1016/j.talanta.2017.08.015] [PMID: 28917728]

[14] Feng C, Pedrero Z, Li P, *et al.* Investigation of Hg uptake and transport between paddy soil and rice seeds combining Hg isotopic composition and speciation Elementa 2016; 2016. [http://dx.doi.org/10.12952/journal.elementa.000087]

[15] Lin Y, Yang Y, Li Y, *et al.* Ultrasensitive speciation analysis of mercury in rice by headspace solid phase microextraction using porous carbons and gas chromatography-dielectric barrier discharge optical emission spectrometry. Environ Sci Technol 2016; 50(5): 2468-76. [http://dx.doi.org/10.1021/acs.est.5b04328] [PMID: 26828416]

[16] Sundberg J, Oskarsson A. Placental and lactational transfer of mercury from rats exposed to methylmercury in their diet: Speciation of mercury in the offspring. J Trace Elem Exp Med 1992; 5: 47-56.

[17] Palkovicova L, Ursinyova M, Masanova V, Yu Z, Hertz-Picciotto I. Maternal amalgam dental fillings as the source of mercury exposure in developing fetus and newborn. J Expo Sci Environ Epidemiol 2008; 18(3): 326-31. [http://dx.doi.org/10.1038/sj.jes.7500606] [PMID: 17851449]

[18] Richardson GM, Wilson R, Allard D, Purtill C, Douma S, Gravière J. Mercury exposure and risks from dental amalgam in the US population, post-2000. Sci Total Environ 2011; 409(20): 4257-68. [http://dx.doi.org/10.1016/j.scitotenv.2011.06.035] [PMID: 21782213]

[19] Vieira SM, de Almeida R, Holanda IBB, *et al.* Total and methyl-mercury in hair and milk of mothers living in the city of Porto Velho and in villages along the Rio Madeira, Amazon, Brazil. Int J Hyg Environ Health 2013; 216(6): 682-9. [http://dx.doi.org/10.1016/j.ijheh.2012.12.011] [PMID: 23340120]

[20] Rebelo FM, Cunha LRD, Andrade PD, Costa Junior WAD, Bastos WR, Caldas ED. Mercury in breast milk from women in the Federal District, Brazil and dietary risk assessment for breastfed infants. J Trace Elem Med Biol 2017; 44: 99-103.

[http://dx.doi.org/10.1016/j.jtemb.2017.06.009] [PMID: 28965609]

[21] Iwai-Shimada M, Satoh H, Nakai K, Tatsuta N, Murata K, Akagi H. Methylmercury in the breast milk of Japanese mothers and lactational exposure of their infants. Chemosphere 2015; 126: 67-72. [http://dx.doi.org/10.1016/j.chemosphere.2014.12.086] [PMID: 25723533]

[22] Miklavčič A, Cuderman P, Mazej D, *et al.* Biomarkers of low-level mercury exposure through fish consumption in pregnant and lactating Slovenian women. Environ Res 2011; 111(8): 1201-7. [http://dx.doi.org/10.1016/j.envres.2011.07.006] [PMID: 21835399]

[23] Vacchina V, Séby F, Chekri R, *et al.* Optimization and validation of the methods for the total mercury and methylmercury determination in breast milk. Talanta 2017; 167: 404-10. [http://dx.doi.org/10.1016/j.talanta.2017.02.046] [PMID: 28340739]

[24] Cerbino MR, Vieira JCS, Braga CP, *et al.* Metalloproteomics approach to analyze mercury in breast milk and hair samples of lactating women in communities of the amazon basin, brazil. Biol Trace Elem Res 2018; 181(2): 216-26. [http://dx.doi.org/10.1007/s12011-017-1057-4] [PMID: 28547667]

[25] Millour S, Noël L, Kadar A, *et al.* Pb, Hg, Cd, As, Sb and Al levels in foodstuffs from the 2nd French total diet study. Food Chem 2011; 126(4): 1787-99. [http://dx.doi.org/10.1016/j.foodchem.2010.12.086] [PMID: 25213958]

[26] Rodushkin I, Engström E, Sörlin D, Baxter D. Levels of inorganic constituents in raw nuts and seeds on the Swedish market. Sci Total Environ 2008; 392(2-3): 290-304. [http://dx.doi.org/10.1016/j.scitotenv.2007.11.024] [PMID: 18155749]

[27] da Silva MJ, Paim APS, Pimentel MF, Cervera ML, de la Guardia M. Determination of total mercury in nuts at ultratrace level. Anal Chim Acta 2014; 838: 13-9. [http://dx.doi.org/10.1016/j.aca.2014.06.049] [PMID: 25064238]

[28] Kannamkumarath SS, Wuilloud RG, Caruso JA. Studies of various elements of nutritional and toxicological interest associated with different molecular weight fractions in Brazil nuts. J Agric Food Chem 2004; 52(19): 5773-80. [http://dx.doi.org/10.1021/jf0496649] [PMID: 15366819]

[29] Pohl P, Bielawska-Pohl A, Dzimitrowicz A, *et al.* Recent achievements in element analysis of bee honeys by atomic and mass spectrometry methods, *TrAC*. Trends Analyt Chem 2017; 93: 67-77. [http://dx.doi.org/10.1016/j.trac.2017.05.009]

[30] Quinto M, Miedico O, Spadaccino G, *et al.* Characterization, chemometric evaluation, and human health-related aspects of essential and toxic elements in Italian honey samples by inductively coupled plasma mass spectrometry. Environ Sci Pollut Res Int 2016; 23(24): 25374-84. [http://dx.doi.org/10.1007/s11356-016-7662-5] [PMID: 27696193]

[31] Stocker A, Schramel P, Kettrup A, Bengsch E. Trace and mineral elements in royal jelly and homeostatic effects. J Trace Elem Med Biol 2005; 19(2-3): 183-9. [http://dx.doi.org/10.1016/j.jtemb.2005.08.004] [PMID: 16325534]

[32] Vieira HP, Nascentes CC, Windmöller CC. Development and comparison of two analytical methods to quantify the mercury content in honey. J Food Compos Anal 2014; 34: 1-6. [http://dx.doi.org/10.1016/j.jfca.2014.02.001]

[33] Pandelova M, Lopez WL, Michalke B, Schramm KW. Ca, Cd, Cu, Fe, Hg, Mn, Ni, Pb, Se, and Zn contents in baby foods from the EU market: Comparison of assessed infant intakes with the present safety limits for minerals and trace elements. J Food Compos Anal 2012; 27: 120-7. [http://dx.doi.org/10.1016/j.jfca.2012.04.011]

[34] Ribeiro RFL, Germano A. Development and validation of a method for the determination of Hg in animal tissues (equine muscle, bovine kidney and swine kidney, and poultry muscle) by direct mercury analysis (DMA). Microchem J 2015; 121: 237-43. [http://dx.doi.org/10.1016/j.microc.2015.03.005]

[35] Guérin T, Chekri R, Chafey C, Testu C, Hulin M, Noël L. Mercury in foods from the first French total diet study on infants and toddlers. Food Chem 2018; 239: 920-5. [http://dx.doi.org/10.1016/j.foodchem.2017.07.039] [PMID: 28873653]

[36] Martins C, Vasco E, Paixão E, Alvito P. Total mercury in infant food, occurrence and exposure assessment in Portugal. Food Addit Contam Part B Surveill 2013; 6(3): 151-7. [http://dx.doi.org/10.1080/19393210.2013.775603] [PMID: 24779897]

[37] Rezende PS, Silva NC, Moura WD, Windmöller CC. Quantification and speciation of mercury in streams and rivers sediment samples from Paracatu, MG, Brazil, using a direct mercury analyzer®. Microchem J 2018; 140: 199-206. [http://dx.doi.org/10.1016/j.microc.2018.04.006]

[38] Zarco-Fernández S, Montalvo-Quirós S, Mancheño MJ, Cámara C, Muñoz-Olivas R. Development of a thiourea derivative polymer combined to a Direct Mercury Analyser for screening and pre-concentration of mercury species. Talanta 2017; 162: 612-7. [http://dx.doi.org/10.1016/j.talanta.2016.10.048] [PMID: 27837879]

[39] Wang B, Feng L, Mao X, *et al.* Direct determination of trace mercury and cadmium in food by sequential electrothermal vaporization atomic fluorescence spectrometry using tungsten and gold coil traps. J Anal At Spectrom 2018; 33: 1209-16. [http://dx.doi.org/10.1039/C8JA00009C]

[40] Martín-Yerga D, Costa-García A. Recent advances in the electrochemical detection of mercury. Cuur Opin Electrochem 2017; 3: 91-6. [http://dx.doi.org/10.1016/j.coelec.2017.06.012]

[41] Tyszczuk-Rotko K, Sadok I, Barczak M. Thiol-functionalized polysiloxanes modified by lead nanoparticles: Synthesis, characterization and application for determination of trace concentrations of mercury(II). Microporous Mesoporous Mater 2016; 230: 109-17. [http://dx.doi.org/10.1016/j.micromeso.2016.04.043]

[42] Veerakumar P, Chen SM, Madhu R, Veeramani V, Hung CT, Liu SB. Nickel nanoparticle-decorated porous carbons for highly active catalytic reduction of organic dyes and sensitive detection of Hg(II) ions. ACS Appl Mater Interfaces 2015; 7(44): 24810-21. [http://dx.doi.org/10.1021/acsami.5b07900] [PMID: 26479076]

[43] Suherman AL, Tanner EEL, Compton RG. Recent developments in inorganic Hg^{2+} detection by voltammetry, *TrAC*. Trends Analyt Chem 2017; 94: 161-72. [http://dx.doi.org/10.1016/j.trac.2017.07.020]

[44] Wang N, Lin M, Dai H, Ma H. Functionalized gold nanoparticles/reduced graphene oxide nanocomposites for ultrasensitive electrochemical sensing of mercury ions based on thymine-mercur--thymine structure. Biosens Bioelectron 2016; 79: 320-6. [http://dx.doi.org/10.1016/j.bios.2015.12.056] [PMID: 26720921]

[45] Lin Y, Peng Y, Di J. Electrochemical detection of Hg(II) ions based on nanoporous gold nanoparticles modified indium tin oxide electrode. Sens Actuators B Chem 2015; 220: 1086-90. [http://dx.doi.org/10.1016/j.snb.2015.06.064]

[46] Isa IM, Sharif SNM, Hashim N, Ghani SA. Amperometric determination of nanomolar mercury(II) by layered double nanocomposite of zinc/aluminium hydroxide-3(4-methoxyphenyl)propionate modified single-walled carbon nanotube paste electrode. Ionics 2015; 21: 2949-58. [http://dx.doi.org/10.1007/s11581-015-1466-3]

[47] Chaiyo S, Chailapakul O, Siangproh W. Highly sensitive determination of mercury using copper enhancer by diamond electrode coupled with sequential injection-anodic stripping voltammetry. Anal Chim Acta 2014; 852: 55-62. [http://dx.doi.org/10.1016/j.aca.2014.09.011] [PMID: 25441879]

[48] Farahi A, Lahrich S, Achak M, Bakasse M, El Mhammedi MA. Effect of KNO_3 to remove silver

interferences in the determination of mercury(II): Application in milk and breast milk samples. Sens Biosensing Res 2015; 4: 90-5.
[http://dx.doi.org/10.1016/j.sbsr.2015.04.004]

[49] Afkhami A, Madrakian T, Sabounchei SJ, Rezaei M, Samiee S, Pourshahbaz M. Construction of a modified carbon paste electrode for the highly selective simultaneous electrochemical determination of trace amounts of mercury(II) and cadmium(II). Sens Actuators B Chem 2012; 161: 542-8.
[http://dx.doi.org/10.1016/j.snb.2011.10.073]

[50] Giacomino A, Ruo Redda A, Squadrone S, *et al.* Anodic stripping voltammetry with gold electrodes as an alternative method for the routine determination of mercury in fish. Comparison with spectroscopic approaches. Food Chem 2017; 221: 737-45.
[http://dx.doi.org/10.1016/j.foodchem.2016.11.111] [PMID: 27979266]

[51] Guerreiro GV, Zaitouna AJ, Lai RY. Characterization of an electrochemical mercury sensor using alternating current, cyclic, square wave and differential pulse voltammetry. Anal Chim Acta 2014; 810: 79-85.
[http://dx.doi.org/10.1016/j.aca.2013.12.005] [PMID: 24439508]

[52] Duarte FA, Soares BM, Vieira AA, *et al.* Assessment of modified matrix solid-phase dispersion as sample preparation for the determination of CH_3Hg^+ and Hg^{2+} in fish. Anal Chem 2013; 85(10): 5015-22.
[http://dx.doi.org/10.1021/ac4002436] [PMID: 23614538]

[53] Chen Q, Lin Y, Tian Y, *et al.* Single-Drop Solution Electrode Discharge-Induced Cold Vapor Generation Coupling to Matrix Solid-Phase Dispersion: A Robust Approach for Sensitive Quantification of Total Mercury Distribution in Fish. Anal Chem 2017; 89(3): 2093-100.
[http://dx.doi.org/10.1021/acs.analchem.6b04753] [PMID: 28208276]

[54] Yavuz E, Tokalıoğlu Ş, Patat Ş. Magnetic dispersive solid phase extraction with graphene/$ZnFe_2O_4$ nanocomposite adsorbent for the sensitive determination of mercury in water and fish samples by cold vapor atomic absorption spectrometry. Microchem J 2018; 142: 85-93.
[http://dx.doi.org/10.1016/j.microc.2018.06.019]

[55] Karadjova I, Dakova I, Yordanova T, Vasileva P. Nanomaterials for elemental speciation. J Anal At Spectrom 2016; 31: 1949-73.
[http://dx.doi.org/10.1039/C6JA00168H]

[56] Hemmati M, Rajabi M, Asghari A. Magnetic nanoparticle based solid-phase extraction of heavy metal ions: A review on recent advances Microchim Acta 2018; 185

[57] Najafi E, Aboufazeli F, Zhad HRLZ, Sadeghi O, Amani V. A novel magnetic ion imprinted nano-polymer for selective separation and determination of low levels of mercury(II) ions in fish samples. Food Chem 2013; 141(4): 4040-5.
[http://dx.doi.org/10.1016/j.foodchem.2013.06.118] [PMID: 23993582]

[58] Seidi S, Fotouhi M. Magnetic dispersive solid phase extraction based on polythiophene modified magnetic graphene oxide for mercury determination in seafood followed by flow-injection cold vapor atomic absorption spectrometry. Anal Methods 2017; 9: 803-13.
[http://dx.doi.org/10.1039/C6AY02900K]

[59] Tadjarodi A, Abbaszadeh A. A magnetic nanocomposite prepared from chelator-modified magnetite (Fe_3O_4) and HKUST-1 (MOF-199) for separation and preconcentration of mercury(II). Mikrochim Acta 2016; 183: 1391-9.
[http://dx.doi.org/10.1007/s00604-016-1770-2]

[60] Zhu S, Chen B, He M, Huang T, Hu B. Speciation of mercury in water and fish samples by HPLC-ICP-MS after magnetic solid phase extraction. Talanta 2017; 171: 213-9.
[http://dx.doi.org/10.1016/j.talanta.2017.04.068] [PMID: 28551131]

[61] Abolhasani J, Hosseinzadeh Khanmiri R, Babazadeh M, Ghorbani-Kalhor E, Edjlali L, Hassanpour A. Determination of Hg(II) ions in sea food samples after extraction and preconcentration by novel

Fe_3O_4@SiO_2@polythiophene magnetic nanocomposite. Environ Monit Assess 2015; 187. [http://dx.doi.org/10.1007/s10661-015-4770-5]

[62] Özdemir S, Mohamedsaid SA, Kılınç E, Soylak M. Magnetic solid phase extractions of Co(II) and Hg(II) by using magnetized *C. micaceus* from water and food samples. Food Chem 2019; 271: 232-8. [http://dx.doi.org/10.1016/j.foodchem.2018.07.067] [PMID: 30236672]

[63] Vereda Alonso E, Guerrero MdelM, Colorado Cueto P, Barreno Benítez J, Cano Pavón JM, García de Torres A. Development of an on-line solid phase extraction method based on new functionalized magnetic nanoparticles. Use in the determination of mercury in biological and sea-water samples. Talanta 2016; 153: 228-39. [http://dx.doi.org/10.1016/j.talanta.2016.03.027] [PMID: 27130113]

[64] Londonio A, Hasuoka PE, Pacheco P, Gil RA, Smichowski P. Online solid phase extraction-HPL--ICP-MS system for mercury and methylmercury preconcentration using functionalised carbon nanotubes for their determination in dietary supplements. J Anal At Spectrom 2018; 33: 1737-44. [http://dx.doi.org/10.1039/C8JA00188J]

[65] López-García I, Vicente-Martínez Y, Hernández-Córdoba M. Determination of ultratraces of mercury species using separation with magnetic core-modified silver nanoparticles and electrothermal atomic absorption spectrometry. J Anal At Spectrom 2015; 30: 1980-7. [http://dx.doi.org/10.1039/C5JA00213C]

[66] Clémens S, Monperrus M, Donard OFX, Amouroux D, Guérin T. Mercury speciation in seafood using isotope dilution analysis: a review. Talanta 2012; 89: 12-20. [http://dx.doi.org/10.1016/j.talanta.2011.12.064] [PMID: 22284453]

[67] Deng D, Zhang S, Chen H, *et al.* Online solid sampling platform using multi-wall carbon nanotube assisted matrix solid phase dispersion for mercury speciation in fish by HPLC-ICP-MS. J Anal At Spectrom 2015; 30: 882-7. [http://dx.doi.org/10.1039/C4JA00436A]

[68] De Quadros DPC, Campanella B, Onor M, Bramanti E, Borges DLG, D'Ulivo A. Mercury speciation by high-performance liquid chromatography atomic fluorescence spectrometry using an integrated microwave/UV interface. Optimization of a single step procedure for the simultaneous photo-oxidation of mercury species and photo-generation of Hg^0. Spectrochim Acta B At Spectrosc 2014; 101: 312-9. [http://dx.doi.org/10.1016/j.sab.2014.09.019]

[69] Li H, Xu Z, Yang L, Wang Q. Determination and speciation of Hg using HPLC-AFS by atomization of this metal on a UV/nano-ZrO_2/HCOOH photocatalytic reduction unit. J Anal At Spectrom 2015; 30: 916-21. [http://dx.doi.org/10.1039/C4JA00455H]

[70] Yin Y, Liang J, Yang L, Wang Q. Vapour generation at a UV/TiO_2 photocatalysis reaction device for determination and speciation of mercury by AFS and HPLC-AFS. J Anal At Spectrom 2007; 22: 330-4. [http://dx.doi.org/10.1039/b614555h]

[71] Cheng H, Chen X, Shen L, Wang Y, Xu Z, Liu J. Ion-pairing reversed-phase chromatography coupled to inductively coupled plasma mass spectrometry as a tool to determine mercurial species in freshwater fish. J Chromatogr A 2018; 1531: 104-11. [http://dx.doi.org/10.1016/j.chroma.2017.11.029] [PMID: 29173958]

[72] Fang Y, Pan Y, Li P, *et al.* Simultaneous determination of arsenic and mercury species in rice by ion-pairing reversed phase chromatography with inductively coupled plasma mass spectrometry. Food Chem 2016; 213: 609-15. [http://dx.doi.org/10.1016/j.foodchem.2016.07.003] [PMID: 27451225]

[73] Castro Grijalba A, Quintas PY, Fiorentini EF, Wuilloud RG. Usefulness of ionic liquids as mobile phase modifiers in HPLC-CV-AFS for mercury speciation analysis in food. J Anal At Spectrom 2018; 33: 822-34.

[http://dx.doi.org/10.1039/C8JA00059J]

[74] Bendicho C, Bendicho-Lavilla C, Lavilla I. Nanoparticle-assisted chemical speciation of trace elements, *TrAC.* Trends Analyt Chem 2016; 77: 109-21. [http://dx.doi.org/10.1016/j.trac.2015.12.015]

[75] Cheng H, Zhang W, Wang Y, Liu J. Graphene oxide as a stationary phase for speciation of inorganic and organic species of mercury, arsenic and selenium using HPLC with ICP-MS detection Microchim Acta 2018; 185

[76] Wu L, Long Z, Liu L, Zhou Q, Lee Y-I, Zheng C. Microwave-enhanced cold vapor generation for speciation analysis of mercury by atomic fluorescence spectrometry. Talanta 2012; 94: 146-51. [http://dx.doi.org/10.1016/j.talanta.2012.03.009] [PMID: 22608427]

[77] Wu P, He L, Zheng C, Hou X, Sturgeon RE. Applications of chemical vapor generation in non-tetrahydroborate media to analytical atomic spectrometry. J Anal At Spectrom 2010; 25: 1217-46. [http://dx.doi.org/10.1039/c003483e]

[78] Chen KJ, Hsu IH, Sun YC. Determination of methylmercury and inorganic mercury by coupling short-column ion chromatographic separation, on-line photocatalyst-assisted vapor generation, and inductively coupled plasma mass spectrometry. J Chromatogr A 2009; 1216(51): 8933-8. [http://dx.doi.org/10.1016/j.chroma.2009.10.061] [PMID: 19913233]

[79] Ai X, Wang Y, Hou X, Yang L, Zheng C, Wu L. Advanced oxidation using $Fe_{(3)}O_{(4)}$ magnetic nanoparticles and its application in mercury speciation analysis by high performance liquid chromatography-cold vapor generation atomic fluorescence spectrometry. Analyst (Lond) 2013; 138(12): 3494-501.
[http://dx.doi.org/10.1039/c3an00010a] [PMID: 23666232]

[80] Zhang WB, Yang XA, Dong YP, Xue JJ. Speciation of inorganic- and methyl-mercury in biological matrixes by electrochemical vapor generation from an L-cysteine modified graphite electrode with atomic fluorescence spectrometry detection. Anal Chem 2012; 84(21): 9199-207. [http://dx.doi.org/10.1021/ac3018923] [PMID: 23035833]

[81] Shi M-T, Yang X-A, Qin L-M, Zhang W-B. Highly efficient electrocatalytic vapor generation of methylmercury based on the gold particles deposited glassy carbon electrode: A typical application for sensitive mercury speciation analysis in fish samples. Anal Chim Acta 2018; 1025: 58-68. [http://dx.doi.org/10.1016/j.aca.2018.04.057] [PMID: 29801608]

[82] Mo J, Li Q, Guo X, Zhang G, Wang Z. Flow injection photochemical vapor generation coupled with miniaturized solution-cathode glow discharge atomic emission spectrometry for determination and speciation analysis of mercury. Anal Chem 2017; 89(19): 10353-60. [http://dx.doi.org/10.1021/acs.analchem.7b02214] [PMID: 28861992]

[83] Covaci E, Senila M, Ponta M, *et al.* Methylmercury determination in seafood by photochemical vapor generation capacitively coupled plasma microtorch optical emission spectrometry. Talanta 2017; 170: 464-72.
[http://dx.doi.org/10.1016/j.talanta.2017.04.036] [PMID: 28501197]

[84] Yoshimoto K, Anh HT, Yamamoto A, *et al.* Simple analysis of total mercury and methylmercury in seafood using heating vaporization atomic absorption spectrometry. J Toxicol Sci 2016; 41(4): 489-500.
[http://dx.doi.org/10.2131/jts.41.489] [PMID: 27432235]

[85] Covaci E, Senila M, Ponta M, Darvasi E, Frentiu M, Frentiu T. Mercury speciation in seafood using non-chromatographic chemical vapor generation capacitively coupled plasma microtorch optical emission spectrometry method – Evaluation of methylmercury exposure. Food Control 2017; 82: 266-73.
[http://dx.doi.org/10.1016/j.foodcont.2017.07.006]

[86] Covaci E, Senila M, Tanaselia C, *et al.* A highly sensitive eco-scale method for mercury determination in water and food using photochemical vapor generation and miniaturized instrumentation for

capacitively coupled plasma microtorch optical emission spectrometry. J Anal At Spectrom 2018; 33: 799-808.
[http://dx.doi.org/10.1039/C8JA00054A]

[87] Çaylak O, Elçi ŞG, Höl A, Akdoğan A, Divrikli Ü, Elçi L. Use of an aminated Amberlite XAD-4 column coupled to flow injection cold vapour generation atomic absorption spectrometry for mercury speciation in water and fish tissue samples. Food Chem 2019; 274: 487-93.
[http://dx.doi.org/10.1016/j.foodchem.2018.08.107] [PMID: 30372969]

[88] Dakova I, Yordanova T, Karadjova I. Non-chromatographic mercury speciation and determination in wine by new core-shell ion-imprinted sorbents. J Hazard Mater 2012; 231-232: 49-56.
[http://dx.doi.org/10.1016/j.jhazmat.2012.06.034] [PMID: 22771349]

[89] Zarco-Fernández S, Mancheño MJ, Muñoz-Olivas R, Cámara C. A new specific polymeric material for mercury speciation: Application to environmental and food samples. Anal Chim Acta 2015; 897: 109-15.
[http://dx.doi.org/10.1016/j.aca.2015.09.016] [PMID: 26515012]

[90] Chen Z, Wang X, Cheng X, Yang W, Wu Y, Fu F. Specifically and visually detect methyl-mercury and ethyl-mercury in fish sample based on DNA-templated alloy Ag-Au nanoparticles. Anal Chem 2018; 90(8): 5489-95.
[http://dx.doi.org/10.1021/acs.analchem.8b01100] [PMID: 29601183]

[91] Costas-Mora I, Romero V, Lavilla I, Bendicho C. In situ building of a nanoprobe based on fluorescent carbon dots for methylmercury detection. Anal Chem 2014; 86(9): 4536-43.
[http://dx.doi.org/10.1021/ac500517h] [PMID: 24678836]

[92] Clémens S, Monperrus M, Donard OFX, Amouroux D, Guérin T. Mercury speciation analysis in seafood by species-specific isotope dilution: method validation and occurrence data. Anal Bioanal Chem 2011; 401(9): 2699-711.
[http://dx.doi.org/10.1007/s00216-011-5040-1] [PMID: 21533797]

[93] Bergquist BA, Blum JD. Mass-dependent and -independent fractionation of hg isotopes by photoreduction in aquatic systems. Science 2007; 318(5849): 417-20.
[http://dx.doi.org/10.1126/science.1148050] [PMID: 17872409]

[94] Feng C, Pedrero Z, Gentès S, *et al.* Specific pathways of dietary methylmercury and inorganic mercury determined by mercury speciation and isotopic composition in zebrafish (Danio rerio). Environ Sci Technol 2015; 49(21): 12984-93.
[http://dx.doi.org/10.1021/acs.est.5b03587] [PMID: 26398726]

[95] Kritee K, Blum JD, Reinfelder JR, Barkay T. Microbial stable isotope fractionation of mercury: A synthesis of present understanding and future directions. Chem Geol 2013; 336: 13-25.
[http://dx.doi.org/10.1016/j.chemgeo.2012.08.017]

[96] Kwon SY, Blum JD, Carvan MJ, *et al.* Absence of fractionation of mercury isotopes during trophic transfer of methylmercury to freshwater fish in captivity. Environ Sci Technol 2012; 46(14): 7527-34.
[http://dx.doi.org/10.1021/es300794q] [PMID: 22681311]

[97] Yin R, Feng X, Meng B. Stable mercury isotope variation in rice plants (*Oryza sativa L.*) from the Wanshan mercury mining district, SW China. Environ Sci Technol 2013; 47(5): 2238-45.
[http://dx.doi.org/10.1021/es304302a] [PMID: 23363238]

[98] Perrot V, Epov VN, Pastukhov MV, *et al.* Tracing sources and bioaccumulation of mercury in fish of Lake Baikal--Angara River using Hg isotopic composition. Environ Sci Technol 2010; 44(21): 8030-7.
[http://dx.doi.org/10.1021/es101898e] [PMID: 20942479]

[99] Perrot V, Pastukhov MV, Epov VN, Husted S, Donard OFX, Amouroux D. Higher mass-independent isotope fractionation of methylmercury in the pelagic food web of Lake Baikal (Russia). Environ Sci Technol 2012; 46(11): 5902-11.
[http://dx.doi.org/10.1021/es204572g] [PMID: 22545798]

[100] Laffont L, Sonke JE, Maurice L, *et al.* Hg speciation and stable isotope signatures in human hair as a tracer for dietary and occupational exposure to mercury. Environ Sci Technol 2011; 45(23): 9910-6. [http://dx.doi.org/10.1021/es202353m] [PMID: 22003970]

[101] Rothenberg SE, Yin R, Hurley JP, *et al.* Stable Mercury Isotopes in Polished Rice (*Oryza sativa L.*) and Hair from Rice Consumers. Environ Sci Technol 2017; 51(11): 6480-8. [http://dx.doi.org/10.1021/acs.est.7b01039] [PMID: 28482656]

[102] Sherman LS, Blum JD, Franzblau A, Basu N. New insight into biomarkers of human mercury exposure using naturally occurring mercury stable isotopes. Environ Sci Technol 2013; 47(7): 3403-9. [http://dx.doi.org/10.1021/es305250z] [PMID: 23463943]

[103] Li M, Sherman LS, Blum JD, *et al.* Assessing sources of human methylmercury exposure using stable mercury isotopes. Environ Sci Technol 2014; 48(15): 8800-6. [http://dx.doi.org/10.1021/es500340r] [PMID: 24967674]

[104] Masbou J, Point D, Sonke JE, *et al.* Hg stable isotope time trend in ringed seals registers decreasing sea ice cover in the alaskan arctic. Environ Sci Technol 2015; 49(15): 8977-85. [http://dx.doi.org/10.1021/es5048446] [PMID: 26132925]

[105] Point D, Sonke JE, Day RD, *et al.* Methylmercury photodegradation influenced by sea-ice cover in Arctic marine ecosystems. Nat Geosci 2011; 4: 188-94. [http://dx.doi.org/10.1038/ngeo1049]

[106] Renedo M, Amouroux D, Duval B, *et al.* Seabird tissues as efficient biomonitoring tools for Hg isotopic investigations: implications of using blood and feathers from chicks and adults. Environ Sci Technol 2018; 52(7): 4227-34. [http://dx.doi.org/10.1021/acs.est.8b00422] [PMID: 29514456]

[107] Renedo M, Amouroux D, Pedrero Z, Bustamante P, Cherel Y. Identification of sources and bioaccumulation pathways of MeHg in subantarctic penguins: a stable isotopic investigation. Sci Rep 2018; 8(1): 8865. [http://dx.doi.org/10.1038/s41598-018-27079-9] [PMID: 29891979]

[108] Cransveld A, Amouroux D, Tessier E, *et al.* Mercury stable isotopes discriminate different populations of european seabass and trace potential Hg sources around europe. Environ Sci Technol 2017; 51(21): 12219-28. [http://dx.doi.org/10.1021/acs.est.7b01307] [PMID: 28942649]

[109] Bérail S, Cavalheiro J, Tessier E, *et al.* Determination of total Hg isotopic composition at ultra-trace levels by on line cold vapor generation and dual gold-amalgamation coupled to MC-ICP-MS. J Anal At Spectrom 2017; 32: 373-84. [http://dx.doi.org/10.1039/C6JA00375C]

[110] Epov VN, Berail S, Jimenez-Moreno M, *et al.* Approach to measure isotopic ratios in species using multicollector-ICPMS coupled with chromatography. Anal Chem 2010; 82(13): 5652-62. [http://dx.doi.org/10.1021/ac100648f] [PMID: 20545335]

[111] Masbou J, Point D, Sonke JE. Application of a selective extraction method for methylmercury compound specific stable isotope analysis (MeHg-CSIA) in biological materials. J Anal At Spectrom 2013; 28: 1620-8. [http://dx.doi.org/10.1039/c3ja50185j]

[112] Entwisle J, Malinovsky D, Dunn PJH, Goenaga-Infante H. Hg isotope ratio measurements of methylmercury in fish tissues using HPLC with off line cold vapour generation MC-ICPMS. J Anal At Spectrom 2018; 33: 1645-54. [http://dx.doi.org/10.1039/C8JA00099A]

[113] Afonso C, Costa S, Cardoso C, *et al.* Evaluation of the risk/benefit associated to the consumption of raw and cooked farmed meagre based on the bioaccessibility of selenium, eicosapentaenoic acid and docosahexaenoic acid, total mercury, and methylmercury determined by an *in vitro* digestion model.

Food Chem 2015; 170: 249-56.
[http://dx.doi.org/10.1016/j.foodchem.2014.08.044] [PMID: 25306342]

[114] Alves RN, Maulvault AL, Barbosa VL, *et al.* Oral bioaccessibility of toxic and essential elements in raw and cooked commercial seafood species available in European markets. Food Chem 2018; 267: 15-27.
[http://dx.doi.org/10.1016/j.foodchem.2017.11.045] [PMID: 29934150]

[115] Bradley MA, Barst BD, Basu N. A review of mercury bioavailability in humans and fish. Int J Environ Res Public Health 2017; 14(2): 14.
[http://dx.doi.org/10.3390/ijerph14020169] [PMID: 28208586]

[116] Girard C, Charette T, Leclerc M, Shapiro BJ, Amyot M. Cooking and co-ingested polyphenols reduce in vitro methylmercury bioaccessibility from fish and may alter exposure in humans. Sci Total Environ 2018; 616-617: 863-74.
[http://dx.doi.org/10.1016/j.scitotenv.2017.10.236] [PMID: 29096961]

[117] Jadán Piedra C, Sánchez V, Vélez D, Devesa V. Reduction of mercury bioaccessibility using dietary strategies. Food Sci Technol 2016; 71: 10-6.
[http://dx.doi.org/10.1016/j.lwt.2016.03.015]

[118] Ouédraogo O, Amyot M. Effects of various cooking methods and food components on bioaccessibility of mercury from fish. Environ Res 2011; 111(8): 1064-9.
[http://dx.doi.org/10.1016/j.envres.2011.09.018] [PMID: 22014585]

[119] He M, Wang WX. Factors affecting the bioaccessibility of methylmercury in several marine fish species. J Agric Food Chem 2011; 59(13): 7155-62.
[http://dx.doi.org/10.1021/jf201424g] [PMID: 21650469]

[120] Ralston NVC, Ralston CR, Raymond LJ. Selenium Health Benefit Values: Updated Criteria for Mercury Risk Assessments. Biol Trace Elem Res 2016; 171(2): 262-9.
[http://dx.doi.org/10.1007/s12011-015-0516-z] [PMID: 26463749]

CHAPTER 9

Process Contaminants

Marta Mesías, Francisca Holgado and **Francisco J. Morales***

Institute of Food Science, Technology and Nutrition, Spanish National Research Council (ICTAN-CSIC), C/ José Antonio Novais, 10. 28040 Madrid, Spain

Abstract: Contaminants are substances that may be present in foods as a result of production, preparation, food formulation, processing, packaging, transport and storage, as well as a result of environmental contaminant. Among them, process contaminants are generated in foods due to chemical reactions occurring during cooking, processing and preservation and are considered to exert adverse toxicological effects in humans. This chapter focuses on some of these process contaminants, specifically on contaminants formed after thermal treatment of foods, such as acrylamide, furan, heterocyclic aromatic amines, chloropropanodiols and their esters, glycidol and glycidyl esters. Heat-generated food contaminants are mostly produced during cooking at high temperatures as a result of Maillard reaction and lipid oxidation, although other non-thermal reactions may also contribute to their formation. Characterization, toxicological considerations, chemical formation, occurrence and exposure are detailed, as well as mitigation strategies applied to prevent their formation and/or reduce and remove from the processed food.

Keywords: Acrylamide, Analysis, Chemical reaction, Cooking, Diet, Exposure, Food safety, Furan, Glycidol esters, Heat, Heterocyclic aromatic amines, Intake, Maillard reaction, Mitigation, 3-monochloropropanediol, Preventive strategies, Process contaminants, Risk, Toxic, Xenobiotics.

INTRODUCTION

General Considerations to Food Safety

Food safety is an activity related to the evaluation of microbiological, chemical and physical hazards that cover handling, preparing and storing food in ways that prevent food-borne illness. In the past, major attention on food safety has been given to food microbiology issues resulting from unexpected and sudden microorganism contamination and outbreak. Since mid of the last century, food

* **Corresponding author Francisco J. Morales:** Institute of Food Science, Technology and Nutrition, Spanish National Research Council (ICTAN-CSIC), C/ José Antonio Novais, 10. 28040 Madrid, Spain; Tel: +34 915492300; E-mail: fjmorales@ictan.csic.es

Belén Gómara & María Luisa Marina (Eds.)
Current and Future Developments in Food Science (Vol. 1)

hygiene does not only address terms of microbiological health hazards,but also chemical food contamination. Chemicals are essentially present in practically every place in the world, including living organisms, and of course in all foods. In foods, these are predominantly proteins, saccharides, lipids, low-molecular organic compounds, minerals and water. The nature of the chemical substances confers their toxicological properties. However, most of the chemicals present in our foods are harmless contributing to the nutritional, technological and sensorial properties of foods, but also being necessary to participate into the physiological reactions in the living organism. In that sense, food contains (naturally, intentionally or additives) a wide range of substances. Among the intentionally added additives, the regulated additives are largely known. Additives are important to increase the nutritional value of foods (vitamins, minerals, amino acids), the sensory properties (pigments, sweeteners, flavoring and its enhancers), and the shelf life of foods (antimicrobials, antioxidants).

Xenobiotics can be described as any foreign chemical in the food that may endanger human safety since they are biochemically active substances. Once they enter into the organism, they can induce or inhibit metabolic pathways, which could be related to enzymes, or transporters expressed in the human host, as well as the microbiota of the gastrointestinal tract. Xenobiotics can be classified into broad categories, according to their relevance in terms of food safety, namely contaminants and chemicals that have been intentionally added to food or raw commodities [1]. Some examples are those potentially present in raw foods (residues of veterinary drugs, environmental pollutants, fertilizers), those intentionally introduced during technological processing, chemicals passed from the packaging and processing equipment, is formed during storage.

The presence of xenobiotics in foods is practically unavoidable. However, the application of food security systems in EU, such as hazard analysis and critical control points (HACCP) and good hygienic practice (GHP), together with a continuous updated legislation, is a guarantee to avoid or even reduce to an acceptable level, the occurrence of most of the harmful xenobiotics to foods [2]. The *Codex Alimentarius*, established in 1963 under FAO/World Health Organization Commission, provides a basis for regulations in order to control the content of hazardous chemicals in foods. Food safety should be a guarantee that no adverse effects occur in humans after food ingestion, ultimately having an impact on human health and wellness.

Risk Assessment Scheme

Hazard control in foods is supervised by supranational official authorities, governmental agencies and the scientific community. There are a number of food safety regulatory bodies and international organizations which are responsible for the food safety, such as *Codex Alimentarius* Commission, World Health Organization (WHO), Food and Agricultural Organization (FAO), World Trade Organization, US Food and Drug Administration (FDA), Food Standards Australia New Zealand, among others. In Europe, the European Food Safety Authority (EFSA) is the keystone of European Union risk assessment regarding food and feed safety that works in close collaboration with national authorities. EFSA, established in 2002, provides independent scientific advice and clear communication on existing and emerging risks. The remit of EFSA concerns the entire food chain covering aspects of human, animal and plant health, and, sometimes, environmental protection. Its focus in human health is on food safety and its scientific advice may contribute to various phases in the policy cycle: reflection, regulation, verification and review. The EU follows the as-low-as reasonably-achievable (ALARA) principle of decision-making for any risk assessment advice. The Scientific Panel of Food Chain Contaminants (CONTAM) would assist EFSA's scientific advice for chemical contaminants.

It is important to describe the differences between hazard and risk as defined by the General Food Law in Europe [3]. Hazard is defined as - a biological, chemical or physical agent in, or condition of, food or feed with the potential to cause an adverse health effect -, while 'risk' means - a function of the probability of an adverse health effect and the severity of that effect, consequential to a hazard.

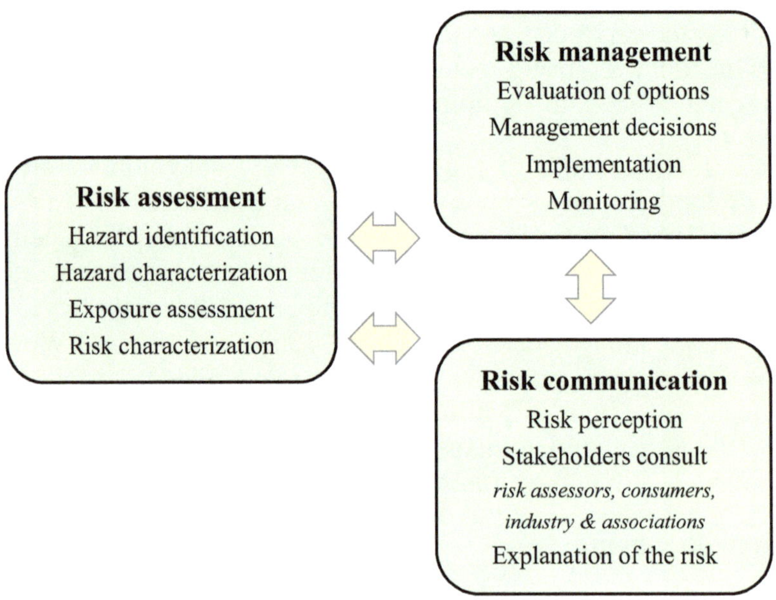

Fig. (1). Risk analysis framework.

According to WHO [4], the framework of risk analysis process, characterized by a hierarchical and stepwise approach, is divided into three interlinked areas, which are *risk assessment*, *risk management* and *risk communication* (Fig. **1**). In Europe, EFSA is responsible for risk assessment assisted by scientific community and food business stakeholders, and the Commission is competent for risk management decisions assisted by EU member states. Risk communication is a shared competence between EFSA and the European Commission [3]. Risk management is the process of weighing policy alternatives in consultation with all stakeholders, considering the outcome of the risk assessment and other factors relevant for the health protection. Risk management actions also take into consideration other factors such as consumer information, prevention of misleading practices, feasibility of controlling a risk, effective risk mitigation actions, socio-economic effects, environmental impact, among others. Risk communication is critical since nowadays, food safety issues are rapidly disseminated to distant regions and countries through websites social networks or broadcast journalist's report. To sum up it is also important to consider a risk/benefit perspective. Benefit-risk assessment is a science-based process intended to estimate the benefits and risks for humans following exposure (or lack of exposure) to a particular food or food component and to integrate them in comparable measures, thus facilitating better-informed decisions by decision-makers [5].

Risk assessment is a scientifically based process to evaluate potential adverse health effects resulting from human exposure to foodborne hazards. An adverse effect is a change in the morphology, physiology, growth, development, reproduction or lifespan of an organism, system or (sub)population that results in an impairment of functional capacity, an impairment of the capacity to compensate for additional stress or an increase in susceptibility to other influences [6]. Adverse effect, as defined by the FDA, is -any biochemical, physiological, anatomical, pathological, and/or behavioral change that results in functional impairment that may affect the performance of the whole organism or reduce the ability of the organism to respond to an additional challenge. The adverse effects can be described in different terms but usually are classified in genotoxicity, carcinogenicity, neurotoxicity, inmunotoxicity, reproductive and developmental toxicity. The risk assessment is performed to assess whether a toxic compound forms a risk for a population. The risk assessment paradigm consists of four sequential steps, which are (i) *hazard identification* of the adverse effects, (ii) *hazard characterization* of the extent, (iii) *exposure assessment* and (iv) *risk characterization* which combine the risk of the adverse health effect in a population or sector of the population at a given exposure level. In other words, it is necessary to compile information on different aspects, such as levels of the chemical compounds in foods and consumption levels of these foods in a certain population at a large, including high end-users or restricted to a relevant group of population such as infants, pregnant women, the elderly or people with specific health conditions.

During the hazard identification, potentially harmful chemicals that may cause adverse health effects and which may be present in a particular food or group of food are identified. Exposure assessment is defined as the qualitative and/or quantitative evaluation of the likely dietary intake of the hazard. In the stage of hazard characterization, the nature of the adverse health effects is evaluated in a qualitative and/or quantitative way (dose-response relationship). During the risk characterization, all the evidence from the previous three steps is combined in order to obtain a risk estimate on the severity to the population [7]. It is necessary to identify the average risk in a certain food, but also under extreme conditions. Risk assessment describes the relationships between the scale of exposure and the frequency of negative health effects, by providing safe exposure limits for chemicals in foods (*i.e.*, maximum residue levels).

Estimation of Food Chemical Intake

Human dietary exposure to chemicals is through ingestion of foods and beverages (water) and reflects an external dose of the chemical entering the body. It is obvious to say that dietary exposure of a certain contaminants in a food which is not consumed does not represent a risk. Data for contaminants and residues may be obtained by monitoring programs, targeted surveys or total diet study approaches focusing on foods that are expected to contain the target contaminants. To determine the population at risk, calculated exposure levels are compared with a relevant toxicological reference value. Models of acute toxicity (*i.e.*, exposure to a single ingestion), are often not useful for hazard identification and risk assessment in relation to foods and food chemicals [8]. Acute toxicity studies are conducted over a period of 24 hours or less, to determine a single maximally tolerated or lethal dose.

There are different models to estimate the dietary exposure to food chemicals; direct method (based on collection of a duplicate diet) and indirect methods (based on food frequency questionnaires). The consumption data is mostly derived from dietary surveys, usually conducted at the national level on a representative sample of individuals covering the general population, as well as critical groups vulnerable or are expected to have exposures significantly different from those of the general population (*i.e.*, infants, children, pregnant women or the elderly). Short-term and long-term assessments are performed depending on the toxicity of the compound and the expected level of exposure (acute or chronic). Short-term exposure assessments relate to the dietary exposure to a harmful substance during a period of 24 hours or less. Long-term exposure relates to exposures over a longer period of time (months or years). Carcinogenicity associated to chronic exposure is evaluated by repeated daily doses for two years in animal (*i.e.,* rodents), which is more or less equivalent to "lifetime" exposure. Repeated daily doses for 14-28 days are used for short-term exposure. The majority of chemical hazards are characterized by potentially long-term effects on human health. Contaminants that accumulate in the organisms are considered as the most dangerous. Standards of exposure can be estimated taking into consideration if the contaminants are potential carcinogens or not [9].

Whatever the method used, Kroes *et al.* [10] summarized that the exposure assessment has to cover three major aspects: (i) how to determine quantitatively the presence of a chemical in individual foods and diets, including its fate during the processes within the food production chain; (ii) how to determine the consumption patterns of the individual foods containing the relevant chemicals; (iii) how to integrate both the likelihood of consumers eating large amounts of the given foods and of the relevant chemical being present in these foods at high levels. Then, the variation of the chemical compounds in food is not routinely considered in the current methods of long-term exposure assessment, such as the point estimate approach, deterministic approach, and the probabilistic approach. The point-estimate approach is based on the selection of a fixed level in the distribution of consumption multiplied by a fixed value chosen from the distribution of concentration. The point estimate approach calculates the dietary intake by multiplying a mean/median or 95th percentile concentration level with a mean consumption level of different foods containing the target substance. It is important to identify a robust and harmonized database containing a harmonized collection of food items. EFSA has consequently launched a series of harmonization activities in the area of food consumption and chemical occurrence data collection to facilitate the collection of high quality data for use in exposure assessments. FoodEx2 is a comprehensive food classification and description system aimed at covering the need to describe foods in data collections across different food safety domains [11]. The point estimate approach does not reflect the exposure of the population but it is considered a conservative scenario. This approach may overestimate exposures when deriving the mean consumption level from food balance sheet data since, for instance, it does not consider the portion of the food already consumed or the waste during food preparation. Although there are some limitations, the point estimate approach is used world-wide as a first screening of step of exposure assessment. Today, probabilistic modelling is an advanced combination method in scenario evaluation that integrates the calculation of the uncertainty and consequently the use of what-if scenarios risk. This quantitative risk assessment is quite interesting since it will give insights in which issues are playing a major role in the contamination and the final exposure (*i.e.,* raw material, formulation, processing, consumer behavior, *etc.*). Probabilistic modelling addresses the variation of the target chemical in the food and a distribution of consumption levels in different scenarios. However, the effects of the mixture or the cocktail of chemical contaminants is a challenge, and therefore new studies of cumulative risk assessment have been triggered [12]. The combined effects of chemical contaminants have caused concerns among policy makers because combined effects can be additive or even higher than effects caused by individually.

Toxicological Considerations

The relationship between the occurrence of a contaminant in foods and its biological response is driven by two aspects; i) toxicokinetics, which relates to the delivery (bioaccesibility/bioavailability, mobility, distribution) of the chemical and its removal (excretion) from the site of action and ii) toxicodynamics relates to the interaction between the chemical and any active metabolites (biometabolisms) at the site of action and the final outcome or toxicological response. Once the chemical contaminant enters in the human body, and is recognized as a non-native molecule, a cascade of metabolic changes is induced. Most of the low-molecular weight contaminants are metabolized by a variety of phase I and phase II enzymes that are present largely in the liver. Phase I metabolism involves the oxidation, reduction or hydrolysis of the xenobiotic with the introduction of groups suitable for subsequent phase II or conjugation reactions [13]. The family of oxidoreductases, such as cytochrome P450 (CYP) enzymes, is responsible for the chemical transformation of the xenobiotic in humans.

Based on the presumed mode of action (non-genotoxic or genotoxic) of the carcinogen, there are two approaches to risk assessment for food chemical contaminants. For non-genotoxic chemicals a threshold of action is assumed from which a no - effect level can be extrapolated experimentally (NOAEL - no observable adverse effect) setting a tolerable daily intake (TDI). They express the amount that can be ingested over a defined period of time without appreciable health risk. They are generally derived by dividing the NOAEL by safety factors (or uncertainty factors) to account for variability within the laboratory animal species and extrapolate results from the most sensitive animal species to humans. In 2009 and revised in 2017, EFSA adopted the use of the benchmark dose (BMD) approach in risk assessment instead of the NOAEL approach, since it makes a more extended use of dose-response data and it allows for a quantification of the uncertainties in the dose-response data. The BMD approach is applicable to all chemicals in food, irrespective of their category or origin for deriving a point of departure (Reference Point) [14, 15].

For genotoxic contaminants mechanisms of which involve chemical interaction of the carcinogen with DNA, there is no threshold and consequently there is no safe level, potentially causing direct adverse effects on DNA. Therefore, no tolerable exposure levels can be set. The genotoxicity should be evaluated using a range of *in vitro* tests for mutation (bacteria), chromosomal damage and changes in chromosome number. Positive results should be confirmed in an *in vivo* genotoxicity study. EFSA developed the MOE (margin-of-exposure) approach as a harmonized methodology for assessing the risk of substances in food and feed, which are genotoxic and carcinogenic to promote useful advice on genotoxic contaminants found in food [16]. MOE approach considers the margin between doses shown in experimental studies to cause adverse effects and the amounts ingested by the population. In general, an MOE of 10,000 or higher, based on a benchmark dose, lower confidence limit for a 10% change in effect ($BMDL_{10}$) from an animal study, might be considered of low concern. This approach is in line with the precautionary principle to foods where there is uncertainty regarding the existence or extent of risks to human health; the institutions may take protective measures without having to wait until the reality and seriousness of those risks become fully apparent [3]. In addition, this evaluation should extrapolate data from animals to humans, considering interspecies and intraspecies variations.

Process Contaminants

Contaminants do not have a specific function in foods or food production, and their presence is considered undesirable. Contaminants are substances that may be present as a result of production (including manage of crops, pre-/post-harvest, animals, *etc.*), preparation, food formulation, processing, packaging, transport and storage, or as a result of environmental contamination. Food packaging materials and containers such as bottles, cups and plates, used to improve food handling and transport, can contain chemical substances such as plastic, elements of which can migrate into the food. Residues are specifically named to plant protection products (*i.e.,* fertilizers, pesticides, biocides) and veterinary drugs (*i.e.,* antibiotics, cocciodiostatics) in foods. Chemical compounds such as metals and nitrates can be present in the soil, water and the atmosphere as a result of some human activities such as farming, industry, or during food production such as high-temperature cooking (*i.e.,* acrylamide) or as a reaction product between ethanol and precursors (cyanide) (*i.e.,* ethyl carbamate produced mainly in stone spirits). Contaminants can also arise from natural toxins produced by fungi, plants and algae, and food contact materials. Fig. **(2)** summarizes the group of chemicals in foods under surveillance by EFSA [7].

Most foods need to be processed to some extent to make them microbiologically safe, palatable, digestible and increase their shelf life. Briefly, the main benefit of food processing is the inactivation of food-borne pathogens (*i.e., Mycobacterium tuberculosis*, *Salmonella*e species, *Staphylococcus aureus*), and the inactivation of natural toxins and enzymes (*i.e.,* trypsin inhibitors in soy). There are concomitant benefits of the processing or cooking: i) enhanced digestibility of foods and bioavailability of nutrients since unfolded denatured proteins are more accessible to digestive enzymes, ii) gelatinization of starch in legumes, cereals and potatoes makes possible its hydrolysis by amylase enzymes. These practices include heating, drying, curing, smoking, fermentation, pasteurization, irradiation, refrigeration, use of additives, bottle-sterilization/canning, high-pressure and aseptic packaging, among other. It is estimated that 80-90% of foods used in home cooking are semi-processed [17].

a. Regulated Food Ingredients

- additives *(i.e. coloring, preservatives, sweeteners, emulsifiers)*
- enzymes *(i.e. asparaginase, ß-galactosidase, pancreatinin)*
- flavouring *(i.e. methanethiol, nonenol, vanillin, thiophene)*
- supplements *(i.e. catechins, isoflavones, caffeine)*

b. Food Chain Residues

- veterinary drugs *(i.e. antibiotics, feed additives)*
- contact materials *(i.e. phthalates)*
- pesticides *(i.e. glyphosate, fipronil)*

c. Contaminants in food

- environmental pollutants
  - brominated flame retardant (i.e. polybrominated diphenyl ethers)
  - dioxins & polychlorinated biphenyls
- metals *(i.e. nickel, mercury, cadmium)*
- mineral oil hydrocarbons
- natural contaminants
  - mycotoxins *(i.e. ochratoxin A, deoxynivalenol or zearalenone)*
  - aflatoxins *(i.e. aflatoxin B1, aflatoxin M1)*
- process contaminants

Fig. (2). Group of chemicals under risk assessment provided by EFSA [7].

Process contaminants are generated in foods because of chemical reactions occurring during cooking and processing, and preservation that are considered to exert adverse toxicological effects in humans (Fig. **3**). In more detail, it is named heat-generated (or thermal-generated) process contaminants that form after thermal/heat treatment of foods, such as acrylamide, furan, heterocyclic aromatic amines, 3-monochloropropanodiol esters and glycidol esters. Heat-generated food contaminants are mostly produced during cooking at high temperatures as a result of Maillard reaction and lipid oxidation. In contrast, there are non-thermal process contaminants such as ethyl carbamate, 3-monochloropropanediol, biogenic amines, n-nitrosamines, and benzene that form by fermentation, acid hydrolysis, oil refining, hydrogenation, chlorination, oxidation or preservation (*i.e.,* curing).

Thermal treatment is the most common process applied to foods to extend shelf-life and increase palatability. Indeed, household food preparation is primarily related to thermal treatment, involving boiling, stewing, frying, baking, grilling, roasting, toasting, microwaving and broiling, using ovens, stoves, toasters or grills. Thermal treatment of foods induces several biological, physical and chemical modifications, leading to sensory, nutritional and textural changes and resulting in enhanced food safety and quality. Thermal food processing plays a pivotal role in the formation of desirable flavors, colors, aromas and textures. Thermal processing partly enhances the nutritional quality, release of bioactive and generation of beneficial components, as well as destruction of negative activity (anti-nutritional substances). However, some undesirable reactions are leading to loss of nutritional value and the formation of potentially mutagenic and carcinogenic molecules [18]. Other newly-formed compounds exert health-promoting effects such as antioxidant and antimutagenic activity.

Heat-induced	**Fermentation**	**Preservation**	**Alcali/acid**
HAAs PAHs MCPD esters Glycidol esters Furan Acrylamide	Ethylcarbamate Biogenic amines	N-nitrosamines Benzene	Lysinoalanine D-amino acids Chloropropanols

Fig. (3). Types of process contaminant in foods. HAAs (heterocyclic aromatic amines), PAH (polycyclic aromatic hydrocarbons), MCPD (monochloropropane-1,2-diol).

During thermal treatments a myriad of compounds are formed, that can be broadly classified according to different chemical classes as alcohols, hydrocarbons, aldehydes, ketones, furans, ketones, pyrazines, pyridines, pyrroles, thiazoles, thiophenes, among others. In 2007, the EU project HEATOX compiled a database of 782 compounds generated as a result of the Maillard reaction and lipid oxidation during thermal treatment and storage [19]. The Maillard reaction is a complex cascade of reactions, involving reducing sugars and amino acids, with a definitive impact on the formation of desirable flavors, aromas, and colored compounds in foods.

The formation of carcinogenic compounds from the Maillard reaction was early confirmed when heterocyclic aromatic amines were isolated from grilled steaks from the reaction of reducing sugars and creatinine. In 2002, it was confirmed that acrylamide forms from the reaction of reducing sugars and asparagine through the Maillard reaction. The potential toxicity of the newly formed compounds was screened by applying an *in-silico* molecular modeling approach that constructs quantitative structure-activity relationship models, whereas 53 compounds were identified as potentially harmful to be both genotoxic and carcinogenic [20]. As classified according to the main functional group, they were pyrazine, thiophene, thiazole, pyrrole, furan, pyridine, oxazole, hydrocarbon and generally described as oxygen-containing compounds. Then, the use of structure-activity relationships in toxicology reveals as a powerful tool since the chemical structure of a compound determines its physical and chemical properties, which are responsible for the biological activity and potential toxicological effects [21].

ACRYLAMIDE

Acrylamide (2-propenamide, C_3H_5NO) (Fig. **4**) is a low molecular weight (71.09 g mol^{-1}), white crystalline, odorless and highly water soluble organic compound used as a monomer in the production of polyacrylamides and with numerous industrial and laboratory applications. Its melting point is 84.5 °C and its boiling point (25 mm Hg) 125 °C (192.6 °C at atmospheric pressure). The ethylenic double bond confers acrylamide electrophilic reactivity, being primarily reactive and promoting the polymerization through radical reactions.

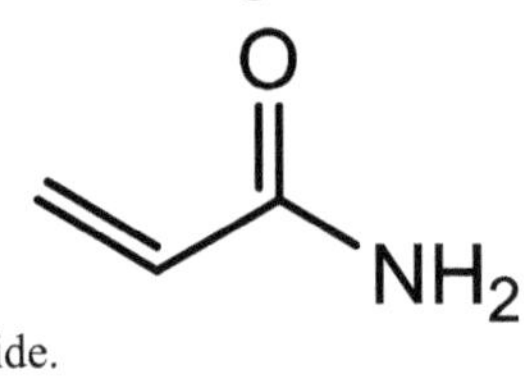

Fig. (4). Molecular structure of acrylamide.

Characterization and Toxicology

Acrylamide has been characterized by its toxicological properties, well studied in animal models, including neurotoxicity, genotoxicity, carcinogenicity and reproductive toxicity. In 1994, the International Agency for Research on Cancer

(IARC) classified acrylamide as a Group 2A carcinogen (probably carcinogenic to humans) [22]. The toxic effects are produced through two ways. Firstly, by the reaction among acrylamide and nucleophilic groups on biological molecules and secondly by the conversion *in vivo* to its metabolite glycidamide, a reactive compound formed through epoxidation of the double bond. After dietary consumption, acrylamide is rapidly absorbed from the gastrointestinal tract and widely distributed to the tissues. In the liver, it is metabolized to glycidamide, which is bound to DNA, having a major role in the genotoxicity of acrylamide in a variety of *in vitro* and *in vivo* test systems. In humans, increasing incidences of endometrial, ovarian, and renal cancer with increased dietary acrylamide intake have been reported [23]. The NOAEL for morphological changes in the nerve systems of rats has been estimated to be 200 μg kg^{-1} body weight (bw) day^{-1} and the TDI 40 μg kg^{-1} bw day^{-1} for neurotoxicity and 2.6 μg kg^{-1} bw day^{-1} for cancer [24]. For genotoxic compounds where no tolerable daily intake is recommended, the MOE approach is used for risk assessment. This value is estimated based on the carcinogenic effect of acrylamide in mammary glands in female rats with a $BMDL_{10}$ of 310 μg kg^{-1} bw day^{-1} and in the Harderian gland of male mice with a $BMDL_{10}$ of 180 g kg^{-1} bw day^{-1} [25].

Chemical Formation in Foods

In 2002, the Swedish National Food Administration and researchers from Stockholm University demonstrated that large amounts of acrylamide were present in certain thermally processed starchy foods, which raised a huge concern about the exposure to this contaminant. Acrylamide is generated in foods when they are subjected to temperatures above 120 °C and in low moisture conditions. It mainly forms through the Maillard reaction between certain amino acids such as asparagine, and reducing sugars (usually fructose and glucose), although other pathways have also been proposed to contribute to the acrylamide formation [26]. Maillard reaction is involved in the appearance of several sensorial aspects of processed foods, such as flavor generation and color development. Therefore, the progress of this reaction is a balance between favorable and unfavorable characteristics.

Occurrence and Exposure

The formation of acrylamide is particularly influenced by the levels of precursors (asparagine and reducing sugars), time and temperature of the process, pH and water activity, among other factors [26]. Acrylamide forms at different levels in numerous baked, fried or roasted carbohydrate-rich foods, highlighting cereals, fried potatoes and coffee as the major sources of exposure to this contaminant. In these foods, the acrylamide formation is promoted by the level of precursors in the raw materials and the high temperatures applied during processing (Fig. **5**).

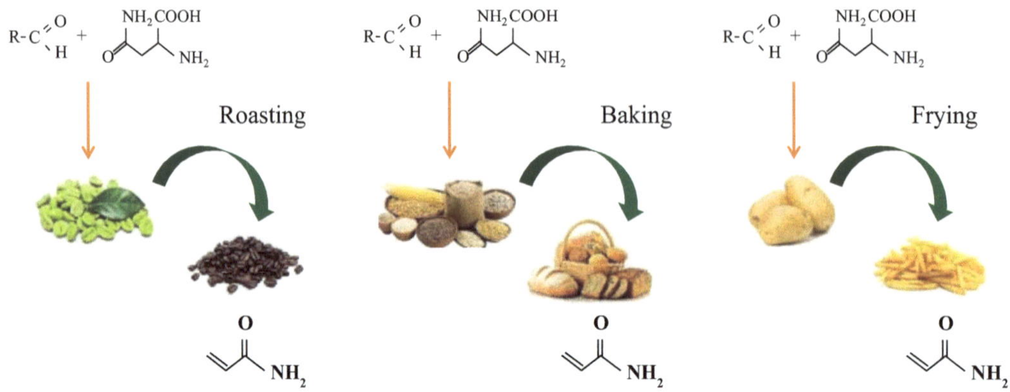

Fig. (5). Formation of acrylamide in roasted coffee, bakery products and fried potatoes.

Factors Affecting the Acrylamide Formation

Acrylamide Precursors

Acrylamide precursors content depends on the type of food raw material, which conditions the development of the reaction in terms of the limiting reactant. For potato products, reducing sugar is the limiting factor of acrylamide forming potential and asparagine is a secondary factor with a lower influence. Levels of precursors in the tuber vary between cultivars and are affected by the variety of potatoes, agronomical factors and growing season. In addition, sugars are also influenced by the storage conditions, since temperatures lower than 6 ºC promotes cold sweetening increasing the levels of glucose and fructose. On the contrary, acrylamide formation in the cereal products is mostly determined by the asparagine content in the flour [27]. Precursors content in cereals also depends on the cultivar and agronomical factors, but varieties and cereal types present less variation and storage does not have any influence. In general, the content of asparagine is higher in whole-grain flour than in the sifted fractions and therefore whole-grain products may contain a higher content of acrylamide. Coffee beans contain sugars and asparagine, which is the amino acid with the second highest concentration. As a consequence, Maillard reaction is promoted during roasting of the coffee beans leading to the formation of acrylamide but also to the production of the typical color and aroma of the coffee. In this case, asparagine is the limiting precursor, that is found in higher amounts in Robusta coffee compared to Arabica coffee, explaining the higher acrylamide amounts in Robusta compared to Arabica coffee [28].

Time and Temperature of the Process

Acrylamide formation in foods also depends on the heating conditions of the process. It is well established that high temperatures during frying, roasting and baking strongly increase acrylamide formation in the processed products. The increase in acrylamide levels with increasing temperature follows an exponential function since that small deviations in temperatures can have large consequences for acrylamide formation. It starts at temperatures slightly above 120 °C and reaches a maximum at around 170-180 °C. However, at higher temperatures acrylamide levels decrease due to a high rate of elimination. This fact is easily observed in coffee, where acrylamide formation reaches the maximum peak some time before the coffee is ready roasted, being the concentration in the roasted coffee somewhat lower than at the peak level. Acrylamide boils at 193 ºC and typical roasting temperature is 210-250 ºC. For that reason, dark-roasted coffee contains less acrylamide than medium-roasted coffee [28]. Not only temperature is crucial, but also the time of heating is important and both factors determine the kinetics of acrylamide formation.

Water Activity and pH

Acrylamide formation is specially affected by the moisture content of the product and by its water activity. Depending on the water content in the foodstuff, water may have an inhibitory or an accelerating impact on the Maillard reaction and, therefore, on acrylamide formation. Acrylamide is formed when water activity is reduced to below 0.8, and in low water activity foods, maximum acrylamide formation takes place when water activity is about 0.4. Additional reduction of water activity tends to decrease the amount of acrylamide. However, it is well known that moisture content is related to thermal processing conditions and it is difficult to separate the effect of water content from that of temperature or time of heating on acrylamide formation [29].

Maillard reaction development is reduced when pH value is decreased and, therefore, acidic conditions minimize acrylamide generation. However, it has been reported that a more alkaline pH does not necessarily imply a higher acrylamide content in the heated products.

Other Factors

Other factors such as baking agents, salts, additives or time fermentation for bakery products [30] and portion size for fried potatoes [31] have also been described to affect the acrylamide formation.

Exposure

EFSA has reported the results of the monitoring acrylamide levels in a total of 10366 foods from 2007 to 2010 [32]. Fig. **(6)** shows the distribution of acrylamide levels in the different foods. Squares represent the content among the 25th and 75th percentiles and bars indicate the maximum levels of acrylamide for each food product. Mean levels ranged from 37 μg kg^{-1} for 'soft bread' to 1504 μg kg^{-1} for 'substitute coffee', with maximum levels for 'substitute coffee' at 7095 μg kg^{-1} and 'potato crisps' at 4804 μg kg^{-1} in 2008 and 2009, respectively.

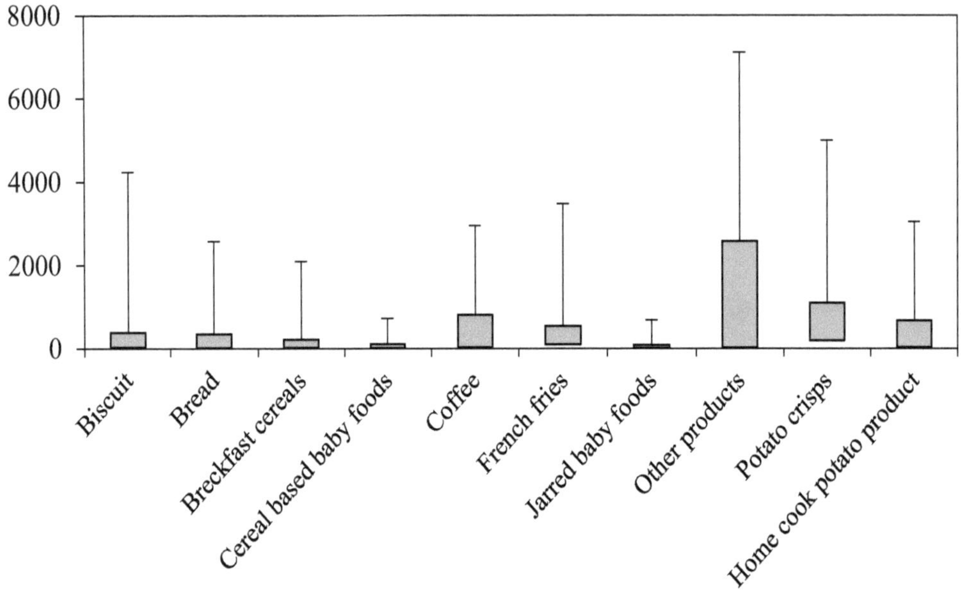

Fig. (6). Acrylamide levels (μg kg^{-1}) in food from monitoring years 2007-2009 [32].

The mean acrylamide intake in Europe has been estimated to range between 0.31 and 1.1 μg kg^{-1} bw day^{-1} for adults (>18 years), between 0.43 and 1.4 μg kg^{-1} bw day^{-1} for adolescents (11-17 years), between 0.70 and 2.05 μg kg^{-1} bw day^{-1} for children (3-10 years) and between 1.2 and 2.4 μg kg^{-1} bw day^{-1} for toddlers (1-3 years). Similarly, the 95th percentile ranged between 0.58-2.3, 0.94-3.1, 1.5-4.2 and 2.4-6.5 μg kg^{-1} bw day^{-1}, for the same groups population. Among the different food sources, the major contributors to the overall acrylamide exposure have been identified to be 'roasted coffee', 'soft bread' and 'fried potatoes' for adults, 'fried potatoes', 'soft bread' and 'biscuits' or 'potato crisps' for adolescents and 'fried potatoes', 'soft bread' or 'unspecified bread' and 'biscuits' for children and toddler population [32].

Mitigation

The knowledge of the harmful effect on human health and the high exposure to this contaminant through the diet led the statement of the Joint Expert Committee of FAO/WHO on Food Additives (JECFA) to reveal that the toxicological threat caused by the intake of acrylamide cannot be excluded [25]. EFSA concluded in 2015 that levels of dietary exposure to acrylamide across age groups indicate a concern due to its carcinogenic effects [33]. For that reason, the implementation of mitigation strategies in order to decrease the formation and consequently the exposure to this contaminant through the diet are necessaries.

Since 2002, the food industry, the academia, the Member States and the EU Commission have undertaken extensive efforts in order to investigate pathways of formation and to reduce the levels of acrylamide in processed foods. On the basis of both results of the investigations and monitoring results obtained over the last years, the European Commission adopted recommendations into the levels of acrylamide in foods, establishing indicative levels for the main food sources of exposure to this contaminant. Member States were recommended to carry out additional surveillance control in cases where the levels of acrylamide in a foodstuff exceed the acrylamide indicative values [34, 35]. In parallel, Food Drink Europe (FDE), in close cooperation with the national food safety authorities and the European Commission, developed a "toolbox", including brochures designed to help food business operators to implement ways relevant for their sector to mitigate acrylamide in food. After all these actions, European Commission established in 2017 mitigation measures for the reduction of acrylamide in foods and identified new benchmark levels for a number of food categories (Table **1**). The regulation for acrylamide in EU entered in force in April 11st 2018 [36]. The Regulation 2158/2017 compels food processors and food business operators in Europe to reduce the presence of acrylamide according to the ALARA principle, establishing measures proportionate to the size and nature of their establishments without compromising the taste, texture and appearance of their products.

Mitigation measures are mainly based on the reduction of acrylamide precursors in raw materials and/or changing some process parameters, although other factors as removing or trapping acrylamide after its formation in foods may also be implemented. Many studies report different mitigation strategies to reduce the formation of acrylamide, some of them are considered by the EU Regulation 2158/2017 [36].

Table 1. Indicative acrylamide values (μg kg^{-1}) according to the recommendations from 2011 and 2013 and benchmark levels for the presence of acrylamide in foodstuffs according to the regulation from 2017.

Food	2011	2013	2017
French Fries (ready-to-eat)	600	600	500
Potato crisps from fresh potatoes and from potato dough	1000	1000	750
Potato-based crackers	-	-	750
Other potato products from potato dough	-	1000	750
Soft bread			
(a) Wheat based bread	150	80	50
(b) Soft bread other than wheat based bread	150	150	100
Breakfast cereals (excl. porridge)			
- bran products and whole grain cereals, gun puffed grain	400	400	300
- wheat and rye based products	400	300	300
- maize, oat, spelt, barley and rice based products	400	200	150
Biscuit and wafers	500	500	350
Crackers with the exception of potato based crackers	500	500	400
Crispbread	500	450	350
Ginger bread	-	1000	800
Products similar to the other products in this category	500	500	300
Roast coffee	450	450	400
Instant (soluble) coffee	900	900	850
Coffee substitutes			
(a) coffee substitutes exclusively from cereals	-	2000	500
(b) coffee substitutes from a mixture of cereals and chicory	-	-	(*)
(c) coffee substitutes exclusively from chicory	-	4000	4000
Baby foods[1], processed cereals based foods for infants and young children excluding biscuits and rusks[2]	80[1]-100[2]	50	40
Biscuits and rusks for infants and young children	250	200	150

(*) Depends on the relative proportion of the ingredients in the final product

Levels of Precursors

A mitigation measure to control the formation of acrylamide is to reduce the levels of precursors (asparagine and reducing sugars) in the raw products. In this sense, an option is the selection of suitable varieties of the raw material with a low content of acrylamide precursors. In the case of potatoes, it is recommended to select potatoes where bruised, spotted or damages are not exceeded and removing the immature tubers. In order to minimize senescent sweetening, potatoes should not be stored and transported at temperatures lower than 6 °C. In addition, in long term stored potatoes, sprouting should be suppressed using appropriate agents. Levels of humidity should also be controlled. For bakery products, the selection of cereals with a low content of asparagine such as rice or maize may be an alternative method to replace partially the wheat flour in a recipe or to consider in new products development.

The sugar content may be decreased by washing, soaking or blanching the raw products before thermal processing whereas asparagine levels can be reduced by pre-treatments with the enzyme asparaginase. Asparaginase converts asparagine into aspartic acid, significantly reducing acrylamide levels and maintaining intact the sensorial attributes of the final product. This strategy has been suggested as an effective technological intervention to mitigate the potential for acrylamide formation in potatoes, cereals and coffee. Fermentation is another tool to lower the occurrence of asparagine in cereal products. In the case of coffee, due to the close relationship between the formation of acrylamide and the development of the desired sensory properties, it is difficult to implement mitigation strategies to reduce acrylamide content in this food. Reduction of precursors could also be reached by using microorganisms able to assimilate free sugars and/or free asparagine, replacing sugars by non reducing sugars or applying genetic approaches. In this respect, numerous patents have been developed [37].

Process Parameters

The heating temperature should be as low as feasibly possible taking into account the specific product and technology applied. Recommendations are focused on keeping the temperature between 160-175 °C when frying and 180-220 °C when using an oven, reducing the cooking time to avoid excessive browning in the finished product. Then temperature and heating times have to be effectively combined to minimize the acrylamide formation but without impairing the taste, texture, color, safety and shelf-life of the products. Moreover, controlling moisture influences the development of Maillard reaction and, therefore, the acrylamide generation. For fried products, the application of vacuum frying may be an alternative process to mitigate acrylamide levels.

Other acrylamide mitigation technologies are based on the addition of acids to reduce the level of pH and limit the reaction. Acrylamide formation exhibits the maximum at pH value of 8 and decreasing this level enhances its elimination and reduces its formation. The addition of free amino acids other than asparagine or from a protein-rich food component to the food matrix has also been reported to strongly reduce the acrylamide content, probably by promoting competing reactions and/or by covalent binding the acrylamide formed. On the other hand, the addition of salts such as NaCl and $CaCl_2$ have been proved to have considerable inhibitory effects on the acrylamide formation and therefore this practice could be another mitigation strategy. In this respect, asparagine can be ionic and electronically associated with cations and early-stage of Maillard reaction can be suppressed. Other additives are also involved in the formation of acrylamide during processing. In bakery products, ammonium bicarbonate is recommended to be replaced by alternative raising agents such as sodium bicarbonate and acidulants or sodium bicarbonate and disodium diphosphates with organic acids or potassium variants.

Size and surface area of the thermal treated foods should also be taken into account since small products lead to promote a higher formation of acrylamide. In addition, color checks on the final cooked products can be performed in order to avoid high levels of acrylamide. Potatoes should be cooked until a golden yellow color and baked products until a lighter color endpoint.

Other Factors Removing or Trapping Acrylamide

Acrylamide can also be removed or trapped once formed. This strategy is mainly focused on coffee samples. It has been reported that when degree of roasting is increased, acrylamide levels are decreased and moreover, acrylamide would be entrapped by the coffee melanoidins. Treatments with supercritical CO_2 and the use of bacterial enzymes may also reduce and control acrylamide content in the roasted coffee (revised by Pedreschi *et al.* [38]).

Methods of Analysis

Since 2002, many analytical methods for acrylamide determination have been developed. Several reviews on the numerous methods for acrylamide detection have been published to date [26, 39, 40]. Due to the sensitivity, selectivity and versatility, the most preferred methods for separation and quantification of acrylamide are liquid chromatography (LC) and high performance liquid chromatography (HPLC) coupled with mass spectrometry. Reliable results can be also obtained by methods based on gas chromatography, with or without derivatization. According to that, official methods of acrylamide determination in foods by isotope dilution liquid or gas chromatography approaches, coupled to tandem mass spectrometry detectors (GC-MS and LC-ESI-MS/MS) have been established by the European Committee of Normalization (CEN, group TC 275) under a specific European Commission Mandate M/463 dated April 30, 2010. These Technical Procedures specify methods for the determination of acrylamide in different matrix, including bakery products, potato products as well as roasted coffee. Among these foods, coffee and its substitute are considered the most difficult matrix to be analyzed due to the presence of interferences that can be co-extracted together with acrylamide and methods have been optimized for the determination. Analysis includes sample extraction in polar solvent (water being mainly used in most published analytical methods), clarification and purification and cleanup step by solid-phase extraction to eliminate interfering compounds. During the extraction step, the internal standard is added with the aim of controlling the recoveries and taking into account possible losses occurring during the sample treatment. GC methods include bromination of acrylamide to be derivatized to 2,3-dibromopropionamide, improving the sensitivity but increasing the complexity of the procedure and the toxicity due to the bromine used. Besides MS detection methods, HPLC-diode-array detection (DAD), HPLC-ultraviolet detection (UV), GC-electron capture detection (ECD) and capillary electrophoresis (CE) have also been employed for the analysis of acrylamide.

Together with standard protocols, other rapid detection methods for acrylamide analysis have been developed. These methods are based on biochemical properties of acrylamide, changes of physiochemical properties of thermally processed foods or biomaterials with high and specific interaction with acrylamide. Since acrylamide formation is correlated with browning levels in processed foods, generated with the advance of Maillard reaction, color and more specifically the redness parameter (a^*) can be used as indicators to evaluate acrylamide concentrations in thermally processed products [41, 42]. Near-infrared spectrometry (NIR) and computer vision are also proposed as methods to monitor acrylamide content in thermally processed foods. Analyses combine pictures of food samples, image analysis and the relationship between acrylamide and browning, allowing improved measurements for uneven surfaces [43]. Browning-based methods are simple procedures, without requiring extraction and purification of the samples. Acrylamide levels can be determined by enzyme-linked immunosorbent assay (ELISA) method through the antigen-antibody reaction. This method offers sensitivity, low cost, short detection time, simplicity and does not require complex sample preparation or expensive equipment. Other novel determinations include supramolecular recognition based methods and electrochemical biosensing procedures. Two types of acrylamide biosensors namely amperometric and potentiometric have been described in the literature. Amperometric biosensors monitor are based on the detection of Hb-acrylamide adducts, whereas potentiometric acrylamide biosensors focus on a direct biochemical interaction between acrylamide and intact bacterial cells [44]. Despite the lower sensitivity and selectivity, fluorescent sensing are alternative methods for the determination of acrylamide in foods. In this case, acrylamide is polymerized and detected by the photophysical property of quantum dots.

FURAN

Furan (C_4H_4O) (Fig. 7) is a small (68.07 g mol^{-1}), heterocyclic organic compound present in the volatile fraction of many thermally processed foods and drinks. It is a colorless liquid, lipophilic and highly volatile, with a boiling point of 31 °C. Its melting point is -85 °C and its boiling point (760 mm Hg) 31.4 °C. Furan is soluble in organic solvents but presents low solubility in water. Due to its volatility, the occurrence of furan is related to the flavor of food products, providing pleasant characteristics. However, it also has been associated with toxic and genotoxic effects.

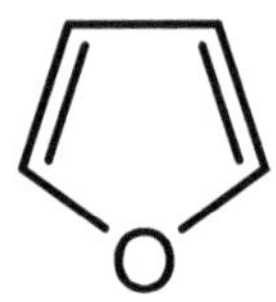

Fig. (7). Molecular structure of furan.

Characterization and Toxicology

Furan has been classified as possibly carcinogenic to humans (Group 2B) by the IARC [22], and FDA and EFSA have confirmed its carcinogenicity and cytotoxicity in animal studies. Furan affects DNA indirectly through a mechanism involving cytotoxicity whereas carcinogenicity is induced through the metabolic activation of cytochrome P450 to cis-2-butene-1,2-diol. According to that, furan exposure indicates a human health concern. Contradictory indications about the genotoxicity of furan have been provided in *in vitro* test.

Furan is hepatotoxic and hepatocarcinogenic in rats and mice and carcinogenicity is considered the critical endpoint for use in human health risk assessment. JECFA established a BMD_{10} of 1.34 mg kg^{-1} bw day^{-1}, based on analysis of liver tumor incidence data from an oral bioassay in mice. According to this information, MOEs between 960 and 480 have been considered for average and high dietary exposures, respectively. The Committee reported that these MOEs indicate a human health concern for furan. On the other hand, the Norwegian Food Safety Authority calculated a $BMDL_{10}$ of 0.14 mg furan kg^{-1} bw day^{-1} based on the incidence of cholangiocarcinomas in rats, resulting in MOEs between 29-13 for 12-months infants and between 74-26 for adults (average and high dietary exposures, respectively) [45].

Chemical Formation in Foods

Furan can be generated in foods from different precursors, including ascorbic acid, amino acids, carbohydrates, unsaturated fatty acids and carotenoids. Formation from ascorbic acid derives from the opening of the hydrolytic ring of ascorbic acid, followed by decarboxylation and carbohydrate degradation mechanisms together with the formation of the intermediary 2-furoic. This compound, in turn, is a furan precursor.

Certain amino acids are capable of forming acetaldehyde and glycolaldehyde, which are precursors of furan through aldol condensation followed by cyclization reactions. These amino acids include serine and cysteine. Other amino acids such as aspartic acid, alanine or threonine can also be involved in the formation of furan, but they need the presence of other reactants. Sugars on their own are able to generate furan, showing relative efficiency following this order of reactivity: D-erythrose > D-ribose > D-sucrose > D-glucose = D-fructose. However, amino compounds also favor the formation of furan from sugars via the Maillard reaction. Regarding fatty acids, it has been reported that only polyunsaturated fatty acids such as linoleic and linolenic acids can generate furan through oxidation mechanisms, whereas monounsaturated acid such as oleic acid, does not form furan [46].

Occurrence and Exposure

Furan is formed in foods during thermal processing. High concentrations have been detected in baked and roasted products, as well as in foods subjected to retorting in cans and jars. Data on furan concentrations in heat treated commercial food products have been collected and reported by EFSA [47]. Fig. (**8**) shows the distribution of furan levels in different groups of foods. These groups are the following: (1) grain and grain-based foods, (2) vegetables and vegetable products, (3) starchy roots and tubers, (4) legumes, nuts and oilseeds, (5) fruit and fruit products, (6) meat and meat products, (7) fish and other seafood, (8) milk and dairy products, (9) sugar and confectionary, (10) animal and vegetable fats and oils, (11) fruit and vegetable juices, (12) non-alcoholic beverages, (13) alcoholic beverage, (14) herbs, spices and condiments, (15) food for infants and small children, (16) composite food, (17) snacks, desserts and other foods. Squares represent the range of the mean values and bars indicate the 95th percentile of furan for each food category. Mean levels ranged from 0.7 $\mu g\ kg^{-1}$ for 'milk and dairy products' and 'animal and vegetable fats and oils' to 4650 $\mu g\ kg^{-1}$ for 'coffee beans', with maximum levels for 'roasted and ground coffee beans', including decaffeinated coffee up to 8700 $\mu g\ kg^{-1}$.

Furan is present in coffee as part of the volatile aroma components generated during roasting; thus, furan concentrations are very low or not detected in green coffee, whereas high concentrations have been observed in roasted ground coffee.

During roasting, furan formation depends on green coffee varieties and roasting conditions [48]. Despite the high furan concentrations in coffee bean, levels decrease from roasting to the cup due to its high volatility and low water solubility, as mentioned later. Mean values of 2446 µg kg^{-1} have been reported for ground roasted coffee, 429 µg kg^{-1} for instant coffee powder and levels around 4.9-31 µg kg^{-1} for coffee drink.

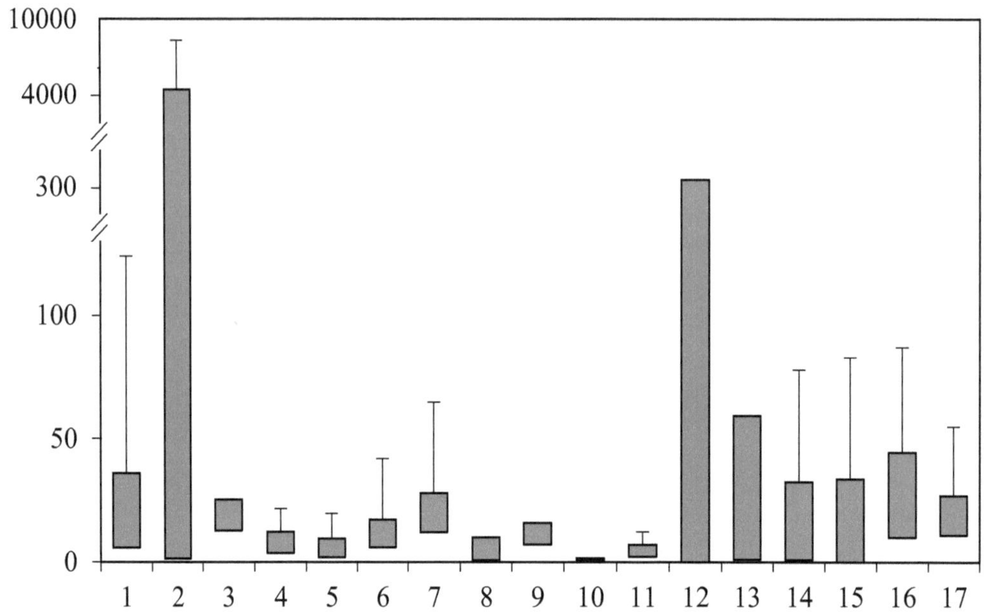

Fig. (8). Furan levels (µg kg^{-1}) in different food groups (based on EFSA [47]).

High furan levels are also found in foods subjected to heat treatment in sealed containers, such as baby foods, soups and other products retorted in cans and jars. In these foods, volatiles are probably trapped in the food container, which avoids the volatilization and consequently increasing the furan content. Levels of furan ranging from 7.7 to 143.0 µg kg^{-1} in fruit and vegetable-based jarred baby foods have been reported by Mesias *et al.* [49]. These authors stated that the higher furan content in vegetable-based baby foods may be associated both with the different way of elaboration of these products and with a greater ascorbic acid degradation by the presence of oxygen in vegetable samples, which is favored by a higher pH in these samples (pH = 5.7) compared to fruit-based baby food (pH = 3.9). Furan is also generated in foods with a low moisture content that are not heated in closed containers, such as cereals, crackers and biscuits or cookies and in other type of foods as soups and soya sauce [47].

Besides commercial products, although in lower levels, furan is also present in foods cooked at home, including toasted, baked, deep-fried and roasted foods. In this regard, concentrations ranging from < 2 to 260 μg kg^{-1} has been reported in toasted bread, from < 2.4 to 27 μg kg^{-1} in baked foods such as pizza, apple cake or breaded products, from 6 to 51 μg kg^{-1} in deep-fried products and around 3-5 μg kg^{-1} in the case of roasted onion. Among these home-cooked foods caramel is highlighted, exhibiting the highest furan levels (51-570 μg kg^{-1}).

Taking into account the furan concentration in the different food categories and considering the dilution factors applied according to the preparation of the foods previous to their consumption, EFSA has estimated the chronic dietary exposure for furan for different age class. Mean dietary exposure (μg kg^{-1} bw day^{-1}) ranges between 0.14-0.99 for infants, 0.22-0.65 for toddlers, 0.19-0.52 for other children, 0.11-0.31 for adolescents, 0.11-0.54 for adults, 0.12-0.61 for elderly and 0.13-0.75 for very elderly [47]. The main contributors to the dietary exposure of infants to furan are foods intended for infants and small children, which account for up to 70% of the exposure. In adults, elderly and very elderly, the group of non-alcoholic beverages was found to be the main source of exposure, including coffee. Composite foods such as ready-to-eat soups and cereal-based dishes as well as grains and grain-based products are other important contributors to the furan exposure throughout all age groups.

Mitigation

Mitigation strategies are focused, on the one hand, on preventive measurements in order to reduce the formation, and, on the other hand, on the removal of the already formed furan from food products.

Preventive Strategies

Furan formation increases with the increase of both temperature and heating time and, therefore, a decrease of the thermal input should be useful to reduce the furan concentration in thermally heated foods. Innovative processing practices as ohmic heating or the vacuum technology have been demonstrated to reduce furan levels in processed foods [50]. Ionization radiation also reduces furan levels of ready-to-eat foods, such as turkey, frankfurters and sausages [51]. On the other hand, changes in the formulation of foods and in consequence in the precursor of furan can also be regarded as a preventive mitigation strategy. This fact could be achieved by precursor removal or substitution, as the use of different sugars which promote to a greater or lesser extent the formation of furan or the addition of food additives and ingredients to avoid this formation, such as phosphate or binary mixtures of ascorbic acid and amino acids, sugars or lipids [50]. Adding antioxidants can also be an efficient strategy to reduce furan levels by inhibiting the formation from thermal oxidation of polyunsaturated fatty acids and carotenoids [52].

Removal Strategies

The high volatility of furan facilitates the reduction of this compound once generated in foods and consequently its exposure. Furan is easily lost by evaporation and therefore cooking or re-heating the foods in an open vessel result in significantly lowering furan levels in comparison with heating in sealed containers. Regarding to coffee, levels of furan are decreased from the bean to the beverage. Thus, through grinding, packed, storage and brewing procedure losses of this compound are produced [48] (Fig. **9**). In addition, the furan concentration in the cup of the coffee brew can be partially evaporated depending on the time of consumption, the stirring of the beverage or the maintenance at different conditions, for instance in a thermo during a workday or in a fridge [53].

Fig. (9). Decrease of furan content in coffee from roasting to the cup (based on Guenther, 2012 [48]).

Methods of Analysis

No official standard methods to determine furan in foods are available and neither there are certified reference materials. Different approaches are used, all of them based on MS detection, quantification using standard additions of deuterated furan (d_4-furan) as an internal standard and separation accomplished by GC. Due to the high volatility of furan, special care should be taken during the analysis. In this respect, to avoid furan losses during sample preparation, sampling and extraction have to be performed under cold conditions. Additional formation of furan from volatile precursors should also be prevented during the analytical procedure.

The most common analysis for determining furan in foods is the headspace-gas chromatography-mass spectrometry (HS-GC-MS) method developed by the US Food and Drug Administration [54]. In this method, samples of food can be weighed directly in the headspace vial, mixed and homogenized with water and then the vial is sealed and fortified with a standard furan solution at concentrations 0.5, 1.0, and 2.0 times the estimated furan content in the food sample. In addition, a standard d_4-furan solution at a concentration similar to that of the estimated furan content in the food sample is added to each vial. This standard addition method is selected for furan quantification in order to avoid matrix effects in the different food samples. Vial is heated to promote volatilization of the analyte in the headspace phase, the headspace gas is sampled and vapor injected into a gas chromatograph. In this analysis, pH of the samples must be taken into account since acidic or basic conditions may affect furan concentrations determined in food samples. The subsequent detection of furan is carried out by mass spectrometry or by nonselective means such as flame ionizing detection. The chromatographic separation is mainly performed on columns allowing an adequate retention and separation of furan from other volatile compounds. GC-MS analysis is used to detect furan and d_4-furan in the selected ion monitoring mode. Furan is identified and quantified using the ions *m/z* 68 and 39 whereas *m/z* 72 and 42 are used for d_4-furan determination.

Furan can also be analyzed by a solid-phase microextraction (SPME) method in combination with GC-MS. SPME uses a fiber coated with a liquid (polymer), a solid (sorbent), or a combination of both. The fiber coating is exposed to the headspace vapors to absorb volatiles (absorption in the case of liquid coatings and adsorption in the case of solid coatings). After that, the SPME fiber is inserted directly into the gas chromatograph for furan desorption and analysis. SPME method allows concentration of the sample, which leads to a higher sensitivity. Proton-transfer reaction mass spectrometry (PTR-MS) could also be used as an analytical method to quantity furan [47].

HETEROCYCLIC AROMATIC AMINES (HAAS)

Heterocyclic aromatic amines are a large group of structurally close genotoxic compounds formed from reactions between creatinine, sugar and amino acids at high temperatures. HAAs are formed in meat, particularly read meat, fish and poultry during pan-frying, roasting/baking, barbecuing, deep-fat frying, smoking and grilling. Different classes of HAAs can be formed from proteinaceous food depending on the temperature.

Characterization and Toxicology

In late 70s, a group of Japanese scientists identified a highly mutagenic class of HAAs in grilled or broiled meat and fish [55]. The *in vitro* mutagenicity associated to HAAs (*i.e.,* imidazoquinolines) is more potent than exerted by the previously known structures of polycyclic aromatic hydrocarbons (*i.e.,* benzo(*a*)pyrene) or mycotoxins (*i.e.,* aflatoxin B1).

Table 2. Chemical structures of different classes of heterocyclic aromatic amines (HAAs).

Type of HAAs	Structure	Abbreviation	CAS number
2-Amino-imidazoquinolines	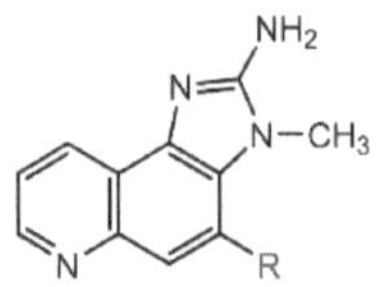	IQ (R: -H) MeIQ (R: $-CH_3$)	76180-96-6 77094-11-2
2-Amino-imidazoquinoxalines	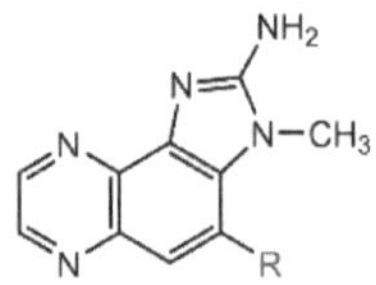	IQx (R: -H) MeIQx (R: $-CH_3$)	108354-47-8 108354-48-9
2-Amino-imidazopyridines	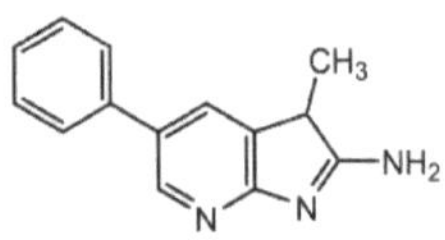	PhIP	105650-23-5
Pyridoindoles (β-Carbolines)	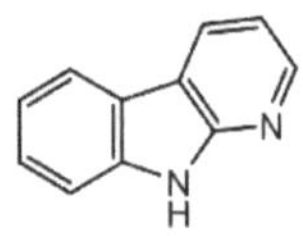	Norharman	244-63-3
Pyridoindoles (γ -Carbolines)	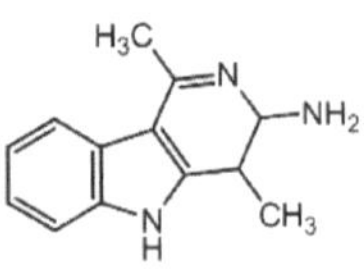	Trp-P-1	62450-06-0

In general, HAAs are formed from thermally processed food heated at temperatures higher than 100 °C, containing nitrogenous compounds, mainly foods of animal origin, such as meat and fish [56]. Furthermore, some structures of HAAs have been identified in wine, beer, and smoked cheese but the source is external to foods, very likely is due to environmental exposure. Since 1977, nearly 25 derivatives of HAAs have been identified from cooked meat or fish. The compounds have been isolated, their structures determined, the mechanisms of formation elucidated by isotope labeling, and the thermal dependence investigated by kinetic modeling. Chemically, HAAs consist of two or three rings with an exocyclic amino group attached to one of the rings. Table **2** describes the main structures of different class of HAAs.

Depending on the dominating chemical structure, HAAs can be divided into: amino-carbolines (*i.e.,* norharman), imidazoquinolines (*i.e.,* IQ, MeIQ), imidazoquinoxalines (*i.e.,* IQx) and imidazopyridines (*i.e.,* PhIP). However, there is another classification according to the temperature needed to their formation. In this sense, HAAs are classified as pyrolytic HAAs and thermic HAAs. The pyrolytic HAAs form at elevated temperatures (> 250 °C) from certain amino acids such as tryptophan, glutamic acid, phenylalanine and ornithine. For instance, the γ-carbolines mutagens (*i.e.,* 2-amino-1, 4-dimethyl-5H-pyridol[4,3-*b*]indole Trp-1) are derived from tryptophan which maintains an indole group in the structure. The thermic HAAs (120-250 °C) are constituted by aminoimidazoarenes, such as imidazoquinolines, imidazoquinoxalines, and imidazopyridines that may be formed from creatine or creatinine, certain free amino acids and sugars *via* the Maillard reaction when meat cooks at mild temperatures (< 250 °C). However, HAAs are not present in uncooked food or during stewing, steaming, and boiling as temperatures are below 120 °C [57].

Once the HAAs are ingested, they are efficiently absorbed, and their bioactivation by the cell machinery takes place. From animal studies, HAAs absorption seems to be almost complete; thus, variations in urinary excretion should provide a measure of interindividual variability in the metabolism of these compounds. Most HAAs are not mutagenic/carcinogenic in their native forms but acquire the capability of forming DNA adducts after metabolic activation. Metabolic activation of the heterocyclic amines predominantly involves CYP1-mediated N-hydroxylation and then O-esterification by phase II enzymes [58]. Then, the first step is the oxidation to hydroxyamine derivatives by cytochrome P450s, *i.e.,* N-hydroxylation by CYP1A2, and subsequent acetylation by N-acetyltransferase type 2 [59]. The genotoxic N-hydroxylation pathway of HAAs amines involves primarily members of the CYP1A subfamily. The activation by CYP1A2 can be induced in humans exposed to a diet rich in HAAs [60]. Examination of the primary oxidative metabolites in a mutagenicity assay such as the Ames *Salmonella typhimurium* test showed that the N-hydroxy metabolites of MeIQx and PhIP were direct-acting mutagens, whereas the ring-hydroxylated products were not [61]. Then, the exocyclic amino group of the imidazo-moiety present in most HAAs is the likely ultimate carcinogen binding to the DNA bases producing DNA adducts through the formation of N-C bonds at guanine bases. DNA adducts have been detected and characterized for all HAA. These amines are potent bacterial and eukaryotic cell mutagens, and all of those tested have been found to induce tumors in laboratory animals. The formation of HAA-DNA adducts is mediated via electrophilic attack of guanine bases at the N-2, N-7, or C-8 position by the HA nitrenium ions. HAAs have been found to be potent carcinogens inducing tumors in liver, lung, haematopoietic system, forestomach, and blood vessels in mice, and colon, small intestine, prostate, mammary gland, hematopoietic system, liver, Zymbal gland, skin, clitoral gland, oral cavity, and urinary bladder in rats (usually, doses between 0.01 and 0.06% in diet for 48-112 weeks) [62]. In addition, further metabolism of the HAAs generates conjugated metabolites that include glucuronides and sulfate and acetyl esters, the esters being reactive derivatives capable of damaging DNA by forming adducts predominantly with guanine bases.

Nowadays, all the HAAs examined have been shown to be carcinogenic in bioassays. Based on animal experiments, IQ has been classified by IARC as a possible human carcinogen (Group 2A) and MeIQ, MeIQx, PhIP, AaC, MeAaC, Trp-P-1, Trp-P-2 and Glu-P-2 as probable human carcinogens (group 2B) [63]. The European Commission has proposed regulations for maximum levels of PhIP and 4,8 - DiMeIQx of 50 μ g kg^{-1} in thermal process flavorings [64]. HAAs can also be classified as polar (IQ-, IQx-, imidazopyridine-type) and non-polar structures which have a common pyridoindole or dipyridoimidazole moiety. The non-polar category is mainly responsible for most of the HAAs associated mutagenic activity detected in food [65]. In summary, N-hydroxylation of the HAAs, the primary metabolic pathway in humans, is also the primary route of HAAs genotoxicity [66].

Chemical Formation in Foods

Various factors influence the formation of HAAs in foods such as temperature, time, method of cooking, acidity, pH, moisture, protein and fat content, and amino acids profile of the protein source, as well as the presence of precursors, enhancers and inhibitors [67]. The class of HAAs aminoimidazoarenes is commonly formed during processing or domestic cooking of foods based on meat and fish. 2-Amino-1-methyl-6-phenylimidazo(4,5-*b*)pyridine (PhIP) is the most abundant mutagenic HAAs found in foods. The main precursors in the raw food are free amino acids, sugars and creatine. Sugars and amino acids react through the Maillard reaction and *via* Streker degradation to form a variety of classes of pyridines and pyrazines. Studies have suggested the involvement of pyridine and pyrazine radicals. However, no clear statement could be made on the role of fat since for one side lipids increase the yield of pyrazines and Strecker aldehydes but there is not effect on the development of mutagenic activity of HAAs. A higher fat content seems to exert a diluting effect of the precursors and contribute to a lower formation of HAAs, but fats also conduct more efficiently the heat into the product giving a local increase of HAAs. However, there are several references in the literature concerning the role of the frying oil in the formation of HAAs during the cooking of meat. Persson *et al.* [68] described that frying oil rich in antioxidants such as virgin olive oil inhibits the formation of HAAs more efficiently than rapeseed oil or refined olive oil. Similarly, an increase in the number of frying cycles leads to an increase in the level and types of HAAs [69]. The sugars occurring naturally in meat are present in approximately half the molar quantity of creatine and amino acids. Meat from horses showed threefold higher glucose levels than beef and formation of HAAs is lower at similar heating conditions. Contrarily, chicken had very low glucose levels, but a similar content of creatine, which increased the concentration of PhIP ten times. Free amino acids are supplied from muscle protein, sugars from glycogen, whereas creatine is an energy metabolite only present in significant levels in muscle cells [62]. Indeed, the addition of creatinine and amino acid form commercial meat flavor as condiment could increase the formation of HAAs during cooking [70]. The creatine cyclizes to creatinine during heating and reacts in an aldol reaction with the pyridine or pyrazine derivatives, respectively, to generate imidazoquinoline (IQ) and imidazoquinoxaline (IQx) as depicted in Fig. (**10**). For example, IQ can be formed from heating of creatinine, glycine and fructose. MeIQx and PhIP result from heating mixtures of creatinine, glucose and glycine, and creatinine, glucose and phenylalanine, respectively. If creatinine is not available, no imidazoquinoline and imidazoquinoxaline are formed. In the case of the PhIP, phenylacetaldehyde is formed from phenylalanine via the Strecker degradation, and reacts in an aldol reaction with creatinine to form an intermediate product. Threonine, glycine, lysine, alanine, and serine lead to MeIQx formation. The formation of amino-carbolines takes place at very high temperatures and depends on amino acids or protein as major precursors. Since creatin(in)e is not required for their formation, they may be produced in foods of animal and plant origins. In the case of formation of β-carbolines, the parent amino acid is tryptophan that reacts with glucose and after oxidation and cyclation forms norhaman.

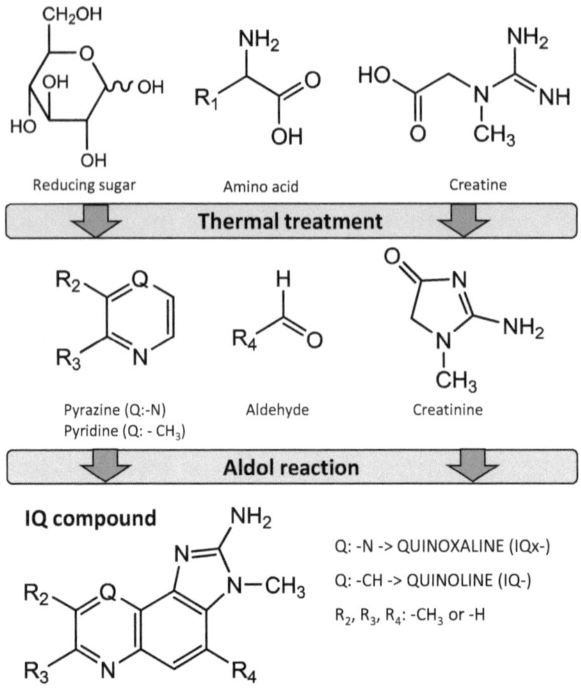

Fig. (10). Mechanism of formation of the heterocyclic aromatic amines classes of IQ- and IQx- from reducing sugars, amino acids and creatine under Maillard reaction and Strecker degradation (adapted from Jagerstad *et al.* [71]).

Occurrence and Exposure

HAAs are present in cooked meat from non-detected to nearly 100 ng g^{-1}, depending of the food (beef, chicken, duck, pork, salmon, etc.), type of preparation (burger, balls, cube, drumsticks, patties, including skin, etc.) and thermal processing applied (pan-frying, deep-frying, baking, roasting, grilling, etc.). The HAAs occurring most often in meat are PhIP, MeIQx, 4,8-DiMeIQx, IQ, MeIQ, and AαC. The use of various preparation methods affects the formation of HAAs differently for different meat products. It has been described that pan-frying and charcoal-grilling produce the highest yield of HAAs at cooking temperatures from 200 °C and above, whereas boiling yields little or no HAAs. Deep-fat frying, roasting, and baking procedure give variable yields. The pH-value also has an impact on the content of HAAs, since HAAs in pork increase about 22% on average (MeIQx - 33%; 4,8-DiMeIQx - 17%; and harman - 9%) in the pale, soft, and exudative meat samples with lower pH values when grilled than the normal muscle at a core temperature of 95 °C [72]. Nevertheless, human exposure to the HAAs, although low, is chronic since meat is consumed daily over a lifetime, but depending upon dietary preferences. Individual's daily exposure to HAAs is essentially zero in the case of vegetarians. Three of these cooked food-derived HAAs, 2-amino-3,8-dimethylimidazo[4,5-*f*]quinoxaline (MeIQx), 2-amino-3,4,8-trimethylimidazo[4,5-*f*]quinoxaline (DiMeIQx), and 2-amino-1-methyl-6-phenylimidazo[4,5-*b*]pyridine (PhIP), are found more frequently than other HAAs. MeIQx, DiMeIQx and PhIP have been detected in beef, lamb, pork, chicken, and fish, especially when chargrilled, fried, or roasted [58]. Large variations are reported in estimated exposure levels, with ranges of daily intake of total HAAs from 160 to 1800 ng $person^{-1}$ day^{-1} [73]. The major source of human exposure to HAAs is through household consumption of cooked meats and fish. Epidemiological results and human daily intake of HAAs obtained from questionnaires show a relationship between the preference for very well-done meat products with increased HAAs levels and an enhanced risk of the incidence of cancer, besides other carcinogens in the diet [74].

Mitigation

In general, the free radical-mediated pathways of HAAs formation can significantly modulate by the addition of compounds which possess antioxidative activity. However, there is not always a direct relationship since potent antioxidant, such as tocopherol, enhances the formation of HAAs. There are also alternative modes of action by scavenging reactive promotors of the reaction which reduce the formation of HAAs. Literature cites the use of different additives (*i.e.,* vitamins) and condiments (*i.e.,* spices and herbs) to mitigate in some extent the formation of HAAs at moderate thermal processing conditions (see for a review [75]). Investigations have been done with spices, herbs, seeds and their extracts rich in antioxidants, mainly polyphenols. It has been described the positive effect of extracts of green tea, grape and pomegranate seeds, apple, apricot, olive, rosemary, thyme, cherry, tomato and oregano. The use of black-, yellow-, red-pepper, onion and garlic powder is more effective when spreading on the surface of the food, for instance when preparing meatballs. Formation of HAAs is an event that occurs at the surface of the product, and then it is better to cook the chicken wings with the skin and then remove it before consumption. It has been also reported the mitigation effect of certain vitamins such as B_1, B_3, B_6, B_7 and C.

Sugar is a precursor of HAAs formation, however adding low levels of sugar (*i.e.,* honey, brown sugar, refined white sugar) to cooked ground meat reduces HAAs formation by inducing competitive reaction with creatine and creatinine [76]. Marinating is an interesting solution to limit the formation of HAAs in meat products as it can act as a barrier to avoid a direct contact of the food product and the heating surface but the response is different according to the HAAs [77]. The marinating in green tea decreases the formation of PhIP and AαC but has no effect in the imidazoquinoxalines.

Since the formation of HAAs has a thermal dependence, a decrease in the cooking temperature or time will lead to an overall decrease in the formation of HAAs [74]. The inhibitory action at the surface of the food is more effective. However, before applying any mitigation strategy, the potential negative modification in the aroma or color of the product should be carefully evaluated.

Methods of Analysis

The determination of HAAs is a challenge since the food matrix is complex (meat, fish) and they are present in very low concentration (few ng g^{-1}). Tandem solid-phase extraction methods have been largely used for the sample preparation of HAAs. The method may be combined with the use of solvents, such as dichloromethane or ethyl acetate depending on the polarity of the target HAAS. It is recommended the use of mixed-mode cartridges with strong cationic exchange sorbent and reverse phase (C_{18}) functionality [78]. In addition, supercritical carbon dioxide with methanol (10%) applying a pressure of 414 bar at 55 °C allowed the isolation of HAAs from solid material with high recoveries of the quinolines and quinoxalines [79]. The most often technique for HAAs quantitation is HPLC-UV and fluorescence detection or UV and MS. After that, depending if searching for a group of HAAs or a wide range of structures it is necessary to apply tandem LC-MS/MS [80].

CHLOROPROPANOLS AND ESTERS, GLYCIDOL AND GLYCIDYL FATTY ACID ESTERS

Chrolopropanols and their fatty esters (chloroesters) are a large family of compounds with a similar core structure substituted with different fatty acids. 3-chloropropane-1,2-diol (3-MCPD), 3-chloropropane-1,2-diol fatty acid esters (3-MCPD-E), glycidol, and glycidyl fatty acid esters (GE) are processing contaminants present in a widespread range of foods and food ingredients. These compounds share similar route of formation and method of analysis, therefore; they are discussed together in this section.

Characterization and Toxicology

First evidence of presence 3-MCPD was in late 70′s in model experiments with lipids hydrolyzed with hydrochloric acid and in acid-hydrolyzed vegetal proteins (HVP) [81]. 3-MCPD forms by reaction of residual oil from vegetal and chloride at high temperatures. 3-MCPD exists as a racemic mixture of *(R)* and *(S)* enantiomers, and its relationship is equimolar in HVP. 3-MCPD is the most abundant of chloropropanols found in foodstuffs. Other isomers as 2-chloropropane-1,3-diol (2-MCPD), and dichloropropanols (*i.e.,* 1,3-dichloropropan-2-ol and 2,3-dichloropropan-1-ol) are found in lower concentrations. Table **3** describes the main chemical structures of different types of chloropropanediols, glycidol and their esters.

In 2002, 3-MCPD-E and to a lesser extent 2-chloropropane-1,3-diol fatty acid esters (2-MCPD-E) were detected in higher contents than those found for 3-MCPD and 2-MCPD in a wide range of thermal processed foods (*i.e.,* cereal-based products, potato-based products, margarines, smoked fish, meat, refined edible oils, fat-based foods). 3-MCPD-E mainly appeared as diesters (85-93%), and esterified with the same or different fatty acids. In the literature, 3-MCPD-E is also named as bound 3-MCPD as opposite to free 3-MCPD. GE are formed in the refining process of oils, specifically during the deodorization step where high temperatures are reached. Palm oil was the first substrate where GE were detected [82]. Nowadays, the high incidence of GE in infant formulas elaborated with refined oils presents high interest, leading to the recent regulation of this compound. This group of products has been under examination by FDA, EFSA, Germany's Federal Institute for Risk Assessment (BfR) in latest years.

Table 3. Chemical structures of different types of chloropropanediols, glycidol and their esters. (R = fatty acid).

Compound	**Structure**	**Abbreviation**	**CAS number**
3-chloropropane-1,2-diol	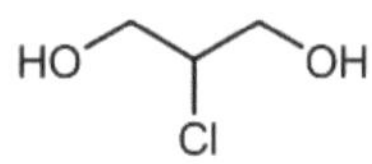	3-MCPD	96-24-2
2-chloropropane-1,3-diol	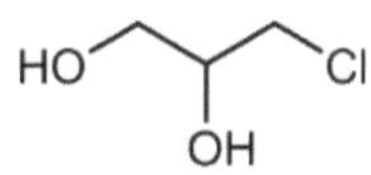	2-MCPD	497-04-01
Glycidol	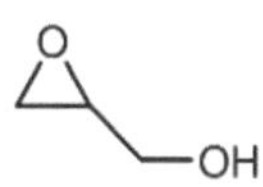	Glycidol	556-52-5
3-MCPD monoester	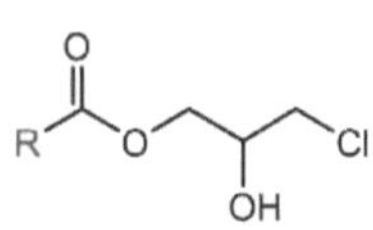	(*i.e.*) 3-MCPD-palmitate (R:-$CH_3(CH_2)_{14}$)	30557-04-1

(Table 3) cont.....

Compound	Structure	Abbreviation	CAS number
3-MCPD diester	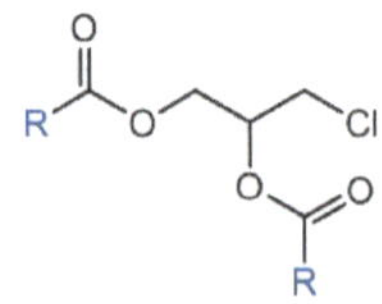	(*i.e.*) 3-MCPD-di-palmitate ($R:-CH_3(CH_2)_{14}$)	51930-97-3
Glycidyl ester	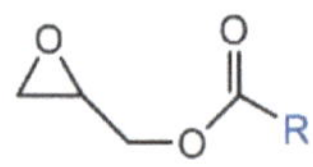	(*i.e.*) glycidyl palmitate ($R:-CH_3(CH_2)_{14}$)	7501-44-2

3-MCPD and its Fatty Acid Esters

The IARC has classified 3-MCPD as a possibly carcinogenic to humans (group 2B) and, for this reason, its content is regulated in a maximum level of 20 μg kg^{-1} in HVP and soy sauce [83, 84]. The oxidative metabolism of 3-MCPD to β-chlorolactaldehyde and β-chlorolactic acid induces toxicity. β-Chlorolactic acid is a major intermediate of 3-MCPD in mammalian species. β-chlorolactate metabolites are supposed to inhibit the glycolysis and cause oxidative stress, which is the mechanism suggested as the possible nephron- and sperm-toxicity. Therefore, kidney and testis are the main target organs affected by 3-MCPD in animal studies (mainly rats) [85].

In vitro studies using an intestinal model showed that 3-MCPD-E are hydrolyzed to 3-MCPD by the pancreatic lipases. The release of 3-MCPD from monoesters was faster than from diesters. Moreover, *in vivo* studies have confirmed the implication of the lipolytic activities in the membranes of intestinal enterocytes [86]. The extent and toxic effects for 3-MCPD-E are similar to those observed for the free 3-MCPD because the esters are cleaved and it is therefore the free form which exerts the toxicity. Consequently, from a toxicological perspective, free and ester forms are considered equivalent on molar base. There is a gap in the assessment of the genotoxicity of 3-MCPD-E since only one *in vivo* study is reported and therefore 3-MCPD-E cannot be classified as genotoxic. Regarding 2-MCPD and 2-MCPD-E, there are few data from *in vitro* genotoxicity studies, and neither *in vivo* with mammalian. The former mechanism through lipases degradation and bioavailability for 3-MCPD is supposed for 2-MCPD-E. Nowadays, it is not possible to undertake the risk characterization for 2-MCPD due to the lack of information [85].

The EFSA CONTAM Panel established a TDI of 0.8 μg kg^{-1} bw day^{-1}, whereas JECFA established a TDI of 4 μg kg^{-1} bw day^{-1}, differences being due to the criteria of application of the BMD approach. After reviewing results from animal studies, the outcome from EFSA was revised in 2017, and the TDI was updated to of 2 μg kg^{-1} bw day^{-1} for 3-MCPD and 3-MCPD-E (expressed as MCPD equivalents) [87].

Glycidol and Glycidyl Esters

The IARC determined that glycidol is a probably carcinogenic to humans (Group 2A) but the carcinogenicity of glycidyl oleate and glycidyl stearate is not yet classified. As 3-MCPD-E, it is supposed that glycidol is released from its parent esters by lipase-catalyzed hydrolysis in the gastrointestinal tract. The epoxide group is very prone to react with nucleophiles. Therefore, cellular nucleophiles biomolecules as glutathione, proteins or nucleic acids can react with glycidol and GE. Glycidol acts as alkylating agent to form DNA adducts in cells. Genotoxicity of glycidol has been examined in *in vitro* and *in vivo* studies. *In vitro* and some *in vivo* data have evidenced the genotixicity of glycidol. The main target organs affected by glycidol in subchronic and chronic toxicity studies are the brain, kidney and testes. Effects include testicular atrophy, antifertility and tumors in different tissues. As it is expected that GE is near totally hydrolyzed in gastrointestinal tract, the same effects can be attributed to GE. Moreover, glycidol was mutagenic in several *in vitro* and *in vivo* tests in rodents. Conversion of 3-MCPD into glycidol, administered in either free or esterified forms, was not observed *in vivo*. The use of human biomarkers as hemoglobin-valina adducts of glycidol (N-(2,3-dihydroxypropyl)-valine) has been proposed as a measure of internal mid-term exposure to glycidol, thus avoiding the inaccuracy of external exposure assessment. Hemoglobin can accumulate adducts of glycidol above 4 months [88]. TDI for glycidol has not been established for the EFSA because of its genotoxic and carcinogenic potential [85].

Chemical Formation in Foods

3-MCPD and, in minor extent 2-MCPD, form in the elaboration of acid-HVP, by the reaction of hydrochloric acid with lipids associated with the proteinaceous cereal materials used in their production at high temperature and pressure. Acid-HPV is used as flavoring ingredient in soya sauce, soups, seasonings and ready-cooked meals. The most important precursors are triacylglycerols (TAG), and to a lesser extent monoacylglycerols (MAG), diacylglycerols (DAG), glycerol and phospholipids. The exact mechanism has not fully elucidated yet. In other processed foods, these compounds are formed from reaction of chloride (present in raw material or added) with acylglycerols or glycerol. 3-MCPD has been found in thermal processed foods and food ingredients no formulated with HVP as cereal-based products, malt-derived ingredients and potato-based, between others. Culinary heat treatments such as baking, roasting or toasting can lead to the formation of these compounds.

MCPDs have also been detected in food products that are not subjected to high temperatures, but in lower contents, such as sausages, salami and cold smoked fish. In sausages and salami, the 3-MCPD can migrate during storage from wet resins based on epichlorohydrin. Regarding meat and fish smoked foods, the 3-MCPD can come from smoke generated from wood during the process and from an ingredient used as an additive that provides smoked flavor to meals "liquid smoke". In 2002, European Commission Directive 2002/72/EC set a limit maximum permitted quantity of the 'residual' epichlorohydrin in the finished material of 1 mg kg^{-1} ; thus 3-MCPD present in food from contact with casings or wrappings should be minimal in foods produced in the EU [89].

The mechanism for MCPDs formation comprises a cyclic acyloxonium ion intermediate [90]. Glycerol is protonated by hydrochloric acid at the hydroxyl groups to form alkyloxonium ions (Fig. **11**). Water is displaced from the primary position hydroxyl alkyloxonium cations forming a racemic mixture of both 3-MCPD enantiomers. The secondary position hydroxyl group alkyloxonium ions dissociate to give water and a carbocation, which reacts with chloride to give 2-MCPD. The formation of MCPDs in foods depends on multiple variables, including fat, salt and water contents. In model thermal processed foods, increasing fat and salt levels, a higher formation of 3-MCPD was found. In absence of water, there is no prior hydrolysis of the acyl group and 3-MCPD is formed by the direct substitution of the glycerol hydroxyl group by chloride ion.

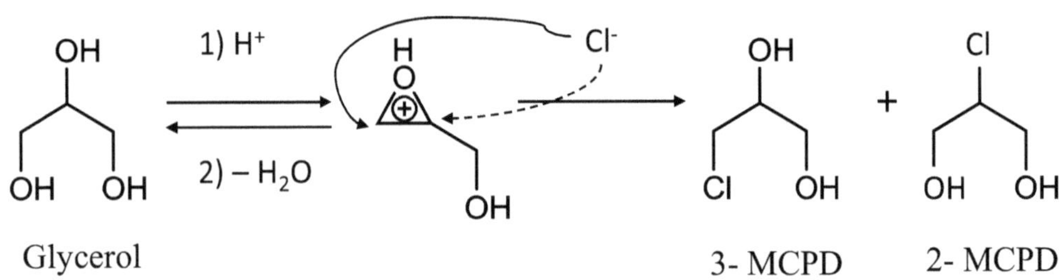

Fig. (11). Formation of chloropropanodiols from glycerol.

3-MCPD-E and in minor extent 2-MCPD-E, are formed during the deodorization step of refining edible oils. The acylglycerols react through mechanisms that include formation *via* an intermediate acyloxonium cation involving a nucleophilic ring-opening substitution reaction by a chlorine anion, or direct nucleophilic substitution of an ester or a hydroxyl group by a chlorine anion (Fig. **12**). At high temperatures, the formation of MCPD-E is accelerated, but the concomitant decomposition rates also increase.

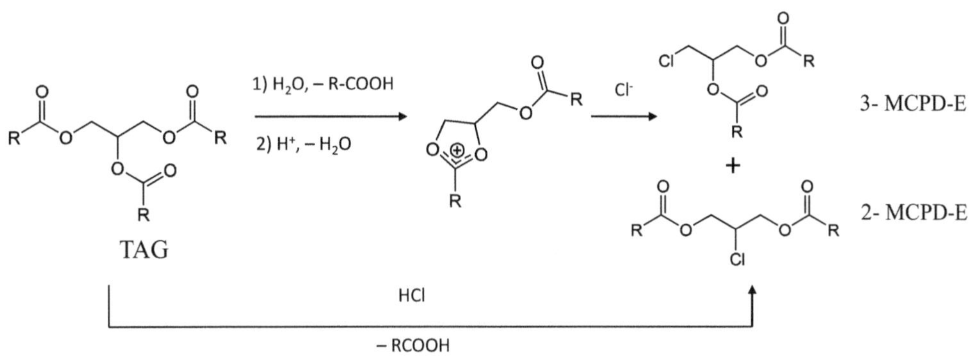

Fig. (12). Formation of esters of MCPD from triacylglycerols (R: fatty acid).

Glycidol and GE were detected for the first time in palm oil, and their presence in other oils is very low [82]. DAG and MAG are the main precursors for GE formation at high temperatures. Palm oil in general has a content of DAGs between 4-12%. GE are mainly generated during refining oil process and specifically in deodorization step where high temperatures are reached (240-270 °C). It is important to highlight that unlike MCPDs-E formation; ion chloride is not involved in GE formation. The mechanisms of formation proposed from DAG are *via* an acyloxonium ion or an intramolecular SN2 reaction [82, 91].

Occurrence and Exposure

The occurrence of 3-MCPD and GE has been roughly classified in literature across different food groups, such as animal and vegetable fats and oils, grain and grain-based products, smoked fish and meat, composite food (herbs, spices and condiments), foods for infant and small children, potato-based products, among others (Table **4**).

A comprehensive food consumption database was applied by EFSA to estimate the chronic dietary exposure to MCPDs and glycidol [85]. The gastrointestinal cleavage of MCPDs-E to their free form increases their intake by a factor 15-20. Chronic dietary exposure to MCPDs and glycidol was assessed as mean and high (95th percentile) exposure across dietary surveys in 23 European countries and 78,990 participants. The consumption data were collected using single or repeated 24 or 48 hours' dietary recalls or dietary records covering from 3 to 7 days per subject. The highest exposure was identified in the population groups of younger age, infants (<12 months old), toddler (≥ 12 to < 36 months old) and other children (≥ 36 months to <10 years old). The mean exposure to MCPDs and glycidol was 0.5-1.5, 0.2-0.7 and 0.3-0.9 μg kg^{-1} bw day^{-1}, respectively. The major contributors were infants- and follow-on formulas, vegetable fats and oils and cookies for infants and vegetable fats and oils, cookies and pastries and cakes for toddlers. In adolescents (≥ 10 to <18 years old), adults (≥ 18 to <65 years old), elderly (≥ 65 to <75 years old) and very elderly (≥ 75 years old) population groups, the mean exposure to 3-MCPD, 2-MCPD and glycidol ranged from 0.2-0.7, 0.1-0.3 and 0.1-0.5 μg kg^{-1} bw day^{-1}, respectively. Margarine and similar, pastries and cakes, fried or baked potato and vegetable fats and oils were the main sources of MCPDs, whereas for glycidol those were fried or roast meat.

Table 4. Occurrence of chloropropanols and glycidyl esters in foods.

Food	Subgroup	3-MCPD Range/Average (mg kg^{-1})	Refs.	GE Range or Average (mg kg^{-1})	Refs.
Oil	Palm oil	0.25-5.77	[92]	0.32-6.30	[82]
		0.11-8.85	[93]	0.30-18.00	[104]
		2.71	[94]	25.60-28.00	[105]
				9.26-9.40	[106]
				30.20	[107]
	Palm olein	1.51-6.59	[93]	15.60	[108]
	Coconut oil	0.025-0.31	[93]	0.5-3	[104]
	Rice oil	0.32	[94]	27.22-28.76	[105]
				33.70	[108]
	Soya oil	0.10	[94]	0.12	[109]
				1.56	[107]
	Sunflower oil	0.30	[95]	0.02-0.90	[110]
	Rapeseed oil	0.48	[95]	0.01-1.10	[110]
Fat	Margarine	0.15-7.7	[96]	3.63	[111]
Infant food	Infant formula	0.011-0.23	[97]	0.16-0.87	[112]
		0.075-0.588	[98]	0.005-0.40	[99]
		0.57-4.1	[96]		
		0.061-0.76	[99]		
	Growing-up	0.062-0.291	[98]		
Meat	Salami	0.88-6.41	[95, 100]		
	Ham	n.d.-2.64	[100]		
	Chicken grilled	0.26-0.74	[97, 100]		

(Table 4) cont.....

Food	**Subgroup**	**3-MCPD Range/Average (mg kg^{-1})**	**Refs.**	**GE Range or Average (mg kg^{-1})**	**Refs.**
Coffee	Coffee	0.1-0.39	[100, 101]		
Cereals	Biscuits	0.249-0.696	[97]		
	Bread toasted	0.06-0.16	[102]		
	Breakfast cereals	0.04-0.888	[97]		
	Crispbread	0.42-0.58	[100]		
	Doughnuts	0.42-1.21	[100]		
Potato products	Crisps	0.048-1.186	[97]		
	Crisps	0.229-1.008	[103]		

Data obtained from infant fed only with infant formulas showed a daily intake of 2.4 μg kg^{-1} bw for 3-MCPD, slightly above of TDI, 1.0 μg kg^{-1} bw for 2-MCPD and 1.9 μg kg^{-1} bw for glycidol [85]. There is not TDI established for glycidol. MOE estimates < 25,000 in the youngest age group population (infants, toddlers and other children). Infants fed only with formulas presented a MOE value of 5,400 for the mean exposure. It is an important point of control since the glycidol exposure could affect in the physical development as a result of its genotoxicity [113]. Despite the relevance of the incidence of 3-MCPD and glycidol in the subpopulation of infants, there are few studies covering infant formulas.

Mitigation

It is difficult to avoid completely the formation of heat-induced contaminants while at meantime still ensuring the quality and safety of the product and maintaining their inherent properties to meet consumer expectations. However, there are three main points where the formation of 3-MCPD-E and GE in vegetable fats and oils can be minimized, in raw material, refining process and final product.

Regarding the raw material, the type of soil, fertilizer, genotype of plant, technique of harvest and the time between ripeness and processing are factors that influence in later formation of these contaminants. The choice of raw materials with a low content of precursors is desirable and to eliminate the critical reactants from the raw oil can lead to a lower formation of 3-MCPD-E after refining process (*i.e.,* the use of fertilizers with a low chloride ion content). In addition, the introduction of a washing step before refining can be applied to reduce the level of chlorine precursors that are water-soluble. Regarding GE, minimizing the time from recollection to processing the oil decreases the enzymatic hydrolysis of the TAG, reducing then the content of DAG.

Refining process implies several steps as degumming, neutralization, bleaching and deodorization. The most critical step between them is the deodorization, where high temperatures are applied. Chemical refining is preferred to physical one since lower temperatures are applied in the deodorization step. However, in some oils as palm oil, physical refining is necessary to be applied in order to eliminate its high content of free fatty acids. The temperature of deodorization has to be as low as possible to get a balance that assesses the complete purification of oil (it ensures that other potential volatile contaminants are safely removed) and minimizes the formation of 3-MCPD-E and related compounds. Furthermore, introduction of additional processing steps can be considered to decrease the content of these contaminants.

Another mitigation strategy is the removal of the esters in the final product (after refining of oil) *i.e.,* adding different adsorbents materials (zeolite, synthetic magnesium silicate) in the refined oil, applying a mild thermal treatment during a short time and then removing them, thus always maintaining the product quality [114]. Improvements of refining conditions with a replacement of the deodorization step by including a short-path distillation have been proposed. This process allows gentle removal of volatile compounds at moderate low temperatures.

In a laboratory-scale study with palm oil using short-path distillation, formation of 3-MCPD was not found in oil but undesirable quality changes were detected in the oil. However, with a post deodorization at 180 °C better results were obtained with an optimal quality of oil and remaining very low levels of 3-MCPD-E and GE [115]. Mitigation efforts have achieved a reduction of above 50% for GE in fats and oils across five years 2010-2015 and reductions of 20-30% of 3-MCPD-E have been achieved in edible oils [85]. However, the efforts for minimizing these contaminants should be intensified. Regarding 3-MCPD, effective modifications have been applied to reduce its formation in acid-HVP soy sauce. FAO has proposed three strategies that imply the careful control of acid hydrolysis, alkaline treatment after acid hydrolysis, and substitution of HCl with sulfuric acid.

Legislation

3-MCPD is regulated for EU Commission under (EC) No 1881/2006 with a maximum level of 20 μg kg^{-1} in HVP and soy sauce [84]. Later, an EU Recommendation 2014/661/EU7 on the monitoring of the presence of MCPDs, MCPDs-E and GE in foods, remarked the importance of collecting data of occurrence and exposure. The Recommendation specifies type of food to be analyzed, sampling procedures and analytical methods for their determination [116]. Regarding 2-MCPD and 2-MCPD-E and 3-MCPD-E no levels are established so far. Maximum levels of GE (expressed as glycidol) in vegetable oils and fats, and infant formulas and foods for special medical purposes intended for infants and young children have been recently regulated (EU 2018/290). This normative has been also included in earlier Commission regulation (EC) No 1881/2006 [84].

Methods of Analysis

There are well-characterized analytical methods to determine free MCPDs in foods [117]. Suitable methods to quantify unstable free glycidol are not described, however GE can be determined in edible oils as join development by American Oil Chemists' Society (AOCS) and Japan Oil Chemists' Society (JOCS) (Official Method Cd 28-10) [118]. Nowadays, there are not official methods to determine MCPDs, glycidol and GE in foods. In 2013, EFSA remarked the importance of the development and establishment of standards analytical methods to determine these compounds with accuracy and precision [119]. Two analytical approaches are applied, as described below.

Direct Determination

The direct determination is based on the quantification of each analyte (3-MCPD, 2-MCPD-E, and GE) without previous chemical transformation. The main advantage of using direct methods is the absence of interferences that can be formed from side chemical reactions (artefacts that can over o underestimate their contents, ex-novo formation). However, they have important drawbacks as the majority presence of TAG and to a lesser extent, DAG in oils. TAG and DAG must be isolated prior analysis for assessment the accuracy and precision of direct methods since of their structural similarity with the target analytes. In addition, many individual standards are necessary for the determination of each compound. The most reliable results are obtained with high number of standards used in the analysis, which implies significant costs and efforts. They are not all commercially available and must be synthesized in-house. Only in relation to 3-MCPD-E, 63 potential esters can be formed from the seven main fatty acids present in oils. Moreover, as LOD can be different in function of each analyte, the sum of LODs for each ester produces higher LODs for GE and 3-MCPD-E than in indirect methods [120].

The first direct methods were developed for the quantification of 3-MCPD in soya sauce, seasonings and protein vegetable hydrolyzed. These methods were based on gas chromatography [121]. The low molecular weight of 3-MCPD difficulties its detection by MS because they cannot be always distinguished from background noise. The high limits of detection of methods based on direct GC were unsuitable to comply with regulation of these compounds, which contents are marked in levels of few mg kg^{-1}. Analyses of 3-MCPD were also performed with HPLC with refractive index detection. Ultraviolet or fluorescence detection cannot be applied because of the absence of chromophore groups in the free forms (3-MCPD, 2-MCPD and glycidol). Capillary electrophoresis with electrochemical detection was applied for soy sauce but providing low sensitivity. In summary, first initial direct methods applied did not provided enough sensitivity for 3-MCPD.

Regarding direct analysis for 3-MCPD-E and GE, numerous research papers have been published in later years due to their extension and importance in refined edible oils and processed oil-based foods. Table **5** summarizes different direct methods applied in oils and foodstuffs. Most of them add an internal isotopically labeled standard for quantification. It should be noted that the application of a pre-treatment with SPE or gel permeation chromatography (GPC) an optimum separation of GE from oil matrix have been achieved following LC-MS [124, 130]. This methodology has been adopted for AOCS and JOCS lead to Official Method Cd 28-10 for determination of GE in oils and fats [118].

Table 5. Direct and indirect methodologies applied for analysis of 3-MCPD-E, 2-MCPD-E and GE in oils and foodstuffs.

DIRECT METHODS					
Analyte	**Matrix**	**Pre-treatment**	**Determination**	**LOD (mg kg^{-1})**	**Refs.**
3-MCPD-E[1] GE[2]	Oil	SPE	LC-MS/MS-ESI	0.06[1] < 0.3[2]	[120]
3-MCPD-E[1] GE[2]	Oil	SPE	LC-TOF-MS-ESI	0.02-0.86[1]* 0.06-0.16[2]*	[109]
GE[2]	Oil	SPE	LC-MS/MS-SIM	1.4-3.7[2]*	[122] [123]
GE[2]	Oil	GPC + SPE	LC-TOF-MS-SIM LC-MS/MS-SIM	< 0.05[2]	[124]
3-MCPD-E[1] GE[2]	Oil		LC-TOF-MS	0.1-1.69[1] 0.07-0.29[2]	[108]
GE[2]	Oil	SPE	GC-MS-SIM	0.01[2]	[107]
INDIRECT METHODS					
Analyte	**Matrix**	**Pre-treatment**	**Determination**	**LOD (mg kg^{-1})**	**REF**
3-MCPD-E[1]	Baby foods	SPE	GC-MS-SIM	0.006[1]	[98]
3-MCPD-E[1] 2-MCPD-E[3] GE[2]	Oil and fats	AC-TR + DER	GC-MS-SIM	0.04[1,3] 0.06[2]	[125]

(Table 5) cont.....

DIRECT METHODS					
Analyte	**Matrix**	**Pre-treatment**	**Determination**	**LOD (mg kg^{-1})**	**Refs.**
3-MCPD-E^1 2-MCPD-E^3 GE^2	Oil and fats	AC-TR + DER	GC-MS-SIM	0.05^1 ≤ 0.025^2	[104]
3-MCPD-E^1 2-MCPD-E^3 GE^2	Oil, margarine, mayonnaise, fish oils	AL-TR+ DER	GC-MS-SIM	0.02^1 0.05^3 0.05^2	[126] [127]
3-MCPD-E^1 GE^2	Fish oils	EN-TR + DER	GC-MS/MS	0.05^1 0.02^2	[128]
3-MCPD-E^1 2-MCPD-E^3 GE^2	Cereal-based products, fish, meat, potato	AL-TR + DER	GC-MS-SIM	7^{1,3} 17^2	[129]

AC-TR: Acid transesterification. AL-TR: Alkaline transesterification. EN-TR: Enzymatic transesterification. DER: Derivatization with PBA. *ng mL^{-1}. ^{1}: 3-MCPD-E. ^{2}: GE. ^{3}: 2-MCPD-E.

Indirect Determination

The indirect methods imply the release of free forms from their respective esters, thus quantify the total MCPDs (as free form) and glycidol. Most of methods apply a transesterification in acid or alkaline medium [110, 125, 129, 131, 132], although the enzymatic hydrolysis has also been applied [102, 126]. The objective is to achieve the complete cleavage of free form and derivatize them prior GC analysis. Table **5** summarizes different indirect methods for determining these compounds. AOCS has validated three indirect methods for their simultaneous analysis suitable for oils: Cd 29a-13 [133], Cd 29b-13 [134], and Cd 29c-13 [135]. The latter is also applicable for oils extracted with caution from processed fat and oil-containing foods. In addition, an extraction method for oil-based emulsions has been recently validated (Cd 30-15) [136]. Cd 30-15 consists on an extraction procedure for isolating MCPDs and GE from oil-based emulsions such as spreads, margarines, dressings and mayonnaise. After extraction, any of them (Cd 29a-13, Cd 29b-13, or Cd 29c-13) can be used to quantify these process contaminants. AOCS proposes the use of these methods to determine these compounds in foods due to the lack of an official method. Furthermore, EU commission also recommends the use of the AOCS Official Methods Cd 29a,b,c-13 not only for analysis of bound MCPDs and glycidol in oils and fats but also in oil- and fat containing foods, with LOQ = 0.1 mg kg^{-1} in the oil/fat fraction and LOQ ≤ 0.01 mg kg^{-1} in foods with ≤ 10% of fat. In the meantime, several collaborative validation methods have been carried out for the determination the content of MCPDs in foods [131, 137].

The differences between these three AOCS methods are the transesterification conditions: the medium (acid or basic), the time (5 or 16 h) and temperature (-22

°C, room temperature, 40 °C). After transesterification, a necessary step of derivatization is carried out for increasing the volatility of these compounds and facilitating their GC-MS detection. In this case, all the methods employ the same derivatization agent, phenylboronic acid (PBA) in aqueous (acetone/water) or organic medium (diethyl ether). Briefly, method Cd 29a-13 is carried out firstly through the conversion of GE to 3-Monobromopropane diol ester (3-MBPD-E) and after a long transesterification in acid medium (16 hours, 40 °C). Method Cd 29b-13 includes two assays 1) conversion of GE to 3-MBPD and 2) determination of MCPDs with an alkaline transesterification (16 hours, -22 °C). The third method Cd 29c-13 is faster but it does not determine 2-MCPD. It also includes two assays 1) conversion of GE to 3-MBPD and 2) determination of 3-MCPD after alkaline transesterification (5 hours, room temperature). The GE content is calculated by difference applying a correction factor.

All of these methods have in common that they have many steps in which chemical transformations are carried out and have to be controlled. Zelinkova *et al.* have recently published an interesting study focused on critical steps of indirect analytical method to quantify fatty acid esters of MCPD and glycidol [132]. Furthermore, a partial conversion of 3-MCPD into glycidol has been found in alkaline transesterification [104, 128, 138]. The method based on acidic transesterification can be proposed as a more robust method [138].

Fig. **(13)** depicts a diagram of analytical procedure of the indirect method. The procedure starts with a differentiation regarding the oil/fat content. If this content is less than 5%, internal standards are added before the extraction of fat/oil from foods. Internal standards are isotopically labelled esters of MCPDs and glycidol. ASE or Pressurized liquid extraction (PLE) is the method preferred by extraction of oil respect to other conventional extraction techniques *i.e.,* Soxhlet. PLE includes several extraction cycles to obtain total oil extraction and therefore high levels of analyte recovery [131]. Another advantage is its high level of automatization, that makes it suitable for routine analysis. The tert-butyl methyl ether is the most used extraction solvent [137, 139]. Although in general the contribution of free form of MCPDs is very low respect their bound form, to ensure the real content of both forms, the determination of free forms also has to be also carried out. In this case, regarding indirect methods, after addition of internal standards, the aqueous phase is recollected and derivatization is carried out. Different derivatization reagents as acetone, heptafluorobutyryl imidazole and PBA have been proposed, the latter being the most commonly used. These agents react with diol groups to form volatile derivatives. However, PBA can cause interference and instrument issues *i.e.,* decrease of the shelf life of chromatographic column. There are two strategies regarding the solvent applied with PBA. PBA can be dissolved in an aqueous medium (acetone/water) [125] or in organic solvents (diethyl ether) [104, 129]. However, derivatization in an organic medium requires another previous extraction step for removal the analytes from the aqueous phase and results in lower sensitivity although the results were comparable [132]. The analysis is carried out in GC-MS, and the quantification is based on the relation of compound/isotope labeled compound.

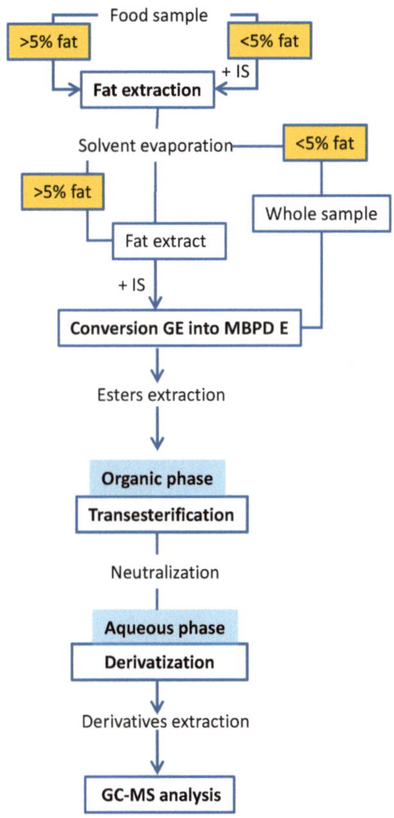

Fig. (13). Simplified diagram of methodology of indirect determination of esters of 3- and 2-MCPD and glycidol. Adapted from Wenzl *et al.* [137].

A study in vegetable oils compared the direct method using liquid chromatography coupled to time of flight mass spectrometry (LC-ToF-MS) with the indirect method through acid transesterification with GC-MS and obtained a good correlation for 3-MCPD and 2-MCPD. Nevertheless, the indirect method presented more advantages for routine use since it requires less analytical standards, facilitating the automation and being applicable to all type of vegetable oils [93]. Furthermore, the most common equipment used in indirect methods is GC-MS, which is more economical feasible than equipment generally required for direct methods.

CONCLUDING REMARKS

The risk analysis of processing contaminants remains a challenging issue since they are formed during the home-cooking or industrial processing of foods usually consumed in our diet. The occurrence of process contaminants is low but their ubiquity in many food commodities increase the exposure and reduce the safety level. This aspect has drawn attention from consumers and regulators in terms of potential human health risk. Particularly, the heat-induced process contaminants are naturally formed that makes any strategy of mitigation more complex since different perspectives should be considered and balanced. The chemical pathways involved in their formation are the same, or at least run parallel, to those routes that lead the formation of other chemicals that impair appreciate taste, flavor and health beneficial properties to food. Additionally, it could be the paradox that actions driven to limiting a specific heat-induced contaminant will increase the formation of other undesirable compounds in the food. This is a limitation in the application of any mitigation strategy since in some cases they lead to changes in the sensorial, nutritional profile, and textural changes that make the alternative process unacceptable to consumers. On the other hand, it is necessary to increase efforts to the development of simple, robust and cost-effective analytical methods to quantify with the adequate level of uncertainty the presence of heat-induced process contaminants in any food matrix. The complexity of matrixes requires a time-consuming sample-preparation step, which is often still the bottleneck of the analytical procedure to reach the adequate limit of detection levels. Identification of novel heat-induced process contaminants will likely occur in coming years, indeed, there are some emerging compounds currently under toxicological evaluation. Since the identification of the HAAs in the 70s of the last century, new compounds with adverse toxicological effects that may pose a risk to human health, such as acrylamide, furan, glycidol, GE, or 3-MCPD-E, have been described in this century. Heat-induced process contaminants are widespread and it is necessary a global risk to risk and risk to benefit evaluation, taking into consideration subgroups of the population which are more susceptible to a higher exposure and unwanted side effects.

CONSENT FOR PUBLICATION

Not applicable.

CONFLICT OF INTEREST

The authors confirm that this chapter contents have no conflict of interest.

ACKNOWLEDGEMENT

This work was funded by the Comunidad of Madrid (Spain) and European funding from FEDER Program (project S2013/ABI-3028, AVANSECAL-CM).

REFERENCES

[1] Dorne JLCM, Bordajandi LR, Amzal B, Ferrari P, Verger P. Combining analytical techniques, exposure assessment and biological effects for risk assessment of chemicals in food. Trends Analyt Chem 2009; 28: 695-707. [http://dx.doi.org/10.1016/j.trac.2009.03.008]

[2] Regulation (EC) No 852/200 of the European Parliament and of the Council of 29 April 2004 on the hygiene of foodstuffs. OJEU 2004; 139: 1-54.

[3] European Parliament and Council Regulation (EC) No. 178/2002 of the European Parliament and of the Council of 28 January 2002 laying down the general principles and requirements of food law, establishing the European Food Safety Authority and laying down procedures in matters of food safety. *Off.* J Eur Comm 2002; 31: 1-24.

[4] WHO (World Health Organization). About Risk Analysis in Food.. http://www.who.int/foodsafety/micro/riskanalysis/en/ [accessed Aug 6, 2018].

[5] Tijhuis MJ, de Jong N, Pohjola MV, *et al.* State of the art in benefit-risk analysis: food and nutrition. Food Chem Toxicol 2012; 50(1): 5-25. [http://dx.doi.org/10.1016/j.fct.2011.06.010] [PMID: 21679741]

[6] IPCS (International Program on Chemical Safety). Environmental Health Criteria 240 Principles and methods for the risk assessment of chemicals in food. Geneva, Switzerland: World Health Organization 2009.

[7] EFSA (European Food Safety Authority). Scientific Opinion on Risk Assessment Terminology. EFSA J 2012; 10: 2664. [http://dx.doi.org/10.2903/j.efsa.2012.2664]

[8] Barlow SM, Greig JB, Bridges JW, *et al.* Hazard identification by methods of animal-based toxicology. Food Chem Toxicol 2002; 40(2-3): 145-91. [http://dx.doi.org/10.1016/S0278-6915(01)00117-X] [PMID: 11893397]

[9] Boobis AR, Doe JE, Heinrich-Hirsch B, *et al.* IPCS framework for analyzing the relevance of a noncancer mode of action for humans. Crit Rev Toxicol 2008; 38(2): 87-96. [http://dx.doi.org/10.1080/10408440701749421] [PMID: 18259981]

[10] Kroes R, Müller D, Lambe J, *et al.* Assessment of intake from the diet. Food Chem Toxicol 2002; 40(2-3): 327-85. [http://dx.doi.org/10.1016/S0278-6915(01)00113-2] [PMID: 11893401]

[11] EFSA (European Food Safety Authority). The food classification and description system FoodEx2 (revision 2). Supporting publication 2015; EN-804: pp. 1-90.

[12] EFSA (European Food Safety Authority). International frameworks dealing with human risk assessment of combined exposure to multiple chemicals. EFSA J 2013; 11: 3313.

[13] Kemper RA, Hayes JA, Bogdanffy MS. Principles and methods of toxicology. Boca Raton, FL: CRC Press 2007; pp. 103-78.

[14] EFSA (European Food Safety Authority). Guidance of the Scientific Committee on a request from EFSA on the use of the benchmark dose approach in risk assessment. EFSA J 2009; 7: 1150.

[15] EFSA (European Food Safety Authority). Update: use of the benchmark dose approach in risk assessment. EFSA J 2017; 15: 4658.

[16] EFSA (European Food Safety Authority). Opinion of the scientific committee on a request from EFSA related to a Harmonised approach for risk assessment of substances which are both genotoxic and carcinogenic. EFSA J 2005; 282: 1-31.

[17] Jongen WMF, Meulenberg MTG. Innovation in Agri-Food Systems Product Quality and Consumer Acceptance. Wageningen: Wageningen Academic Publishers 2005. [http://dx.doi.org/10.3920/978-90-8686-666-3]

[18] van Boekel M, Fogliano V, Pellegrini N, *et al.* A review on the beneficial aspects of food processing. Mol Nutr Food Res 2010; 54(9): 1215-47. [http://dx.doi.org/10.1002/mnfr.200900608] [PMID: 20725924]

[19] Chaudhry Q, Cotterill J, Watkins R. Acrylamide and Other Health Hazardous Compounds in Heat - Treated Foods. UK: Woodhead Publishing 2006; pp. 132-60. [http://dx.doi.org/10.1533/9781845692018.1.132]

[20] Cotterill JV, Chaudhry MQ, Matthews W, Watkins RW. *In silico* assessment of toxicity of heat-generated food contaminants. Food Chem Toxicol 2008; 46(6): 1905-18. [http://dx.doi.org/10.1016/j.fct.2008.01.030] [PMID: 18334278]

[21] McKinney JD, Richard A, Waller C, Newman MC, Gerberick F. The practice of structure activity relationships (SAR) in toxicology. Toxicol Sci 2000; 56(1): 8-17. [http://dx.doi.org/10.1093/toxsci/56.1.8] [PMID: 10869449]

[22] IARC (Internacional Agency for Research on Cancer). Monographs on the evaluation of carcinogen risk to humans: some industrial chemicals No 60. Lyon, France: International Agency for Research on Cancer 1994.

[23] Hogervorst J, Schouten L, van den Brandt P. Reaction on Gargas *et al.*: Acrylamide: consideration of species differences and nonlinear processes in estimating risk and safety for human ingestion. Food Chem Toxicol 2009; 47(11): 2871-2. [http://dx.doi.org/10.1016/j.fct.2009.08.004] [PMID: 19682529]

[24] Tardiff RG, Gargas ML, Kirman CR, Carson ML, Sweeney LM. Estimation of safe dietary intake levels of acrylamide for humans. Food Chem Toxicol 2010; 48(2): 658-67. [http://dx.doi.org/10.1016/j.fct.2009.11.048] [PMID: 19948203]

[25] FAO/WHO Experts Committee on Food Additives (JECFA). Rome,: Summary and conclusions of the sixtyfourth meeting of the Joint FAO/WHO Experts Committee on Food Additives (JECFA 2005.

[26] Keramat J, LeBail A, Prost C, Soltanizadeh N. Acrylamide in Foods: Chemistry and Analysis. A Review. Food Bioprocess Technol 2011; 4: 340-63. [http://dx.doi.org/10.1007/s11947-010-0470-x]

[27] Yaylayan VA, Stadler RH. Acrylamide formation in food: a mechanistic perspective. J AOAC Int 2005; 88(1): 262-7. [PMID: 15759750]

[28] Bagdonaite K, Derler K, Murkovic M. Determination of acrylamide during roasting of coffee. J Agric Food Chem 2008; 56(15): 6081-6. [http://dx.doi.org/10.1021/jf073051p] [PMID: 18624446]

[29] Ahrné L, Claes-Göran A, Floberg P, Rosén J, Lingnert H. Effect of crust temperature and water content on acrylamide formation during baking of white bread: steam and falling temperature baking. Lebensm Wiss Technol 2007; 40: 1708-15. [http://dx.doi.org/10.1016/j.lwt.2007.01.010]

[30] Mesias M, Morales FJ. Acrylamide in foods Analysis, content and potential health effects. London, UK: Academic Press 2016; pp. 131-57.

[31] Luning P, Sanny M. Acrylamide in foods Analysis, content and potential health effects. London, UK: Academic Press 2016; pp. 159-79.

[32] EFSA (European Food Safety Authority). Results on acrylamide levels in food from monitoring years 2007-2009. EFSA J 2011; 4: 2133.

[33] EFSA (European Food Safety Authority). Scientific opinion on acrylamide in food. EFSA J 2015; 13: 4104.
[http://dx.doi.org/10.2903/j.efsa.2015.4104]

[34] EC (European Commission). Commission recommendation of 10.1.2011 on investigations into the levels of acrylamide in food. 2011 https://ec.europa.eu/food/sites/food/files/safety/docs/cs_contaminants_catalogue_acrylamide_recommendation_10012011_food_en.pdf [(accessed Aug 2, 2018).];

[35] EC (European Commission). Commission recommendation of 8 November 2013 on investigations into the levels of acrylamide in food. OJUE, 2013; L 301: 15-7.

[36] EC (European Commission). Comission regulation (EU) 2017/2158 of 20 November 2017 establishing mitigation measures and benchmark levels for the reduction of the presence of acrylamide in food OJUE, 2017; L 304: 24-44.

[37] Mariotti S, Pedreschi F, Carrasco JA, Granby K. Patented techniques for acrylamide mitigation in high-temperature processed foods. Recent Pat Food Nutr Agric 2011; 3(3): 158-71.
[http://dx.doi.org/10.2174/2212798411103030158] [PMID: 21846322]

[38] Pedreschi F, Mariotti MS, Granby K. Current issues in dietary acrylamide: formation, mitigation and risk assessment. J Sci Food Agric 2014; 94(1): 9-20.
[http://dx.doi.org/10.1002/jsfa.6349] [PMID: 23939985]

[39] Morales FJ, Mesias M. Coffee in health and disease prevention. London, UK: Academic Press 2015; pp. 1013-21.
[http://dx.doi.org/10.1016/B978-0-12-409517-5.00111-X]

[40] Tekkeli SEK, Önal C, Önal A. A review of current methods for the determination of acrylamide in food products. Food Anal Methods 2012; 5: 29-39.
[http://dx.doi.org/10.1007/s12161-011-9277-2]

[41] Gökmen V, Senyuva HZ. A generic method for the determination of acrylamide in thermally processed foods. J Chromatogr A 2006; 1120(1-2): 194-8.
[http://dx.doi.org/10.1016/j.chroma.2006.01.084] [PMID: 16464456]

[42] Mesias M, Delgado-Andrade C, Holgado F, Morales FJ. Acrylamide content in French fries prepared in households: A pilot study in Spanish homes. Food Chem 2018; 260: 44-52.
[http://dx.doi.org/10.1016/j.foodchem.2018.03.140] [PMID: 29699680]

[43] Gökmen V, Mogol BA. Computer vision based image analysis for rapid detection of acrylamide in heated foods. Qual Assur Saf Crops Foods 2010; 2: 203-7.
[http://dx.doi.org/10.1111/j.1757-837X.2010.00072.x]

[44] Batra B, Pundir CS. Acrylamide in foods Analysis, content and potential health effects. London, UK: Academic Press 2016; pp. 497-505.

[45] VKM (Norwegian Scientific Committee for Food Safety). Risk assessment of furan exposure in the Norwegian population; Opinion of the Panel on Food Additives, Flavorings, Processing Aids, Materials in Contact with Food and Cosmetics and the Panel on Contaminants of the Norwegian Scientific Committee for Food safety. 2012; 107.

[46] Becalski A, Seaman S. Furan precursors in food: a model study and development of a simple headspace method for determination of furan. J AOAC Int 2005; 88(1): 102-6.
[PMID: 15759732]

[47] EFSA (European Food Safety Authority). Risks for public health related to the presence of furan and methylfurans in food. EFSA J 2017; 15: 5005.

[48] Guenther H. Coffee Emerging health effects and disease prevention. USA: JohnWiley & Sons, Inc.

and the Institute of Food Technologists: Iowa 2012; pp. 307-18.

[49] Mesias M, Guerra-Hernández E, García-Villanova B. Furan content in Spanish baby foods and its relation with potential precursors. CYTA J Food 2013; 11: 1-6. [http://dx.doi.org/10.1080/19476337.2012.669797]

[50] Anese M, Suman M. Mitigation strategies of furan and 5-hydroxymethylfurfural in food. Food Res Int 2013; 51: 257-64. [http://dx.doi.org/10.1016/j.foodres.2012.12.024]

[51] Fan X, Sommers CH. Effect of gamma radiation on furan formation in ready-to-eat products and their ingredients. J Food Sci 2006; 71: C407-12. [http://dx.doi.org/10.1111/j.1750-3841.2006.00136.x]

[52] Seok YJ, Her JY, Kim YG, *et al.* Furan in thermally processed foods - A review. Toxicol Res 2015; 31: 241-53. [http://dx.doi.org/10.5487/TR.2015.31.3.241] [PMID: 26483883]

[53] Mesias M, Morales FJ. Reliable estimation of dietary exposure to furan from coffee: An automatic vending machine as a case study. Food Res Int 2014; 61: 257-63. [http://dx.doi.org/10.1016/j.foodres.2013.07.054] [PMID: 28412000]

[54] US FDA (US Food and Drug Administration). Determination of furan in foods Department of Health and Human Services FDA. Rockville, MD: US Food and Drug Administration 2004.

[55] Nagao M, Honda M, Seino Y, Yahagi T, Sugimura T. Mutagenicities of smoke condensates and the charred surface of fish and meat. Cancer Lett 1977; 2(4-5): 221-6. [http://dx.doi.org/10.1016/S0304-3835(77)80025-6] [PMID: 45723]

[56] Skog K, Solyakov A, Arvidsson P, Jägerstad M. Analysis of nonpolar heterocyclic amines in cooked foods and meat extracts using gas chromatography-mass spectrometry. J Chromatogr A 1998; 803(1-2): 227-33. [http://dx.doi.org/10.1016/S0021-9673(97)01266-1] [PMID: 9604333]

[57] Persson E, Sjoholm I, Skog K. Heat and mass transfer in chicken breasts - effect on PhIP formation. Eur Food Res Technol 2002; 214: 455-9. [http://dx.doi.org/10.1007/s00217-001-0462-1]

[58] Gooderham NJ, Murray S, Lynch AM, *et al.* Food-derived heterocyclic amine mutagens: variable metabolism and significance to humans. Drug Metab Dispos 2001; 29(4 Pt 2): 529-34. [PMID: 11259345]

[59] Boobis AR, Lynch AM, Murray S, *et al.* CYP1A2-catalyzed conversion of dietary heterocyclic amines to their proximate carcinogens is their major route of metabolism in humans. Cancer Res 1994; 54(1): 89-94. [PMID: 8261468]

[60] Dingley KH, Curtis KD, Nowell S, Felton JS, Lang NP, Turteltaub KW. DNA and protein adduct formation in the colon and blood of humans after exposure to a dietary-relevant dose of 2-amino-1-methyl-6-phenylimidazo[4,5-b]pyridine. Cancer Epidemiol Biomarkers Prev 1999; 8(6): 507-12. [PMID: 10385140]

[61] Zhao K, Murray S, Davies DS, Boobis AR, Gooderham NJ. Metabolism of the food derived mutagen and carcinogen 2-amino-1-methyl-6-phenylimidazo(4,5-b)pyridine (PhIP) by human liver microsomes. Carcinogenesis 1994; 15(6): 1285-8. [http://dx.doi.org/10.1093/carcin/15.6.1285] [PMID: 8020169]

[62] Jägerstad M, Skog K. Genotoxicity of heat-processed foods. Mutat Res 2005; 574(1-2): 156-72. [http://dx.doi.org/10.1016/j.mrfmmm.2005.01.030] [PMID: 15914214]

[63] IARC (International Agency For Research On Cancer). IARC monographs of the evaluation of the carcinogenic risk of chemicals to humans Some naturally occurring substances: food items and constituents, heterocyclic aromatic amines and mycotoxins. Lyon: World Health Organization

International Agency for Research on Cancer 1993; p. 56.

[64] EFSA (European Food Safety Authority). Guidance on the data required for the risk assessment of flavourings to be used in or on foods. EFSA J 2010; 8: 1623. [http://dx.doi.org/10.2903/j.efsa.2010.1623]

[65] Sugimura T, Wakabayashi K, Nakagama H, Nagao M. Heterocyclic amines: Mutagens/carcinogens produced during cooking of meat and fish. Cancer Sci 2004; 95(4): 290-9. [http://dx.doi.org/10.1111/j.1349-7006.2004.tb03205.x] [PMID: 15072585]

[66] Gibis M. Heterocyclic aromatic amines in cooked meat products: Causes, Formation, Occurrence, and Risk assessment. Compr Rev Food Sci Food Saf 2016; 15: 269-302. [http://dx.doi.org/10.1111/1541-4337.12186]

[67] Rahman U, Sahar A, Khan MI, Nadeem M. Production of heterocyclic aromatic amines in meat: Chemistry, health risks and inhibition. A review. Lebensm Wiss Technol 2014; 59: 229-33. [http://dx.doi.org/10.1016/j.lwt.2014.06.005]

[68] Persson E, Graziani G, Ferracane R, Fogliano V, Skog K. Influence of antioxidants in virgin olive oil on the formation of heterocyclic amines in fried beefburgers. Food Chem Toxicol 2003; 41(11): 1587-97. [http://dx.doi.org/10.1016/S0278-6915(03)00190-X] [PMID: 12963012]

[69] Wang Y, Hui T, Zhang YW, *et al.* Effects of frying conditions on the formation of heterocyclic amines and trans fatty acids in grass carp (*Ctenopharyngodon idellus*). Food Chem 2015; 167: 251-7. [http://dx.doi.org/10.1016/j.foodchem.2014.06.109] [PMID: 25148986]

[70] Bordas M, Moyano E, Puignou L, Galceran MT. Formation and stability of heterocyclic amines in a meat flavour model system. Effect of temperature, time and precursors. J Chromatogr B Analyt Technol Biomed Life Sci 2004; 802(1): 11-7. [http://dx.doi.org/10.1016/j.jchromb.2003.09.024] [PMID: 15035992]

[71] Jägerstad M, Skog K, Grivas S, Olsson K. Formation of heterocyclic amines using model systems. Mutat Res 1991; 259(3-4): 219-33. [http://dx.doi.org/10.1016/0165-1218(91)90119-7] [PMID: 2017209]

[72] Polak T, Dosler D, Zlender B, Gasperlin L. Heterocyclic amines in aged and thermally treated pork longissimus dorsi muscle of normal and PSE quality. *Food Sci. Technol.* Lebensm Wiss Technol 2009; 42: 504-13. [http://dx.doi.org/10.1016/j.lwt.2008.09.014]

[73] Knize MN. Acrylamide and other hazardous compounds in heat treated foods. Cambridge: Woodhead 2006; pp. 231-46. [http://dx.doi.org/10.1533/9781845692018.2.231]

[74] Gibis M, Kruwinnus M, Weiss J. Impact of different pan-frying conditions on the formation of heterocyclic aromatic amines and sensory quality in fried bacon. Food Chem 2015; 168: 383-9. [http://dx.doi.org/10.1016/j.foodchem.2014.07.074] [PMID: 25172724]

[75] Rannou C, Laroque D, Renault E, Prost C, Sérot T. Mitigation strategies of acrylamide, furans, heterocyclic amines and browning during the Maillard reaction in foods. Food Res Int 2016; 90: 154-76. [http://dx.doi.org/10.1016/j.foodres.2016.10.037] [PMID: 29195868]

[76] Shin HS, Ustunol Z. Influence of honey-containing marinades on heterocyclic aromatic amine formation and overall mutagenicity in fried beef steak and chicken breast. J Food Sci 2004; 2004(69): 147-53. [http://dx.doi.org/10.1111/j.1365-2621.2004.tb13350.x]

[77] Hasnol NDS, Jinap S, Sanny M. Effect of different types of sugars in a marinating formulation on the formation of heterocyclic amines in grilled chicken. Food Chem 2014; 145: 514-21. [http://dx.doi.org/10.1016/j.foodchem.2013.08.086] [PMID: 24128508]

[78] Messner C, Murkovic M. Evaluation of a new model system for studying the formation of heterocyclic amines. J Chromatogr B Analyt Technol Biomed Life Sci 2004; 802(1): 19-26. [http://dx.doi.org/10.1016/j.jchromb.2003.11.015] [PMID: 15035993]

[79] Thiebaud HP, Knize MG, Kuzmicky PA, Felton JS, Hsieh DP. Mutagenicity and chemical analysis of fumes from cooking meat. J Agric Food Chem 1994; 42: 1502-10. [http://dx.doi.org/10.1021/jf00043a021]

[80] Fu Y, Zhao G, Wang S, *et al.* Simultaneous determination of fifteen heterocyclic aromatic amines in the urine of smokers and nonsmokers using ultra-high performance liquid chromatography-tandem mass spectrometry. J Chromatogr A 2014; 1333: 45-53. [http://dx.doi.org/10.1016/j.chroma.2014.01.057] [PMID: 24529957]

[81] Velísek J, Davídek J, Hajslová J, Kubelka V, Janícek G, Mánková B. Chlorohydrins in protein hydrolysates. Z Lebensm Unters Forsch 1978; 167(4): 241-4. [http://dx.doi.org/10.1007/BF01135595] [PMID: 716635]

[82] Weißhaar R, Perz R. Fatty acid esters of glycidol in refined fats and oils. Eur J Lipid Sci Technol 2010; 112: 158-65. [http://dx.doi.org/10.1002/ejlt.200900137]

[83] IARC (International Agency for Research on Cancer). 3-monochloro-1,2-propanediol. Some chemicals present in industrial and consumer products, food and drinking-water. IARC Monographs 2012; 101: 349-74.

[84] Commission Regulation (EC) No. 1881/2006 of 19 December 2006 setting maximum levels for certain contaminants in foodstuffs. OJ L 2006; 364: 5.

[85] EFSA (European Food Safety Authority). Scientific opinion on Risks for human health related to the presence of3- and 2-monochloropropanediol (MCPD), and their fatty acid esters, and glycidyl fatty acid esters in food. EFSA J 2016; 14: 4426.

[86] Buhrke T, Frenzel F, Kuhlmann J, Lampen A. 2-Chloro-1,3-propanediol (2-MCPD) and its fatty acid esters: cytotoxicity, metabolism, and transport by human intestinal Caco-2 cells. Arch Toxicol 2015; 89(12): 2243-51. [http://dx.doi.org/10.1007/s00204-014-1395-3] [PMID: 25354796]

[87] EFSA (European Food Safety Authority). Scientific opinion on Update of the risk assessment on 3-monochloropropane diol and its fatty acid esters. EFSA J 2018; 16: 5083.

[88] Hielscher J, Monien BH, Abraham K, Jessel S, Seidel A, Lampen A. An isotope-dilution UPLC-MS/MS technique for the human biomonitoring of the internal exposure to glycidol via a valine adduct at the N-terminus of hemoglobin. J Chromatogr B Analyt Technol Biomed Life Sci 2017; 1059: 7-13. [http://dx.doi.org/10.1016/j.jchromb.2017.05.022] [PMID: 28554062]

[89] Commission Directive 2002/72/EC. Relating to plastic materials and articles intended to come into contact with foodstuffs. 6th August. 2002https://eur-lex.europa.eu/lega--content/EN/TXT/HTML/?uri=LEGISSUM:l21301&from=EN [(accessed Aug 1, 2018).];

[90] Collier PD, Cromie DDO, Davies AP. Mechanism of formation of chloropropanols present in protein hydrolysates. J Am Oil Chem Soc 1991; 68: 785-90. [http://dx.doi.org/10.1007/BF02662173]

[91] Cheng W, Liu G, Wang L, Liu Z. Glycidyl fatty acid esters in refined edible oils: a review on formation, occurrence, analysis, and elimination methods. Compr Rev Food Sci Food Saf 2017; 16: 263-81. [http://dx.doi.org/10.1111/1541-4337.12251]

[92] Razak RAA, Ainie K, Siew WL, *et al.* Detection and monitoring of 3-monochloropropane-1,2-diol (3-MCPD) esters in cooking oils. Food Control 2012; 25: 355-60. [http://dx.doi.org/10.1016/j.foodcont.2011.10.058]

[93] Dubois M, Tarres A, Goldmann T, Empl AM, Donaubauer A, Seefelder W. Comparison of indirect and direct quantification of esters of monochloropropanediol in vegetable oil. J Chromatogr A 2012; 1236: 189-201.
[http://dx.doi.org/10.1016/j.chroma.2012.03.009] [PMID: 22444428]

[94] Yamazaki K, Ogiso M, Isagawa S, Urushiyama T, Ukena T, Kibune N. A new, direct analytical method using LC-MS/MS for fatty acid esters of 3-chloro-1,2-propanediol (3-MCPD esters) in edible oils. Food Addit Contam Part A Chem Anal Control Expo Risk Assess 2013; 30(1): 52-68.
[http://dx.doi.org/10.1080/19440049.2012.713031] [PMID: 22909228]

[95] Zelinková Z, Svejkovská B, Velísek J, Dolezal M. Fatty acid esters of 3-chloropropane-1,2-diol in edible oils. Food Addit Contam 2006; 23(12): 1290-8.
[http://dx.doi.org/10.1080/02652030600887628] [PMID: 17118872]

[96] Weisshaar R. Fatty acid esters of 3-MCPD: overview of occurrence and exposure estimates. Eur J Lipid Sci Technol 2011; 113: 304-8.
[http://dx.doi.org/10.1002/ejlt.201000312]

[97] FSA (Food Standards Agency). Survey of process contaminants in retail foods - Year 3: 2009. Food Standards Agency Food Survey Information Sheet 03/10 , 2010 http://www.food.gov.uk/multimedia/pdfs/fsis0310.pdf [(accessed Jul 25, 2018).];

[98] Zelinková Z, Doležal M, Velıšek J. Occurrence of 3-chloropropane-1,2-diol fatty acid esters in infant and baby foods. Eur Food Res Technol 2009; 228: 571-8.
[http://dx.doi.org/10.1007/s00217-008-0965-0]

[99] Leigh J, MacMahon S. Occurrence of 3-monochloropropanediol esters and glycidyl esters in commercial infant formulas in the United States. Food Addit Contam Part A Chem Anal Control Expo Risk Assess 2017; 34(3): 356-70.
[http://dx.doi.org/10.1080/19440049.2016.1276304] [PMID: 28004609]

[100] Svejkovská B, Novotný O, Divinová M, Réblová Z, Doležal M, Velıšek J. Esters of 3-chloropropan--1,2-diol in foodstuffs. Czech J Food Sci 2004; 22: 190-6.
[http://dx.doi.org/10.17221/3423-CJFS]

[101] Doležal M, Chaloupská M, Divinová M, Svejkovská B, Velıšek J. Occurrence of 3- chloropropane-1,2-diol and its esters in coffee. Eur Food Res Technol 2005; 221: 221-5.
[http://dx.doi.org/10.1007/s00217-004-1118-8]

[102] Hamlet CG, Sadd PA. Chloropropanols and their esters in cereal products. Czech J Food Sci 2004; 22: 259-626.
[http://dx.doi.org/10.17221/10676-CJFS]

[103] Zelinková Z, Doležal M, Velıšek J. 3-Chloropropane-1,2-diol fatty acid esters in potato products. Czech J Food Sci 2009; 27: 421-4.
[http://dx.doi.org/10.17221/1065-CJFS]

[104] Kuhlmann J. Determination of bound 2,3-epoxy-1-propanol (glycidol) and bound monochloropropanediol (MCPD) in refined oils. Eur J Lipid Sci Technol 2011; 113: 335-44. [http://dx.doi.org/10.1002/ejlt.201000313]

[105] Shiro H, Kondo N, Kibune N, Masukawa Y. Direct method for quantification of glycidol fatty acid esters in edible oils. Eur J Lipid Sci Technol 2011; 113: 356-60.
[http://dx.doi.org/10.1002/ejlt.201000395]

[106] Blumhorst MR, Collison MW, Cantrill R, *et al.* Collaborative study for the analysis of glycidyl fatty acid esters in edible oils using LC-MS. J Am Oil Chem Soc 2013; 90: 493-500.
[http://dx.doi.org/10.1007/s11746-012-2187-7]

[107] Steenbergen H, Hrnčiřík K, Ermacora A, de Koning S, Janssen HG. Direct analysis of intact glycidyl fatty acid esters in edible oils using gas chromatography-mass spectrometry. J Chromatogr A 2013; 1313: 202-11.

[http://dx.doi.org/10.1016/j.chroma.2013.06.056] [PMID: 23891377]

[108] Haines TD, Adlaf KJ, Pierceall RM, Lee I, Venkitasubramanian P, Collison MW. Direct determination of MCPD fatty acid esters and glycidyl fatty acid esters in vegetable oils by LC-TOFMS. J Am Oil Chem Soc 2011; 88(1): 1-14.
[http://dx.doi.org/10.1007/s11746-010-1732-5] [PMID: 21350591]

[109] Hori K, Koriyama N, Omori H, Kuriyama M, Arishima T, Tsumura K. Simultaneous determination of 3-MCPD fatty acid esters and glycidol fatty acid esters in edible oils using liquid chromatography time-of-flight mass spectrometry. Lebensm Wiss Technol 2012; 48: 204-8.
[http://dx.doi.org/10.1016/j.lwt.2012.03.014]

[110] Kuhlmann J. Analysis and occurrence of dichloropropanol fatty acid esters and related process-induced contaminants in edible oils and fats. Eur J Lipid Sci Technol 2016; 118: 382-95.
[http://dx.doi.org/10.1002/ejlt.201400518]

[111] Küsters M, Bimber U, Reeser S, Gallitzendörfer R, Gerhartz M. Simultaneous determination and differentiation of glycidyl esters and 3-monochloropropane-1,2-diol (MCPD) esters in different foodstuffs by GC-MS. J Agric Food Chem 2011; 59(11): 6263-70.
[http://dx.doi.org/10.1021/jf200493b] [PMID: 21545150]

[112] Wöhrlin F, Fry H, Lahrssen-Wiederholt M, Preiß-Weigert A. Occurrence of fatty acid esters of 3-MCPD, 2-MCPD and glycidol in infant formula. Food Addit Contam Part A Chem Anal Control Expo Risk Assess 2015; 32(11): 1810-22.
[http://dx.doi.org/10.1080/19440049.2015.1071497] [PMID: 26179516]

[113] Jędrkiewicz R, Kupska M, Głowacz A, Gromadzka J, Namieśnik J. 3-MCPD: A worldwide problem of food chemistry. Crit Rev Food Sci Nutr 2016; 56(14): 2268-77.
[http://dx.doi.org/10.1080/10408398.2013.829414] [PMID: 25830907]

[114] Strijowski U, Heinz V, Franke K. Removal of 3-MCPD esters and related substances after refining by adsorbent material. Eur J Lipid Sci Technol 2011; 113: 387-92.
[http://dx.doi.org/10.1002/ejlt.201000323]

[115] Pudel F, Benecke P, Vosmann K, Matth B. 3-MCPD- and glycidyl esters can be mitigated in vegetable oils by use of short path distillation. Eur J Lipid Sci Technol 2016; 118: 396-405.
[http://dx.doi.org/10.1002/ejlt.201500095]

[116] Commission Recommendation of 10 September 2014 on the monitoring of the presence of 2 and 3-monochloropropane-1,2-diol (2 and 3-MCPD), 2- and 3-MCPD fatty acid esters and glycidyl fatty acid esters in food. OJEU, 2014; L 271: 93-35.

[117] Lee BQ, Khor SM. 3-Chloropropane-1,2-diol (3-MCPD) in soy sauce: a review on the formation, reduction, and detection of this potential carcinogen. Compr Rev Food Sci Food Saf 2015; 14: 48-66.
[http://dx.doi.org/10.1111/1541-4337.12120]

[118] AOCS/JOCS (American Oil Chemists' Society/Japan Oil Chemists' Society). Glycidyl fatty acid esters in edible oils. Joint AOCS/JOCS Official Method Cd 28-10, 2012.

[119] Analysis of occurrence of 3-monochloropropane-1,2-diol (3-MCPD) in food in Europe in the years 2009-2011 and preliminary exposure assessment. EFSA J 2013; 11: 3381.
[http://dx.doi.org/10.2903/j.efsa.2013.3381]

[120] MacMahon S, Mazzola E, Begley TH, Diachenko GW. Analysis of processing contaminants in edible oils. Part 1. Liquid chromatography-tandem mass spectrometry method for the direct detection of 3-monochloropropanediol monoesters and glycidyl esters. J Agric Food Chem 2013; 61(20): 4737-47.
[http://dx.doi.org/10.1021/jf4005803] [PMID: 23590632]

[121] Kissa E. Determination of 3-chloropropanediol and related dioxolanes by gas chromatography. J Chromatogr A 1992; 605: 134-8.
[http://dx.doi.org/10.1016/0021-9673(92)85038-U]

[122] Masukawa Y, Shiro H, Nakamura S, *et al.* A new analytical method for the quantification of glycidol

fatty acid esters in edible oils. J Oleo Sci 2010; 59(2): 81-8.
[http://dx.doi.org/10.5650/jos.59.81] [PMID: 20103980]

[123] Masukawa Y, Shiro H, Kondo N, Kudo N. Generalized method to quantify glycidol fatty acid esters in edible oils. J Am Oil Chem Soc 2011; 88: 15-21.
[http://dx.doi.org/10.1007/s11746-010-1645-3]

[124] Dubois M, Tarres A, Goldmann T, Loeffelmann G, Donaubauer A, Seefelder W. Determination of seven glycidyl esters in edible oils by gel permeation chromatography extraction and liquid chromatography coupled to mass spectrometry detection. J Agric Food Chem 2011; 59(23): 12291-301.
[http://dx.doi.org/10.1021/jf2028347] [PMID: 22017647]

[125] Ermacora A, Hrncirik K. A novel method for simultaneous monitoring of 2-MCPD, 3-MCPD and glycidyl esters in oils and fats. J Am Oil Chem Soc 2013; 90: 1-8.
[http://dx.doi.org/10.1007/s11746-012-2132-9]

[126] Miyazaki K, Koyama K. An improved enzymatic indirect method for simultaneous determinations of 3-MCPD esters and glycidyl esters in fish oils. J Oleo Sci 2017; 66(10): 1085-93.
[http://dx.doi.org/10.5650/jos.ess17082] [PMID: 28924085]

[127] Miyazaki K, Koyama K. Application of indirect enzymatic method for determinations of 2-/3-MC-D-Es and Gly-Es in foods containing fats and oils. J Am Oil Chem Soc 2016; 93: 885-93.
[http://dx.doi.org/10.1007/s11746-016-2833-6]

[128] Garballo-Rubio A, Soto-Chinchilla J, Moreno A, Zafra-Gómez A. A novel method for the determination of glycidyl and 3-monochloropropanediol esters in fish oil by gas chromatography tandem mass spectrometry. Talanta 2017; 165: 267-73.
[http://dx.doi.org/10.1016/j.talanta.2016.12.060] [PMID: 28153252]

[129] Samaras VG, Giri A, Zelinkova Z, Karasek L, Buttinger G, Wenzl T. Analytical method for the trace determination of esterified 3- and 2-monochloropropanediol and glycidyl fatty acid esters in various food matrices. J Chromatogr A 2016; 1466: 136-47.
[http://dx.doi.org/10.1016/j.chroma.2016.08.071] [PMID: 27623063]

[130] Musukawa Y, Shiro H, Kondo N, Kudo N. Generalized method to quantify glycidol fatty acid esters in edible oils. J Am Oil Chem Soc 2011; 88: 15-21.
[http://dx.doi.org/10.1007/s11746-010-1645-3]

[131] Fry H, Schödel C, These A, Preiß-Weigert A. Collaborative study for the determination of 3-MCPD- and 2-MCPD- fatty acid esters in fat containing foods First Collaborative Study Part II Method Validation and Proficiency Test BfR Wissenschaft. Berlin, Germany: Federal Institute for Risk Assessment 2013.

[132] Zelinkova Z, Giri A, Wenzl T. Assessment of critical steps of a GC/MS based indirect analytical method for the determination of fatty acid esters of monochloropropanediols (MCPDEs) and of glycidol (GEs). Food Control 2017; 77: 65-75.
[http://dx.doi.org/10.1016/j.foodcont.2017.01.024] [PMID: 28674464]

[133] AOCS (American Oil Chemists Society). 2- and 3-MCPD fatty acid esters and glycidol fatty acid esters in edible oils and fats by acid transesterification. AOCS Official Method Cd 29a-13 2013.

[134] AOCS (American Oil Chemists Society). Determination of bound monochloropropanediol-(MCPD-) and bound 2,3-epoxy-1-propanol (glycidol-) by gas chromatography/mass spectrometry (GC/MS). AOCS Official Method Cd 29b-13 2013.

[135] AOCS (American Oil Chemists Society). Fatty-acid-bound 3-chloropropane- 1,2-diol (3-MCPD) and 2,3-epoxi-propane-1-ol (glycidol), determination in oils and fats by GC/MS (differential measurement). AOCS Official Method Cd 29c-13 2013.

[136] AOCS (American Oil Chemists Society). Analysis of 2- and 3-MCPD fatty acid esters and glycidyl fatty acid esters in oil-based emulsions. AOCS Method Cd 30-15, 2015.

[137] Wenzl T, Samaras V, Giri A, Buttinger G, Karasek L, Zelinkova Z. Development and validation of analytical methods for the analysis of 3-MCPD (both in free and ester form) and glycidyl esters in various food matrices and performance of an ad-hoc survey on specific food groups in support to a scientific opinion on comprehensive risk assessment on the presence of 3-MCPD and glycidyl esters in food. European Commission Joint Research Centre Institute for Reference Materials and Measurements, 2015 http://publications.jrc.ec.europa.eu/repository/handle/JRC95649 [(accessed Jul 1, 2018)];

[138] Hrncirik K, Zelinkova Z, Ermacora A. Critical factors of indirect determination of 3- chloropropane-1,2-diol esters. Eur J Lipid Sci Technol 2011; 113: 361-7. [http://dx.doi.org/10.1002/ejlt.201000316]

[139] Bratinova S, Buttinger G, Karasek L. Determination of the MCPD fatty acid esters and glycidyl fatty acid esters in food. Report on the collaborative trial organized by the EURL-PAH and Process Contaminants. European Commission Joint Research Centre Institute for Reference Materials and Measurements, 2017. https://ec.europa.eu/jrc/en/publication/determination-mcpd-fatty-acid-esters-and-glycidyl-fatty-acid-esters-food-report-collaborative-trial [(accessed Jul 19, 2018)];

CHAPTER 10

Mycotoxins

Yelko Rodríguez-Carrasco[1,*] and **Alberto Ritieni**[2,*]

[1] *University of Valencia, Department of Food Chemistry and Toxicology, Av/ Vicent A. Estellés, s/n 46100 Burjassot, Valencia, Spain*

[2] *Università di Napoli Federico II, Department of Pharmacy, Via D. Montesano, 49 - 80131Napoli, Italy*

Abstract: Mycotoxins are secondary metabolites produced by fungal species which can usually be found in foodstuffs. The effects of some food-borne mycotoxins are acute, symptoms of severe illness appearing very quickly. Other mycotoxins occurring in food have longer term chronic or cumulative effects on health, including the induction of cancers and immune deficiency. Thus, Regulation (EC) 1881/2006, partially amended by other Regulations, set maximum contents of some mycotoxins in different foodstuffs allowing to evaluate risks and take actions to protect public health. In this chapter, mycotoxins with significant health and food production impact are discussed by considering the following items: chemical structure, conditions of their production, occurrence in food, maximum limits, toxicity and analytical methods. The chapter also includes the exposure assessment approach to these food contaminants, their metabolism and the proposed biomarkers in the literature. A final remark about the toxicogenomic approach is also included in the chapter as a future trend in the study of mycotoxins.

Keywords: Aflatoxins, *Alternaria* toxins, Biomarkers, Chromatography, Emerging fusariotoxins, Exposure assessment, Food, Fumonisins, Mass spectrometry, Metabolism, Metabolites, Mycotoxins, Ochratoxin A, Occurrence, Patulin, Trichothecenes, Toxicity, Zearalenone.

GENERAL INTRODUCTION

Preamble

The term mycotoxin derived from Greek words mikes (fungus) and toxicum (poison).

* **Corresponding author Yelko Rodríguez Carrasco and Alberto Ritieni:** University of Valencia, Department of Food Chemistry and Toxicology, Av/ Vicent A. Estellés, s/n - 46100 Burjassot, Valencia, Spain; Tel: +34 963543047; Fax: +34 963544284; E-mail: yelko.rodriguez@uv.es and Università di Napoli Federico II, Department of Pharmacy, Via D. Montesano, 49 - 80131 Napoli, Italy; Tel: +39 081678652; Fax: +39 081678610;
E-mail: alberto.ritieni@unina.it, respectively

Belén Gómara & María Luisa Marina (Eds.)
Current and Future Developments in Food Science (Vol. 1)

Mycotoxins are defined as secondary fungal metabolites, low molecular weight (MW <800 Da), produced by filamentous fungi whose ingestion, inhalation or dermal absorption can cause various diseases and disorders known as mycotoxicosis, the severity of which depends on the toxicity of mycotoxin, the degree of exposure, age and nutritional status of both human and animal [1, 2].

The oldest mycotoxicosis described in human is the Ergotism, a condition described in the Middle Age and caused by the consumption of plant products contaminated by ergot alkaloids produced by the fungus *Claviceps purpurea* which triggered several epidemics that devastated Western Europe [3]. The most important case of human mycotoxicosis by trichothecenes was described as endemic in parts of Russia in 1932 causing high mortality rates (60% of those affected). Initially it was thought that the disease had an infectious origin and could even be due to a vitamin deficiency so confused with diseases such as scarlet fever, diphtheria, pellagra and even scurvy, but was finally in 1943 when it was named as "leukopenia toxic hemorrhagic", better known as ATA (*Alimentary toxic aleukia*) caused by contamination of crops by T-2 toxin; a toxin produced by *Fusarium sporotrichoides* [4]. The discovery of aflatoxins in the 60s of the 20th century marked a turning point in the study of Mycotoxicology, when thousands of young turkey and other birds died in the England because of an illness which was coined as "turkey disease X" due to the consumption of peanut flour contaminated by toxins from *Aspergillus flavus* [3, 4].

Since then, there have been significant amount of research conducted to determine the presence and toxicity of mycotoxins in various food matrices and which have driven the development of strategies for detoxification of these toxic compounds to ensure food safety [5].

OVERVIEW

Among the major mycotoxin producers molds are included those within the genera *Aspergillus*, *Penicillium*, *Fusarium*, *Alternaria* or *Claviceps*. Those fungi under certain conditions (temperature, humidity, water activity, pH, substrate composition, etc.) can colonize and subsequently contaminate with mycotoxins food and feed (Table **1**) [1].

These fungi are widely distributed worldwide. *Aspergillus* genera is usually isolated in tropical area, whereas *Fusarium* and *Penicillium* are mainly found in cold climates and in temperate zones, respectively. Nonetheless, there is not a clear pattern at this moment due to the climate change [6, 7]. Fig. (**1**) shows the global map of mycotoxin occurrence and risk in different regions.

Table 1. Conditions of temperature and water activity (aw) for the growth of the main mycotoxigenic fungi and the production of their toxins.

Species	Growth		Toxin Production		
	Temperature	Minimum aw	Mycotoxin	Minimum aw	Optimal aw
Aspergillus flavus	24 °C to 37 °C	0.78	AFs	0.80-0.82	0.95-0.98
Aspergillus parasiticus		0.80		0.83	0.98
Aspergillus ochraceus	24 °C to 37 °C	0.77	OTA	0.80-0.90	0.95 to 0.99
Penicillium verrucosum	20 °C to 32 °C	0.80		0.83-0.85	0.90-0.99
Penicillium expansum	23 °C to 27 °C	0.83	PAT	-	0.98
Fusarium verticillioides	25 °C to 37 °C	0.90	FBs	-	0.97
Fusarium proliferatum		0.90		-	0.97
Fusarium sporotrichioides	15 °C to 27 °C	0.90	ZON and TCs	0.95	0.97 to 0.99
Fusarium graminearum		0.90		0.95	0.97 to 0.99
Fusarium culmorum		0.90		0.95	0.97 to 0.99

AFs: aflatoxins; OTA: ochratoxin A; PAT: patulin; FBs: fumonisins; ZON: zearalenone; TCs: trichothecenes

The invasion by these fungi may occur during the pre-harvest (field) or post-harvest stages (storage, transport and processing), causing both serious economic losses and health problems among humans and livestock [8]. It has to be highlighted that a same toxigenic strain can produce various mycotoxins, and one mycotoxin can be synthesized by different fungi. These metabolites have different chemical structures and biological activities. According to the literature, there have been described about 400 mycotoxins, being the most important due to their adverse health effects on human and animals the followings: aflatoxins, ochratoxin A, fumonisins, trichothecenes, zearalenone and patulin [9].

The presence of mycotoxins along the food chain remains a major public health problem [10 - 12]. In this sense, the Food and Agriculture Organization of the World Health Organization (FAO/WHO) has estimated that at least 25% of the world's crops are contaminated with mycotoxins [13]. Table **2** shows the most frequent combination of contaminated food and type of mycotoxin. In this line, mycotoxins are within the food contaminants with the highest number of notifications according to the annual reports published by the Rapid Alert System for Food and Feed (RASFF) [14]. This trend has been maintained over time adjusting to the natural fluctuations of these contaminants. Table **3** shows the total notifications of mycotoxins since 2004. Table **4** shows a comparison of the number of notifications recorded in recent years for mycotoxins, pathogenic microorganisms and pesticide residues.

Fig. (1). Incidence of the main mycotoxins according to regions. Reproduced with permission from [7].

Table 2. Foods most frequently contaminated by mycotoxins and main producing species.

Mycotoxin	Gender	Species	Foods most Frequently Contaminated
Aflatoxins	*Aspergillus*	*A. flavus, A. parasiticus, A. bombycis, A. pseudotamarii, A.ochraceoroseus*	Groundnut, corn, wheat, rice, rapeseed, cottonseed, nuts, eggs, milk, cheese, figs
Ochratoxin A	*Aspergillus*	*A. ochraceus, A. melleus, A. niger, A.sulphureus, A. carbonarius*	Wheat, barley, oats, corn, dry beans, peanuts, cheese, coffee, raisins, grapes, dried fruits, wine
Patulin	*Penicillium*	*P. verrucosum*	Fruit, especially apples, fruit juice
	Aspergillus	*A. clavi, A. terreus*	
	Penicillium	*P. expansum, P. carneum, P. paneum, P. sclerotigenum*	
	Paecylomyces	*P. varioti*	
	Byssochlamys	*B. fulva, B. nívea*	

(Table 2) cont.....

Mycotoxin	Gender	Species	Foods most Frequently Contaminated
Fumonisins	*Fusarium*	*F. verticillioides, F. proliferatum, Anthophilum F., F. Dlamini, F. napiforme*	Corn (bran, flour, breakfast cereals, infant formulas, popcorn, roasted corn, sweet corn), beer, asparagus, dried figs
Trichothecenes type A	*Fusarium*	*F. sporotrichioides, poae F., F. equiseti*	Wheat, corn, barley, oats, bread, breakfast cereals, beer
Trichothecenes type B	*Fusarium*	*F. graminearum, F. culmorum*	
Zearalenone	*Fusarium*	*F. graminearum, F. culmorum, F. Cerealis, F. semitectum*	Corn, oats, sorghum, wheat
Enniatins, beauvericin	*Fusarium*	*F. avenaceum, F. sporotrichioides, F. Poea, F. langsethiae*	Wheat, corn, barley, oats, bread, breakfast cereals, beer
Alternaria toxins	*Alternaria*	*A. alternata, A. tenuissima, A. arborescens, A. radicina, A. brassicae, A. brassicicola, A. infectoria*	Cereals, tomatoes, apples, grapes, cucumbers, tangerines, oil crops, oilseeds, melons

Table 3. Evolution of total notifications of mycotoxins according to the RASFF's annual reports.

Year	2004	2005	2006	2007	2008	2009	2010	2011	2012	2013	2014	2015	2016	2017
Notifications	880	996	878	760	933	669	688	635	538	405	383	495	551	581

Table 4. Comparison of total notifications according to hazards (RASFF).

Hazards	**Year**						
	2011	**2012**	**2013**	**2014**	**2015**	**2016**	**2017**
Mycotoxins	635	538	405	383	495	551	581
Pathogenic microorganisms	599	592	774	782	745	685	1073
Pesticide Residues	363	447	452	435	405	253	338

MAIN MYCOTOXINS

Mycotoxins Produced by *Aspergillus* and/or *Penicillium* Fungi

Aflatoxins

Aflatoxins (AFs) are polycyclic unsaturated organic compounds which are very closely related to pentacyclic compounds, widely known for their carcinogenic potential. These compounds usually consist of a coumarin nucleus flanked with a highly reactive bifuran ring system on one side and a six membered lactone ring on the other side.

Production

AFs are mainly produced by strains of *Aspergillus flavus* and *Aspergillus parasiticus*. The most important AFs are called B1, B2, G1, G2 and M1. AFM1 is not directly produced by any AFs producing fungi but it is a metabolic derivative of AFB1 [15].

Presence in Food

The main food invaded by fungi of the genus *Aspergillus* and subsequent aflatoxin production are: peanuts, corn, cotton seed and pistachios. AFs have also been detected in other matrices such as oilseeds (sunflower and soybean), cereals, almonds, hazelnuts, raisins, figs, coffee, cocoa and some spices like paprika and pepper [16, 17].

Toxicology

Symptoms of acute aflatoxicosis include hepatocellular hemorrhagic necrosis, edema and lethargy. It also presents cytotoxic effects by inducing lipid peroxidation and causing oxidative damage in hepatocytes. Symptoms of chronic poisoning by aflatoxins is not very specific and includes decreased body weight, decreased production of eggs and milk as well as increased susceptibility to various infectious diseases due to the immunosuppressive effects caused by these toxins. AFs presented carcinogenic and mutagenic effects. Therefore, the International Agency for Research on Cancer (IARC) has classified AFs in group 1 (carcinogenic to humans agents) [18].

Ochratoxin A

Ochratoxin A (OTA) is a pentaketide derived from the dihydrocoumarins family coupled to β-phenylalanine.

Production

The main OTA producing species are *Aspergillus ochraceus* and *Penicillium verrucosum*. Other fungi such as *Aspergillus niger* and *Aspergillus carbonarius* are also relevant OTA producing species according to the matrix or geographical area [19].

Presence in Food

The main sources of this food contaminant are cereals such as wheat, barley, maize or rice and rice products, alcoholic beverages such as beer and wine (especially red wine), and other matrices such as grape juice, raisins, vegetables, coffee, cocoa and nuts [16].

Toxicology

Symptoms of poisoning OTA includes nephrosis and liver necrosis with multifocal hemorrhages in major organs [19]. Regarding chronic effects, OTA is classified as hepatotoxic agent, nephrotoxic, neurotoxic, immunosuppressive, cytotoxic and genotoxic. OTA was classified in group 2B as possible human carcinogen by IARC [18].

Patulin

Patulin (PAT) is a furopyran and lactone that is (2H-pyran-3(6H)-ylidene) acetic acid which is substituted by hydroxy groups at positions 2 and 4 and in which the hydroxy group at position 4 has condensed with the carboxy group to give the corresponding bicyclic lactone.

Production

Patulin is mainly produced by *Penicillium expansum*; however, PAT also may be produced by other species of *Penicillium*, *Aspergillus*, *Byssochlamys* and *Paecylomyces*.

Presence in Food

All species previously mentioned have the ability to grow on fruits, vegetables, cereals, cheese and grass silage and corn for animal feed. Among them, fruits, particularly apple and their derivatives (such as juice, puree and compote), are considered the main source of PAT [16, 17].

Toxicology

The effects of acute intoxication by PAT include nausea, vomiting and other gastrointestinal symptoms that are associated with kidney damage. It has been observed that the PAT has immunotoxic properties at high concentrations. Chronic ingestion of high doses of PAT produces very serious immune, gastrointestinal and neurotoxic disorders [3]. However, the IARC shows that there is no evidence to show a carcinogenic effect of this mycotoxin neither in experimental animals nor in humans so, it was classified within the group 3 [20].

Mycotoxins Produced by *Fusarium* Fungi

Fumonisins

Fumonisins (FBs) are polyketides and thereby members of the largest class of fungal secondary metabolites. They contain 2-propane-1,2,3-tricarboxylic acid side chains esterified to an aminopolyol backbone. According to the chemical groups present along the linear chain, fumonisin can be classified into four series: fumonisins A, B, C and P. However, toxins belonging to the B series, and in the following decreasing order: B1, B2, B3 and B4 are those that are isolated mostly in foods.

Production

Fumonisins are produced by various *Fusarium* strains, mainly *Fusarium verticillioides* and *Fusarium proliferatum*. The most important is the fumonisin FB1. It can constitute up to 70% of all FBs in food [21].

Presence in Food

The main sources of FBs are cereals and cereal derived products, being corn the most susceptible food matrix to be infested by these fungi [16]. Other matrices in which the presence of FBs have been detected includes feed or beer, as a result of the use of semolina or corn syrup as technological coadjutant.

Toxicology

The high structural similarity of the FBs with sphingolipids determines its mechanism of action which is based on competitive inhibition of ceramide synthase enzyme causing disturbance in ceramide's metabolism and sphingolipids [22]. The symptoms of acute FBs poisoning include abdominal pain and diarrhea. At chronic level, there is a probable association between consumption of food contaminated by FBs and esophageal and liver cancer. Based on that, FBs were classified in Group 2B by IARC [18].

Trichothecenes

Trichothecene (TCs) mycotoxins are a family of tetracyclic sesquiterpenoid substances comprising over 200 compounds of widely varying toxicity. The epoxy group at C-12 and C-13 is considered essential for toxicity. TCs are classified in types A–D according to their structure. Main type A TCs include T-2 toxins and HT-2 toxin, diacetoxyscirpenol (DAS), neosolaniol (NEO), etc. and position have a different C-8 ketone functional group, while the type B characterized by having a ketone group in position C-8. Type B TCs encompass as deoxynivalenol (DON) and nivalenol (NIV) and acetylated compounds: 3-acetyl (3-ADON) [1].

Production

Fusarium sporotrichioidies, *Fusarium poae* and *Fusarium equiseti* are the main producers of type A TCs. *Fusarium graminearum* and *Fusarium culmorum* are the most important producers of type B TCs [15].

Presence in Food

Cereals, particularly wheat but also corn, barley, oats or rice among others, are the main sources of TCs in the diet. Mycotoxin content varies according to location in the grain, being higher in the outer parts of the cereal. Among the TCs, the most commonly found in food include DON, T-2 and HT-2. The latter exhibit toxicity ten times higher compared to DON; however, its impact on food is lower. TC contamination is also commonly found in cereal-based products such as bread, breakfast cereals, baby food, beer etc [16, 17, 23].

Toxicology

Symptoms associated with acute toxicity include nausea, vomiting, loss of appetite, weight loss and diarrhea. Ingestion of high doses can cause severe lymphoid tissue, injury intestinal bleeding and cell destruction in epithelial mucous membranes of the intestine [24]. TCs are also potent inhibitors of protein synthesis in eukaryotes and have been linked to diseases such as alimentary toxic aleukia (ATA) [25]. All TCs are classified by IARC (DON, NIV, FUS-X and T-2) within group 3 [18].

Zearalenone

Zearalenone (ZON) is a non-steroidal estrogenic mycotoxin. Chemically it is a derivative of β-resorcylic acid lactone.

Production

The main zearalenone producing fungi are *Fusarium graminearum*, *F. culmorum*, *F. semitectum* and *F. Cerealis* [15]. The main ZON producing species are similar to those that produce TCs, and growth conditions of the fungus and mycotoxin production are identical. Therefore, the simultaneous presence of TCs and ZON in various food matrices is commonly reported.

Presence in Food

Particularly maize and derived products are susceptible to be infested by those fungi and subsequent production of ZON. ZON contamination have also been reported in other matrices such as barley, oats, sorghum, rice and derived products detected [17].

Toxicology

ZON toxicity and some of its metabolites, especially α-ZOL is associated with reproductive problems in some animals and possibly humans, since they have the ability to bind competitively with the estrogen receptors. Recently, a study has shown that ZON may stimulate the growth of mammary carcinomas in humans; nonetheless, IARC does not have sufficient scientific evidence and this toxin and its metabolites have been classified in Group 3 [18, 26].

Emerging Fusarium Mycotoxins

The emerging *Fusarium* mycotoxins are beauvericin (BEA), enniatins (ENN) and fusaproliferin (FUS). ENNs and BEA belong to a group of cyclic hexadepsipetides composed of alternating residues of three *N*-methyl amino acids, commonly valine, leucine and isoleucine, and three hydroxyl acids, typically hydroxyisovaleric acid [27].

Production

These toxins are produced by various *Fusarium* species, such as *F. avenaceum*, *F. sporotrichioides*, *F. Poea* or *F. langsethiae* among others.

Presence in Food

The before mentioned fungi are able to colonize and subsequently produce mycotoxins in food matrices highly consumed worldwide as cereals and their derivatives. The presence of these food contaminants has also been detected in other matrices such as nuts [25]. In Mediterranean countries, unprocessed and processed grains containing ENNs and BEA in a concentration range from few μg kg^{-1} to over mg kg^{-1} levels were observed [27]. Unlike other *Fusarium* mycotoxins, no limits have been set for ENNs up to know.

Toxicology

ENNs have ionophoric properties through which these toxins are capable of promoting the formation of stable lipophilic complexes with essential cations mono and divalent (Ca^{2+}, Na^{+}, K^{+}) and their transport through biological membranes, disrupting normal physiological concentrations. This ionophoric behavior seems to be related to their wide range of biological activity. *In vitro* studies have demonstrated their antimicrobial, insecticidal and phytotoxic effect. Furthermore, ENNs exerted a remarkable cytotoxic effect in several human and animal cells lines at very low micromolar range as recently reviewed in literature [28]. The observed difference of *in vitro* and *in vivo* toxicity might have its origin in their low bioavailability [29 - 35].

Mycotoxins Produced by *Alternaria* Fungi

The major *Alternaria* toxins belong to the chemical groups Dibenzo-α-pyrones, which include alternariol (AOH) and alternariolmonomethyl ether (AME) and cyclic tetrapeptides represented by tentoxin (TEN). Emerging *Alternaria* mycotoxins mainly belong to five chemical classes: pyranones or benzopyrones (altenuene (ALT); altenuisol (AS); altenusin (ALN)), amine/amide metabolites (tenuazonic acid (TeA); altersetin (ALS)), perylenequinones (altertoxins (ATXs), alterperylenol or alteichin (ALTCH), and stemphyltoxin (STE)), and anthraquinones (Macrosporin A, Altersolanol (As-A)) [36].

Production

The most common *Alternaria* species include *A. alternata*, *A. tenuissima*, *A. arborescens*, *A. radicina*, *A. brassicae*, *A. brassicicola*, and *A. infectoria*. *Alternaria* species has been shown to be a significant contaminant of refrigerated fruits, vegetables, and stored feedstuffs [4]. *A. alternata* is the most important within the genus as regards to mycotoxin contamination of fruits and vegetables.

Presence in Food

Alternaria species can colonize a wide range of plants including cereals, tomatoes, apples, grapes, oil crops, oilseeds, sunflower seeds, oranges, lemons, melons, cucumbers, cauliflowers, peppers, and tangerines [36, 37].

Toxicology

Toxicological data are limited to the above-mentioned *Alternaria* mycotoxins. They have been related to a range of pathologies from hematological disorders to esophageal cancer. They are also mutagenic, clastogenic, and estrogenic in microbial and mammalian cell systems and tumorigenic in rats [36, 37].

LEGISLATION

Because of the growing concern in recent years in food safety for the presence of mycotoxins in foods, health authorities have recognized the potential risks to the health of humans and animals posed by the contamination of food by mycotoxins. Consequently, the risk assessment has led the Joint FAO/WHO Expert Committee on Food Additives (JECFA) to estimate the dose considered safe, such as the tolerable daily intake (TDI) or in case of shortage of reliable data on the consequences of human exposure, the provisional estimated tolerable daily intake (PTDI) or provisional tolerable weekly intake (PTWI). Currently the European Food Safety Authority (EFSA) has established tolerable intakes for OTA, PAT, the sum of FBs, ZON, NIV, the sum of DON and its acetylated forms (3-ADON and 15-ADON) and the sum of the HT-2 toxin and T-2 (Table **5**). For aflatoxins, they exposure must be as low as reasonably achievable due to the act that they are both genotoxic and carcinogenic substances and therefore a tolerable intake cannot be established. On the other hand, not only maximum permissible limits for some mycotoxins in various food matrices have been established but also protocols concerning sampling and specifications required for analytical methods as stated in Regulation (EC) 401/2006 [38] partially amended by Regulation (EU) 178/2010 [39].

Table 5. Proposed tolerable intakes for the main mycotoxins.

Mycotoxin	Tolerable Intake	Reference
OTA	0.12 μg kg^{-1} bw weekly	[40]
PAT	0.4 μg kg^{-1} bw daily	[41]
Σ (FB1+FB2+B3)	2 μg kg^{-1} bw daily	[42]
Σ (DON+3-ADON+15-ADON)	1 μg kg^{-1} bw daily	[43]
NIV	1.2 μg kg^{-1} bw daily	[44]
Σ (HT-2+T-2)	0.1 μg kg^{-1} bw daily	[45]
ZON	0.25 μg kg^{-1} bw daily	[46]

In 2006, entered into force the current Regulation 1881/2006 [47] as regards mycotoxins in foodstuff setting maximum content of AFB1, B2, G1, G2 and M1 in nuts, spices, cereals and cereal products, such as baby food set, as well as in milk and infant formula. Similarly the maximum levels of OTA were also fixed in cereals and cereal products, wine, grape juice, raisins and spices. As regards PAT, Regulation sets maximum allowable limits in fruit juices, products derived from apples, fermented beverages and baby food. As far as *Fusarium* mycotoxins are concerned, Regulation set maximum levels of DON, ZON, and FBs mainly in cereals and derivatives (Table **6**).

So far, Regulation 1881/2006 of the Commission of 19 December 2006 has been partially amended by:

- Regulation 1126/2007 of the Commission of 28 September 2007 [48] as regards *Fusarium* toxins in corn and corn products,
- Regulation 105/2010 of the Commission of 5 February 2010 [49] as regards OTA in certain products,
- Regulation 165/2010 of the Commission of 26 February 2010 [50] as regards the contents of the AFs on certain products,
- Regulation 1058/2012 of the Commission of 12 November 2012 [51] as regards the content of AFs in figs.

Table 6. Maximum limits of mycotoxins according to foodstuffs.

Mycotoxins	Food Products	Maximum Level (μg kg^{-1})
AFB1	Processed cereals and baby foods for infants and young children, dietary foods for special medical purposes, cereals and cereal products, dried fruits, nuts, spices and peanuts	0.1 – 8
AFM1	Infant formulas and follow-on formulas, dietary foods for special medical purposes, raw milk, heat-treated milk and milk for the manufacture of dairy products	0.025 – 0.050
Sum of AFs	Cereals and cereal products, nuts, nuts, coffee and peanuts	4 – 15
OTA	Processed cereals and baby foods for infants and young children, dietary foods for special medical purposes, cereals and cereal-based products, grape juice, wine, coffee and raisins	0.5 – 10
Sum of FBs	Foods made from corn and baby foods for infants and young children, corn-based food and unprocessed maize	200 – 2000
PAT	Fruit juices and solid products made from apple, fermented drinks derived from apples or containing apple juice, nectars, baby foods other than processed cereal-based foods for infants and young children	10 – 50
DON	Processed cereals and baby foods for infants and young children, some cereals and cereal products, unprocessed cereals	200 – 1750
ZON	Corn-based foods for infants and children, foods made from cereals, other than maize for infants and children, corn and corn-based products, corn and other cereals unprocessed	20 – 200

MYCOTOXIN ANALYSIS

Food safety is a growing concern for consumers, governments and producers due to the globalization of markets where food is produced and distributed worldwide. With food scares on the increase the Authority has a vital role to play in ensuring both the safety of food and consumer confidence in the food chain. In this regard, the establishment of regulatory limits has driven the development and improvement of analytical procedures for the identification and quantification of mycotoxins in food, feed and biological matrices. Association of Official

Analytical Chemists (AOAC) recommended standardized methods for mycotoxins analysis which have been validated in interlaboratory studies showing a high degree of accuracy and precision; however, some of those methods provide insufficient detection limits to reach the demands for measurement at low levels (Table 7) [52]. Several review articles have compiled the methods of analysis published in the last decade for the determination of mycotoxins [53 - 56].

Table 7. Official methods for the determination of mycotoxins (modified of [52]).

Mycotoxin	Method	Product	LOD (μg kg^{-1})
AFs	990.17	Corn	1.5 - 10
OTA	991.44	Corn and barley	10
PAT	995.10	Apple juice	20
FBs	995.15	Corn	100 - 500
DON	986.18	Wheat	350
ZON	994.01	Wheat	1000

Sampling

The great heterogeneity in the distribution of toxigenic fungi greatly hinders mycotoxin analysis, and specifically the sampling stage. To overcome that, the European Union (EU) has established methods of sampling and analysis for the official control of the levels of mycotoxins in foodstuffs in Regulation (EC) No 401/2006 [38].

Extraction and Purification

Analytical methods based on chromatographic or immunoassay techniques typically require a solvent extraction to release the mycotoxin from the matrix, and a subsequent step of purification to reduce the matrix effect. Among the most common solvents for the extraction of mycotoxins are mixtures of acetonitrile:water and methanol:water. The addition of salts and pH changes can improve the extraction efficiency.

In the analytical method, the clean-up is the separation of mycotoxins of interest of other co-extracted compounds. There are several techniques of extraction and purification used for mycotoxin analysis. Currently there is a growing interest in the simplification and miniaturization of analytical techniques, thereby reducing the amount of sample used in the extraction and consumption of organic solvents. Therefore, some research groups have developed alternative sample preparation procedures based on microextraction techniques mainly aimed to determine mycotoxins and their metabolites not only in foodstuffs but also in biological samples (urine, tissues, hair, *etc.*).

Confirmation Techniques

Determining mycotoxins levels requires analytical procedures both sensitive and selective enough to detect these compounds to below the permitted maximum levels. Chromatographic techniques coupled to mass spectrometry are the choice for the identification and quantification of mycotoxins. These techniques allow reaching enough sensitivity and selectivity to fulfill the requirements set in Regulation.

Chromatographic Techniques

Gas Chromatography

Gas chromatography (GC) is a technique based on the separation of compounds based on their volatility and affinity for the stationary phase. After derivatization of the free hydroxyl groups of the mycotoxins to trimethylsilyl ethers forms (TMS), heptafluorobutyric esters (HFB) or trifluoroacetyl esters (TFA), they become volatile enough to GC analysis. The more sensitive and specific GC detectors are Electron Capture Detector (ECD) and mass spectrometry (MS).

Scott *et al.*, 1981 [57] performed the first application of GC-MS for analysis of DON in wheat and other cereals with satisfactory results. Currently, this technique is performed for analysis of both trichothecenes type A and B. In Table **8** are shown some GC-MS methods published in the last years for determination of mycotoxins.

Table 8. GC-MS multi-mycotoxin methods published in last years.

Year	Matrix	GC-MS		Reference
		Detector	LOD/LOQ (µg kg^{-1} or µg L^{-1})	
2011	barley	Q	0.013-1.91 / 0.05-8	[58]
2012	corn	Q	14-65 / 42-196	[59]
2012	semolina	QqQ	*n.m.* / 1.25-10	[60]
2012	corn	IT	70 / 120	[61]
2013	corn, wheat	Q	0.04-10 / 0.1-30	[62]
2013	wheat, rice, corn	QqQ	0.6-5 / 1.25-10	[63]
2014	urine	QqQ	0.12-4 / 0.25-8	[64]
2015	oils	QqQ	0.01-0.06 / 0.03-0.2	[65]

(Table 8) cont.....

Year	Matrix	GC-MS		Reference
		Detector	LOD/LOQ (µg kg^{-1} or µg L^{-1})	
2015	beer	QqQ	0.05-8 / 0.10-16	[66]
2016	murine macrophage ANA-1 cell	ToF/MS	*n.m.*	[67]
2017	barley and derived products	QqQ	1.6-70 / 3.1-80	[68]
2018	ready-to-eat food	QqQ	0.04-1.5 / 0.13-5	[69]
2018	chicken liver	QqQ	0.05-0.15 / 0.25-0.75	[70]

n.m. not mentioned; Q: single quadrupole; QqQ: triple quadrupole; IT: ion trap; ToF: time-of-flight

Liquid Chromatography

Liquid chromatography (LC) is the preferred technique in mycotoxin analysis because of its versatility and wide applicability. For mycotoxin analysis, LC is used in reverse phase (RP). The most common is the phase octadecylsilica, used along with mixtures of polar solvents as mobile phases (*e.g.*, water, methanol and acetonitrile). For the study of compounds with similar chemical structure and polarity isocratic elution is preferred while the gradient elution is useful for the study of mycotoxins of different polarity.

The association between liquid chromatography and mass spectrometry became possible with the development of atmospheric pressure ionization (API) techniques. API techniques used in LC-MS are electrospray and APCI (atmospheric pressure chemical ionization).

The first application of the LC-MS for the AFs analysis was performed by Rao *et al.*, in 1973 [71] and since then numerous studies have been published for the determination of mycotoxins with physicochemical properties very different [72]. An important advance in LC was the development of ultra-high performance liquid chromatography (UHPLC). An ultra-high pressure system allows using of small particle-packed columns with small diameter, which has a positive effect on both system efficiency and analysis time. Some of the methods published in the last years for the determination of mycotoxins by LC-MS are shown in Table **9**.

Table 9. LC-MS multi-mycotoxin methods published in last years.

Year	Matrix	LC-MS		Reference
		Detector	LOD/LOQ (µg kg^{-1} or µg L^{-1})	
2010	corn, wheat, barley	ToF	5-50 / 10-100	[73]
2010	corn, wheat	QTRAP	0.2-80 / 0.1-120	[74]
2011	baby food, milk	QqQ	2-9 / 6-25	[75]

(Table 9) cont.....

Year	Matrix	LC-MS		Reference
		Detector	LOD/LOQ (µg kg^{-1} or µg L^{-1})	
2011	cereals and cereal products	QqQ	0.001 / 0.005	[76]
2012	corn	QqQ	0.04-3.4 / 0.1-11	[77]
2013	nuts and dried fruits	QqQ	0.02-0.15 / 0.1-0.5	[78]
2013	lotus seed	QqQ	0.003-0.007 / 0.01-0.02	[79]
2014	baby food	QqQ	0.05-5.5	[80]
2014	wine	QqQ	0.4-200 / 0.2-200	[81]
2015	milk	QqQ	0.05-2 / *n.m.*	[82]
2016	biological samples	QqQ	n.m. / 0.05-0.15	[83]
2017	milk	QqQ	0.020-10.140 / *n.m.*	[84]
2017	plant-based beverages	QqQ	*n.m.* / 0.05-15	[85]
2018	urine	Q-Orbitrap HRMS	0.0005 / 0.001	[86]
2018	milk	Q-Orbitrap HRMS	0.001 / 0.002	[87]
2019	soy-based burgers	Q-Orbitrap HRMS	0.01-0.1 / 0.2-5	[88]

n.m. not mentioned; QqQ: triple quadrupole; QTRAP: triple quadrupole linear ion trap; ToF: time-of-flight; Q-Orbitrap HRMS: Q exactive hybrid quadrupole-orbitrap mass spectrometer

Mass Spectrometry

Mass spectrometry is a technique based on sample ionization and separation and recording of ions produced by mass-charge (*m/z*) ratio, in a vacuum system. The main distinction between different mass spectrometers is in the type of analyzer used. Analyzers most relevant for analysis of food contaminants include ion trap (IT), quadrupole (Q) and triple quadrupole (QqQ), time of flight (ToF) and Orbitrap. There are two types of tandem (MS/MS) in the space (triple quadrupole and hybrid quadrupole time of flight) and tandem in time (ion trap), and combinations of both types of tandem in time and space (linear trap or hybrid ion trap-time of flight). The combination of several analyzers allows obtaining structural information and increasing the specificity of quantitative determinations.

QqQ analyzers can monitor a specific transition (Selected Reaction Monitoring, SRM) so it will increase significantly the signal/noise ratio (Signal to Noise, S/N) as well as increasing sensitivity. Working with specific transitions results in highly selective methods and can be used with confirmatory purposes.

The development of technology has enabled to develop analytical procedures based on mass spectrometry (low and high resolution) for the simultaneous determination of mycotoxins. In this sense the high resolving power contributed by chromatographic techniques with the high selectivity and sensitivity provided by the mass spectrometer, converts coupling GC-MS or LC-MS in highly powerful analytical tools for the identification and quantification of mycotoxins. The recent development of high-resolution mass spectrometry (HRMS) employing Orbitrap technology has made possible to achieve higher resolving power, higher mass accuracy, high sensibility and complementary structural information if compared with other MS detectors. Furthermore, HRMS enables a fully untargeted measurement with the ability to retrospectively detect additional compounds in the raw data, which were not anticipated to be of interest at the time of sample analysis. Hence, the high sensibility and sensitivity of the instrument allow identifying also trace levels (ng kg^{-1}) of both target analytes and non-target substances.

However, the wide chemical diversity of toxins and the different concentrations found in food, feed and biological samples, make difficult their simultaneous determination. Apart from this, the acquisition and subsequent maintenance of equipment, especially high resolution mass spectrometry instruments, can make large financial investments for laboratories.

EXPOSURE ASSESSMENT

In the White Paper on Food Safety [89] is stated that risk analysis should be the basis of food safety through its three components:

- risk assessment (scientific advice and data analysis)
- risk management (regulation and control)
- and risk communication.

Toxicological risk assessment produced by mycotoxins seeks to identify and estimate the probability in which their exposure produce adverse health effects. The methodology for evaluation is approached from four stages:

- hazard identification,
- hazard characterization,
- exposure assessment
- and risk characterization,

the latter being the integration of all the information collected in the previous steps in order to estimate the likelihood of adverse health effects in the population as a result of exposure.

Traditionally, studies assessing exposure to food contaminants were carried out by combining data consumption of certain foods with data on the concentration of toxic substances in them, and taking into account the body weight of the population under study. However, it notes that risk assessment is subjected to uncertainties related to the data and selecting the appropriate model. In this regard, it is essential to have reliable data on food intake in order to carry out a proper assessment of exposure from the concentrations of chemicals in food. It is important to get detailed data on food consumption by the average consumer and different population groups to assess exposure, especially vulnerable groups [90]. In this sense, total diet study (TDS) can be a complementary approach to traditional monitoring and surveillance programs, which is designed to provide a solid basis for calculating population dietary exposure and assessing potential impact on public health. A TDS includes the selection of foods based on food consumption data to represent a large portion of a typical diet, their preparation to food as consumed and the subsequent pooling of related foods before analysis.

In foodstuffs, can also be found conjugated mycotoxins, in which the toxin is usually bound to a more polar substance like glucose. These conjugated forms are known as masked mycotoxins. Nowadays, there is a growing interest related to the determination of these hidden forms of mycotoxins in food and feed commodities. They represent an additional potential risk for the consumer. These forms may be hydrolyzed to their precursors in the digestive tract, and may present toxic effects comparable to the parent compound [91].

Therefore, in order to minimize the uncertainty associated with the assessment of exposure through food analysis, in recent years important efforts have been carried out to assess the exposure to mycotoxins through the development biomarkers of exposure which is measured by biological fluids and tissues and can be correlated with exposure to a specific compound [92 - 95].

METABOLISM AND BIOMARKERS

From a toxicological point of view, the use of biological markers or biomarkers, have as main objectives to measure the exposure to a xenobiotic agent causing disease and to predict toxic response that could occur [96]. It implies that the toxic compound must be bioaccessible and bioavailable. The bioaccessibility refers to the amount of toxin released from the food matrix in the gastrointestinal tract and therefore it can be absorbed. Bioavailability refers to the bioavailable fraction of toxic reaching the systemic circulation from the gastrointestinal tract and is available to exert adverse effects. Beyond these concepts, it has to be also considered the metabolism and biotransformation of the toxic compound. The characterization of metabolites is of important consideration in the safety and risk evaluation of food contaminants because they can influence the overall toxic profile.

Both the sensitivity and specificity are essential parameters in the development of biomarkers [97]. Several biomarkers of exposure for some of the mycotoxins of greatest toxicological concern have been proposed in literature. For example, for AFB1 is proposed the urinary evaluation of AFM1 and the adduct 8,9-dihydro-8-(N^7-guanyl)-9-hydroxy-AFB1, products of their metabolism mediated by cytochrome P450 (CYP) as well as the albumin-AFB1 adduct in plasma [98, 99].

With respect to OTA, its concentration in plasma and the concentration of OTA and its metabolite ochratoxin alpha (OTA-α) in urine are the most used biomarkers for assessing exposure to this mycotoxin [100]. However, it is still necessary to establish the relationship between the concentration of this toxin in physiological fluids and the daily intake [101].

As far as the biomarkers of exposure to *Fusarium* toxins are concerned, the sphinganine/sphingosine (Sa/So) ratio in urine or serum has been proposed as a biomarker to evaluate exposure to fumonisins. However, a study published by Westhuizen *et al.*, 2010 [102] concluded that there is no clear association between this ratio and exposure to fumonisin. On the other hand, urinary levels of FB1 in free form showed a positive correlation with the intake of this mycotoxin [103]. As regards the toxicokinetics of DON, it is not yet fully resolved for humans. Currently it is known that there are two major metabolites in mammals: de-epoxydeoxynivalenol (DOM-1), product detoxification generated by the intestinal microbiota, and glucuronides of DON (DON-GlcAs) result of phase II metabolism by the action of uridine diphosphate-glucuronosyltransferase (UGT) [104 - 106]. Several studies have concluded that DOM-1 metabolite is not a majority in humans and their relationship as a biomarker study is under evaluation [104]. In contrast, several studies conducted in humans have shown that a high percentage of DON (between 70-90%) is excreted by the kidneys as DON-GlcAs [106]. In this line Meky *et al.* 2003 [107] developed and validated a biomarker of exposure to DON in urine based on enzymatic hydrolysis of DON-GlcAs with β-glucuronidase. Hence total DON (sum of DON in free form + DON released upon hydrolysis) can be easily obtained. The validated method was correlated with DON intake via consumption of cereals and cereal-based foods. Subsequently, Warth *et al.*, 2011 [108] developed a methodology to quantify directly in urine the major DON-GlcAs, namely deoxynivalenol-3-O-glucuronide (DON-3-GlcA) and deoxynivalenol-15-O-glucuronide (DON-15-GlcA), without requiring the enzymatic hydrolysis. Recently, new DON-GlcAs forms have also been identified (deoxynivalenol-7-O-glucuronide and deoxynivalenol-8-O-glucuronide) [109, 110]. On the other hand, some studies have suggested that 3-ADON and 15-

ADON probably are rapidly deacetylated *in vivo* to DON as it has been observed in pigs [111].

Concerning ZON there is little literature as regards the metabolic pathway of this mycotoxin in humans. The main biotransformation ZON in animals include hydroxylation which leads to the formation of metabolites α-ZOL and β-ZOL and conjugation with glucuronic acid (ZON-GlcAs) [112]. There have also been described other ZON metabolites, namely ZAN, α-ZAL and β-ZAL. However, these compounds *per se* have also been detected in cereals [113]. In terms of toxicity, the estrogenic activity of α-ZOL is ten times higher than ZON, while β-ZOL has a lower affinity for the receptor [114]. Several studies have shown that glucuronidation is a major route of detoxification of ZON and their metabolites in humans [115, 116]. Consequently, the evaluation of ZON-GlcAs urine could be evaluated as a potential biomarker of exposure to ZON.

As regards the emerging *Fusarium* mycotoxins, there is currently few data regarding their detection and their biotransformation products in tissues and biological fluids, complicating the evaluation of the potential risk related to the exposure to them. Recent studies have reported that phase II metabolism was most likely irrelevant for enniatins [29, 30]. On contrary, main phase I Enn B and Enn B1 metabolites were identified and structurally characterized by liquid chromatography (LC) coupled to: ion trap mass spectrometry (ITMS), multiple-stage mass fragmentation (MS^n), time-of-flight mass spectrometry (Q/ToF) and high resolution mass spectrometry (HRMS) using an Orbitrap mass spectrometer. The identified metabolites were the result of hydroxylation, carboxylation and N-demethylation processes [29, 31, 86].

FUTURE TRENDS

The embracement of big data by geneticians and bioinformaticians has marked a turning point for toxicologist. Toxicogenomics is one of the most innovative omics to date which aims to elucidate the molecular mechanisms of toxicity and modes of action by analyzing chemical induced changes in gene expression patterns [117]. Toxicogenomic approach could be used to understand the complex effects of chemicals and to assist in health risk assessment, and it has already generated interesting results for predicting compound genotoxicity and carcinogenicity [118]. Microarrays and next generation sequencing (NGS) techniques will become the gold standard for understanding molecular mechanisms of toxicity. The latter, are revolutionizing genomic/proteomic studies providing high-throughput datasets with high precision and accuracy [119 - 121].

CONSENT FOR PUBLICATION

Not applicable.

CONFLICT OF INTEREST

The authors confirm that this chapter contents have no conflict of interest.

ACKNOWLEDGEMENTS

Declared none.

REFERENCES

[1] Bennett JW, Klich M. Mycotoxins. Clin Microbiol Rev 2003; 16(3): 497-516. [http://dx.doi.org/10.1128/CMR.16.3.497-516.2003] [PMID: 12857779]

[2] Hussein HS, Brasel JM. Toxicity, metabolism, and impact of mycotoxins on humans and animals. Toxicology 2001; 167(2): 101-34. [http://dx.doi.org/10.1016/S0300-483X(01)00471-1] [PMID: 11567776]

[3] Soriano JM. Micotoxinas en alimentos. España 2007.

[4] Ramos AJ. Micotoxinas y micotoxicosis. España: AMV Ediciones 2011.

[5] Garcia D, Ramos AJ, Sanchis V, Marín S. Predicting mycotoxins in foods: a review. Food Microbiol 2009; 26(8): 757-69. [http://dx.doi.org/10.1016/j.fm.2009.05.014] [PMID: 19835759]

[6] Paterson RRM, Lima N. How will climate change affect mycotoxins in food? Food Res Int 2010; 43: 1902-14. [http://dx.doi.org/10.1016/j.foodres.2009.07.010]

[7] Biomin. https://www.biomin.net/en/blog-posts/2017-biomin-mycotoxin-survey-results/ (last accessed: 09/10/1018).

[8] FAO Manual on the application of the HACCP system in mycotoxin prevention and control. FAO Food Nutr Pap 2001; 73: 1-114.

[9] Bräse S, Encinas A, Keck J, Nising CF. Chemistry and biology of mycotoxins and related fungal metabolites. Chem Rev 2009; 109(9): 3903-90. [http://dx.doi.org/10.1021/cr050001f] [PMID: 19534495]

[10] Richard JL. Some major mycotoxins and their mycotoxicoses - an overview. Int J Food Microbiol 2007; 119(1-2): 3-10. [http://dx.doi.org/10.1016/j.ijfoodmicro.2007.07.019] [PMID: 17719115]

[11] Shephard GS. Determination of mycotoxins in human foods. Chem Soc Rev 2008; 37(11): 2468-77. [http://dx.doi.org/10.1039/b713084h] [PMID: 18949120]

[12] Serrano AB, Font G, Ruiz MJ, Ferrer E. Co-occurrence and risk assessment of mycotoxins in food and diet from Mediterranean area. Food Chem 2012; 135(2): 423-9. [http://dx.doi.org/10.1016/j.foodchem.2012.03.064] [PMID: 22868109]

[13] Food and Agriculture Organization, World Health Organization (FAO /WHO). Safety Evaluation of Certain Mycotoxins in Food. Geneva, Switzerland: World Health Organization 2001; pp. 281-387.

[14] RASFF, Rapid Alert System for Food and Feed. https://ec.europa.eu/food/safety/rasff_en (last accessed, 25/09/2018).

[15] Tantaoui A. Les mycotoxines Production, transfert et dans les denrées alimentaires persistance. Rabat: Ed. Fenex 2014.

[16] Sweeney MJ, Dobson AD. Mycotoxin production by *Aspergillus, Fusarium* and *Penicillium* species. Int J Food Microbiol 1998; 43(3): 141-58.
[http://dx.doi.org/10.1016/S0168-1605(98)00112-3] [PMID: 9801191]

[17] Weidenbörner M. Mycotoxins in foodstuffs. Ed. Springer, Bonn, Germany 2014.

[18] IARC (International Agency for Research on Cancer). Some naturally occurring substances: Constituents and food items, heterocyclic aromatic amines and mycotoxins. IARC Monogr Eval Carcinog Risks Hum 1993; 56: 245.

[19] Kuiper-Goodman T, Scott PM. Risk assessment of the mycotoxin ochratoxin A. Biomed Environ Sci 1989; 2(3): 179-248.
[PMID: 2692617]

[20] IARC (International Agency for Research on Cancer). Monographs on the evaluation of carcinogenic risk of chemicals to humans. Lyon, France 1986; p. 40.

[21] Rheeder JP, Marasas WFO, Vismer HF. Production of fumonisin analogs by *Fusarium* species. Appl Environ Microbiol 2002; 68(5): 2101-5.
[http://dx.doi.org/10.1128/AEM.68.5.2101-2105.2002] [PMID: 11976077]

[22] Soriano JM, González L, Catalá AI. Mechanism of action of sphingolipids and their metabolites in the toxicity of fumonisin B1. Prog Lipid Res 2005; 44(6): 345-56.
[http://dx.doi.org/10.1016/j.plipres.2005.09.001] [PMID: 16266752]

[23] Eudes F, Comeau A, Rioux S, Collin J. Impact of trichothecenes on *Fusarium* head blight [*Fusarium graminearum*] development in spring wheat (*Triticum aestivum*). Can J Plant Pathol 2001; 23: 318-22.
[http://dx.doi.org/10.1080/07060660109506948]

[24] Creppy EE. Update of survey, regulation and toxic effects of mycotoxins in Europe. Toxicol Lett 2002; 127(1-3): 19-28.
[http://dx.doi.org/10.1016/S0378-4274(01)00479-9] [PMID: 12052637]

[25] Marin S, Ramos AJ, Cano-Sancho G, Sanchis V. Mycotoxins: occurrence, toxicology, and exposure assessment. Food Chem Toxicol 2013; 60: 218-37.
[http://dx.doi.org/10.1016/j.fct.2013.07.047] [PMID: 23907020]

[26] Zinedine A, Soriano JM, Moltó JC, Mañes J. Review on the toxicity, occurrence, metabolism, detoxification, and intake of zearalenone regulations: An oestrogenic mycotoxin. Food Chem Toxicol 2007; 45: 1-18.
[http://dx.doi.org/10.1016/j.fct.2006.07.030] [PMID: 17045381]

[27] Jestoi M. Emerging *fusarium*-mycotoxins fusaproliferin, beauvericin, enniatins, and moniliformin: a review. Crit Rev Food Sci Nutr 2008; 48(1): 21-49.
[http://dx.doi.org/10.1080/10408390601062021] [PMID: 18274964]

[28] Prosperini A, Berrada H, Ruiz MJ, *et al.* A review of the mycotoxin enniatin B. Front Public Health 2017; 5: 304.
[http://dx.doi.org/10.3389/fpubh.2017.00304] [PMID: 29201864]

[29] Fæste CK, Ivanova L, Uhlig S. *In vitro* metabolism of the mycotoxin enniatin B in different species and cytochrome p450 enzyme phenotyping by chemical inhibitors. Drug Metab Dispos 2011; 39(9): 1768-76.
[http://dx.doi.org/10.1124/dmd.111.039529] [PMID: 21622627]

[30] Fraeyman S, Devreese M, Antonissen G, De Baere S, Rychlik M, Croubels S. Comparative oral bioavailability, toxicokinetics, and biotransformation of enniatin B1 and enniatin B in broiler chickens. J Agric Food Chem 2016; 64(38): 7259-64.
[http://dx.doi.org/10.1021/acs.jafc.6b02913] [PMID: 27632250]

[31] Ivanova L, Uhlig S, Devreese M, Croubels S, Fæste CK. Biotransformation of the mycotoxin enniatin B1 in pigs: A comparative *in vitro* and *in vivo* approach. Food Chem Toxicol 2017; 105: 506-17. [http://dx.doi.org/10.1016/j.fct.2017.04.041] [PMID: 28472676]

[32] Manyes L, Escrivá L, Serrano AB, *et al.* A preliminary study in Wistar rats with enniatin A contaminated feed. Toxicol Mech Methods 2014; 24(3): 179-90. [http://dx.doi.org/10.3109/15376516.2013.876135] [PMID: 24329503]

[33] Ruiz MJ, Franzova P, Juan-Garcia A, Font G. Toxicological interactions Between the beauvericin mycotoxins, deoxynivalenol and T-2 toxin in CHO-K1 cells *in vitro.* Toxicon 2011; 58: 315-26. [http://dx.doi.org/10.1016/j.toxicon.2011.07.015] [PMID: 21821061]

[34] Mallebrera B, Font G, Ruiz MJ. Disturbance of antioxidant capacity produced by beauvericin in CHO-K1 cells. Toxicol Lett 2014; 226(3): 337-42. [http://dx.doi.org/10.1016/j.toxlet.2014.02.023] [PMID: 24613870]

[35] Prosperini A, Juan-Garcia A, Font G, Ruiz MJ. Reactive oxygen species and mitochondrial apoptosis involvement in damage in Caco-2 cells induced by enniatins A, A1, B and B1. Toxicol Lett 2013; 222: 36-44. [http://dx.doi.org/10.1016/j.toxlet.2013.07.009] [PMID: 23867914]

[36] Fraeyman S, Croubels S, Devreese M, Antonissen G. Emerging *Fusarium* and *Alternaria* mycotoxins: occurrence, toxicity and toxicokinetics. Toxins (Basel) 2017; 9(7): 228. [http://dx.doi.org/10.3390/toxins9070228] [PMID: 28718805]

[37] Escriva L, Oueslati S, Font G, Manyes L. Alternaria Mycotoxins in Food and Feed: An Overview. J Food Qual. 2017; p. 20. Article ID 1569748

[38] EC No 401/2006, COMMISSION REGULATION (EC) No 401/2006 of 23 February 2006 laying down the methods of sampling and analysis for the official monitoring of the levels of mycotoxins in foodstuffs (Text with EEA relevance), 2006.

[39] Commission Regulation (EU) No 178/2010, Commission Regulation (EU) No 178/2010 of 2 March 2010 amending Regulation (EC) No 401/2006 as regards groundnuts (peanuts), other oilseeds, tree nuts, apricot kernels , liquorice and vegetable oil (Text with EEA relevance), 2010.

[40] EFSA Panel on Contaminants in the Food Chain (CONTAM), Review of the Scientific Panel on contaminants in the food chain [CONTAM] related to Ochratoxin A in food, 2006.

[41] EFSA Report of experts participating in Task 3.2.8 Assessment of dietary intake of Patulin by the population of EU Member States, 2002.

[42] Scientific Committee on Food, Updated review of the Scientific Committee on Food on Fumonisin B1, B2 and B3, 2003.

[43] SCF. Review of the Scientific Committee on Food on Fusarium toxins.Part 6: Group evaluation of T-2 toxin, HT-2 toxin, nivalenol and deoxynivalenol. EFSA J 2011; 9(6): 2197.

[44] EFSA Panel on Contaminants in the Food Chain (CONTAM). Scientific Opinion on Risks for animals and public health related to the presence of nivalenol in food and feed. EFSA J 2013; 11(6): 3262. [http://dx.doi.org/10.2903/j.efsa.2013.3262]

[45] SCF. Scientific Opinion on the Risks for animals and public health related to the presence of T-2 and HT-2 toxin in food and feed. EFSA J 2011; 9(12): 2481. [http://dx.doi.org/10.2903/j.efsa.2011.2481]

[46] SCF. Scientific Opinion on the Risks for public health related to the presence of zearalenone in food. EFSA J 2011; 9(6): 2197. [http://dx.doi.org/10.2903/j.efsa.2011.2197]

[47] EC No 1881/2006 COMMISSION REGULATION (EC) No 1881/2006 of 19 December 2006 setting maximum levels for contaminants in foodstuffs Certain (Text with EEA relevance), 2006.

[48] EC No 1126/2007, COMMISSION REGULATION (EC) No 1126/2007 of 28 September 2007 amending Regulation (EC) No 1881/2006 setting maximum levels for contaminants in foodstuffs Certain as regards Fusarium toxins in maize and maize products (Text with EEA relevance), 2007.

[49] COMMISSION REGULATION (EU) No 105/2010, COMMISSION REGULATION (EU) No 105/2010 of 5 February 2010 amending Regulation (EC) No 1881/2006 setting maximum levels for contaminants in foodstuffs Certain as regards Ochratoxin A (Text with EEA relevance), 2010.

[50] COMMISSION REGULATION (EU) No 165/2010, COMMISSION REGULATION (EU) No 165/2010 of February 26 2010am ending Regulation (EC) No 1881/2006 setting maximum levels for contaminants in foodstuffs as Certain regards aflatoxins (Text with EEA relevance), 2010.

[51] COMMISSION REGULATION (EU) No 1058/2012, Commission Regulation (EU) No 1058/2012 of 12 November 2012 amending Regulation (EC) No 1881/2006 as regards maximum levels for aflatoxins in dried figs (Text with EEA relevance), 2012.

[52] Pohland AE, Trucksess MW. Mycotoxin Method Evaluation – An Introduction.Methods in Molecular BiologyMycotoxin Protocols. Humana Press Inc.Totowa, New Jersey (USA)20011571310.

[53] Turner NW, Bramhmbhatt H, Szabo-Vezse M, Poma A, Coker R, Piletsky SA. Analytical methods for determination of mycotoxins: An update (2009-2014). Anal Chim Acta 2015; 901: 12-33. [http://dx.doi.org/10.1016/j.aca.2015.10.013] [PMID: 26614054]

[54] Köppen R, Koch M, Siegel D, Merkel S, Maul R, Nehls I. Determination of mycotoxins in foods: current state of analytical methods and limitations. Appl Microbiol Biotechnol 2010; 86(6): 1595-612. [http://dx.doi.org/10.1007/s00253-010-2535-1] [PMID: 20339844]

[55] Pereira VL, Fernandes JO, Cunha SC. Mycotoxins in foodstuffs and related cereals: A review on recent occurrence and methods of analysis. Trends Food Sci Technol 2014; 36: 96-136. [http://dx.doi.org/10.1016/j.tifs.2014.01.005]

[56] Righetti L, Dall'Asta C. New Analytical Frontiers in Mycotoxin Research. Reference Module in Food Science 2018.

[57] Scott PM, Lau PY, Kanhere SR. Gas chromatography with electron capture and mass spectrometric detection of deoxynivalenol in wheat and other grains. J Assoc Off Anal Chem 1981; 64(6): 1364-71. [PMID: 7309655]

[58] Ibáñez-See M, Lopez E, Gonzalez-Penas E. Simultaneous determination of type-A and type-B trichothecenes in samples by GC-MS barley. Food Control 2011; 22: 1428-34. [http://dx.doi.org/10.1016/j.foodcont.2011.03.004]

[59] Ferreira I, Fernandes JO, Cunha SC. Optimization and validation of a method based on QuEChERS procedure and gas chromatography-mass spectrometry for the determination of multi-mycotoxins in popcorn. Food Control 2012; 27: 188-93. [http://dx.doi.org/10.1016/j.foodcont.2012.03.014]

[60] Rodríguez-Carrasco Y, Berrada H, Font G, Mañes J. Multi-mycotoxin analysis in wheat semolina using an acetonitrile-based extraction procedure and gas chromatography-tandem mass spectrometry. J Chromatogr A 2012; 1270: 28-40. [http://dx.doi.org/10.1016/j.chroma.2012.10.061] [PMID: 23182289]

[61] Tran ST, Smith TK, Girgis GN. A survey of free and conjugated deoxynivalenol in the 2008 corn crop in Ontario, Canada. J Sci Food Agric 2012; 92(1): 37-41. [http://dx.doi.org/10.1002/jsfa.4674] [PMID: 21997524]

[62] Gregori R, Meriggi P, Pietri A, Formenti S, Baccarini G, Battilani P. Dynamics of fungi and related mycotoxins in silo storage during cereal bags. Food Control 2013; 30: 280-7. [http://dx.doi.org/10.1016/j.foodcont.2012.06.033]

[63] Rodríguez-Carrasco Y, Moltó JC, Berrada H, Mañes J. A survey of trichothecenes, zearalenone and patulin in milled grain-based products using GC-MS/MS. Food Chem 2014; 146: 212-9.

[http://dx.doi.org/10.1016/j.foodchem.2013.09.053] [PMID: 24176334]

[64] Rodríguez-Carrasco Y, Moltó JC, Mañes J, Berrada H. Development of a GC-MS/MS strategy to determine 15 mycotoxins and metabolites in human urine. Talanta 2014; 128: 125-31. [http://dx.doi.org/10.1016/j.talanta.2014.04.072] [PMID: 25059139]

[65] Qian M, Zhang H, Wu L, Jin N, Wang J, Jiang K. Simultaneous determination of zearalenone and its derivatives in edible vegetable oil by gel permeation chromatography and gas chromatography-triple quadrupole mass spectrometry. Food Chem 2015; 166: 23-8. [http://dx.doi.org/10.1016/j.foodchem.2014.05.133] [PMID: 25053023]

[66] Rodríguez-Carrasco Y, Fattore M, Albrizio S, Berrada H, Mañes J. Occurrence of *Fusarium* mycotoxins and their dietary intake through beer consumption by the European population. Food Chem 2015; 178: 149-55. [http://dx.doi.org/10.1016/j.foodchem.2015.01.092] [PMID: 25704695]

[67] Ji J, Zhu P, Pi F, *et al.* GC-TOF/MS-based metabolomic strategy for combined toxicity effects of deoxynivalenol and zearalenone on murine macrophage ANA-1 cells. Toxicon 2016; 120: 175-84. [http://dx.doi.org/10.1016/j.toxicon.2016.08.003] [PMID: 27530666]

[68] Juan C, Berrada H, Mañes J, Oueslati S. Multi-mycotoxin determination in barley and derived products from Tunisia and estimation of their dietary intake. Food Chem Toxicol 2017; 103: 148-56. [http://dx.doi.org/10.1016/j.fct.2017.02.037] [PMID: 28249780]

[69] Carballo D, Moltó JC, Berrada H, Ferrer E. Presence of mycotoxins in ready-to-eat food and subsequent risk assessment. Food Chem Toxicol 2018; 121: 558-65. [http://dx.doi.org/10.1016/j.fct.2018.09.054] [PMID: 30266314]

[70] Mahmoud AF, Escrivá L, Rodríguez-Carrasco Y, Moltó JC, Berrada H. Determination of trichothecenes in chicken liver using gas chromatography coupled with triple-quadrupole mass spectrometry. Lebensm Wiss Technol 2018; 93: 237-42. [http://dx.doi.org/10.1016/j.lwt.2018.03.043]

[71] Rao GHR, Anders MW. Aflatoxin detection by high-speed liquid chromatography and mass spectrometry. J Chromatogr A 1973; 84(2): 402-6. [http://dx.doi.org/10.1016/S0021-9673(01)91723-6] [PMID: 4745807]

[72] Krska R, Welzig E, Berthiller F, Molinelli A, Mizaikoff B. Advances in the analysis of mycotoxins and its quality assurance. Food Addit Contam 2005; 22(4): 345-53. [http://dx.doi.org/10.1080/02652030500070192] [PMID: 16019804]

[73] Zachariášová M, Lacina O, Malachova A, *et al.* Novel approaches in analysis of Fusarium mycotoxins in cereals employing ultra performance liquid chromatography coupled with high resolution mass spectrometry. Anal Chim Acta 2010; 662(1): 51-61. [http://dx.doi.org/10.1016/j.aca.2009.12.034] [PMID: 20152265]

[74] Capriotti AL, Foglia P, Gubbiotti R, Roccia C, Samperi R, Laganà A. Development and validation of a liquid chromatography/atmospheric pressure photoionization-tandem mass spectrometric method for the analysis of mycotoxins subjected to commission regulation (EC) No. 1881/2006 in cereals. J Chromatogr A 2010; 1217(39): 6044-51. [http://dx.doi.org/10.1016/j.chroma.2010.07.018] [PMID: 20728087]

[75] Beltran E, Ibáñez M, Sancho JV, Cortes MÁ, Yusa V, Hernandez F. UHPLC-MS/MS determination of aflatoxins highly sensitive, the aflatoxin M1 metabolite and Ochratoxin A in baby food and milk. Food Chem 2011; 126: 737-44. [http://dx.doi.org/10.1016/j.foodchem.2010.11.056]

[76] Tam J, Pantazopoulos P, Scott PM, Moisey J, Dabeka RW, Richard IDK. Application of isotope dilution mass spectrometry: determination of ochratoxin A in the Canadian Total Diet Study. Food Addit Contam Part A Chem Anal Control Expo Risk Assess 2011; 28(6): 754-61. [http://dx.doi.org/10.1080/19440049.2010.504750] [PMID: 21623499]

[77] Varga E, Glauner T, Köppen R, *et al.* Stable isotope dilution assay for the accurate determination of mycotoxins in maize by UHPLC-MS/MS. Anal Bioanal Chem 2012; 402(9): 2675-86. [http://dx.doi.org/10.1007/s00216-012-5757-5] [PMID: 22293971]

[78] Tolosa J, Font G, Mañes J, Ferrer E. Nuts and dried fruits: Natural occurrence of emerging *Fusarium* mycotoxins. Food Control 2013; 33: 215-20. [http://dx.doi.org/10.1016/j.foodcont.2013.02.023]

[79] Liu S, Qiu F, Kong W, Wei J, Xiao X, Yang M. Development and validation of an accurate and rapid LC-ESI-MS/MS method for the simultaneous quantification of aflatoxin B1, B2, G1 and G2 in lotus seeds. Food Control 2013; 29: 156-61. [http://dx.doi.org/10.1016/j.foodcont.2012.05.069]

[80] Juan C, Raiola A, Mañes J, Ritieni A. Presence of mycotoxin in commercial infant formulas and baby foods from Italian market. Food Control 2014; 39: 227-36. [http://dx.doi.org/10.1016/j.foodcont.2013.10.036]

[81] Pizzutti IR, de Kok A, Scholten J, *et al.* Development, optimization and validation of a multimethod for the determination of 36 mycotoxins in wines by liquid chromatography-tandem mass spectrometry. Talanta 2014; 129: 352-63. [http://dx.doi.org/10.1016/j.talanta.2014.05.017] [PMID: 25127606]

[82] Wang X, Li P. Rapid screening of mycotoxins in liquid milk and milk powder by size-exclusion automated SPE-UPLC-MS/MS and quantification of matrix effects over the whole chromatographic run. Food Chem 2015; 173: 897-904. [http://dx.doi.org/10.1016/j.foodchem.2014.10.056] [PMID: 25466104]

[83] Rodríguez-Carrasco Y, Heilos D, Richter L, *et al.* Mouse tissue distribution and persistence of the food-born fusariotoxins Enniatin B and Beauvericin. Toxicol Lett 2016; 247: 35-44. [http://dx.doi.org/10.1016/j.toxlet.2016.02.008] [PMID: 26892719]

[84] Flores-Flores ME, González-Peñas E. An LC-MS/MS method for multi-mycotoxin quantification in cow milk. Food Chem 2017; 218: 378-85. [http://dx.doi.org/10.1016/j.foodchem.2016.09.101] [PMID: 27719924]

[85] Miró-Abella E, Herrero P, Canela N, *et al.* Determination of mycotoxins in plant-based beverages using QuEChERS and liquid chromatography-tandem mass spectrometry. Food Chem 2017; 229: 366-72. [http://dx.doi.org/10.1016/j.foodchem.2017.02.078] [PMID: 28372187]

[86] Rodríguez-Carrasco Y, Izzo L, Gaspari A, Graziani G, Mañes J, Ritieni A. Urinary levels of enniatin B and its phase I metabolites: First human pilot biomonitoring study. Food Chem Toxicol 2018; 118: 454-9. [http://dx.doi.org/10.1016/j.fct.2018.05.052] [PMID: 29802944]

[87] Rodríguez-Carrasco Y, Izzo L, Gaspari A, Graziani G, Mañes J, Ritieni A. Simultaneous Determination of AFB1 and AFM1 in Milk Samples by Ultra High Performance Liquid Chromatography Coupled to Quadrupole Orbitrap Mass Spectrometry. Beverages 2018; 4: 43. [http://dx.doi.org/10.3390/beverages4020043]

[88] Rodríguez-Carrasco Y, Castaldo L, Gaspari A, Graziani G, Ritieni A. Development of an UHPLC--Orbitrap HRMS method for simultaneous determination of mycotoxins and isoflavones in soy-based burgers. Lebensm Wiss Technol 2019; 99: 34-42. [http://dx.doi.org/10.1016/j.lwt.2018.09.046]

[89] White Paper on Food Safety Commission of the European Communities. Brussels, COM (99) 719 final, 12 January 2000.

[90] Joint FAO/WHO International Program on Chemical Safety, in Anonymous Principles and Methods for the Risk Assessment of Chemicals in Food. Chapter 6. Dietary exposure assessment of chemicals in food, 2009.

[91] Berthiller F, Crews C, Dall'Asta C, *et al.* Masked mycotoxins: a review. Mol Nutr Food Res 2013; 57(1): 165-86.
[http://dx.doi.org/10.1002/mnfr.201100764] [PMID: 23047235]

[92] Ediage EN, Di Mavungu JD, Song S, Wu A, Van Peteghem C, De Saeger S. A direct assessment of mycotoxin biomarkers in human urine samples by liquid chromatography tandem mass spectrometry. Anal Chim Acta 2012; 741: 58-69.
[http://dx.doi.org/10.1016/j.aca.2012.06.038] [PMID: 22840705]

[93] Warth B, Sulyok M, Fruhmann P, *et al.* Development and validation of a rapid multi-biomarker liquid chromatography/tandem mass spectrometry method to assess human exposure to mycotoxins. Rapid Commun Mass Spectrom 2012; 26(13): 1533-40.
[http://dx.doi.org/10.1002/rcm.6255] [PMID: 22638970]

[94] Shephard GS, Burger HM, Gambacorta L, *et al.* Multiple mycotoxin exposure determined by urinary biomarkers in rural subsistence farmers in the former Transkei, South Africa. Food Chem Toxicol 2013; 62: 217-25.
[http://dx.doi.org/10.1016/j.fct.2013.08.040] [PMID: 23985452]

[95] Rodríguez-Carrasco Y, Mañes J, Berrada H, Font G. Preliminary estimation of deoxynivalenol excretion through a 24 h pilot study. Toxins (Basel) 2015; 7(3): 705-18.
[http://dx.doi.org/10.3390/toxins7030705] [PMID: 25723325]

[96] Sandra S, Arango V. Biomarkers for risk assessment on human health, *Rev. Fac. Nac.* Salud Pública 2012; 30: 75-82.

[97] Crews H, Alink G, Andersen R, *et al.* A critical assessment of some biomarker approaches linked with dietary intake. Br J Nutr 2001; 86 (Suppl. 1): S5-S35.
[http://dx.doi.org/10.1079/BJN2001337] [PMID: 11520422]

[98] Mykkänen H, Zhu H, Salminen E, *et al.* Fecal and urinary excretion of aflatoxin B1 metabolites (AFQ1, AFM1 and AFB-N7-guanine) in young Chinese males. Int J Cancer 2005; 115(6): 879-84.
[http://dx.doi.org/10.1002/ijc.20951] [PMID: 15723309]

[99] Kensler TW, Roebuck BD, Wogan GN, Groopman JD. Aflatoxin: a 50-year odyssey of mechanistic and translational toxicology. Toxicol Sci 2011; 120 (Suppl. 1): S28-48.
[http://dx.doi.org/10.1093/toxsci/kfq283] [PMID: 20881231]

[100] Muñoz K, Blaszkewicz M, Degen GH. Simultaneous analysis of ochratoxin A and its major metabolite ochratoxin alpha in plasma and urine for an advanced biomonitoring of the mycotoxin. J Chromatogr B Analyt Technol Biomed Life Sci 2010; 878(27): 2623-9.
[http://dx.doi.org/10.1016/j.jchromb.2009.11.044] [PMID: 20031488]

[101] Gilbert J, Brereton P, MacDonald S. Assessment of dietary exposure to ochratoxin A in the UK using a duplicate diet approach and analysis of urine and plasma samples. Food Addit Contam 2001; 18(12): 1088-93.
[http://dx.doi.org/10.1080/02652030110070030] [PMID: 11761119]

[102] van der Westhuizen L, Shephard GS, Rheeder JP, Burger H. Single fumonisin exposure levels and sphingoid based in rural populations consuming maize in South Africa. Food Chem Toxicol 2010; 48: 1698-703.
[http://dx.doi.org/10.1016/j.fct.2010.03.047] [PMID: 20363278]

[103] Shephard GS, Van Der Westhuizen L, Sewram V. Biomarkers of exposure to fumonisin mycotoxins: a review. Food Addit Contam 2007; 24(10): 1196-201.
[http://dx.doi.org/10.1080/02652030701513818] [PMID: 17886192]

[104] Turner PC, Hopton RP, Lecluse Y, White KLM, Fisher J, Lebailly P. Determinants of urinary deoxynivalenol and de-epoxy deoxynivalenol in male farmers from Normandy, France. J Agric Food Chem 2010; 58(8): 5206-12.
[http://dx.doi.org/10.1021/jf100892v] [PMID: 20349912]

[105] Turner PC, Hopton RP, White KLM, Fisher J, Cade JE, Wild CP. Assessment of deoxynivalenol metabolite profiles in UK adults. Food Chem Toxicol 2011; 49(1): 132-5. [http://dx.doi.org/10.1016/j.fct.2010.10.007] [PMID: 20934480]

[106] Turner PC, White KLM, Burley VJ, *et al.* A comparison of deoxynivalenol intake and urinary deoxynivalenol in UK adults. Biomarkers 2010; 15(6): 553-62. [http://dx.doi.org/10.3109/1354750X.2010.495787] [PMID: 20572795]

[107] Meky FA, Turner PC, Ashcroft AE, *et al.* Development of a urinary biomarker of human exposure to deoxynivalenol. Food Chem Toxicol 2003; 41(2): 265-73. [http://dx.doi.org/10.1016/S0278-6915(02)00228-4] [PMID: 12480302]

[108] Warth B, Sulyok M, Berthiller F, *et al.* Direct quantification of deoxynivalenol glucuronide in human urine as biomarker of *Fusarium* mycotoxin exposure to the deoxynivalenol. Anal Bioanal Chem 2011; 401: 95-200. [http://dx.doi.org/10.1007/s00216-011-5095-z] [PMID: 21604166]

[109] Warth B, Sulyok M, Fruhmann P, *et al.* Assessment of human deoxynivalenol exposure using an LC-MS/MS based biomarker method. Toxicol Lett 2012; 211(1): 85-90. [http://dx.doi.org/10.1016/j.toxlet.2012.02.023] [PMID: 22429874]

[110] Maul R, Warth B, Kant JS, *et al.* Investigation of the hepatic glucuronidation pattern of the *Fusarium* mycotoxin deoxynivalenol in various species. Chem Res Toxicol 2012; 25(12): 2715-7. [http://dx.doi.org/10.1021/tx300348x] [PMID: 23106612]

[111] Eriksen GS, Pettersson H, Lindberg JE. Absorption, metabolism and excretion of 3-acetyl DON in pigs. Arch Tierernahr 2003; 57(5): 335-45. [PMID: 14620907]

[112] Pfeiffer E, Hildebrand A, Mikula H, Metzler M. Glucuronidation of zearalenone, zeranol and four metabolites *in vitro*: formation of glucuronides by various microsomes and human UDP-glucuronosyltransferase isoforms. Mol Nutr Food Res 2010; 54(10): 1468-76. [http://dx.doi.org/10.1002/mnfr.200900524] [PMID: 20397195]

[113] Mizutani K, Nagatomi Y, Mochizuki N. Metabolism of zearalenone in the course of beer fermentation. Toxins (Basel) 2011; 3(2): 134-41. [http://dx.doi.org/10.3390/toxins3020134] [PMID: 22069701]

[114] Malekinejad H, Maas-Bakker R, Fink-Gremmels J. Species differences in the hepatic biotransformation of zearalenone. Vet J 2006; 172(1): 96-102. [http://dx.doi.org/10.1016/j.tvjl.2005.03.004] [PMID: 15907386]

[115] de Andrés F, Zougagh M, Castañeda G, Ríos A. Determination of zearalenone and its metabolites in urine samples by liquid chromatography with electrochemical detection using a carbon nanotube-modified electrode. J Chromatogr A 2008; 1212(1-2): 54-60. [http://dx.doi.org/10.1016/j.chroma.2008.09.112] [PMID: 18945434]

[116] Solfrizzo M, Gambacorta L, Visconti A. Assessment of multi-mycotoxin exposure in southern Italy by urinary multi-biomarker determination. Toxins (Basel) 2014; 6(2): 523-38. [http://dx.doi.org/10.3390/toxins6020523] [PMID: 24476712]

[117] Li Z, Long Y, Zhong L, *et al.* RNA sequencing provides insights into the toxicogenomic response of ZF4 cells to methyl methanesulfonate. J Appl Toxicol 2016; 36(1): 94-104. [http://dx.doi.org/10.1002/jat.3147] [PMID: 26011545]

[118] Wilson VS, Keshava N, Hester S, *et al.* Utilizing toxicogenomic data to understand chemical mechanism of action in risk assessment. Toxicol Appl Pharmacol 2013; 271(3): 299-308. [http://dx.doi.org/10.1016/j.taap.2011.01.017] [PMID: 21295051]

[119] Caiment F, Gaj S, Claessen S, Kleinjans J. High-throughput data integration of RNA-miRN--circRNA reveals novel insights into mechanisms of benzo[a]pyrene-induced carcinogenicity. Nucleic Acids Res 2015; 43(5): 2525-34.

[http://dx.doi.org/10.1093/nar/gkv115] [PMID: 25690898]

[120] Manyes L, Escrivá L, Font G, Ruiz MJ. Transcriptomic changes after exposure to Enniatin B in Jurkat human T lymphoblastoid cells. Toxicol Lett 2018; 280: S294.

[121] Escrivá L, Jennen D, Caiment F, Manyes L. Transcriptomic study of the toxic mechanism triggered by beauvericin in Jurkat cells. Toxicol Lett 2018; 284: 213-21. [http://dx.doi.org/10.1016/j.toxlet.2017.11.035] [PMID: 29203277]

CHAPTER 11

Biogenic Amines

Gianni Sagratini*, **Giovanni Caprioli, Massimo Ricciutelli** and **Sauro Vittori**

University of Camerino, School of Pharmacy, Via S. Agostino 1, 62032 Camerino (MC), Italy

Abstract: Biogenic amines (BAs) are basic molecules present in food formed by decarboxylation of aminoacids of proteins. They have a particular profile from a toxicological point of view, and the intake of food with high presence of BAs can generate various problems and allergic responses. Due to the importance of their toxicological aspects, BAs are considered as an important indicator of freshness and quality of food, through the evaluation of specific indices that take into account their concentration in food, *i.e.*, Biogenic Amine Index (BAI) or the ratio spermidine/spermine (SPD/SPM). Many foods can be contaminated by the high levels of BAs as meat, cheese, fish, beer, wine and baby foods, and no regulation exists by EFSA or FDA except for histamine in fish. The analytical methodologies used for the detection of the BAs in food are normally based on a primary step of sample preparation (extraction and purification) and then on a second step of instrumental analysis that uses high performance liquid chromatography (HPLC) or gas chromatography (GC) coupled to various detectors as diode array detector (DAD), fluorescence detector (FD), mass spectrometry (MS) and tandem mass spectrometry (MS/MS). Also capillary electrophoresis (CE) has been used for the analysis of BAs in food. This chapter describes an overview on the presence of BAs in foods and the most important analytical strategies for their analysis and detection.

Keywords: Biogenic amines, CE, DAD, Derivatization, FD, GC, HPLC, MS proteic food, Sample preparation, Shelf life markers.

INTRODUCTION

Biogenic amines (BAs) are basic molecules of low molecular mass present in living organisms and, hence, in food [1]. Based on their chemical structures, they are classified into three categories: (1) aromatic, as histamine, tryptamine, tyramine, 2-phenylethylamine, (2) aliphatic diamines, as cadaverine and putrescine, and (3) aliphatic polyamines, as spermine and spermidine (Fig. **1**) [2].

BAs are biologically active molecules that are involved in many cellular functions; monoamines play an important role in neurotransmission and the reg-

* **Corresponding author Gianni Sagratini:** University of Camerino, School of Pharmacy, Via S. Agostino 1, 62032 Camerino (MC), Italy; Tel: +39 0737 402238; E-mail: gianni.sagratini@unicam.it

Belén Gómara & María Luisa Marina (Eds.)
Current and Future Developments in Food Science (Vol. 1)

ulation of blood pressure, polyamines are essential for cellular proliferation and differentiation as they participate in the synthesis of DNA, RNA, and proteins [3].

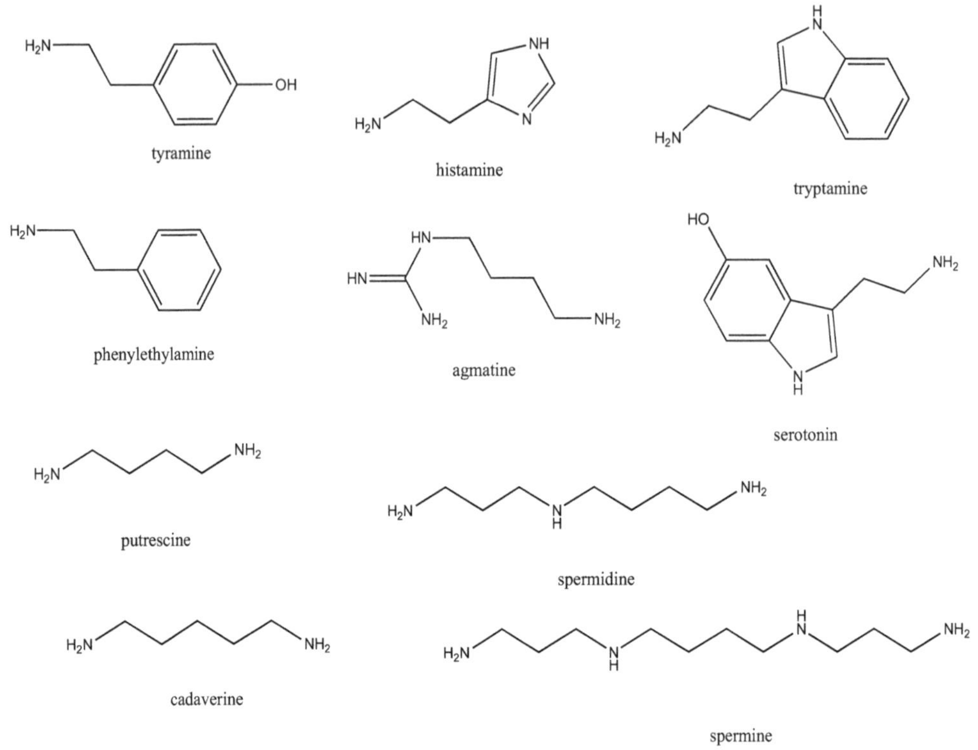

Fig. (1). Structures of the most relevant biogenic amines occurring in food.

BAs are produced in food of proteic origin by three possible mechanisms: (a) decarboxylation of aminoacids (promoted by the decarboxylase enzymes present in various microorganisms) [4], normal cellular metabolism of tissues [5], amination or transamination of aldehydes and ketones [4].

BAs are important from a toxicological point of view, because the intake of food with high concentrations of BAs can generate migraine, headaches, gastrointestinal disorders, and allergic responses. Histamine poisoning produces effects on cardiovascular system such as low blood pressure [6], while tyramine causes allergic skin reactions and increasing blood pressure by releasing noradrenaline from the sympathetic nervous [5]. Other amines, such as spermidine or spermine, have also been associated with the development of food allergy. In normal conditions, the human body can detoxify histamine and tyramine coming from foods by acetylation and oxidation mediated by the enzymes monoamine oxidase (MAO), diamine oxidase (DAO), and polyamine oxidase (PAO) [5]. However, if these detoxifying mechanisms are upset because there is a lack of aminooxidases, BAs increased their concentration in the body and could cause serious toxicological problems. Putrescine and cadaverine, although not considered toxic individually, can increase the effect of histamine and tyramine by interacting with aminooxidases and decreasing with the detoxifying mechanism [7]. It is really very difficult to establish limits of toxicity of BAs in food, because their effect does not depend on their presence alone but is also influenced by other molecules and by the ability of the detoxifying mechanisms.

Literature reports that BAs are potential precursors for the formation of carcinogenic N-nitroso compounds. The reaction of primary amines and nitrosating-agents produces alkylating species, which can react with other food components by generating toxic compounds [8]. The secondary amines such as agmatine, spermine, spermidine and others can react with nitrile and produce the nitrosamines, while tertiary amines produce a range of labile N-nitroso derivatives [9].

Due to the importance of their toxicological aspects, BAs are considered as an important indicator of freshness and quality of food [10]. Various indices can be used for evaluating the quality of fresh food, firstly Biogenic Amine Index (BAI) that is the sum (mg kg^{-1}) of putrescine, cadaverine, histamine and tyramine, then the ratio between spermidine and spermine (SPD/SPM) and the total sum of analyzed BAs. Also the Chemical Quality Index (CQI) has been taken into account for evaluating the quality of food, in particular fish; it is calculated by the sum of the concentration of putrescine + cadaverine + histamine divided by spermine + spermidine + 1. In particular, a CQI between 0 and 1 indicates good quality tuna, between 1 and 10, borderline, and, higher than 10, decomposed [11].

Although BAs have been described as potential toxic compounds, the maximum histamine level is only regulated in fishery products, at 50 mg kg^{-1} by the US Food and Drug Administration (FDA), and at 100 mg kg^{-1} by the European Community [12]. However, the European Food Safety Authority (EFSA) produce a scientific opinion where it describes the risks related to the intake of BAs in fermented products [13]. This document highlights the importance of controlling these molecules in food, and the need to validate analytical methods for their detection. On the other hand, some European countries recommended fixed limits for histamine in wine [Germany (2 mg L^{-1}), Belgium (5-6 mg L^{-1}), and France (8 mg L^{-1})] [14, 15].

Food safety has promoted more research in the field of BAs in the last few years, and thus many countries are involved in the monitoring of BA occurrence in food (Fig. **2**) [16].

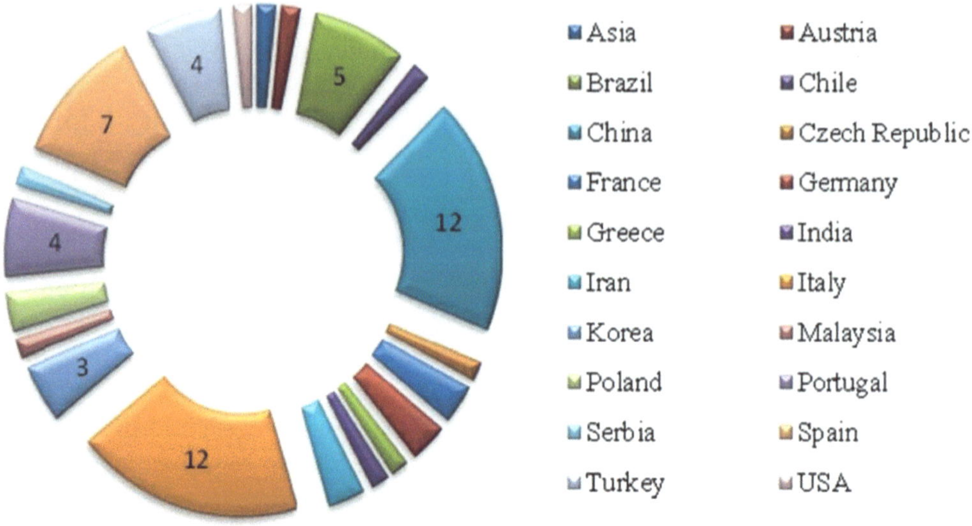

Fig. (2). Percentage breakdown of biogenic amine researches related to food products published since 2010 categorized with the geographical region of the analyzed samples. Reprinted with permission from [16].

In theory, all foods that contain free amino acids or protein and are exposed to conditions enabling microbial or biochemical activity, can produce BAs. These amines generally occur in fermented products such as cheese, wine, beer, but they can be present in fresh food as fish and meat. It is important to underline that the BAs content can vary depending on several extrinsic and intrinsic factors during the manufacturing process, such as the ripening conditions, formulations, physico-chemical parameters as well as microflora development and its decarboxylase activity [16].

Here we report the most important food in which it is possible to detect BAs at relevant levels.

Meat

From a nutritional point of view, meat represents an excellent source of macro and micronutrients, but it is perishable and very susceptible to chemical and physical modifications. The different nutrients composition of meat makes it an ideal environment for the growth and proliferation of microorganisms and common food-borne pathogens. In particular, both red meat (bovine) and white meat (chicken) are particularly sensitive to protein degradation that occurs under appropriate specific conditions, and increasing, in this way, the level of BAs during the time of storage [17]. Many *Enterobacteriaceae*, *Pseudomonas spp.*, and certain *Lactobacillus*, *Enterococcus*, *Staphylococcus spp.* and *Brochothrix thermosphacta* are particularly involved in BAs formation in meat [18]. The most important BAs in meat and meat products are putrescine, tyramine, cadaverine, and also histamine. The only amines present at significant levels in fresh meat are spermidine and spermine. Some amines such as tyramine, putrescine and cadaverine can be formed during the storage of meat. Sirocchi *et al.* [10] reported the presence of spermidine and spermine in Italian fresh chicken meat at a concentration of 17.60 and 109.16 mg kg $^{-1}$, respectively. Lazaro *et al.* [19] analyzed the content of tyramine up to 356.8 mg kg^{-1}, moreover the presence of putrescine cadaverine, spermine and spermidine in chicken samples from Brazil. The highest concentration of BAs in meat and meat products has been reported in sausages from Greece. Papavergou *et al.* [20] described the occurrence of cadaverine, putrescine and histamine in concentrations up to 689.89 mg kg^{-1} in fermented sausages.

Seafood

Seafood is a specific food that can be related to several biological, chemical, and physical hazards, the most common of which are biotoxins, pathogenic bacteria, chemical contaminants and BAs. The most frequent food intoxication caused by BAs involves histamine. Histamine poisoning is caused by eating seafood, mainly scombroid fish such as mackerel (*Scomber japonicus*), tuna (*Thunnus spp.*), mahimahi (*Coryphaena hippurus*), sardine (*Sardina pilchardus*) and bluefish (*Pomatomus saltratrix*) [21]. The most important BAs in seafood are tyramine, histamine, putrescine, tryptamine and cadaverine. These are synthesized by their respective amino acids tyrosine, histidine, ornithine, tryptophan, and lysine. Spermine and spermidine are formed by putrescine [22]. Sagratini *et al.* reported that histamine was detected in 5 of 8 Spanish fish samples in a concentration range of 0.73-0.92 mg kg^{-1}, that is very far from the maximum level established by European Union (100 mg kg^{-1}) or FDA (50 mg kg^{-1}). Moreover, spermine was the amine found in 7 of 8 samples at highest levels, with an average concentration of 3.28 mg kg^{-1} and in a range of 1.14-5.16 mg kg^{-1}; spermidine was found in all 8 samples with an average concentration of 2.07 mg kg^{-1} in a range of 0.25-5.56 mg kg^{-1} [23]. Bilgin *et al.* detected the mean concentration of cadaverine (up to 122.18 $\pm$ 68 mg kg^{-1}) in canned tuna and sardines from Turkey, tryptamine (up to 190.61 $\pm$ 7.67 mg kg^{-1}) in marinated anchovies, histamine in maximum concentration (up to 110.33 $\pm$ 9.87 mg kg^{-1}) and putrescine (up to 116.53 $\pm$ 2.90 mg kg^{-1}) in canned tuna [24]. One study was conducted in Malaysian shrimp sauce detecting tryptamine, putrescine and tyramine up to 57.9 mg L^{-1}, followed by 2-phenylethylamine, histamine and spermidine (up to 5.4 mg L^{-1}) [25].

Cheese

High concentrations of BAs can be found in fermented foods such as cheese and dairy products due to microbial activity. During cheese ripening, free amino acids may be converted into BAs mostly affected by bacterial decarboxylases of a contaminating microflora [26]. It is important to underline that the technology of ripening may be crucial for the development of BAs, and elevated ripening temperatures of cheese can generate greater production of BAs (*e.g.* tyramine, putrescine and cadaverine) [27]. Mayer *et al.* reported that 13.8% of cheese samples from Austria revealed a histamine content of 100 mg kg^{-1}, and 22.4% had a tyramine content of 100 mg kg^{-1}. Moreover, 8.6% of samples had a putrescine or cadaverine content higher than 100 mg kg^{-1}. The total concentration of BAs in two cheese samples was around 1940 mg kg^{-1} [28]. Valsamaki *et al.* reported that the total BAs content in mature cheese from Greece (Feta) was 330 mg kg^{-1} and, after a 120-day storage, 617 mg kg^{-1}. Tyramine and putrescine were the most detected BAs in mature samples (69.7% and 71.2% at 60 and 120 days, respectively) [29]. In general, among different types of cheese; cadaverine the most detected BA in high concentrations. The presence of cadaverine in Formaggio di Fossa ranged up to 1303 ± 5.02 mg kg^{-1} [30] and in Cabrales 774.51 mg kg^{-1} from Italy [31]. A very high concentration of cadaverine was found in Otlu peynior cheese from Turkey (1844.5 mg kg^{-1}) [32].

Beer and Liqueurs

Beer is the fifth most consumed beverage in the world and is produced mainly by mixing malted barley and hops and selected yeast strains, such as *Saccharomyces cerevisiae* (top fermenting), *Saccharomyces carlsbergensis* with wild yeast and lactic acid bacteria. In beers, BAs are produced by microbial contamination, mainly bacterial, *e.g.* histamine, tyramine and cadaverine [33]. BAs present in raw materials (malted barleys, yeasts and hop) are represented by spermidine, phenylethylamine, and spermine. From a toxicological point of view, beers could increase the adverse effects of BAs due to the presence of ethanol, which reduces/inhibits the activity of monoamine oxidase and diamine oxidase, enzymes responsible of BA metabolism. Other compounds such as trimethylamine and acetaldehyde could increase the histamine toxicity [34]. Pradenas *et al.* analyzed 101 beers from Chile and reported that BAs content in beer samples ranged from 0.53 to 85.04 mg L^{-1}, and the beers produced by microbreweries (19.13 ± 16.92 mg L^{-1}) showed a significant higher biogenic amines content (P ¼ 0.0021) than beers produced by macrobreweries (9.65 ± 4.50 mg L^{-1}). Putrescine was the most important BA found in all kind of beers [35]. Almeida *et al.*, 2012 reported the analysis of eighteen BAs in beer samples from Portugal [33]. The most dominant ones were putrescine, tyramine and dimethylamine with concentrations up to

12.777 mg L^{-1}.

Liqueurs are flavored sweetened spirit beverages (15-30% alcohol; >10% sugar) normally consumed in small quantities. Although liqueurs are not strictly fermented beverages, the alcohol used in their production is always obtained by fermentation and further distillation of crop products, so liqueurs can contain BAs. Cunha *et al.* analyzed 27 liqueur samples from Portugal, including 11 fruit liqueurs, 8 herbs liqueurs, 5 coffee liqueurs, 2 honey liqueurs, and 1 milk liqueur. The volatile amines methylamine, ethylamine, and dimethylamine, and the non-volatile amines morpholine, cadaverine, and histamine, were the BAs detected and, in only 15 out of 27 samples, putrescine showed the mean content of 0.38 mg L^{-1} [36].

Wine

Wine is one of the most famous and investigated beverages, important not only for nutritional and healthy properties but also for its hedonistic significance. Literature reports that more than 15 amines have been identified in wine and their total concentration has been reported to range from a few ng L^{-1} to about 50 mg L^{-1}, depending on the type of the wine.

According to the study performed by Soufleros *et al.* [37], their concentration is low after alcoholic fermentation, and increases in most wines during malolactic fermentation. Amines such as methylamine, ethylamine, phenylethylamine, isoamylamine already present in grape must be produced and degraded during vinification [38]. It is widely known that several bacteria are implicated in BAs production in wine including *Lactobacillus, Pediococcus, Oenococcus spp.* and *Leuconostoc*. Literature reports that, in white wines the range of BAs ranged from n.d. to 12.8 mg L^{-1}. High concentrations were found in Italian wines for histamine, putrescine, tyramine and ethylamine while in Portuguese wines only histamine presented in 8.94 mg L^{-1}. Higher concentrations were found for histamine 23.1 ± 2.2 mg L^{-1} in Portugal, in Italy 10.8 mg L^{-1}, in France 14.05 mg L^{-1} and in different EU countries (Italy, France, Germany, Switzerland, Austria and Spain) 16.2 mg L^{-1} [16].

Baby Foods

Adverse reactions among infants and young children could appear after consumption of food containing small amounts of BAs. The study of Czajkowska-Mysłek *et al.* revealed the first assessment of BAs occurrence in ready-to-eat vegetable without/with fish, meat and fruit baby products intended for the youngest consumers. The content of BAs was evaluated in 68 commercial baby foods produced by 10 leading manufacturers available in Poland. The total

BA contents in analyzed foods were obtained in the range of 1283-101421 ng g^{-1}. The highest concentration of histamine (2375 ng g^{-1}) was found in the sample with spinach, tyramine (1667 ng g^{-1}) in fruit sample with banana [39].

EXTRACTION AND PURIFICATION OF BAS IN FOOD

BAs are present in complex matrices as food and their extraction has to consider nutritional and chemical properties of food.

Solid samples such as fish, meat, cheese, and their derivatives, are extracted through the use of acids, essentially hydrochloric acid (HCl), trichloroacetic acid (TCA) or perchloric acid (PCA) [25]. Some authors have compared the extraction efficiency of different acids for different amines but they obtained different results. A summary of the different extraction methods was reported in Table **1.**

Eerola *et al.* [40] extracted 2 g of dry sausage with 10 mL of 0.4 M perchloric acid by using an Ultraturrax and then directly derivatized the sample. The same was performed by Draisci *et al.* [41] that extracted kiwi, cheese (Emmenthal), fish (pilchard) and dry fermented sausages with 0.375 M perchloric acid and then defatted the aqueous solution (only for fish, meat and cheese) with hexane (2:1 v/v). Innocente *et al.* [42] extracted seven BAs from cheese by using 20 mL of 0.1 M hydrochloric acid and then homogenized the slurry in a Politron homogenizer before derivatization with dansyl chloride. The use of HCl for cheese was also confirmed by Moret *et al.* [43] that reported that 0.1 M HCl appeared to be suitable for the analysis of cheese, however, it was not a good choice for fish or meat. This can be explained by sample turbidity and the presence of poorly-resolved interfering peaks. In fact, pH is an important parameter exerting the greatest influence on the partition equilibrium of amines between the aqueous and organic solutions. They highlighted that 5% of thrichloracetic acid (TCA) represented a better choice for fish and meat. Also Lange *et al.* [44] studied the suitability of different solvents for amine extraction from food such as 0.1 M HCl, two different concentrations (5%, 10%) of trichloroacetic acid (TCA), methanol in water (50 and 75%), and 100% methanol. They highlighted that with 100% methanol the lowest signal/noise ratios and the highest peaks could be observed. An extraction with 10% TCA was less useful because of high signal/noise ratio and low peak resolution. On the other hand, the use of 5% TCA resulted in the best peak resolutions. Therefore, 5% TCA was used normally for extraction of the food samples, except for cheese in which HCl remained the favorable one. The use of TCA 5% was also the best choice for Sagratini *et al.* [23] that set up a new analytical method for analyzing eight underivatized biogenic amines in fish. BAs were extracted from 5 g of homogenized fresh or frozen fish, for one minute with 15 mL of TCA 5% using an Ultraturrax and repeating thrice the extraction.

Similar extraction procedures with the same extraction solvent were followed by Sirocchi *et al.* [10] that quantified eight biogenic amines in meat wrapped in a new active packaging system and by Sirocchi *et al.* that extracted ten biogenic amines from the same matrix. Among acid, also acetic acid can be used. In fact, Molognoni *et al.* [45] described a new and fast multiresidues method for the analysis of meat and fish products by liquid chromatography-tandem mass spectrometry (LC-MS/MS). They extracted six biogenic amines, among a total of 17 analytes, by solid-liquid extraction (SLE): from 2 g of food matrix they performed the extraction with 10 mL of methanol:acetonitrile:water 45:45:10 acidified with 0.1% acetic acid and homogenize the sample with Ultraturrax.

Other alternatives to SLE including sample clean-up and/or sample enrichment techniques are supercritical fluid extraction (SFE) [46] or solid phase microextraction (SPME) [47]. These techniques can be applied to different kind of samples either naturally liquid, as biological fluids or beverages or after acidic extraction from solid material. SFE has been applied mainly for aromatic amine extraction [48]. SPME has been applied to aromatic amines [49, 50] and aliphatic amines [51]. Saaid *et al.* [25] extracted ten BAs from different food samples by a hollow fiber liquid-phase microextraction (LPME) followed by *in situ* derivatization using dansyl chloride. Conditions for LPME were optimized such as type of organic solvent (dihexyl ether), temperature (70 °C), extraction time (30 min), stirring speed (1600 rpm) and salt addition (no NaCl) that influence the extraction efficiency. This method was successfully developed for the high-performance liquid chromatography-ultraviolet (HPLC-UV) determination of the BAs (tryptamine, putrescine, cadaverine, histamine, tyramine, spermidine) in shrimp sauce and tomato ketchup samples.

Liquid samples such as fruit juices and alcoholic beverages are usually extracted in alkaline condition independently of their pH values or if they are aliphatic or aromatic, thus often samples are alkalinized in the sample preparation step. Busto *et al.* [52] analyzed sixteen amines in wines with a fully automated derivatization step by means of an injection programme. The injection system mixes the reagents automatically: the alkalinity needed to perform the derivatization was obtained by using the borate buffer (pH 8.8) and then the derivative reagent was formed by adding acetonitrile to the 6-aminoquinolyl-N-hydroxy-succinimidyl carbamate. Dang *et al.* [53] analyzed histamine in red and white wine without sample extraction and/or purification as red wine was diluted 1:5 with mobile phase meanwhile the white wine sample was injected directly for LC-MS analysis. Also Cai *et al.* [54] detected only histamine in beer by directly injected samples in LC-MS. Lee *et al.* [55] developed a fast and robust method for the extraction of 9 BAs in Korean turbid rice wine. They performed a deproteinization step with sulfosalicylic acid, then samples were derivatized with danshyl chloride in alkaline condition.

These traditional extraction techniques such as liquid-liquid extraction (LLE) are still applied in food samples treatment but they displayed some disadvantages as they need a conspicuous amount of solvents, often hazardous (such as toluene, chloroform, dichloromethane, ethyl acetate or *iso*-hexane) and they are time consuming procedures. For these reasons, new LLE-based techniques have evolved by using liquid-phase micro-extractions (LPME) [56, 57]. Aflaki *et al.* [56] extracted nine BAs from non-alcoholic beers: the sample was saturated by solid sodium chloride, adjusted to pH=12.0 using sodium hydroxide and extracted with n-butanol–chloroform (1:1, v/v). Afterwards, in another study [57] the same authors performed the benzoyl chloride derivatization of BAs with a low amount of reagents and they perform sample preparation in significantly less time than other procedures. In fact, they added 0.5 g polyvinylpyrrolidone (PVP) to 15 mL of food sample in a beaker; the mixtures were stirred for 10 min on a magnetic stirrer and were then centrifuged before derivatization and analysis. Other BA extraction techniques from beverages are based on the LLE, such as salting-out assisted liquid-liquid extraction (SALLE) [58, 59]. Ramos *et al.* [58] developed a novel method for the determination of seven BAs in wines, which includes salting-out assisted liquid-liquid extraction, derivatization with dansyl chloride followed by analysis by using HPLC with fluorimetric detection. In their procedures, 2 mL of a wine sample were mixed with 1 mL of acetonitrile and 1 mL of dansylchloride and to ensure a correct derivatization step, wine samples were previously mixed with phosphate buffer before HPLC analysis. Instead, Jain *et al.* [59] proposed a novel methodology for the quantification of biogenic amines in fruit juices and alcoholic beverages involving reaction of BAs with 1-naphthylisothiocyanate, extraction of 1-naphthylthiourea derivatives with acetonitrile and when solvents phase separation occurred using ammonium sulfate, a process called salting-out assisted liquid-liquid extraction. Other BA extraction techniques from beverages based on the LLE include dispersive liquid-liquid microextraction based on solidification of floating organic droplets (DLLME-SFO) [60], and vortex assisted surfactant-enhanced emulsification liquid-liquid microextraction (VSLLME) [61]. In the first study, Jia *et al.* [60] analyzed nine BAs in wines. The analytes were firstly derivatized in situ with benzoyl chloride and then extracted by DLLME-SFO, before LC–UV analysis. Donthuan *et al.* [61] developed an easy, fast and green method based on modified dispersive liquid–liquid microextraction and HPLC analysis for the simultaneous quantification of five BAs in fermented food. BAs were derivatized with 9-fluorenylmethyl chloroformate, extracted by vortex-assisted surfactant-enhanced emulsification liquid-liquid microextraction (acetonitrile), and then analyzed by HPLC.

As reported above, amine extraction from liquid samples has been commonly carried out by using liquid-liquid extraction despite the disadvantages associated such as matrix interference, emulsion formation, large volumes of hazardous solvents and time. In addition, solid matrix is critical and the main issue is to remove interferences from the sample matrix, improving the detection limits, and enhancing the separation and detection performance. For this reason, both solid and liquid samples extracted can be further purified by solid phase extraction (SPE) to retain either the analytes or the interferences [62]. For instance, amines can be fixed efficiently by anion exchange as they display positive charge at acid pH values. On the other side, reversed-phase C18, polymeric, and anion-exchange cartridges can be used for sample cleanup to retain and, thus, remove interferences from polyphenols, acids, and other organic compounds. Sagratini *et al.* [23] developed a method based on LC-MS/MS using a triple quadrupole (QqQ) analyzer for the analysis of eight biogenic amines in fish tissues. After acidic extraction, solid phase extraction clean-up was performed with STRATA X cartridge. The cartridge was conditioned with methanol followed by water, then sample with a pH adjusted to 11 with NH_4OH 28% loaded on the cartridges. Finally, analytes were eluted from the STRATA X sorbents with a mixture of methanol/acetic acid (99:1, v/v). Sirocchi *et al.* [10] quantified eight biogenic amines in meat wrapped in a new active packaging system. In this case, the aqueous residue was purified on a Strata C18-E cartridge. The cartridge was activated with acetonitrile, conditioned with water, then the aqueous sample residue was loaded onto the cartridge. In this case, elution was performed using acetonitrile. Jia *et al.* [63] developed a method for the simultaneous quantification of seven biogenic amines in Korean wines. The BAs were pre-column derivatized with dansyl chloride and then extracted by a solid phase extraction (SPE) cartridge They used Hydrophilic-Lipophilic-Balanced (HLB) SPE cartridges that were activated with methanol followed by pure water. Then, 1.6 mL of the derivatized solution was loaded onto the cartridge. The dansylated amines were finally eluted with 1 mL of acetonitrile. As mentioned above, strong cationic exchange (SCX) cartridges can be used for the extraction and purification of aliphatic and BAs. Busto *et al.* [64] described a method for the quantification of the most important BAs in wines and solid phase extraction (SPE) with SCX cartridges was used prior to derivatization to improve the selectivity and the sensitivity of the method. The cartridges were activated with methanol and 0.1 mM HCI. Then, sample was passed through the conditioned cartridge and amines finally eluted with two fractions of 75 mM borate in MeOH 50%. Jurado-Sanchez *et al.* [65] set-up a continuous SPE procedure for the pre-concentration of various groups of amines (*i.e.*, aliphatic amines, anilines, chloroanilines and N-nitrosamines) from water samples. For this reason, they compared several sorbent materials such as conventional (LiChrolut EN, Oasis HLB, RP-C18 and graphitized carbon black), and non-conventional sorbents (C60 and C70 fullerenes, and nanotubes) in order to choose the best one in terms of universality, sensitivity and precision for amine retention. In this sense, the LiChrolut EN column shows the highest enrichment factors as it has a breakthrough volume up to 300 mL compared to those of Oasis HLB, which only tolerate volumes up to 180 mL and as LiChrolut EN displayed higher recovery it was used in the application of the proposed method.

After the purification step, biogenic amines are commonly analyzed by using HPLC-DAD, HPLC-FLD and HPLC-MS. Due to their low volatility and lack of chromophores, biogenic amines are mostly analyzed by LC with pre and post column derivatization and UV-visible or fluorescence detection [66]. The most common reagents used for the preparation of fluorescent derivatives are o-phthalaldehyde (OPA) [67, 68], benzoyl chloride [69], dansyl chloride [10, 42, 70] and 6-aminoquinolyl-N-hydrosysuccinimidyl carbamate (AQC) [71]. Other newly used derivatives are 3-(4-chlorobenzoyl)-quinoline-2-carboxaldehyde [72] and succinimidylferrocenyl propionate [73].

OPA has the disadvantage that it reacts only with primary amines. On the other side, dansyl chloride reacts both with primary and secondary amines, leading to more stable products with respect to those formed with OPA [74]. Liu *et al.* [75] quantified six BAs in sausage and cheese by using off-line derivatization with dansyl chloride, 9-fluorenylmethoxycarbonyl chloride, benzoyl chloride and dansyl chloride, before HPLC-UV detection. The results showed that both the off-line 9-fluorenylmethoxycarbonyl and dansyl derivatization were not suitable for HPLC analysis of BAs when batch injection was used because the derivatives were instable, whereas both the off-line dansyl and benzoyl derivatization were suitable for HPLC analysis of BAs when batch injection was used, but the latter needed to maintain the derivatives at 4 °C to ensure that benzoylated tyramine was not degraded when waiting for the analysis [75]. However, derivation reactions are time consuming and they might lead to low recovery, analyte loss and contamination; nowadays LC coupled to MS offers a great potential because it has the advantage that a less consuming sample preparation is needed, at the same time avoiding the derivatization process.

ANALYTICAL TECHNIQUES IN BAS ANALYSIS

The analytical determination of BAs in different matrices is very relevance in different fields like food safety, process control in food industry and diagnosis of diseases and food poisonings (episodes of human intoxication). Since the first description by Antoni van Leeuwenhoek in 1678 of crystals that were much later identified as polyamine salts, a lot of analytical methods have been developed; the most used techniques are the High Performance Liquid Chromatography (HPLC), the Gas Chromatography (GC) and Capillary Electrophoresis (CE). However, other alternative methods have been developed like biosensors and ELISA test.

Table 1. Different extraction methods used for the analysis of BAs.

Matrix	BAs extracted	Extraction solvent	Extraction method	Purification	Derivatization	Ref.
Meat	spermidine, spermine, cadaverine, putrescine, histamine, tyramine, tryptamine, phenylethylamine	TCA 5%	Solid/liquid extraction with Ultraturrax	Yes with SPE C18	Dansyl chloride	[10]
Fish	spermidine, spermine, cadaverine, putrescine, histamine, tyramine, tryptamine, phenylethylamine	TCA 5%	Solid/liquid extraction with Ultraturrax	Yes with SPE STRATA X	Dansyl chloride for HPLC-FLD analyses No for HPLC-MS/MS analyses	[23]
Shrimp sauce and tomato ketchup	spermidine, putrescine, cadaverine, 2-phenylethylamine, histamine, tryptamine, tyramine	Dihexyl ether	Hollow fiber based liquid-phase microextraction (LPME)	No	Dansyl chloride	[25]
Sausages	phenylethylamine, histamine, serotonin, cadaverine, spermine, spermidine, putrescine, tryptamine, tyramine	Perchloric acid 0.4 M	Solid/liquid extraction with Ultraturrax	No	Dansyl chloride	[40]
Kiwi, cheese, fish and dry fermented sausages	putrescine, cadaverine, tyrosine, tyramine, histamine, histidine, spermidine	Perchloric acid 0.375 M	Solid/liquid extraction with homogenizer	Defattening with hexane	No	[41]
Cheese	tryptamine, phenylethylamine, putrescine, cadaverine, histamine, tyramine, spermine, spermidine, proline	Hydrochloric acid 0.1 M	Solid/liquid extraction with homogenizer	No	Dansyl chloride	[42]
Cheese, meat and fish	tryptamine, phenylethylamine, putrescine, cadaverine, histamine, tyramine, spermine, spermidine	HCl 0.1 M for cheese TCA 5% for fish and meat	Solid/liquid extraction with homogenizer	No	Dansyl chloride	[43]

(Table 1) cont.....

Matrix	BAs extracted	Extraction solvent	Extraction method	Purification	Derivatization	Ref.
Fish, Cheese, Meat products, Vegetarian products	histamine, tyramine, spermidine, spermine, cadaverine, putrescine, agmantine	HCl 0.1 M, TCA 5%, TCA 10%, methanol 50%, methanol 75%, methanol 100%.	Solid/liquid extraction with homogenizer	No clean-up for capillary electrophoresis measurement Yes with SPE C18 or either liquid liquid extraction (LLE) with butanol-heptane for HPLC analyses	No for capillary electrophoresis measurement with o-phthaldialdehyde (OPA) for HPLC	[44]
Meat and fish	histamine, tyramine, cadaverine, putrescine, spermidine, spermine,	Methanol:acetonitrile:water 45:45:10 acidified with 0.1% acetic acid	Solid/liquid extraction with homogenizer	No	No	[45]
Wine	ethanolamine, methylamine, histamine, ethylamine, pyrrolidine, isopropylamine, propylamine, tyramine, isobutylamine, putrescine, cadaverine, phenethylamine, 3-methylbutylamine, amylamine and hexylamine	Acetonitrile	Liquid/liquid extraction	No	6-aminoquinolyl-N-hydroxy-succinimidyl carbamate	[52]
Wine	histamine	No	Direct injection	No	No	[53]
Beer	histamine	No	Direct injection	No	No	[54]
Rice wine	agmatine, phenylethylamine, tyramine, putrescine, cadaverine, histamine, tryptamine, spermidine, spermine	No	No	Deproteinization with sulfosalicylic acid	Dansyl chloride	[55]
Non-alcoholic beers	histamine, tyramine, cadaverine, putrescine, spermidine, spermine, tryptamine, agmantine, phenylethylamine,	n-butanol–chloroform (1:1, v/v) at pH 12 using sodium hydroxide	Liquid-phase micro-extractions (LPME)	No	Benzoyl chloride	[56]
Non-alcoholic beers	histamine, tyramine, cadaverine, putrescine, spermidine, spermine, tryptamine, agmantine, phenylethylamine,	Treatment with polyvinylpyrrolidone (PVP) to remove polyphenols and directly derivatized	Liquid-phase micro-extractions (LPME)	No	Benzoyl chloride	[57]

(Table 1) cont.....

Matrix	BAs extracted	Extraction solvent	Extraction method	Purification	Derivatization	Ref.
Wine	histamine, putrescine, cadaverine, phenylethylamine, methylamine, dimethylamine, iso-pentylamine, spermidine, ethylamine	Acetonitrile	Salting-out assisted liquid-liquid extraction (SALLE)	No	Dansyl chloride	[58]
Fruit juices and beers	tyramine, putrescine, tryptamine, 2-phenylethylamine, spermidine, methyl amine, cysteamine	Acetonitrile	Salting-out assisted liquid-liquid extraction (SALLE)	No	1-naphthylisothiocyanate	[59]
Wine	putrescine, cadaverine, 1,3-diaminopropane, tryptamine, phenylethylamine, spermidine, spermine, histamine, tyramine	Acetonitrile	Dispersive liquid-liquid microextraction based on solidification of floating organic droplets (DLLME-SFO)	No	Benzoyl chloride	[60]
Wine, beer and fermented fish	histamine, tyramine, cadaverine, tryptamine, spermidine	Acetonitrile	Vortex assisted surfactant-enhanced emulsification liquid-liquid microextraction (VSLLME)	No	9-fluorenylmethyl chloroformate	[61]
Wine	tyramine, tryptamine, putrescine, 2-phenylethylamine, cadaverine, histamine, spermidine	No	No	Yes with HLB SPE	Dansyl chloride	[63]
Wine	Methylamine, histamine, ethylamine, tyramine, putrescine, cadaverine, p-phenethylamine, 3-methylbutylamine	No	No	Yes with SCX SPE	6-aminoquinolyln-hydroxy-succinimidyl carbamate	[64]
Water	Various groups of amines (*i.e.*, aliphatic amines, anilines, chloroanilines and N-nitrosamines)	No	No	Yes/ LiChrolut EN, Oasis HLB, RP-C18 and graphitized carbon black	No	[65]
Meat	spermidine, spermine, cadaverine, putrescine, histamine, tyramine, tryptamine, phenylethylamine, serotonin, agmantine	TCA 5%	Solid/liquid extraction with Ultraturrax	No	No	[70]

Liquid Chromatography (LC)

As previously said, the BAs are very polar molecules with low molecular mass classifiable in three different types: aromatic, (histamine, tryptamine, tyramine, 2-phenylethylamine), aliphatic diamines, (cadaverine and putrescine), and aliphatic polyamines (spermine and spermidine). For these characteristics, their determination is not simple both in terms of chromatographic separation and in terms of detection. The need to determine BAs even at low concentration and in very complex matrices, such as food, leads to derivatization of the BAs. In fact, the derivatives have a minor polarity that improves the chromatographic separation by reversed phase techniques and an excellent absorption property in the ultraviolet, the visible and the fluorescence wavelength ranges [71]. The methods using fluorogenic or chromogenic dyes are nowadays the most important and used methods (more than 80%), thanks to their good selectivity and excellent sensitivity. The most popular probes suitable to react with the amino group is the dansyl chloride (reacting with both primary and secondary amines) but also other derivatization agents are used. The derivatization process can be done in pre-column or in post-column reactions (Fig. **3**).

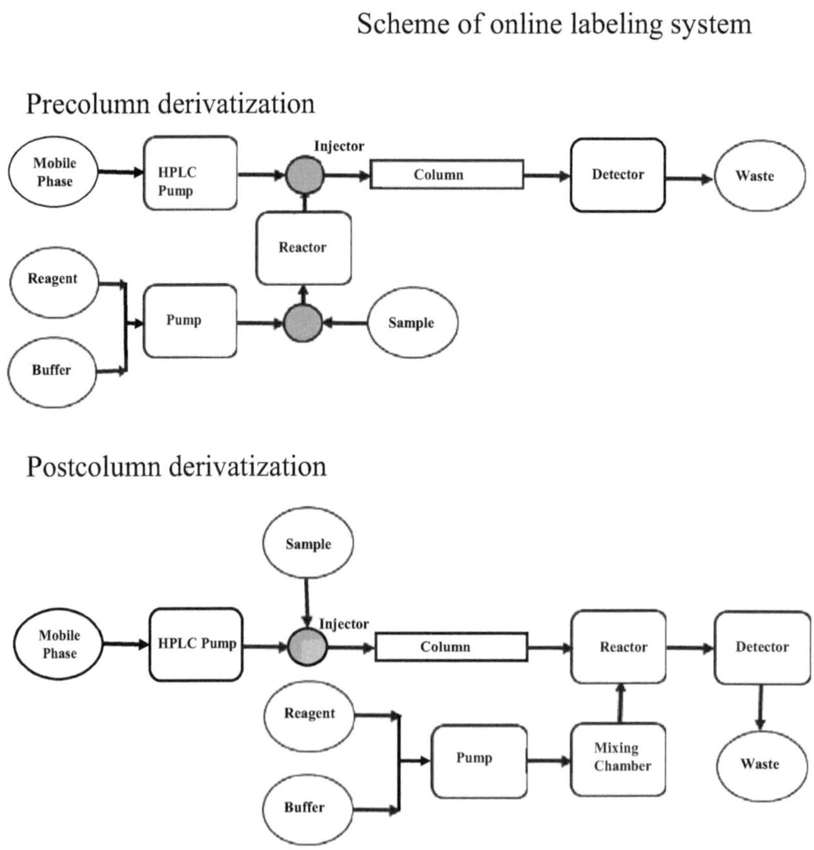

Fig. (3). Scheme of online labeling system: **(a)** precolumn and **(b)** postcolumn derivatization.

Normally, pre-column labeling is developed in off-line mode in glass vials heating to mild temperature to accelerate the reaction. About the chromatographic column, the most common methods employ C18 phases, but also different reversed phases like Kinetex core-shell PFP [76] and Synergy Hydro by Phenomenex [70] or others octadecylsilane stationary phases functionalized for the separation of polar compounds. These phases are particularly useful for analysis of underivatized BAs in chromatographic methods hyphenated with MS techniques. The evolution of chromatography systems has led recently at a new technique called UHPLC or UPLC (Ultra High Performance Liquid Chromatography): column efficiency increases with reduction in particle size of stationary phase, so a reduction of the particles from 3-5 μm, used in conventional analytical column, to the sub-2 μm in UHPLC involves an improvement of peak resolution. Furthermore, the UHPLC offers an increase of sensitivity and analysis speed and a reduction of analysis time (from 20-30 minutes to 2-10 minutes) and solvent flow (from 1-2 mL min^{-1} to 250-750 μL min^{-1}). UHPLC works at higher pressure than HPLC (15,000-18,000 psi vs 5000-6000 psi) and all parts of the system are able to withstand such conditions; on the other hand, UHPLC technologies and their materials are more complex but, at same time, more expensive. Spectrophotometric and fluorescence detectors are extensively used for detection of derivatives of BAs both in research area and quality control applications; as earlier said, pre-column derivatization with dansyl chloride is used because gives an excellent sensitivity as evidenced from Preti *et al.* They achieved LOD values in the range from 2 μg L^{-1} (PUT) to 23 μg L^{-1} (AGM) using fluorescence detection with UHPLC in wine and fruit nectar samples [77]. Others researchers obtained a minor, but always good, sensitivity by using UV-Vis spectrophotometric detection, *i.e.*, from 13 μg L^{-1} to 112 μg L^{-1} [78, 79]. Furthermore, both detectors are stable, robust and easy to use and widely diffused in analytical laboratories, so they represent the most important analysis systems. An alternative is the evaporative light scattering detector (ELSD) which is known as universal detector as its detection sensitivity is independent of the compound's spectral properties and based upon the absolute quantity of the compound. ELSD has been applied for underivatized BAs determination in cocoa-based products with LOD values in the range from 10 to 20 μg L^{-1} [80]. Some drawbacks are associated with the derivatization such as analyte loss, possibility of contamination and time-consuming analysis. HPLC and UHPLC coupled with MS techniques are suitable to detect underivatized amines although normally the methods including a derivatization step in order to improve the peak resolution and sensitivity. The polar nature both of the underivatized and derivatized BAs allows the use of atmospheric pressure chemical ionization (APCI) and electrospray ionization (ESI) ion sources though the last one is the most used. Regarding the mass analyzers, the triple quadrupole (QqQ) and ion trap (ITD) give a good performance. For instance, Sirocchi *et al.* [70] detected ten BAs in meat samples by HPLC-MS/MS working in multiple reaction monitoring (MRM) (Fig. **4**) both in ITD and QqQ mass analyzers. In terms of analytical performance, the latter was the most sensitive and reproducible; the range of the correlation coefficients of the calibration curves of the analyzed compounds was 0.987-0.999, and the limits of detection were in the range of 2-100 μg L^{-1}.

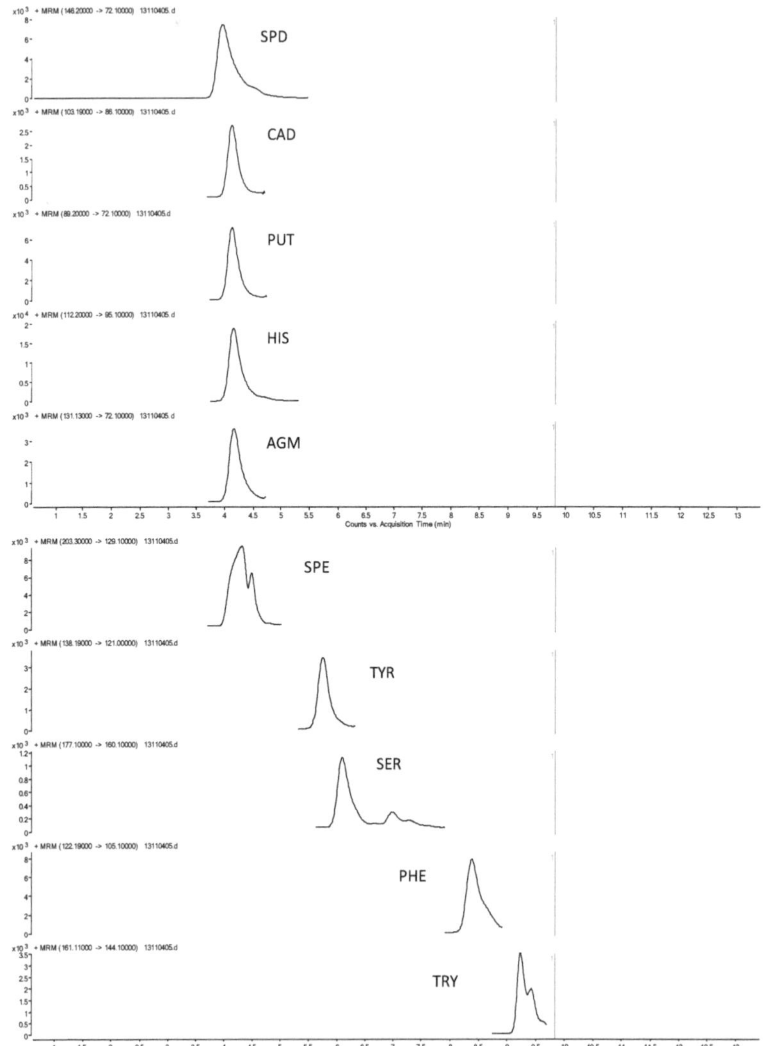

Fig. (4). HPLC-MS/MS chromatogram (QqQ) acquired in MRM mode of a spiked meat sample at concentration of 0.25 mg kg^{-1}. Single BAs are reported in the figure. Spermidine (SPD), cadaverine (CAD), putrescine (PUT), histamine (HIS), agmatine (AGM), spermine (SPE), tyrosine (TYR), serotonine (SER), phenylethylamine (PHE), tryptamine (TRY).

Another way to analyze the BAs in food matrices based on the use of a stable isotope-coded derivatization and LC-MS/MS analysis was developed by Mazzotti *et al.* [81]. N-Hydroxysuccinimidyl ester of d0/d4-4-methoxybenzoic acid, which mainly functionalized primary amines, was utilized to identify and characterize molecular structure of BAs by MS/MS in various foodstuffs. The derivatives obtained provided selective cleavage at the binding site between the reagent and the amine in ESI-MS/MS experiments and produced a characteristic fragment derived from the reagent moiety, allowing the simultaneous determination of labeled amines under MS/MS conditions with LOD values in the range from 10 to 20 μg L^{-1}. Besides, Lee *et al.* [82] used LC-ESI-MS with the Linear Ion trap (QTrap instrument) for MS/MS determination of dansyl derivatives of BAs in cheonggukjan (food made from fermented soybeans) with LOD values in the range from 10 to 40 μg L^{-1}. Nowadays, the high resolution mass spectrometers (HRMS) are emerging as one of the most powerful instruments for identification and confirmation of the food contaminants. It is possible to divide the HRMS instruments in four types: magnetic sector, quadrupole-time of flight (q-TOF), Orbitrap and Fourier transform ion cyclotron resonance (FTICR). The most frequently used with LC methods are q-TOF and Orbitrap analyzers and many methods have already been developed with these two important analytical techniques. Jia *et al.* [60] obtained LOD from 3 μg L^{-1} to 15 μg L^{-1} in bokbunja wines with q-TOF technologies while Self *et al.* [83] obtained LOD from 20 μg L^{-1} to 250 μg L^{-1} in tuna samples. The high resolution, together with the accuracy, lead to identification of a considerable variety of contaminants in food, not only the targeted ones but also untargeted (unknown) contaminants. In conclusion, both spectrophotometric or fluorimetric detectors and mass spectrometer detectors demonstrate to be effective for the determination of BAs in food samples. However, MS detectors, although more expensive, have some advantages such as the possibility to work on bare amines avoiding the derivatization step and to have more specificity besides having the possibility to do untargeted analysis.

Gas Chromatography (GC)

This technique is not as normally employed as LC techniques for determination of BAs in food samples but some studies are reported on fermented beverages offering an alternative to LC methods. To increase the volatility of the analytes a derivatization is necessary. Different type of derivatizing reagents can be applied but the most popular are alkylated chloroformates. The reaction occurs at room temperature, in a short time and in aqueous basic solution followed by extraction procedure. Recently, some works used isobutyl chloroformate derivatives coupled with SPME [84] or DLLME on fish or alcoholic beverage samples [33, 85]. The GC, especially if combined with MS, represent a valid technique for BAs analysis giving a good reproducibility and LODs of the same order of LC techniques.

Capillary Electrophoresis (CE)

CE represents a family of performing techniques to analyze the BAs in the food samples despite its sensitivity is lower than other techniques as those previously shown. Reliability, rapidity, low reagent consumption and easy for use making it a very useful analytical way to determine the BAs. It possible classify CE in the following types: Capillary zone electrophoresis (CZE), capillary gel electrophoresis (CGE), capillary isoelectric focusing (CIEF), capillary isotachophoresis (CITP) and micellar electrokinetic chromatography (MEKC). Detection methods as fluorescence, absorbance, electrochemical and MS can be coupled with CE techniques. To improve the sensitivity, some analytical strategies are used. For example, off-line pre-column derivatization (o-phthalaldehyde, dansylchloride, dabsylchloride, β-naphthol, Fluorescein isothiocyanate isomer) is the most common application. Furthermore, some innovative on-line pre-concentration preceded by a step involving clean-up procedures (addition of polyvinylpyrrolidone followed by filtration) have contributed to enhance the sensitivity of methods. An application of this analytical approach in wine and beer samples, using an MS/MS technique, reported values of LOD in the range of 1-2 μg L^{-1} [86]. Ginterova *et al.* [87] developed a method without derivatization by using on-line sample pre-concentration with an automated on-line combination of capillary isotachophoresis-capillary zone electrophoresis (cITP-CZE) equipped with photometric detection at 280 nm. This method has been used with success to analyze the histamine, 2-phenylethylamine and tyramine in wine samples with the following LOD values: were 0.35 mg L^{-1} for histamine, 0.33 mg L^{-1} for 2-phenylethylamine and 0.37 mg L^{-1} for tyramine. An isotachophoresis (CITP) method was setup by Jastrzebska *et al.* [88] without needing derivatization. They determined the BAs in a 14-minutes run with LODs better than a compared method with HPLC-UV and dansyl derivatives. Finally, a new and fast separation (LODs 0.42-1.26 nM) of seven BAs from wine samples was performed by Uzasci *et al.* [89] using nonionic micellar electrokinetic chromatography (MEKC) method through fluorescein isothiocyanate isomer (FITC) coupled to laser-induced fluorescence detector (LIF).

Others Methodologies

LC, GC and CE are among the most important instrumental methods for precise quantitative analysis of BAs, but these methods are time-consuming and they require considerable skill. Furthermore, these instrumentations have high costs of purchase and maintenance. Biosensors are based on biological recognition through enzyme specificity and they are a good and cheap alternative to the traditional methods of analytes determination. These methods can be applied for routine analysis with semi-quantitative and quantitative determination of biogenic amines. They do not require a special instrumentation, and do not need sample clean-up and derivatization procedures. Besides, an important advantage of this technique is that it may be used for in situ analysis [74]. Most biosensors use amine oxidases (AO) which catalyze the oxidative deamination of primary amines, diamines and substituted amines to produce aldehyde, ammonia and hydrogen peroxide [90]. Normally, biosensors are able to detect just an amine or an overall total amount of BAs. Exceptionally, a dual enzymatic sensor was developed to simultaneously determine histamine and putrescine [91]. Alternatively, a sensor with molecularly imprinted nanoparticles with a high affinity for histamine has developed by Bazosabal *et al.* [92]. Regarding transduction systems, electrochemical (potentiometric or amperometric) and optical sensors have been employed. Furthermore, Ramon-Marquez *et al.* have developed a novel optical sensor using nanotechnology. This sensor detects tryptamine by using the intrinsic phosphorescence property and its application on beer samples reaches LOD of 6 $\mu g\ L^{-1}$ [93]. Finally, another way for the determination of BAs is ELISA (Enzyme-Linked Immuno Sorbent Assay) test. It has good precision and sensitivity but it can be time-consuming and expensive if applied to a small number of samples [94]. Mainly it is applied for the determination of histamine in fish but there are also some applications in wine matrices [95]. Recently, a sensitive method has been developed by Luo *et al.* [96] for wine matrices. The LOD for histamine is similar to those in LC methods (93.6 $\mu g\ L^{-1}$).

CONCLUDING REMARKS

BAs represent an important tool of food analysis due to the role that they have both as chemical markers for establishing the quality of food and for the importance that they possess from a toxicological point of view. Many foods, especially fermented, can contain high level of these compounds and actually there are no regulations that establish their maximum limits except for histamine. Methodologies that use chromatographic techniques as HPLC or GC coupled to various detectors as MS have been developed, but new rapid and sensitive analytical approaches are needed for investigating the level of BAs and hence the safety of food.

CONSENT FOR PUBLICATION

Not applicable.

CONFLICT OF INTEREST

The authors confirm that this chapter contents have no conflict of interest.

ACKNOWLEDGEMENTS

Declared none.

REFERENCES

[1] Ruiz-Capillas C, Jiménez-Colmenero F. Biogenic amines in meat and meat products. Crit Rev Food Sci Nutr 2004; 44(7-8): 489-99. [http://dx.doi.org/10.1080/10408690490489341] [PMID: 15969322]

[2] Smith TA. Amines in food. Food Chem 1980; 6: 169-200. [http://dx.doi.org/10.1016/0308-8146(81)90008-X]

[3] Sentellas S, Núñez Ó, Saurina J. Recent advances in the determination of biogenic amines in food samples by (U)HPLC. J Agric Food Chem 2016; 64(11): 7667-78. [http://dx.doi.org/10.1021/acs.jafc.6b02789] [PMID: 27689967]

[4] Karovicova J, Kohaidova Z. Biogenic amines in food. Chem Pap 2005; 59: 70-9.

[5] Bardòcz S. 30 Polyamines in food and their consequences for food quality and human health. Trends Food Sci Technol 1995; 6: 341-6. [http://dx.doi.org/10.1016/S0924-2244(00)89169-4]

[6] Stratton JE, Hutkins RW, Taylor SL. Biogenic amines in cheese and other fermented food. A review. J Food Prot 1991; 54(6): 460-70. [http://dx.doi.org/10.4315/0362-028X-54.6.460] [PMID: 31051616]

[7] Rice SL, Eitenmiller RR, Koehler PE. Biologically active amines in food: a review. J Milk Food Technol 1976; 39: 353-8. [http://dx.doi.org/10.4315/0022-2747-39.5.353]

[8] Karovičá J, Kohajdová Z. Biogenic amines in food. Chem Pap 2005; 59: 70-9.

[9] Giurfa M. Associative learning: the instructive function of biogenic amines. Curr Biol 2006; 16(20): R892-5. [http://dx.doi.org/10.1016/j.cub.2006.09.021] [PMID: 17055975]

[10] Sirocchi V, Caprioli G, Cecchini C, *et al.* Biogenic amines as freshness index of meat wrapped in a new active packaging system formulated with essential oils of Rosmarinus officinalis. Int J Food Sci Nutr 2013; 64(8): 921-8. [http://dx.doi.org/10.3109/09637486.2013.809706] [PMID: 23815565]

[11] Silva CMG, Gloria MBA. Bioactive amines in chicken breast and thigh after slaughter and during storage at 4±1 °C and in chicken based meat products. Food Chem 2002; 78: 241-8. [http://dx.doi.org/10.1016/S0308-8146(01)00404-6]

[12] Bedia-Erim F. Recent analytical approaches to the analysis of biogenic amines in food samples. Trends Analyt Chem 2013; 52: 239-47. [http://dx.doi.org/10.1016/j.trac.2013.05.018]

[13] EFSA Panel on Biological Hazards (BIOHAZ). Scientific opinion on risk based control of biogenic amine formation in fermented foods. EFSA J 2011; 9: 93.

[14] Bauza T, Blaise A, Daumas F, Cabanis JC. Determination of biogenic amines and their precursor amino acids in wines of the Valle du Rhone by highperformance liquid chromatography with precolumn derivatization and fluorimetric detection. J Chromatogr A 1995; 707: 373-9. [http://dx.doi.org/10.1016/0021-9673(95)00318-H]

[15] Landete JM, Ferrer S, Polo L, Pardo I. Biogenic amines in wines from three Spanish regions. J Agric Food Chem 2005; 53(4): 1119-24. [http://dx.doi.org/10.1021/jf049340k] [PMID: 15713028]

[16] Papageorgiou M, Lambropoulou D, Morrison C, Kłodzińska E, Namieśnik J, Płotka-Wasylka J. Literature update of analytical methods for biogenic amines determination in food and beverages. Trends Analyt Chem 2018; 98: 128-42.
[http://dx.doi.org/10.1016/j.trac.2017.11.001]

[17] Vinci G, Antonelli ML. Biogenic amines. Quality index of freshness in red and white meat. Food Control 2002; 13: 519-24.
[http://dx.doi.org/10.1016/S0956-7135(02)00031-2]

[18] Galgano F, Favati F, Bonadio M, Lorusso V, Romano P. Role of biogenic amines as index freshness in beef meat packed with different biopolymeric materials. Food Res Int 2009; 42: 1147-52.
[http://dx.doi.org/10.1016/j.foodres.2009.05.012]

[19] Lazaro CA, Conte-Júnior CA, Canto AC, *et al.* Biogenic amines as bacterial quality indicators in different poultry meat species. Lebensm Wiss Technol 2015; 60: 15-21.
[http://dx.doi.org/10.1016/j.lwt.2014.09.025]

[20] Papavergou EJ, Savvaidis IN, Ambrosiadis IA. Levels of biogenic amines in retail market fermented meat products. Food Chem 2012; 135(4): 2750-5.
[http://dx.doi.org/10.1016/j.foodchem.2012.07.049] [PMID: 22980868]

[21] Al Bulushi I, Poole S, Deeth HC, Dykes GA. Biogenic amines in fish: roles in intoxication, spoilage, and nitrosamine formation--a review. Crit Rev Food Sci Nutr 2009; 49(4): 369-77.
[http://dx.doi.org/10.1080/10408390802067514] [PMID: 19234946]

[22] Biji KB, Ravishankar CN, Venkateswarlu R, Mohan CO, Gopal TKS. Biogenic amines in seafood: a review. J Food Sci Technol 2016; 53(5): 2210-8.
[http://dx.doi.org/10.1007/s13197-016-2224-x] [PMID: 27407186]

[23] Sagratini G, Fernández-Franzón M, De Berardinis F, Font G, Vittori S, Mañes J. Simultaneous determination of eight underivatised biogenic amines in fish by solid phase extraction and liquid chromatography-tandem mass spectrometry. Food Chem 2012; 132(1): 537-43.
[http://dx.doi.org/10.1016/j.foodchem.2011.10.054] [PMID: 26434328]

[24] Bilgin B, Gençcelep H. Determination of biogenic amines in fish products. Food Sci Biotechnol 2015; 24: 1907-13.
[http://dx.doi.org/10.1007/s10068-015-0251-4]

[25] Saaid M, Saad B, Ali AS, Saleh MI, Basheer C, Lee HK. In situ derivatization hollow fibre liquid-phase microextraction for the determination of biogenic amines in food samples. J Chromatogr A 2009; 1216(27): 5165-70.
[http://dx.doi.org/10.1016/j.chroma.2009.04.091] [PMID: 19481215]

[26] Combarros-Fuertes P, Fernández D, Arenas R, Diezhandino I, Tornadijo ME, Fresno JM. Biogenic amines in Zamorano cheese: factors involved in their accumulation. J Sci Food Agric 2016; 96(1): 295-305.
[http://dx.doi.org/10.1002/jsfa.7093] [PMID: 25641417]

[27] Samková E, Dadáková E, Pelikánová T. Changes in biogenic amine and polyamine contents in smear-ripened cheeses during storage. Eur Food Res Technol 2013; 237: 309-14.
[http://dx.doi.org/10.1007/s00217-013-1993-y]

[28] Mayer HK, Fiechter G, Fischer E. A new ultra-pressure liquid chromatography method for the determination of biogenic amines in cheese. J Chromatogr A 2010; 1217(19): 3251-7.
[http://dx.doi.org/10.1016/j.chroma.2009.09.027] [PMID: 19781707]

[29] Valsamaki K, Michaelidou A, Polychroniadou A. Biogenic amine production in Feta cheese. Food Chem 2000; 71: 259-66.
[http://dx.doi.org/10.1016/S0308-8146(00)00168-0]

[30] Mascaro N, Stocchi R, Ricciutelli M, *et al.* Biogenic amine content and chemical and physical features of Italian formaggio di Fossa. Ital J Food Saf 2010; 1: 49-53.

[http://dx.doi.org/10.4081/ijfs.2010.8.49]

[31] Restuccia D, Spizzirri UG, Puoci F, *et al.* A new method for the determination of biogenic amines in cheese by LC with evaporative light scattering detector. Talanta 2011; 85(1): 363-9. [http://dx.doi.org/10.1016/j.talanta.2011.03.080] [PMID: 21645711]

[32] Andiç S, Gençcelep H, Tunçtürk Y, Köse S. The effect of storage temperatures and packaging methods on properties of Motal cheese. J Dairy Sci 2010; 93(3): 849-59. [http://dx.doi.org/10.3168/jds.2009-2413] [PMID: 20172204]

[33] Almeida C, Fernandes JO, Cunha SC. A novel dispersive liquid-liquid microextraction (DLLME) gas chromatography-mass spectrometry (GC-MS) method for the determination of eighteen biogenic amines in beer. Food Control 2012; 25: 380-8. [http://dx.doi.org/10.1016/j.foodcont.2011.10.052]

[34] Cardona-Galvez M, Gonzalez-Domínguez A. Intoxicacion histamínica o escombriosis en pescados. Alimentaria 2005; 365: 68-73.

[35] Pradenas J, Galarce-Bustos O, Henríquez-Aedo K, Mundaca-Uribe R, Aranda M. Occurrence of biogenic amines in beers from Chilean market. Food Control 2016; 70: 138-44. [http://dx.doi.org/10.1016/j.foodcont.2016.05.043]

[36] Cunha SC, Lopes R, Fernandes JO. Biogenic amines in liqueurs: Influence of processing and composition. J Food Compos Anal 2017; 56: 147-55. [http://dx.doi.org/10.1016/j.jfca.2016.11.016]

[37] Soufleros E, Barrios ML, Bertrand A. Correlation between the content of biogenic amines and other wine compounds. Am J Enol Vitic 1998; 49: 266-78.

[38] Lonvaud-Funel A. Biogenic amines in wines: role of lactic acid bacteria. FEMS Microbiol Lett 2001; 199(1): 9-13. [http://dx.doi.org/10.1111/j.1574-6968.2001.tb10643.x] [PMID: 11356560]

[39] Czajkowska-Mysłek A, Leszczyńska J. Risk assessment related to biogenic amines occurrence in ready-to-eat baby foods. Food Chem Toxicol 2017; 105: 82-92. [http://dx.doi.org/10.1016/j.fct.2017.03.061] [PMID: 28366843]

[40] Eerola S, Hinkkanen R, Lindfors E, Hirvi T. Liquid chromatographic determination of biogenic amines in dry sausages. J AOAC Int 1993; 76(3): 575-7. [PMID: 8318851]

[41] Draisci R, Giannetti L, Boria P, Lucentini L, Palleschi L, Cavalli S. Improved ion chromatography-integrated pulsed amperometric detection method for the evaluation of biogenic amines in food of vegetable or animal origin and in fermented foods. J Chromatogr A 1998; 798(1-2): 109-16. [http://dx.doi.org/10.1016/S0021-9673(97)01198-9] [PMID: 9542132]

[42] Innocente N, Biasutti M, Padovese M, Moret S. Determination of biogenic amines in cheese using HPLC technique and direct derivatization of acid extract. Food Chem 2007; 101: 1285-9. [http://dx.doi.org/10.1016/j.foodchem.2005.12.026]

[43] Moret S, Conte LS. High-performance liquid chromatographic evaluation of biogenic amines in foods. An analysis of different methods of sample preparation in relation to food characteristics. J Chromatogr A 1996; 729(1-2): 363-9. [http://dx.doi.org/10.1016/0021-9673(95)00961-2] [PMID: 9004961]

[44] Lange J, Thomas K, Wittmann C. Comparison of a capillary electrophoresis method with high-performance liquid chromatography for the determination of biogenic amines in various food samples. J Chromatogr B Analyt Technol Biomed Life Sci 2002; 779(2): 229-39. [http://dx.doi.org/10.1016/S1570-0232(02)00372-0] [PMID: 12361737]

[45] Molognoni L, Daguer H, de Sá Ploêncio LA, De Dea Lindner J. A multi-purpose tool for food inspection: Simultaneous determination of various classes of preservatives and biogenic amines in meat and fish products by LC-MS. Talanta 2018; 178: 1053-66.

[http://dx.doi.org/10.1016/j.talanta.2017.08.081] [PMID: 29136796]

[46] Garrigós MC, Reche F, Pernías K, Sánchez A, Jiménez A. Determination of some aromatic amines in finger-paints for children's use by supercritical fluid extraction combined with gas chromatography. J Chromatogr A 1998; 819(1-2): 259-66.
[http://dx.doi.org/10.1016/S0021-9673(98)00432-4] [PMID: 9781420]

[47] Theodoridis G, Koster EHM, de Jong GJ. Solid-phase microextraction for the analysis of biological samples. J Chromatogr B Biomed Sci Appl 2000; 745(1): 49-82.
[http://dx.doi.org/10.1016/S0378-4347(00)00203-6] [PMID: 10997704]

[48] Less M, Schmidt TC, von Löw E, Stork G. Gas chromatographic determination of aromatic amines in water samples after solid-phase extraction and derivatization with iodine. II. Enrichment. J Chromatogr A 1998; 810(1-2): 173-82.
[http://dx.doi.org/10.1016/S0021-9673(98)00232-5] [PMID: 9691297]

[49] Wu YC, Huang SD. Solid-phase microextraction coupled with high-performance liquid chromatography for the determination of aromatic amines. Anal Chem 1999; 71: 310-8.
[http://dx.doi.org/10.1021/ac980614a]

[50] DeBruin LS, Josephy PD, Pawliszyn JB. Solid-phase microextraction of monocyclic aromatic amines from biological fluids. Anal Chem 1998; 70(9): 1986-92.
[http://dx.doi.org/10.1021/ac9709413] [PMID: 9599591]

[51] Pan LJ, Chong M, Pawliszyn J. Determination of amines in air and water using derivatization combined with solid-phase microextraction. J Chromatogr A 1997; 773: 249-60.
[http://dx.doi.org/10.1016/S0021-9673(97)00179-9]

[52] Busto O, Guasch J, Borrull F. Determination of biogenic amines in wine after precolumn derivatization with 6-aminoquinolyl-N-hydroxysuccinimidyl carbamate. J Chromatogr A 1996; 737: 205-13.
[http://dx.doi.org/10.1016/0021-9673(96)00022-2]

[53] Dang A, Pesek JJ, Matyska MT. The use of aqueous normal phase chromatography as an analytical tool for food analysis: determination of histamine as a model system. Food Chem 2013; 141(4): 4226-30.
[http://dx.doi.org/10.1016/j.foodchem.2013.06.005] [PMID: 23993609]

[54] Cai J, Li M, Xiong X, Fang X, Xu R. Detection of histamine in beer by nano extractive electrospray ionization mass spectrometry. J Mass Spectrom 2014; 49(1): 9-12.
[http://dx.doi.org/10.1002/jms.3315] [PMID: 24446257]

[55] Lee S, Yoo M, Shin D. The identification and quantification of biogenic amines in Korean turbid rice wine, Makgeolli by HPLC with mass spectrometry detection. Lebensm Wiss Technol 2015; 62: 350-6.
[http://dx.doi.org/10.1016/j.lwt.2015.01.016]

[56] Aflaki F, Ghoulipour V, Saemian N, Sheibani S. Biogenic amine contents in non-alcoholic beers: screening and optimization of derivatization. Food Anal Methods 2014; 7: 713-20.
[http://dx.doi.org/10.1007/s12161-013-9746-x]

[57] Aflaki F, Ghoulipour V, Saemian N, Salahinejad M. A simple method for benzoyl chloride derivatization of biogenic amines for high performance liquid chromatography. Anal Methods 2014; 6: 1482-7.
[http://dx.doi.org/10.1039/c3ay41830h]

[58] Ramos RM, Valente IM, Rodrigues JA. Analysis of biogenic amines in wines by salting-out assisted liquid-liquid extraction and high-performance liquid chromatography with fluorimetric detection. Talanta 2014; 124: 146-51.
[http://dx.doi.org/10.1016/j.talanta.2014.02.026] [PMID: 24767456]

[59] Jain A, Gupta M, Verma KK. Salting-out assisted liquid-liquid extraction for the determination of biogenic amines in fruit juices and alcoholic beverages after derivatization with 1-

naphthylisothiocyanate and high performance liquid chromatography. J Chromatogr A 2015; 1422: 60-72.
[http://dx.doi.org/10.1016/j.chroma.2015.10.036] [PMID: 26518497]

[60] Jia S, Ryu Y, Kwon SW, Lee J. An in situ benzoylation-dispersive liquid-liquid microextraction method based on solidification of floating organic droplets for determination of biogenic amines by liquid chromatography-ultraviolet analysis. J Chromatogr A 2013; 1282: 1-10.
[http://dx.doi.org/10.1016/j.chroma.2013.01.041] [PMID: 23415468]

[61] Donthuan J, Yunchalard S, Srijaranai S. Vortex-assisted surfactant-enhanced-emulsification liquid-liquid microextraction of biogenic amines in fermented foods before their simultaneous analysis by high-performance liquid chromatography. J Sep Sci 2014; 37(21): 3164-73.
[http://dx.doi.org/10.1002/jssc.201400570] [PMID: 25142110]

[62] Molins-Legua C, Campins-Falco P. Solid phase extraction of amines. Anal Chim Acta 2015; 546: 206-20.
[http://dx.doi.org/10.1016/j.aca.2005.05.021]

[63] Jia S, Kang YP, Park JH, Lee J, Kwon SW. Determination of biogenic amines in Bokbunja (Rubus coreanus Miq.) wines using a novel ultra-performance liquid chromatography coupled with quadrupole-time of flight mass spectrometry. Food Chem 2012; 132(3): 1185-90.
[http://dx.doi.org/10.1016/j.foodchem.2011.11.069] [PMID: 29243599]

[64] Busto O, Miracle M, Guasch J, Borrull F. Solid phase extraction of biogenic amines from wine before chromatographic analysis of their AQC derivatives. J Liq Chromatogr Relat Technol 1997; 20: 743-55.
[http://dx.doi.org/10.1080/10826079708014139]

[65] Jurado-Sánchez B, Ballesteros E, Gallego M. Comparison of several solid-phase extraction sorbents for continuous determination of amines in water by gas chromatography-mass spectrometry. Talanta 2009; 79(3): 613-20.
[http://dx.doi.org/10.1016/j.talanta.2009.04.035] [PMID: 19576420]

[66] Hernández-Orte P, Peña-Gallego A, Ibarz MJ, Cacho J, Ferreira V. Determination of the biogenic amines in musts and wines before and after malolactic fermentation using 6-aminoquinolyl-N-hydroxysuccinimidyl carbamate as the derivatizing agent. J Chromatogr A 2006; 1129(2): 160-4.
[http://dx.doi.org/10.1016/j.chroma.2006.06.111] [PMID: 16876810]

[67] Moret S, Smela D, Populin T, Conte LS. A survey on free biogenic amine content of fresh and preserved vegetables. Food Chem 2005; 89: 355-61.
[http://dx.doi.org/10.1016/j.foodchem.2004.02.050]

[68] Salazar MT, Smith TK, Harris A. High-performance liquid chromatographic method for determination of biogenic amines in feedstuffs, complete feeds, and animal tissues. J Agric Food Chem 2000; 48(5): 1708-12.
[http://dx.doi.org/10.1021/jf990893i] [PMID: 10820083]

[69] Balamatsia CC, Paleologos EK, Kontominas MG, Savvaidis IN. Correlation between microbial flora, sensory changes and biogenic amines formation in fresh chicken meat stored aerobically or under modified atmosphere packaging at 4 °C: Possible role of biogenic amines as spoilage indicators. Antonie van Leeuwenhoek. J Microbiol 2006; 89: 9-17.
[PMID: 16528580]

[70] Sirocchi V, Caprioli G, Ricciutelli M, Vittori S, Sagratini G. Simultaneous determination of ten underivatized biogenic amines in meat by liquid chromatography-tandem mass spectrometry (HPLC-MS/MS). J Mass Spectrom 2014; 49(9): 819-25.
[http://dx.doi.org/10.1002/jms.3418] [PMID: 25230178]

[71] Peña-Gallego A, Hernández-Orte P, Cacho J, Ferreira V. Biogenic amine determination in wines using solid-phase extraction: a comparative study. J Chromatogr A 2009; 1216(15): 3398-401.
[http://dx.doi.org/10.1016/j.chroma.2009.01.106] [PMID: 19232626]

[72] Zhang N, Wang H, Guo XF, Zhang HS. Determination of primary biogenic amines in various food products using isocratic HPLC with 3-(4-chlorobenzoyl)-quinoline-2-carboxaldehyde as a new precolumn fluorogenic reagent. J Sep Sci 2008; 31(15): 2804-12. [http://dx.doi.org/10.1002/jssc.200800247] [PMID: 18668510]

[73] Bomke S, Seiwert B, Dudek L, Effkemann S, Karst U. Determination of biogenic amines in food samples using derivatization followed by liquid chromatography/mass spectrometry. Anal Bioanal Chem 2009; 393(1): 247-56. [http://dx.doi.org/10.1007/s00216-008-2420-2] [PMID: 18843485]

[74] Onal A. A review: Current analytical methods for the determination of biogenic amines in foods. Food Chem 2007; 103: 1475-86. [http://dx.doi.org/10.1016/j.foodchem.2006.08.028]

[75] Liu SJ, Xu JJ, Ma CL, Guo CF. A comparative analysis of derivatization strategies for the determination of biogenic amines in sausage and cheese by HPLC. Food Chem 2018; 266: 275-83. [http://dx.doi.org/10.1016/j.foodchem.2018.06.001] [PMID: 30381186]

[76] Berbegal C, Pardo I, Ferrer S. The use of core-shell high-performance liquid chromatography column technology to improve biogenic amine quantification in wine. J Sci Food Agric 2016; 96(5): 1556-61. [http://dx.doi.org/10.1002/jsfa.7249] [PMID: 25959778]

[77] Preti R, Antonelli ML, Bernacchia R, Vinci G. Fast determination of biogenic amines in beverages by a core-shell particle column. Food Chem 2015; 187: 555-62. [http://dx.doi.org/10.1016/j.foodchem.2015.04.075] [PMID: 25977063]

[78] Kantor A, Kacaniova M, Pachlova V. Biogenic amines content in different wine samples. J Microbiol Biotechnol Food Sci 2015; 4: 37-40. [http://dx.doi.org/10.15414/jmbfs.2015.4.special1.37-40]

[79] Simat V, Dalgaard P. Use of small diameter column particles to enhance HPLC determination of histamine and other biogenic amines in seafood. Lebensm Wiss Technol 2010; 44: 399-406. [http://dx.doi.org/10.1016/j.lwt.2010.08.011]

[80] Spizzirri UG, Parisi OI, Picci N, Restuccia D. Applicationof LC with evaporative light scattering detector for biogenic amines determination in fair trade cocoa-based products. Food Anal Methods 2016; 9: 2200-9. [http://dx.doi.org/10.1007/s12161-016-0398-5]

[81] Mazzotti F, Di Donna L, Napoli A, *et al.* N-hydroxysuccinimidyl p-methoxybenzoate as suitable derivative reagent for isotopic dilution assay of biogenic amines in food. J Mass Spectrom 2014; 49(9): 802-10. [http://dx.doi.org/10.1002/jms.3417] [PMID: 25230176]

[82] Lee S, Eom HS, Yoo M, Cho Y, Shin D. Determination of biogenic amines in Cheonggukjang using ultra high pressure liquid chromatography coupled with mass spectrometry. Food Sci Biotechnol 2011; 20: 123-9. [http://dx.doi.org/10.1007/s10068-011-0017-6]

[83] Self RL, Wu WH, Marks HS. Simultaneous quantification of eight biogenic amine compounds in tuna by matrix solid-phase dispersion followed by HPLC-orbitrap mass spectrometry. J Agric Food Chem 2011; 59(11): 5906-13. [http://dx.doi.org/10.1021/jf200455r] [PMID: 21534596]

[84] Huang J, Gan N, Lv F, Cao Y, Ou C, Tang H. Environmentally friendly solid-phase microextraction coupled with gas chromatography and mass spectrometry for the determination of biogenic amines in fish samples. J Sep Sci 2016; 39(22): 4384-90. [http://dx.doi.org/10.1002/jssc.201600893] [PMID: 27753266]

[85] Płotka-Wasylka J, Simeonov V, Namieśnik J. An in situ derivatization - dispersive liquid-liquid microextraction combined with gas-chromatography - mass spectrometry for determining biogenic

amines in home-made fermented alcoholic drinks. J Chromatogr A 2016; 1453: 10-8.
[http://dx.doi.org/10.1016/j.chroma.2016.05.052] [PMID: 27237593]

[86] Daniel D, Dos Santos VB, Vidal DTR, do Lago CL. Determination of biogenic amines in beer and wine by capillary electrophoresis-tandem mass spectrometry. J Chromatogr A 2015; 1416: 121-8.
[http://dx.doi.org/10.1016/j.chroma.2015.08.065] [PMID: 26362807]

[87] Ginterová P, Marák J, Staňová A, Maier V, Sevčík J, Kaniansky D. Determination of selected biogenic amines in red wines by automated on-line combination of capillary isotachophoresis-capillary zone electrophoresis. J Chromatogr B Analyt Technol Biomed Life Sci 2012; 904: 135-9.
[http://dx.doi.org/10.1016/j.jchromb.2012.07.018] [PMID: 22884886]

[88] Jastrzębska A, Piasta A, Szłyk E. Simultaneous determination of selected biogenic amines in alcoholic beverage samples by isotachophoretic and chromatographic methods. Food Addit Contam Part A Chem Anal Control Expo Risk Assess 2014; 31(1): 83-92.
[http://dx.doi.org/10.1080/19440049.2013.855326] [PMID: 24350674]

[89] Uzasçi S, Baskan S, Bedia L, Erim F. Biogenic amines in wines and pomegranate molasses-a non-ionic micellar electrokinetic chromatography Assay with laser-induced fluorescence detection. Food Anal Methods 2012; 5: 104-8.
[http://dx.doi.org/10.1007/s12161-011-9220-6]

[90] Alonso-Lomillo MA, Domínguez-Renedo O, Matos P, Arcos-Martínez MJ. Disposable biosensors for determination of biogenic amines. Anal Chim Acta 2010; 665(1): 26-31.
[http://dx.doi.org/10.1016/j.aca.2010.03.012] [PMID: 20381686]

[91] Henao-Escobar W, Del Torno-de Román L, Domínguez-Renedo O, Alonso-Lomillo MA, Arcos-Martínez MJ. Dual enzymatic biosensor for simultaneous amperometric determination of histamine and putrescine. Food Chem 2016; 190: 818-23.
[http://dx.doi.org/10.1016/j.foodchem.2015.06.035] [PMID: 26213043]

[92] Basozabal I, Guerreiro A, Gomez-Caballero A, Aranzazu Goicolea M, Barrio RJ. Direct potentiometric quantification of histamine using solid-phase imprinted nanoparticles as recognition elements. Biosens Bioelectron 2014; 58: 138-44.
[http://dx.doi.org/10.1016/j.bios.2014.02.054] [PMID: 24632140]

[93] Ramon-Marquez T, Medina-Castillo AL, Fernandez-Gutierrez A, Fernandez-Sanchez JF. Novel optical sensing film based on a functional nonwoven nanofibre mat for an easy, fast and highly selective and sensitive detection of tryptamine in beer. Biosens Bioelectron 2016; 79: 600-7.
[http://dx.doi.org/10.1016/j.bios.2015.12.091] [PMID: 26761616]

[94] Alves RC, Barroso MF, González-García MB, Oliveira MBPP, Delerue-Matos C. New trends in food allergens detection: towards biosensing strategies. Crit Rev Food Sci Nutr 2016; 56(14): 2304-19.
[http://dx.doi.org/10.1080/10408398.2013.831026] [PMID: 25779935]

[95] Marcobal A, Polo MC, Martín-Alvarez PJ, Moreno-Arribas MV. Biogenic amine content of red Spanish wines: comparison of a direct ELISA and anHPLC method for the determination of histamine in wines. Food Res Int 2005; 38: 387-94.
[http://dx.doi.org/10.1016/j.foodres.2004.10.008]

[96] Luo L, Xu ZL, Yang JY, *et al.* Synthesis of novel haptens and development of an enzyme-linked immunosorbent assay for quantification of histamine in foods. J Agric Food Chem 2014; 62(51): 12299-308.
[http://dx.doi.org/10.1021/jf504689x] [PMID: 25417820]

SUBJECT INDEX

A

B

Belén Gómara & María Luisa Marina (Eds.)
Current and Future Developments in Food Science (Vol. 1)

E

F

G

H

I

J

L

M

Q

R

S

www.ingramcontent.com/pod-product-compliance
Lightning Source LLC
LaVergne TN
LVHW070943210926
848482LV00002B/79